**On Being Adjacent to Historical Violence**

# On Being Adjacent to Historical Violence

---

Edited by
Irene Kacandes

DE GRUYTER

ISBN 978-3-11-075326-4
e-ISBN (PDF) 978-3-11-075329-5
e-ISBN (EPUB) 978-3-11-075335-6

**Library of Congress Control Number: 2021948043**

**Bibliographic information published by the Deutsche Nationalbibliothek**
The Deutsche Nationalbibliothek lists this publication in the Deutsche Nationalbibliografie;
detailed bibliographic data are available on the Internet at http://dnb.dnb.de.

Cover image: German Troops Crossing the Polish Border, 1939 © Viktor Witkowski
Typesetting: Integra Software Services Pvt. Ltd.
Printing and binding: CPI books GmbH, Leck

www.degruyter.com

Dedicated to the memory of Susanne and Half Zantop
and of all individuals whose lives
were cut short by the violent actions of other human beings.

# Contents

## Part Three: **Journeys**

Irene Kacandes
# On Being Adjacent to Historical Violence: Introduction

## Overview

This book offers to academic and general public readers timely reflections about our relationships to violence. Taking cues from the self-reflexivity, themes, and subject matters of Holocaust, queer, and Black studies, this large group of diverse intellectuals wrestles with questions that connect past, present, and future: where do I stand in relation to violence? What is my attitude toward that adjacency? Whose story gets to be told by whom? What story do I understand this image to be telling? How do I co-witness to another's suffering? How do I honor the agency and resilience of family members or historical personages? How do past violence and injustice connect to and affect the present? In smart, self-conscious, passionate, and often painfully beautiful prose, cultural practitioners, historians, and cultural studies scholars explore such questions, inviting readers to do the same. By making available compelling examples of thinkers performing their own work within the cauldron of crises that came to a boil in 2020, *On Being Adjacent to Historical Violence* models some strategies and lines of thought for moving forward with hope.

## The Moment

*April 2020 to April 2021.* This is the year in which this project was conceived and carried out. It is a period saturated with disease of many natures. Covid-19 may well end up being the problem that remains most associated with this time frame. The new virus and the pandemic it caused certainly succeeded in bringing the whole world to an almost complete standstill for a while, and huge resources of money and effort were required to bring the disease somewhat under control. However, the consequences of the diseases of systemic racism, ecological devastation, poverty, and political polarization seeded centuries prior by colonialism, chattel slavery, imperialism, and industrial capitalism were similarly brought into the consciousness of more people in this same period than perhaps ever before. Particularly in the United States, news and images of police killings of Black Americans were relentless throughout this year. The murder of George Floyd by Derek Chauvin on 20 May 2020 became known to the whole world through the

https://doi.org/10.1515/9783110753295-001

video filmed at the scene of the crime by teenager Darnella Frazier. Images of traumatized would-be migrants to Europe from Africa and Asia rescued from near-drowning in the Mediterranean and of would-be migrants from Central America at the southern border of the United States also became quotidian. For many believers in democracy, the scenes of insurgents storming the US Capitol building on 6 January 2021 to disrupt the verification of a fairly executed election will not soon be forgotten. What was perhaps the most novel about this period for the ordinary citizen was the increasing evidence of how these scourges, historical and contemporary, were connected. A closer consideration of the pandemic provides evidence.

While many commentators originally pronounced this new corona virus as the "Great Equalizer" because of how widely and quickly it spread geographically, infecting individuals of all kinds – old, young, and middle-aged; rich and poor; famous and anonymous; those seemingly in the bloom of good health and those not – it didn't take Covid-19 long to make clear to us that any insistence on "sameness" obscured the cruel reality of our wildly *unequal* chances of being exposed to the virus in the first place, getting really sick from it, and, yes, dying. To understand the inequality, the connection of scourges must be noticed. Already by Spring 2020 it became clear to those paying attention that it mattered whether you were in China, Germany, the USA, or Brazil, the inner city or the countryside, in a democracy or in a dictatorship. It mattered whether you had access to clean water and healthcare. It mattered whether your circumstances allowed you or forced you to observe social distancing. Whether you were housed or unhoused. Whether you lived with others or alone. Were incarcerated or free. It mattered whether you risked exposure to the virus every day by tending to the sick and elderly, washing hospital linens, collecting garbage, stocking shelves or ringing up groceries, delivering packages, burying the dead. In many societies in the contemporary world such service jobs near the bottom of the economic ladder are disproportionately held by members of racial and ethnic minorities. It mattered whether people doing these jobs had adequate protective equipment. To be sure, populations in many places quickly began showing their thanks for "essential workers" with nightly applause or posted signs of gratitude, and yet it took a very long time to actually give those individuals the resources they needed to keep safe. Untold numbers are still waiting for protective equipment.

A similar unevenness and belatedness of protection is being revealed as I write this with regard to world-wide access to the vaccines that were developed and have proven themselves to be effective in cutting death rates.

It's the death rates that force us to see more profound connections than "just" direct exposure due to jobs. To take the case of the United States, by early April 2020, the *Washington Post* analyzed available data, discovering that majority Black counties had three times the rate of infections and almost six times the rate of deaths as counties that were majority white.[1] There is a plethora of statistics for the United States and elsewhere that shows similar discrepancies for those on the social or economic bottom from those at the top.

In addition to proximity to jobs that bring them in contact with the virus, members of racial and ethnic minorities suffer at a higher rate than the general population conditions such as diabetes, heart disease, and lung disease that render them more vulnerable to this new respiratory disease. These conditions themselves are often triggered by environmental degradation to the places where minority communities live and by other quotidian stresses of discrimination in its many forms. The pandemic has also shown racism's reach to be insidious, because even when minority individuals and the poor make it to a hospital, not all hospitals are equally staffed and equipped; potentially fatal too is unacknowledged bias in healthcare providers that can often lead to undertreatment or lack of treatment altogether.[2] The case of Dr. Susan Moore, a Black American, brought home to many the lethalness of implicit bias. This medical doctor knew what treatments she could be getting when she was hospitalized with Covid-19 in Indiana, and yet when she requested additional pain medication, her white attending physician downplayed her pain and suggested she would be discharged. Dr. Moore posted a video to social media where she poignantly explained what was happening to her and concluded that "if I was white, I wouldn't have to go through that." After her video testimony went viral – a metaphor that seems even more apt since the beginning of the pandemic – Dr. Moore's physical pain was subsequently addressed in the hospital, but she was soon sent home and died only two weeks after her post. Commenting on Dr. Moore's experiences, Dr. Christina Council, a Black physician in Maryland stated, "Sometimes when we think about medical bias it seems so far removed. We can sit there and say, 'OK, it can happen to someone that may be poorer.' But when you actually see it happen to

---

1 Reis Thibault, Andrews Ba Trans, Vanessa Williams, "The Coronavirus is infecting and killing black Americans at an alarmingly high rate," 7 April 2020, *The Washington Post*, accessed 29 April 2021, https://www.washingtonpost.com/nation/2020/04/07/coronavirus-is-infecting-killing-black-americans-an-alarmingly-high-rate-post-analysis-shows/?arc404=true.

2 Bias in healthcare was implicitly cited in the April 2020 *Washington Post* analysis and more explicitly in Sujata Gupta, "Why African-Americans may be especially vulnerable to COVID-19," *Science Times*, 10 April 2020, accessed 29 April 2021, https://www.sciencenews.org/article/co ronavirus-why-african-americans-vulnerable-covid-19-health-race; and "Why is the Pandemic Killing So Many Black Americans?" *The Daily*, 20 May 2020, accessed 29 April 2021, https://www.nytimes.com/2020/05/20/podcasts/the-daily/black-death-rate-coronavirus.html.

a colleague and you're seeing her in the hospital bed literally pleading for her life, it just hits a different way and really hits home and says, 'Wow, we need to do something.'"[3]

I share Dr. Council's reaction and so have people around the world. In the past few years, millions of us have viewed probably thousands of videos and still images of individuals pleading for their own lives and the lives of their families: parents of children starving to death in Yemen, parents and children screaming while being separated by US Border agents, adult children pressing their hands to hospital and nursing home windows begging their parents to hang on, willing that their loved ones realize they're not alone; George Floyd and shamefully almost countless other people of color pleading for their lives in the presence of the police. By virtue of viewing those images and reading information about them, we are, whether we choose it consciously or not, in some relation to the violence depicted there.

We need to do something. Oh, yes, we do. The problem for many of us is that the issues seem so complex, so entangled that it's easy to hesitate, even to feel paralyzed in the face of that complexity and the current plethora of injustices and inequalities. Perhaps we can unfreeze ourselves by being reminded of the advice from a victim of an earlier pandemic, Xolani Nkosi Johnson, a Zulu boy who died of inherited AIDS at the age of 12 on 1 June 2001, weighing just 20 pounds. Nkosi used to tell interviewer Jim Wooten of ABC, who quickly became an admirer and a friend of this young AIDS advocate:

Do all you can with what you have in the time you have in the place you are.[4]

Right. We must summon our courage and exercise our imagination to bear witness to and actually do something about the tragedies unfolding in front of us. Like 17-year-old Darnella Frazier filming the arrest and murder of George Floyd. Like 15-year-old Greta Thunberg staging a one-girl protest in front of the Swedish Parliament Building. Like Alicia Garza, Patrisse Cullors, and Opal Tometi founding #BlackLivesMatter in response to the 2013 acquittal of the murderer of Trayvon Martin. Like millions of people going into the streets to protest George Floyd's murder.

---

**3** All facts and quotations from John Eligon, "Black Doctor Dies of Covid-19 After Complaining of Racist Treatment," *New York Times,* 23 December 2020, accessed 29 April 2021, https://www.nytimes.com/2020/12/23/us/susan-moore-black-doctor-indiana.html.
**4** As quoted in Sam Roberts, "Captivated by the Courageous Spirit of a Small Child Who Was Larger Than Life," which appeared in the print version of the *New York Times* on 4 December 2004, accessed 29 April 2021, https://www.nytimes.com/2004/12/04/books/captivated-by-the-courageous-spirit-of-a-small-child-who-was-larger.html.

What can *I* do with all I have in this time and place? I am a writer, a scholar and teacher on trauma and specifically the Holocaust and Neonazism. And I have a lot of smart friends. One year ago, I asked them to help me create a response to what was unfolding in front of us. They said yes. They would help.

This book is our response. It is not a guidebook to civil disobedience or protest. In fact, it's not a guidebook to anything in particular beyond careful thinking. The North American so-called culture wars have exacerbated suspicion about intellectuals, about people who are paid to think and to teach others how (not what!) to think. But I and the team behind this book submit that the kind of "thinking through" that is often only possible when trying to express a train of thought in writing is something that is sorely needed right now if we hope to understand and take action on intricate problems. In the pages that follow you will find a collection of short think pieces about being in relation to violence. To put it otherwise, this is a collection of what some individuals decided to think about in this year of connected crises, of what those individuals were capable of thinking about in relation to their own previous work and in relation to discovering and demonstrating connections between types of violence in the past and the present, whether private presents or the more public one I've been describing above. In the closing pages of this volume, you will find a transcript of these writers thinking out loud together, an activity that, when exercised with careful and respectful listening, can similarly capture complexity. To be sure, reading the transcript of a discussion has a different density and rhythm than reading the composed thought of a single person. We include both to reach the maximum number of people.

## The Group

There are thirty-four contributors to this volume of thirty-two essays (one was written by three people). They are trained in and draw on a number of traditional disciplines: history, cultural studies, literary and critical theory, folklore, sociology, political science, conversation analysis, disability studies, art history, life writing. And they combine these fields in creative and productive ways to produce interdisciplinary approaches to their topics. Although most of these individuals work in a university environment, academics are joined here by an artist-filmmaker, a psychiatrist, and a fiction writer. The scholars among us range in rank and experience from one individual finishing her Ph.D. through all the tenure-track ranks to illustrious emeriti. I contacted these individuals because I knew them to be thinking about topics that seemed connected in my mind to the nexus of problems we were dealing with in society at large, but also because I knew them to be good writers. For me, the good writer is one who self-consciously reflects on the act of writing itself, not so much getting each word or sentence right, though that's important, too, but rather registering and judging the work that any given piece of writing can and does do.

These individuals I realized would be thinking hard about the catastrophes we were all living through and were also deeply committed to the political, in the sense of connecting up many aspects of any culture or society. Further, these scholar-friends had long made a commitment to interrogating their own practice and considering their intellectual work as needing to make a difference to the society in which it is produced. To my thinking already in the early months of the pandemic, it became quite evident that everyone, starting with myself, had to do their work and live their lives differently if we were going to emerge out of these crises into a more just period of world history. Though staying "socially distanced" quickly became a dictum in 2020, I work best when in dialog with others. I needed to connect to these folks to figure out my own way forward and how that path might help others. "Connect" we did: through our well-practiced use of email and on that new platform, at least new for many of us: Zoom.

In a way not necessarily signposted in the volume itself, these contributors represent intellectual networks. Many of us knew each other from four seminars within the general fields of Holocaust and German interdisciplinary studies that had either occurred or were proposed before the general shutdowns of 2020. Leslie Morris and Karen Remmler had planned the first of these already in late 2017. "Private Matters: Expanding the Margins of the *Lebenslauf* [cv]" was a seminar conducted over three mornings at the 2018 conference of the German Studies Association (GSA) and aspired to share work that explores "the shifts in academic writing that have created a space for interweaving life stories with analytical work" (from the call for submissions). A second seminar, "Probing the Limits of Holocaust Memoir," followed soon after, organized by Leslie Morris for the 2018 Holocaust Educational Foundation's biannual conference on "Lessons and Legacies of the Holocaust"; participants were invited to: "integrate our work as scholars of the Holocaust with our private lives in families that might include Holocaust survivors, perpetrators, or those who have been marked indelibly by historical trauma" (also from a call for submissions). A follow-up seminar was planned for the 2020 GSA conference organized again by Morris and Remmler, "Genealogies of Self-Reflection: Writing in the Wake of Trauma" (and took place on-line in October 2020); and yet another follow-up at the Holocaust conference was organized by me with the title "Opting to Affiliate with Atrocity" and to take place in November 2020 (originally cancelled and now postponed to 2022).

Some scholarly work that was inflecting the emphases of these seminars included older works like Saul Friedlander's edited volume *Probing the Limits of Representation* (1992), Alice Kaplan's *French Lessons: A Memoir* (1993), Susan R. Suleiman's *Budapest Diary* (1996), Mark Roseman's *A Past in Hiding: Memory and Survival in Nazi German* (2000), Irene Kacandes's *Daddy's War. A Paramemoir* (2009), and Marianne Hirsch and Leo Spitzer's *Ghosts of Home: The Afterlife of Czernowitz in Jewish Memory* (2010). More recent references included Christina Sharpe's *In the Wake: On Blackness and Being* (2016), Nora Krug's *Heimat/Belonging: A German Reckons with History and Home* (2018), Michael Rothberg's *The Implicated Subject: Beyond Victims and Perpetrators* (2019), and Angelika Bammer's *Born After: Reckoning with the German Past* (2019).

The interrogation and integration of public and private lives in scholarly work was the red thread through these conferences and indeed is one of the red threads of this volume. Another commonality with many of the books listed above is a commitment to addressing a broader public than just the strictly academic one to be found, for instance, in those conferences where we typically share our work.

As the circle of contributors expanded, so did our references to foundational and recent work in feminist and queer studies, Black studies, including Black German studies, certain branches of historical work, narratology, and experimental life writing. Roland Barthes, Audre Lorde, Eve Sedgwick, bell hooks, Patricia Williams, W. G. Sebald, Lisa Lowe, Judith Butler, Saidiya Hartman, and the team behind the groundbreaking *Farbe Bekennen* (*Showing Our Colors: Afro-German Women Speak Out*) are not always mentioned by name in these essays, and yet are nonetheless inspiring what is. Finally, a text that many of us read and put us into a certain framework and frame of mind in terms of topics and also thinking carefully about them, was the blog created by the Katz Center of the University of Pennsylvania called, "Knowing the Victim? Reflections on Empathy, Analogy, and Voice from the Shoah to the Present." As convener David Myers positioned that exchange: "The Holocaust and the BLM movement share the problem of knowing another's experience. Judith Butler, Cheryl Greenberg, Marianne Hirsch, and Robin D. G. Kelley tackle the core epistemological and moral question of whether we can know another's experience, and what is at stake in our answer."[5]

---

5 David Myers, "Knowing the Victim? Reflections on Empathy, Analogy, and Voice from the Shoah to the Present," 25 June 2020, accessed 30 April 2021, https://katz.sas.upenn.edu/resources/blog/knowing-victim-reflections-empathy-analogy-and-voice-shoah-present?fbclid=IwAR0e1ONuQbu5DCFvmNkyfvp3MRr3U74dJIC0COX7-0lR9ZKuEdJEcMtNX1E.

Creating connections btween histories that are often kept separate, figuring out reparative gestures, and making room for futurity were goals this team articulated, goals that led me to yet other scholars whose work I did not previously know but whose ways of thinking and areas of expertise were obviously going to contribute richly to the project. Not all those contacted could or wanted to reach back, of course; my ambitious schedule to get this volume out fast was a deterrent to signing on for many, and thus my desire for even more diverse topics, approaches, and identity positions of writers than that with which I started went partially unfulfilled. Still, in this book readers will travel to France, Greece, the Asian Caribbean, the African diaspora, Native American reservations and Indigenous territories, and Southeast Asia, in addition to Central and Eastern Europe, the regions plowed so regularly in German and Holocaust studies in general and in many of the contributors' previous publications in particular. Time periods discussed range from the eighteenth to the twenty-first centuries. Connecting German studies to other fields, and Germany to other places has been a longtime goal of my teaching, writing, and editorial work, as has been making clear that cultural studies is not concerned only with recent history. To put it more squarely in the framework of this book: the present is inflected and constituted by many pasts, those more recent and more distant, those of nearer and farther places.

There are two other points about the contributors that I'd like to mention here. In addition to various intellectual networks that overlapped to bring us into the same publishing space, it seems important to me to point out affective networks as evidence of a way in which the professional and the personal simply cannot be separated. Not every contributor might agree with the way I see these affective networks, and since that's not exactly my point, I'll just spell out the one that cannot be teased apart from my own development as a scholar, teacher, writer, and human being. Contributing to this volume are my teacher and graduate school mentor at Harvard, Susan R. Suleiman; my fellow graduate student in the program in Comparative Literature there, Bettina Brandt; my colleague from my first academic post at the University of Texas at Austin, Ann Cvetkovich; a student from my first independently taught Holocaust course (also at Texas) Erin McGlothlin; and my (past and present) Dartmouth colleagues and collaborators Marianne Hirsch, Leo Spitzer, Susannah Heschel, and Viktor Witkowski. Each of these individuals knows about, indeed lived through, with, alongside, adjacent to me during at least one personal crisis when it was not clear to me that I could move forward: the enmeshment of the personal and the professional indeed. It would be hopelessly solipsistic if I mentioned this only for my own gratification and the opportunity to express thanks for this and other affective networks connecting me to people in this book.

Rather, I want to use this example to make a second point, that is, to suggest that 2020–2021 engendered personal, professional, and political crises for *innumerable* human beings. Some of those human beings managed those crises partly by connecting to new kindred souls, or to borrow a phrase from my friend Ann Cvetkovich, by becoming part of a new tribe. With regard to the contributors to this project, the new group is certainly not completely cohesive, and, in any case, we are not a group that ever got to be together in one physical space. Some of our needs were and remain different. Still, a shared goal was trying to talk together for the purposes of understanding better the crises we were experiencing and creating something new to address them. The form, frequency, and timing of gathering for this project also varied for the specific individuals involved. Starting with a colloquy in summer 2020 convened by me, but then continuing on, propelled by whatever the various need for contact was and how it could best be addressed, people met in dyads and in smaller and larger groups. I was at first a little surprised and then delighted to learn that various configurations of contributors were choosing to communicate with one another after an initial meeting and right up to the last meeting convened by me on 20 March 2021. Almost everybody got involved in that last group effort in one manner or another, as you can see when you turn to the "Conversation" at the end of this book. The voluntary decision to opt to affiliate with one another at all in a moment when individuals were facing so many private and professional challenges and obligations seems noteworthy to me in and of itself. However, additionally, in coming together, we figured out what this book could be about. In other words, I may have started the process rolling, and most pages have been written by a single person, but this volume represents a group conception.

# The Concepts

## "On being adjacent to historical violence"

This phrase seemed to make sense immediately to those who heard it in 2020 – to individuals invited to contribute and to the director and assistant director of Dartmouth College's Leslie Center for the Humanities who granted us funds to run the initial conversations mentioned above. (I mention these funds because one of the issues of summer 2020 in my estimation was individuals ubiquitously not being adequately compensated for their efforts, and I did not want to create another example of that. Some contributors who consider themselves well-paid wanted to refuse the proffered honorarium, small though it was, out of similar considerations; I urged them to accept it and then donate it to contingency faculty or graduate students if they so chose.)

Proposed by cultural and queer studies scholar Ann Cvetkovich, originally as a title for her own contribution to this volume, the phrase "on being adjacent to historical violence" certainly made immediate sense to me. After hearing several objections to titles I had originally proposed for the volume – mainly because of concern about concepts like "atrocity" and "affiliating with suffering" that seemed to limit the object of study or the possibilities for hope and resilience – I asked Ann if we could use her proposed title for that of the whole book. She had already come up with a new focus for her own contribution and said yes. Here's what Ann originally explained to me about this phrase:

> I'd like to think about the term "adjacent" as a way to describe juxtaposition, or the concept of "sideways/aslant" that is discussed in queer theory, or in your/Goethe's notion of "elective affinity," which you have used before for our work and the links between our projects.[6] And I'm using "historical violence" to avoid other categories such as "trauma" or "genocide" that immediately assume certain genealogies or critical debates. Because my work so often involves scenes of violence that aren't always immediately visible/discernible as such – structural violence/ordinary violence/racism – I favor vocabulary that keeps open that sense of uncertainty about how to describe what the thing is.[7]

Despite what she insisted was its impromptu description, I think Ann's explication is quite refined. I'll add to it rather than qualify it by simulating an exercise I often conduct with my students at the beginning of an academic term: comment on each aspect of the title I've given any particular class.

## "On Being"

Our titular concept begins with unadorned existence. It foregrounds the various ways in which so many people around the world have earnestly struggled to simply continue to be, to stay alive during this crisis year. It evokes individuals whose livelihoods completely evaporated due to their workplaces being shuttered or to the more critical priority of caregiving to children or elders that made it impossible

---

6 See "Part Five" of *Eastern Europe Unmapped: Borders and Peripheries*, eds. Irene Kacandes and Yuliya Komska (New York: Berghahn, 2017), 227–268, where Ann's and my essays appear under the section title "Elective Affinities," a phrase I borrowed from one of the English translations of J. W. von Goethe's novel *Wahlverwandtschaften* (1809).
7 Private email from Ann Cvetkovich to Irene Kacandes.

to work outside the home or to wars encroaching and basic sustenance becoming even more scarce or to literally trying to stay alive in prisons where individuals could not protect themselves from others potentially carrying the virus or to those who put their hands on the steering wheel or in the air to try to continue to exist when the police pulled out their guns. I want us to honor all that effort to be. This book is about Being.

I endorse this phrase also for its echoing of the title of the popular radio and media project launched by Krista Tippett, the "On Being Project," the stated goal of which is: "Pursuing deep thinking and moral imagination, social courage and joy, to renew inner life, outer life, and life together." The six grounding virtues of the project seem particularly relevant to moving forward from our complex crisis year: "words that matter, hospitality, humility, patience, generous listening, adventurous civility." The project identifies these as "spiritual technologies" and "tools for the art of living."[8]

Finally, to my mind, the idea of "being" also leaves open the question of agency. When I first contemplated this project, I thought specifically about the ethical decision to choose to have a relation to another human being against whom violence was being or had been perpetrated. I still consider making such a choice an important option. However, as I talked to contributors about their essay topics, it became clear to me that they also wanted to talk about situations where there is no choice; birth, chance, or external circumstances not of one's making, often draw one into the vortex of violence.

## "Adjacent"

Webster's offers the following glosses for adjacent: "not distant; nearby; having a common endpoint or border; immediately preceding or following." As we've seen above, Ann chose "adjacent" as the descriptor for two or more things, ideas, persons, concepts, and methodologies being in a relation of juxtaposition, sideways, aslant. So, adjacent means "in some kind of proximity," but perhaps not in some fixed, orderly or predictable way. In her essay in this volume, Ann Cvetkovich speaks of her "practice of adjacency – of setting seemingly disparate things alongside one another – as a form of queer method." Adam Z. Newton's essay proposes a "poetics of adjacency," defined as a "skirting the edge of plausible

---

**8** Quotes taken from the "On Being" website, accessed 30 April 2021, that can be found at https://onbeing.org.

deniability perhaps, or else bridging the ambiguous space between encroaching shadow and sanguine overlay." And Mita Choudhury, to offer a third engagement with the term, suggests that empathy can foster adjacency even to those quite distant from us in time and experience.

## "Historical Violence"

Ann's original reflection on this phrase concerns using it as a way to avoid deploying words like "trauma" or "genocide" that bring with them academic debates that she's not necessarily interested in engaging. She also suggests that those words wouldn't necessarily connote the kinds of everyday and non-spectacular violation that her work often concerns. Many essays in this book *do* engage with aggressions of gargantuan proportions, one might say "historic," like those committed during the mid-twentieth century against the disabled and those labeled racially inferior by the Nazis, starting with Jews, Roma-Sinti and Slavic peoples. But I endorse the word "violence" as evoking a greater range of types of violation.

As for "*historical* violence," its multivalence is important to this book. On the one hand, many essays take up violence committed in the past, like the extermination of European Jewry already referenced, and also slavery (Ellis), indentured servitude (Goffe), crimes against Indigenous peoples (Cvetkovich and Riegert), genocidal killings in Southeast Asia (Newton) and rape in the eighteenth century (Choudhury), to name just some. On the other hand, different treatments producing suffering – one might call them less violent or quieter – are also considered in this book, like being treated as suspicious as a Black person in a predominantly white world (Layne) or being discriminated against or spoken for (Shuman, Bohmer, Niyitunga) or being uprooted by emigration (Ellis). What runs through essentially all the essays is the way violence from the past inflects the present. Violence always has a history.

## Co-Witnessing

There's one more concept that I see as central to the work this book is accomplishing, even if it is not signaled explicitly in the title of the volume. In my own professional and ethical framework, the contributors are all performing acts of "co-witnessing."

I first proposed the term "co-witnessing" decades ago to describe the activity of supporting those who have experienced a trauma or other injustice and who, because of that traumatization or by virtue of being incarcerated or dead – or fictional – cannot help themselves. Over my career of thinking through this term, I have moved from applying it originally to prose fiction to exploring what it can mean in analyzing memoir, film, and historical and contemporary events.[9]

My current description of the co-witness can be summarized briefly like this:

- The co-witness is someone who is usually not present at the infliction of a crime or injustice or who is not the direct target of it.
- The co-witness has a desire to think about others, to listen respectfully to their stories or to help research them, to learn their details and complexities.
- The co-witness repeats others' stories without appropriating them.
- The co-witness exercises a high level of self-consciousness about their activity of co-witnessing.

Though only a few contributors reference my framework of "co-witnessing," I believe all the contributors as the writers of their essays display the above characteristics. I also believe that these are attitudes, ways of being, behaviors that can help us move through these calamitous times.

Can such actions as co-witnessing do anybody any good? Well, in the case of injustice that causes individuals' deaths, obviously it won't help the dead. However, following important thinkers such as legal theorist Martha Minow and feminist psychiatrist Judith Herman, I consider it important to co-witness to all situations of injustice or violation even when doing so will not help individual victims, because whose suffering we decide we care about reveals who *we* are as a society. Co-witnessing reveals the values we hold.[10] Being explicit about the connection between our actions and our values strikes me as particularly necessary to creating a more just world.

In my now decades-long trajectory of thinking about co-witnessing, I have come to believe more and more firmly, that while telling/repeating the story in a non-appropriative way is itself a valuable action, it is also critical that co-witnesses learn from historical and contemporary co-witnesses about what

---

**9** See Irene Kacandes, *Talk Fiction: Literature and the Talk Explosion* (Lincoln: University of Nebraska Press, 2001); *Daddy's War. Greek American Stories. A Paramemoir* (Lincoln: University of Nebraska Press, 2009); "'Die Ungnade der späten Geburt': Challenges for Central Europeans in the Twenty-First Century," *German Studies Review* 40, no. 2 (2017): 389–405; and "How Co-Witnessing Could Transform the Post-Pandemic World," in *13 Perspectives on the Pandemic: Thinking in a State of Exception*, ed. Rabea Rittgerodt (Berlin: de Gruyter Verlag, 2020).
**10** Martha Minow, "Surviving Victim Talk," *UCLA Law Review* 40 (1993): 1445. Judith Lewis Herman, M. D., *Trauma and Recovery* (New York: Basic 1997, orig. pub. 1992).

additional types of actions might be taken to rectify an injustice or to prevent one from happening or from continuing to happen. In other words, I firmly believe that the most complete acts of co-witnessing involve learning from what others have done, adding those actions to one's personal, conscious repertoire of responses to injustice, and executing an action from that repertoire or a similar one when the circumstances seem to fit. Launching and curating a Facebook group called "Doing Something" that gathers and shares what individuals have actually done and are doing in the face of a huge array of injustices in our world, I extend my concept of co-witnessing – extend it both in terms of whom I hope to reach and in terms of taking the concept beyond "telling."[11] If readers of this volume add to their repertoires from cues found in its essays (or in that Facebook group), as have witnesses using their cell phones to record police action, that would be a wonderful result in my view.

As I have watched friends and strangers bewildered and understandably floundering in the face of the enormous challenges of 2020, I have thought a lot about the role co-witnessing can play. I even published an audacious if not hopelessly optimistic think piece suggesting that co-witnessing could transform the post-pandemic world.[12] In the course of putting together this volume, I have been involved in hundreds of discussions, and I have realized that I sometimes play a co-witnessing role to contributors' acts of co-witnessing. That has been enormously humbling. I thank all contributors for trusting me enough to let me help them execute the co-witnessing that only they know how to do.

## The Essays

As I hope is already clear by this point, *On Being Adjacent to Historical Violence* is not a traditional academic anthology of essays. It is rather a set of think pieces, reflections to spur thought, essays that reveal thinkers at work: some longer, some shorter; some more academic-sounding, some more intimate; all self-reflective.

While all writers take up the volume's titular concepts in one way or another, the book is divided into three main sections: "Connections"; "Families"; and "Journeys." Those essays that foreground more explicitly connections between

---

**11** The "Doing Something" home page can be found at https://www.facebook.com/groups/458897688025950.
**12** Kacandes, "How Co-Witnessing Could Transform."

different chronologies or catastrophes can be found in the first section of the book. Others exploring more centrally adjacencies created by family bonds are found in the second section. A third section groups essays that describe actual and metaphorical journeys. To be sure, many essays could have appeared in more than one section, and I anticipate that readers will enjoy traveling through this book in a number of different ways from the one I have put forth here. To hint at some hopscotching on subjects that themselves echo with today's challenges: Remmler is interested in the unburied and the improperly buried (in the "Families" section), a topic that must be addressed in Prager's reflection on teaching images of atrocity (in the "Journeys" section). A reader could follow writers' concerns about implications of using the label "victim" in Hirsch ("Connections"), Spitzer and Kuehne ("Families"), and Breger ("Journeys"), and compare their comments to Choudhury's decision to embrace the victim perspective ("Journeys").

Other methods for traveling through the volume could involve writers' decisions that I suggest parallel some of the decisions facing many people as a result of our catastrophic year. It seems almost too obvious to remind here that the nexus of race, privilege, and discrimination has been and continues to be on many individuals' minds and in the public discourse, as it has foregrounded this introduction. Essentially all essays take up this topic, with many addressing it quite explicitly (e.g., Shuman/Bohmer/Nigitunga, Layne, Cvetkovich, Heschel, Zhang, Goffe, Kacandes, Riegert, Ellis). To take another contemporary issue: with death so ubiquitous and yet unpredictable, many individuals have decided not to put off things they perhaps long wished to do but have never done. In terms of this book, several contributors decided to write about the self in a way that they or their disciplines had not given permission to do previously (e.g., Layne, Matzen, Zhang, Flescher, Kuehne, Buerkle, Riegert, and Ellis), or on a topic they'd not taken up before (e.g., Herzog on disability, Heschel on the position of German Jews who converted to Christianity, Kligerman on losing a brother, or Newton on leaving academia). Others take up pieces of family history they'd meant to research or reflect on further at some point or even to entertain the idea of ceasing to engage with family history (Hirsch, Witkowski, Suleiman, Spitzer, Morris, Goffe, Grossmann). Also echoing with issues that have been foregrounded by this past year, some contributors turn to their personal entanglements with the elders they've come in contact with professionally – or is that personally? (Bergen, McGlothlin, Roseman, Grossmann). Some essays reflect most explicitly on the ethics of telling others' stories (Shuman/Bohmer/Niyitunga, Brandt). And certainly not finally but last in this list, some reflect on past decisions, pronouncements, and omissions that they want to correct, foreground, or repair (Cvetkovich, Kacandes, Gross, Choudhury). To be sure and once again, many essays perform more than one of these types of work.

# The Cover

We are fortunate to have an accomplished active artist (in many genres) as part of our group. I thank Viktor Witkowski for giving permission to use one of his images on the cover of this book: "*German Troops Crossing the Polish Border (1939)*," created in 2006. In addition to referencing the violence of the Second World War, a topic that comes up explicitly or implicitly in the majority of essays in this volume, it seems to me that the painting foregrounds the central goal of our book to echo and elucidate themes and actions relevant to the present in which it was created. Although I refer readers of this introduction to Witkowski's own explanation about the genesis of this powerful work in his poignant essay "Borderlands," I can offer a preview by remarking that it is based on an image of the German invasion into Poland that was *completely staged* by a group of German photographers; the scene they set up was also filmed and included as part of the newsreels shown in cinemas back in Germany. Even though the invasion and its consequences were quite real, *the image that communicated the news was faked*. Viktor's own version of this famous image changes it, most obviously through its medium of painting, a medium that necessarily reveals the role of the artist rather than reinforcing the illusion of reality, and also by not including the German soldiers playing at invading Poland for the famous – or rather infamous photo. By doing so, the artist could shift what the painting foregrounds: in Viktor's own words: "By removing the agents of violence from my painting and rendering the border post as unfinished (by revealing the raw linen of the canvas), I turn away from the historic event and the declaration of war it illustrates. My focus is on the border as a site of trauma and transformation: borders that are built, crossed, conquered, torn down, and sometimes forgotten" (Witkowski in this volume). Another important accomplishment of this painting is the way it connects the personal and the political, since the artist's own family history is deeply inflected by its mixed German-Polish heritage and by multiple crossings of borders. By deleting some of the details of the original image, Witkowski succeeds in creating a more abstract image that further allows us to make connections between a specific history – whether of the creation of the photo in 1939 and the Germans it inspired to wage war or of the artist's own experiences of borders – and other histories, emotions, and experiences of borders in viewers' minds or in this book.

# The Acknowledgements

Dartmouth College has been my base for a rich and productive intellectual life for more than a quarter century. While the campus has been the site for me of many enormously generative relationships for which I am profoundly grateful, I want to acknowledge here that it sprawls over land that belonged to the Abenaki peoples. I am indebted to my colleagues who teach about the Abenaki and other Indigenous peoples, not only about their suffering at the hands of European invaders but also about the fullness of their culture and ways of life – in the past and present. I also want to thank all individuals, including the students in one of the courses I was teaching while editing this volume, Women's, Gender and Sexuality Studies 10, who chose to teach themselves about Dartmouth's relationship to Native Americans and to chattel slavery of Africans.

I offer cheerful thanks to Dartmouth College's Leslie Center for the Humanities, its Director Rebecca Biron and Assistant Director Andrea Tarnowski for their early and enthusiastic intellectual and financial support of this project. Mary Fletcher and Erin E. Bennett offered administrative and technical support that was invaluable. A. Christopher Strenta provided needed advice. As a then complete neophyte to "Zoom," I want to thank my former students Savannah M. Cochran '20 and Mary E. Tobin '20 for their good humor in letting me convene a practice colloquium with them before the first big day arrived to speak with the group of contributors. Three 2019 graduates of Dartmouth's Comparative Literature M.A. program, Alba Elliott, Marcus Mamourian, and Stephen Sudia, not only provided technical help during the colloquia but also took notes and created summaries of the discussions that were enormously helpful to all of us, contributing profoundly to the cohesion of the volume, especially for those who were unable to attend all sessions. I want to offer deep thanks to Gail M. Patten, a former administrator of the History Department at Dartmouth College, who took on the huge task of formatting and proofreading this sprawling project. Just knowing she would handle those onerous necessities allowed me a freedom to focus on my tasks.

Though they will remain anonymous, I want to thank the external reviewers of this project. Again, knowing they would be vetting our efforts and offering advice allowed me to relax. Their comments were invaluable to all of us.

I have had a wildly productive and enjoyable experience working with de Gruyter Verlag in general and with my former and current de Gruyter in-house editors: Heiko Hartmann, Susanne Rade, Manuela Gerlof, Lydia White, Anja Michalski, Myrto Aspioti, and Stella Diedrich. I thank them all for their initial and continuing support of me, of the book series that I edit at de Gruyter, and especially for their enthusiasm and assistance with this project.

Mention of editing and editors brings me to the penultimate issue of this foreword: I specifically asked contributors not to edit out signs – not all signs, anyway – of what they were feeling and thinking that came most directly out of the experiences of spring, summer and fall 2020 before first drafts were due. In this manner, I believe *On Being Adjacent to Historical Violence* serves as a valuable record of what thoughtful individuals were capable of articulating and wanted to articulate at that particular moment in history. This is also why I pushed contributors so hard to complete their essays quickly. And they did. Thank you!

This book is dedicated to the memory of Susanne and Half Zantop and of all individuals whose lives were cut short by the violent actions of other human beings. In the spirit of Xolani Nkosi Johnson, may the memory of all they could do with what they had in the time that they had in the places that they lived be eternal!

3 May 2021                                                    Lebanon, NH, USA

## Works Cited

Bammer, Angelika. *Born After: Reckoning with the German Past*. New York and London: Bloomsbury Academic, 2019.

"Doing Something." (homepage) https://www.facebook.com/groups/458897688025950.

Eligon, John. "Black Doctor Dies of Covid-19 After Complaining of Racist Treatment." *New York Times*, 23 December 2020. Accessed 29 April 2021. https://www.nytimes.com/2020/12/23/us/susan-moore-black-doctor-indiana.html.

Friedlander, Saul, ed. *Probing the Limits of Representation. Nazism and the "Final Solution."* Cambridge, MA: Harvard University Press, 1992.

Gupta, Sujata. "Why African-Americans may be especially vulnerable to COVID-19," *Science Times*, 10 April 2020. Accessed 29 April 2021. https://www.sciencenews.org/article/coronavirus-why-african-americans-vulnerable-covid-19-health-race.

Hirsch, Marianne, and Leo Spitzer. *Ghosts of Home: The Afterlife of Czernowitz in Jewish Memory*. Berkeley: University of California Press, 2010.

Kacandes, Irene. *Daddy's War. Greek American Stories. A Paramemoir*. Lincoln: University of Nebraska Press, 2009.

Kacandes, Irene. "How Co-Witnessing Could Transform the Post-Pandemic World." #10. In *13 Perspectives on the Pandemic: Thinking in a State of Exception*. Edited by Rabea Rittgerodt. Berlin: de Gruyter Verlag, 2020.

Kacandes, Irene. *Talk Fiction: Literature and the Talk Explosion*. Lincoln: University of Nebraska Press, 2001.

Kacandes, Irene. "'Die Ungnade der späten Geburt': Challenges for Central Europeans in the Twenty-First Century. The 2016 Presidential Address." *German Studies Review* 40, no. 2 (2017): 389–405.

Kacandes, Irene, and Yuliya Komska, eds. *Eastern Europe Unmapped: Borders and Peripheries*. New York: Berghahn, 2017.

Kaplan, Alice Yaeger. *French Lessons. A Memoir*. Chicago: University of Chicago Press. 1993.

Krug, Nora. *Belonging (Heimat): A German Reckons with History and Home*. New York: Scribner, 2018.

Myers, David. "Knowing the Victim? Reflections on Empathy, Analogy, and Voice from the Shoah to the Present." 25 June 2020. Accessed 30 April 2021. https://katz.sas.upenn.edu/resources/blog/knowing-victim-reflections-empathy-analogy-and-voice-shoah-present?fbclid=IwAR0e1ONuQbu5DCFvmNkyfvp3MRr3U74dJIC0COX7-0lR9ZKuEdJEcMtNX1E.

"On Being." Accessed 30 April 2021. https://onbeing.org.

Roberts, Sam. "Captivated by the Courageous Spirit of a Small Child Who Was Larger Than Life" which appeared in the print version of the *New York Times* on 4 December 2004. Accessed 29 April 2021. https://www.nytimes.com/2004/12/04/books/captivated-by-the-courageous-spirit-of-a-small-child-who-was-larger.html.

Roseman, Mark. *A Past in Hiding: Memory and Survival in Nazi Germany*. New York: Picador, 2000.

Rothberg, Michael. *The Implicated Subject. Beyond Victims and Perpetrators*. Stanford, CA: Stanford University Press, 2019.

Sharpe, Christina. *In the Wake: On Blackness and Being*. Durham: Duke University Press, 2016.

Thibault, Reis, Andrews Ba Trans, and Vanessa Williams, "The Coronavirus is infecting and killing black Americans at an alarmingly high rate." 7 April 2020. *The Washington Post*. Accessed 29 April 2021. https://www.washingtonpost.com/nation/2020/04/07/coronavirus-is-infecting-killing-black-americans-an-alarmingly-high-rate-post-analysis-shows/?arc404=true.

"Why is the Pandemic Killing So Many Black Americans?" *The Daily*. 20 May 2020. Accessed 29 April 2021. https://www.nytimes.com/2020/05/20/podcasts/the-daily/black-death-rate-coronavirus.html.

Part One: **Connections**

# Marianne Hirsch
# Debts

It was possible that I did owe something to my own family and the families of my friends. That is, to tell their stories as simply as possible, in order, you might say, to save a few lives. Grace Paley

Every day my inbox bursts with messages from the Czernowitz listserv.[1] Mostly, I send them off to the trash-bin over my first coffee. I thought I was done with Czernowitz. I spent more than two decades researching, writing, speaking and publishing about my strong personal connection to a place that can no longer be found in any contemporary atlas. I read memoirs and historical accounts, a pored over old maps and postcards and stared at faces in faded photos, trying to imagine and to animate a world that had long ago been destroyed and dispersed. Together with Leo Spitzer, I wrote about the rich Habsburg and interwar Romanian Jewish life that thrived there, and about the persecution, deportation and survival of the city's Jews during the Second World War.[2] I was able to visit my parents' childhood apartments in crumbling buildings in what is now Chernivtsi in Ukraine, and to retrace their ghetto experiences, in their company. I even traveled to Transnistria, a region now in Ukraine and Moldova to which tens of thousands of Jews from Czernowitz/Cernăuți and Greater Romania were deported and where they suffered starvation, disease and mass murder at the hands of Nazi and Romanian perpetrators. I went there twice, even though my parents had been among the fortunate third of the city's Jews who were able to evade deportation. I had to see and touch not just what they suffered, but also what they feared. These experiences fueled my scholarship and lent personal and political urgency to my writing. And now, seventy-five years after the end of what to me will always be "the war," my parents, my aunts and uncles, along with nearly all their contemporaries have died. I would have liked to think that I have paid my debts to this past by listening to them and by retelling their stories "as simply as possible." By trying to ensure that

---

1 See http://czernowitz.ehpes.com/.
2 Marianne Hirsch and Leo Spitzer, *Ghosts of Home: The Afterlife of Czernowitz in Jewish Memory* (Berkeley: University of California Press, 2010).

---

**Note:** An earlier and shorter version of this article appeared in *a/b: Auto/Biography Studies* 32, no. 2 (2017).

https://doi.org/10.1515/9783110753295-002

the particular Jewish history of this border region, largely missing from most historical accounts of the Holocaust, will not sink into oblivion, I had hoped, in Grace Paley's words, "to save a few lives."

The legacies of genocidal violence, however, are not so neatly put aside. They leave their traces and shadows for decades and generations to come. That's what hit me when, one morning a couple of years ago, I was about to delete yet another message from the Czernowitz listserv. This one was from Stephen Winter of New York,

> My mother, Blanka, turned 93 today. Once in a while her memory gets confused
> Her only wish to me today was to go back to Transnistria and die to be with her brother Leopold
> This made me think . . . no one has ever apologized to her for their cruelty!
> What a load Transnistria has been on my 62 years of life.
> I think she deserves at least 1 apology!

This was quickly followed by a response from Miriam Süss from Melbourne:

> Stephen, as I read your post and tears run down my cheeks, I hear the voice of my 97-year-old mother, who at least once a week cries over her losses and the terrible suffering in Bershad Transnistria.
> A shadow over all our childhoods and indeed our adult lives.

And then a message by Sally Bendersky from Santiago:

> I connect with you, Miriam. . . . My mother passed away several years ago, but the memory of her suffering and the impossibility of enjoying her life is still here.

Who is there who might apologize to 93-year-old Blanka Winter and to her son? Hers is a history that has never been properly acknowledged. In a context in which historical transmission is short-circuited by shifts in national borders, the realignment of political orientations, and the contestation, erasure and forgetting of histories even as devastating as the Romanian Holocaust in Cernăuți and Transnistria, survivors transmit more than memories of wartime suffering to their descendants. Their anxieties and needs, their trauma and mourning, are compounded by the limited possibilities of recognition that exist after decades of impunity, political denial, and contestation. It's these incommensurate affects I hear in the brief outcries from around the globe, written in shorthand and, for most, in a foreign language, forging a digital community of what Eva Hoffman has termed the "postgenerations."[3] That community shares the many

---

**3** Eva Hoffman, *After Such Knowledge: Memory, History and the Legacy of the Holocaust* (London: Public Affairs, 2005).

dimensions of postmemory I've myself experienced and analyzed in the art and writing of those who came "after": mourning for a lost world, the impulse to repair the loss and heal to those who have suffered it, anger about the absence of public recognition, frustration in the face of our own ignorance and impotence.[4] In short, the inheritance of a trauma that survives the survivors, overwhelming the present and hijacking the future.

As familial and affiliative descendants of survivors, we continue to feel the need to research and to tell their stories of injustice and persecution, as well as to reclaim moments of courage and resistance on their part. But as the survivors have themselves now, mostly, left our midst, shouldn't we also ask ourselves what more we can *do* with this persisting legacy in our own present? How can we envision a future that recalls past crimes without being paralyzed by them, and without perpetuating a culture of fear and denial, of nationalism and ethnocentrism that was responsible for those crimes in the first place?

I believe that the structure of postmemory applies to co-witnesses not only of past trauma, but also of contemporary catastrophic events occurring in proximate or distant parts of the world.[5] Might our own subject position as descendants of the Holocaust, in particular, move us to be responsive to the violent histories of others, whether past or present? As the geographical reach of the Czernowitz listserv shows, survivors from Nazi-occupied Europe were dispersed throughout the world, and, despite the early resolve of "never again," we, as their descendants, have witnessed injustice, violence and genocide again and again. Even as postgeneration Jews have inherited, viscerally, what it feels like to be deemed "other," undesirable, and unworthy of life, we have also witnessed and been implicated in the "othering" of our neighbors and compatriots. Particularly, the evolution of the Israeli state, in large part created as a refuge for European Holocaust survivors, has raised the question of how the victims can themselves become victimizers of Palestinian populations who have been expelled from their homes, whose lands have been expropriated and who have been living under decades of violent occupation.

As I witness the alarming growth of nationalism and ethnocentrism, racism and anti-Semitism, often based on past grievances, I am keenly aware of how the memory of painful pasts is often mobilized as an alibi for persecution, exclusion and violence. The memory of the Holocaust is regularly used by the state of Israel and by other countries to instill fear and xenophobia. As descendants of Holocaust

---

**4** Marianne Hirsch, *The Generation of Postmemory: Writing and Visual Culture After the Holocaust* (New York: Columbia University Press, 2012).

**5** On co-witnessing, see Irene Kacandes, "How Co-Witnessing Could Transform the Post-Pandemic World," #10, in *13 Perspectives on the Pandemic: Thinking in a State of Exception*, ed. Rabea Rittgerodt (Berlin: de Gruyter Verlag, 2020).

survivors, I believe we need to resist such instrumentalization and to refuse to allow the suffering of our ancestors to be used to inflict further violence. This is what I feel I owe my parents now, and it is a debt as weighty as the injunction to tell their stories, and one that might be more intractable.

For retrospective or distant co-witnesses, the challenge is to allow ourselves to be vulnerable to what Susan Sontag has called "the pain of others," to recognize the entanglements of our connective histories while, at the same time, resisting an appropriative form of identification and empathy between self and other, past and present.[6] It is to mobilize the vulnerability we have inherited as postgenerations to serve as a platform of attunement and connectivity that reaches beyond identity and ethnicity, in favor of solidarity, co-witnessing and co-resistance.

The afterlives of violent histories are anything but simple. I would like to envision a day when we could emerge from our traumatic legacies feeling lighter because we have paid our debts – not by ceasing to revisit the ghosts of our own past, but by feeling that mobilizing them in our writing and our work has opened a door to a more connective present and future.

# Works Cited

Hirsch, Marianne. *The Generation of Postmemory: Writing and Visual Culture After the Holocaust*. New York: Columbia University Press, 2012.

Hirsch, Marianne, and Leo Spitzer. *Ghosts of Home: The Afterlife of Czernowitz in Jewish Memory*. Berkeley: University of California Press, 2010.

Hoffman, Eva. *After Such Knowledge: Memory, History and the Legacy of the Holocaust*. London: Public Affairs, 2005.

Kacandes, Irene. *Talk Fiction: Literature and the Talk Explosion*. Lincoln: University of Nebraska Press, 2001.

Kacandes, Irene. "How Co-Witnessing Could Transform the Post-Pandemic World." #10. In *13 Perspectives on the Pandemic: Thinking in a State of Exception*. Edited by Rabea Rittgerodt. Berlin: de Gruyter Verlag, 2020.

Paley, Grace. "Debts," *The Collected Stories*. New York: Farrar, Straus & Giroux, 1994.

Sontag, Susan. *Regarding the Pain of Others*. New York: Farrar, Straus & Giroux, 2001.

---

6 Susan Sontag, *Regarding the Pain of Others* (New York: Farrar, Straus & Giroux, 2001).

Amy Shuman, Carol Bohmer, Eric Niyitunga

# Telling Our Own Stories and Speaking on Behalf of Others

Whose stories are told and retold? And who can tell them? Who should not tell them? What stories cannot be told, and when, or to whom? What are the obligations to tell, and what are the restrictions? When we speak on behalf of others, how do we avoid silencing or exploiting them? And if people do speak in their own voices, who is listening, and what difference does it make? These are, of course, familiar questions with many answers. Here, we consider the obligations and objections to tell and retell as central to questions about adjacency. Objections are made from a particular position, as a feminist, as a disability rights activist, as a parent, as a Holocaust survivor, as a child of a Holocaust survivor, or as the person with first-hand experience. These "objections *as*" are one kind of adjacency. Obligations also reveal adjacency, especially the obligation to tell *as* a witness.

Our discussion includes both objections to telling and obligations to tell, but we do not regard these as ends of a continuum. Further complicating these categories are the "undertold," the stories that are difficult to tell; these stand in contrast to the "overtold," the stories that become larger than life and serve as allegories for life experience. Undertold and overtold are problematic categories since they imply a value judgment, as if the undertold should be told and the overtold should not. Both are entangled in adjacency; the value judgments often imply that someone is far from or close to the experiences described. In this essay, we are interested in how proximity and/or adjacency to the events recounted is a part of the mandate to tell or the critique of telling. The overtold usually refers to stories that have traveled so far from the tellers who experienced the events recounted that they have either lost a foothold of significance or have acquired new interpretations that are unrecognizable to or even reprehensible to people who consider themselves to be close to the events. The undertold includes the difficulties people face in recounting stories about atrocities that are narratively without precedent for them and invites the question, how do people narrate the inexplicable?

The category we are describing as undertold has been explored extensively, especially in Holocaust literature and other accounts of narrating trauma. Narrating trauma as part of an historical or legal record is complicated not only by questions of what is unspeakable and unrepresentable but also by the fact that, as Marianne Hirsch and Leo Spitzer observe "Mute testimony, deep embodied

https://doi.org/10.1515/9783110753295-003

memory, is not verifiable."[1] Here, with particular attention to questions of adjacency, we also discuss undertold stories as a dimension of what Marco Jacquemet describes as "asymmetrical encounters,"[2] in which tellers' and listeners' different expectations produce significant misunderstandings.

The category we are describing as overtold encompasses both retellings of personally important stories to already familiar listeners and tellings in new contexts to new and distant audiences. Overtold stories often sidestep questions of what is tellable and who can tell what to whom and, in recontextualizing and repositioning narrative events, sometimes introducing new interpretations. Challenging an ethics of adjacency, some overtellings, especially those told by people who did not experience the events they describe, impose different understandings that can be harmful, stigmatizing, and exploitative to the people whose stories are retold. Inevitably, stories do travel to new and unfamiliar audiences; in fact, such narrations are crucial for remembering past atrocities. This is a contested ground we face in considering adjacency (Shuman, 2006).

To address the complex dynamics of retelling others' narratives, we ask, how do new available narratives emerge, and how does this inexplicability translate in retellings and in objections to retelling? When does telling contribute to safety or recovery and when does it place people in danger? Further, we discuss adjacency as a dimension of the experiences people describe including the complex relationships with allies, advocates, and authorities, both benefactors and persecutors. These complex relationships often compromise any easy classification of hero and villain, friend, and enemy, confidant and betrayer, advocate and gatekeeper.

Our explorations begin with our own collaboration. Amy Shuman, a folklorist and narrative scholar, has been collaborating with Carol Bohmer, a lawyer and sociologist, for the past twenty years to help political asylum seekers with the application process in the U.S. and U.K. and to write about how the process works. Eric Niyitunga, a refugee from Burundi, now a senior lecturer at the University of Johannesburg, South Africa, first encountered Carol Bohmer at King's College, London, where he was her student and where they forged a close relationship of support and trust.[3]

---

1 Marianne Hirsch and Leo Spitzer, "The witness in the archive: Holocaust studies/memory studies," *Memory Studies* 2.2 (2009): 151–170. Here, 159.

2 Marco Jacquemet,"Crosstalk 2.0: Asylum and Communicative Breakdowns," *Text & Talk* 31, no. 4 (2011): 475–497, here 477.

3 Carol Bohmer and Amy Shuman, *Rejecting Refugees: Political Asylum in the 21st Century* (New York: Routledge, 2007) and *Political Asylum Deceptions: The Culture of Suspicion* (New York: Springer, 2017).

Each of us has our own personal histories of adjacency. We begin with Amy Shuman's experiences as an ally and advocate for her disabled son, and then our discussion turns to Eric Niyitunga's experiences as a refugee. Dr. Niyitunga told his story to Drs. Bohmer and Shuman soon after arriving in London; within his longer account of many years of fleeing danger, he told one particularly remarkable story. It is a story of a reversal of fortunes in which his life depended on being recognized as a friend by enemy forces. In our initial conversations, he was very selective about which struggles he described and glossed over many of his terrifying experiences in many refugee camps. In our recent conversations, he said "I'm ready to talk about anything because by opening up, you can also help other people who are refugees or asylum seekers." We write about the injustices of the refugee camps elsewhere.[4] Also, we note that disability is itself a significant dimension of refugee camps, and to some extent, we can regard most refugees as experiencing disability, if defined as the experience of social as well as embodied exclusions and stigma, in addition to assessments of who is deserving of a livelihood. Dr. Niyitunga's personal story is a story about adjacency to violence; it is also a story about the recognition of mutual interdependence, a topic that makes it tellable and intelligible beyond the characters involved. Before turning to it, we discuss objections to particular kinds of disability narratives, and the ways that we are implicated in those objections, as an important dimension of understanding adjacency.

Disability rights discourses about *not* telling other people's stories stand in stark contrast to the obligation, even plea, to speak on behalf of others who have been victims of genocide or other atrocities. For example, the poet Derek Mahon includes the following lines in his well-known poem "A Disused Shed in Co. Wexford," written in 1973, a year after Bloody Sunday:

They are begging us, you see, in
    their wordless way,
To do something, to speak on
    their behalf
Or at least not to close the door
    again

The plea for someone to speak is often drowned out in current debates about the representation of suffering by outsiders. (And the term outsider or other is often vague in these debates.) Objecting to the protests against white artist Dana Schutz'

**4** Eric Niyitunga, "Impact of Global War on Terror on the Rights of Refugees and Asylum Seekers in the Horn of Africa – An Analysis of the Somali Refugees and Asylum Seekers in Kenya," *Africa Insight* 45, no. 3 (2015): 46–62.

painting of Emmett Till, Coco Fusco questions the position that, "Whites aren't allowed to depict black suffering." The current controversy about the postponement of the Philip Guston retrospective show reflects similar objections.[5] When is it acceptable for people to represent the suffering of others? In her defense of Schutz' painting, Fusco points out that Emit Till's mother "wanted the *world* to see what those men had done to her son" (Fusco's emphasis), an argument in defense of Schutz's painting. Anna De Fina and others complicate the question of how both narrators (and artists) and the critics who object to their representations make identity claims, both about themselves and about their membership in particular groups.[6] Fusco's larger point is that, "A reasoned conversation about how artists and curators of all backgrounds represent collective traumas and racial injustice would, in an ideal world, be a regular occurrence in art museums and schools."[7] If we are to have this conversation, we need to consider both how some stories become tellable, iconic, representations of injustice and how the tellings can be understood as responses to solicitations, both implicating tellers and listeners and drawing on preconceived, assumed relations of adjacency.

In contrast to those who beg us to speak on their behalf, disability rights activists caution against those who control the stories, the decisions, the policies, and the modes of discourse about disability. They insist on alternatives to the medical, educational, or other "expert" narratives that portray them as impaired and that promote the unquestioned privileging of able-bodied neurotypical lives.

In disability rights discourses, the mantra "nothing about us without us," refers to the many situations when people with disabilities have not been included in conversations and decisions about them.[8] Speaking for oneself is a way of controlling knowledge, but the issue is not primarily about the epistemological accuracy of accounts; instead, it's about the dominance of particular kinds of accounts, records, documents, or portrayals that determine who counts as a human.

---

5 Alex Greenberger, "Controversial Philip Guston Show Postponement Met with Shock and Anger from Art Community," *ARTnews*, Sept. 28 (2020), https://www.artnews.com/art-news/news/philip-guston-retrospective-postponement-reactions-1234571844/.
6 Anna De Fina, "Discourse and Identity," *Discourse Studies: A Multidisciplinary Introduction* 263 (2011): 263–282.
7 Coco Fusco, "Censorship, Not the Painting, Must Go: On Dana Schutz's Image of Emmett Till," *HyperAllergic*, accessed 27 March 2017, https://hyperallergic.com/368290/censorship-not-the-painting-must-go-on-dana-schutzs-image-of-emmett-till/ (2017).
8 James I. Charlton, *Nothing About Us Without Us: Disability Oppression and Empowerment* (Berkeley: University of California Press, 2000).

In her widely-read essay "The Problem of Speaking for Others," Linda Martin Alcoff outlines both the presumptions of speaking as an expert and the objections to being spoken for by "others." Her essay recommends attention to the "particular power relations and discursive effects involved" in speaking for others.[9] As she points out, speaking for others often reinscribes hierarchies. Almost as an afterthought, in the final lines of her essay, she makes a nod to the possible benefits of advocacy. This final point directs us to one of the stickiest dimensions of the disability rights mantra, the responsibilities of allies and advocates.

Parents of people with disabilities often necessarily speak for their children, especially when for whatever reason, they cannot speak for themselves. However, even if parents are necessary advocates, they are not necessarily welcome advocates. Their advocacy can come at a cost to people with disabilities, as evidenced by many of the memoirs written by parents who describe their experiences as traumatic, sometimes only initially traumatic, but in any case, consigning the label of tragedy to the lives of people with disabilities. The questions at stake in these objections are, Whose life? Whose tragedy? As we discuss below, some of the same issues arise in questions about who has the right to represent genocide and other atrocities.

Amy Shuman has considered the problem of speaking for someone else as a parent of her thirty-four-year-old son, Colin, who has an intellectual disability and a pronounced language disability and who is nonetheless a great communicator who enjoys telling stories. Like others, she and her son co-tell his stories.[10] With little explanation, saying only that he finds a particular topic embarrassing or that he wants to tell a particular story, he is very selective about the stories he wants to tell. His objections and preferences do not address either accuracy or the topic of the story but instead concern the relationships conveyed by the stories and the ways that the story positions him in relation to the characters in the story and his listeners. Michael Bamberg identifies three kinds of positioning in narrative: 1) the ways that characters within the story are positioned in relation to each other; 2) the ways that the listeners are positioned; 3) the ways that tellers are positioned in relation to themselves.[11] (We will return to this model below and suggest two additional categories.) Colin's cooperatively told narratives correspond to these

**9** Linda Martin Alcoff, "The Problem of Speaking for Others," in *Who Can Speak: Authority and Critical Identity*, eds. J. Roof and R. Wiegman (Urbana: University of Illinois Press, 1995), 111.

**10** Charles Goodwin, "A Competent Speaker Who Can't Speak: The Social Life of Aphasia," *Journal of Linguistic Anthropology* 14, no. 2 (2004): 151–170.

**11** Michael Bamberg, "Positioning Between Structure and Performance," *Journal of Narrative and Life History* 7, nos. 1–4 (1997): 337.

categories; once he establishes a story in his repertoire, he wants to tell it to everyone in his life; in fact, telling it is a way of establishing relationships with others, no matter whether or not they are familiar with the characters in the story.[12] He describes the actions, but not the motivations, of the characters in the story. He is especially interested in telling stories that position him positively, but he is often satisfied to just claim, "I was there," even if, for example, in stories about his birth, he is not recounting his own memories. Both co-witnessing, in which the listener/witness can serve as a conduit for memories[13] and cooperative telling involve interdependency, a kind of adjacency that invokes ethics and sometimes includes advocacy. Linguist Val Williams argues for cooperative telling, or what she calls "inclusive conversation" to provide an opportunity for people with intellectual disabilities to be heard (2011). Parents' stories in general point to the question of whether advocacy confers the right to tell a story, and whether advocates can claim it also their story.

Disability scholars Faye Ginsburg and Rayna Rapp provide a helpful model for understanding relationships among advocates, allies, and people with disabilities. They observe "the cultural work performed by the circulation of kinship narratives through various public media as an essential element in the refiguring of the body politic as envisioned by advocates of both disability and reproductive rights."[14] Disability is the subject of competing discourses in which biomedical narratives often exclude or render invisible the personal stories of people with disabilities and their families. Ginsburg and Rapp are particularly interested in the ways that conversations about reproductive choice sit at the nexus of the very different narratives about genetic information and about parental choices. As they note, these narratives depend upon but offer very different assessments of "dependency versus autonomy; intimacy versus authority; the acceptance of caretaking versus its rejection; normative cultural scripts versus alternative, more inclusive "rewritings."[15] As they point out, these different assessments are not merely a matter of perspective. The narratives characterize what Ginsburg and Rapp call "anomalous

---

12 Amy Shuman and Katharine Young, "The Body as Medium: a Phenomenological Approach to the Production of affect in Narrative," in *The Edinburgh Companion to Contemporary Narrative Theories,* eds. Z. Dinnen and R. Warhol (Edinburgh: Edinburgh University Press, 2018): 399–416.

13 Irene Kacandes, *Daddy's War: Greek American Stories* (Lincoln: University of Nebraska Press, 2009).

14 Rayna Rapp and Faye D. Ginsburg, "Enabling Disability: Rewriting Kinship, Reimagining Citizenship," *Public Culture* 13, no. 3 (2001): 535.

15 Ginsburg and Rapp, "Enabling Disability," 536.

children" (who are unlike other family members) in larger religious, political, medical, and economic discourses, resulting in a "burgeoning public circulation of intimate disability stories."[16] Reframed for our discussion, an "anomaly" is a person whose adjacency is in question; anomalous children are both excluded and included.

We all participate in the public circulation of competing discourses. The questions are, how do these discourses position us differently? How do they foster alliances? Disability rights activists acknowledge the alliances fostered by parents' narratives about their children's disabilities but also object to the ways that parents' narratives perpetuate stereotypes about people with disabilities as tragic. Alison Piepmeier, who, like Amy Shuman is the parent of a child with cognitive disabilities, describes a problematic pattern in parents' memoirs: "Many of the memoirs I have read reinforce and thereby strengthen our culture's dehumanizing stereotypes that surround and define disability. Through their use of grief, their emphasis on a limited medicalized model, and their framing of the child's disabilities, these memoirs often represent the child not as a person but as a problem with which the parents have had to grapple."[17] The parents are not distant observers accused of voyeuristically exploiting their children. Unlike the protests against Schutz' painting, the parents are allies and advocates, adjacent to their children's experiences but, according to their critics, imposing an unwanted narrative. The parents' memoirs participate in a network in which some readers/parents experience affirmation and recognition. They often foster the creation of virtual networks, in which individuals create bonds with strangers who have had similar experiences. Ginsburg and Rapp refer to this as "mediated kinship,"[18] which, they argue, has the capacity to provide counter-narratives that contribute to changes in the cultural imagination. Importantly, these new kinship relationships, like many other newly forged networks, are formed in response to exigency and exist in relation to the unequal, unjust, non-reciprocal power relations that, in this case, victimize people with disabilities and their families.

These virtual networks and new kinship formations often describe entering an unfamiliar world, outside of expectations and, in the case of disability, stigmatized and made even more unfamiliar by being categorized as "not normal." Many people with disabilities, and many parents, including Amy Shuman, do

---

**16** Ginsburg and Rapp, "Enabling Disability," 537.
**17** Alison Piepmeier, "Saints, Sages, and Victims: Endorsement of and Resistance to Cultural Stereotypes in Memoirs by Parents of Children with Disabilities," *Disability Studies Quarterly* 32, no. 1 (2012): np.
**18** Ginsburg and Rapp, "Enabling Disability," 550.

not find any recognition in the typical parent memoir and instead regard them as yet another zone of unfamiliarity. There are exceptions, of course, that do not rely on the overdetermined narrative of tragedy and grief followed by acceptance and that instead refuse the categories of normal.[19] Parents' narratives about their children's disabilities are an example of the overtold not only because they can insist on overdetermined narrative of but also because they can promote what disability rights activists call "inspiration porn," stories that perpetuate stigma at the same time as they convey a seemingly positive narrative of overcoming difficulties.[20] The celebration of often symbolic accomplishments obscures the injustices and oppression that prevent people with disabilities from having access to resources and participating in civic life.

If disability narratives are overtold, refugee narratives are undertold. As Michael Jackson says, "Many writers have noted that one of the most arresting things about refugee stories – and, more generally, the stories of people in crisis, in torture, and in flight – is that life ceases to be narratable."[21] Like the competing disability discourses, political asylum narratives are also contentious, representing the very different interests, agendas, and knowledge of the asylum seekers and the immigration officials. Both have connections to entrenched historical narratives, though those histories are completely different. And in both cases, of course, neither side is homogenous. Their similarity is in the ways that personal stories converge with political public narratives as part of a continuous feedback loop: individuals looking for ways to understand their experiences find recognition, or create counter-narratives to the public narratives, and the public narratives draw on personal narratives in widening examples of experiences. Some of those personal narratives become iconic and influential, even reshaping paradigms.

Eric Niyitunga's story begins when he was kidnapped from his family home in Burundi by Hutu militia and held captive in the forest for three months because he refused to join the army. As he explained to his captors, he could not join them for two reasons, first, because his faith as a Seventh Day Adventist prevented him from killing and second, because if he joined the militia, he would be

---

**19** Michael Bérubé, *Life as We Know It: A Father, a Family, and an Exceptional Child* (New York: Vintage, 1998) and *The Secret Life of Stories: From Don Quixote to Harry Potter, How Understanding Intellectual Disability Transforms the Way We Read* (New York: New York University Press, 2018).
**20** Stella Young, "Inspiration Porn and the Objectification of Disability," TEDxSydney. Retrieved from http://tedxsydney.com/site/item.cfm (2014).
**21** Michael Jackson, *The Politics of Storytelling: Variations on a Theme by Hannah Arendt* (Copenhagen: Museum Tusculanum Press, 2013), 91.

killing members of his mother's, Tutsi, tribe. As the child of a Hutu father and a Tutsi mother, Dr. Niyitunga was, like many people in his community, half-caste. He explains that the community was not divided by either language or cultural practices, and inter-marriage was common. The Civil War, beginning in 1993 and lasting decades, disrupted those relationships. Everyone was asked to take sides (one kind of solicitation), and Dr. Niyitunga refused. In the forest, he managed to eat because the captives prepared food for the militia, but otherwise he lived in great hardship, for example with no opportunity to bathe for those three months. In response to his explanation for why he couldn't join the militia, one of his captors beat him, saying, "Dialogue can't work because this government has refused, and we must fight against it with weapons." Dr. Niyitunga responded "I cannot. Kill me or don't kill me. I am in your custody. At your mercy. I will never take a gun and fight."

After two months, a convoy came from Tanzania with 300 soldiers. They were, by then, near the Democratic Republic of Congo, about a month's walk from the Tanzania border. And then, in an instant, Dr. Niyitunga's fortunes changed, when the leader of the convoy recognized Dr. Niyitunga. They had gone to school together, and during that time, the leader, who was not a good student, depended on Dr. Niyitunga, who even took his exams for him. Dr. Niyitunga said, "I would do two exams, one with my name and then an exam for him." He changed the name and handwriting. And now, in the forest, their dependency was reversed.

The leader of the convoy said, "Ah Eric, why are you here? Who brought you here?"

Dr. Niyitunga told him, and the leader arranged for his release and his return to his home. In his narrative, Dr. Niyitunga describes their exchange. The convoy leader said, "I just need to know the commander who got you here." (He was a general who has since become one of the key prominent leaders of Burundi.) This former classmate, who knew that Dr. Niytiunga's refusal to join the militia was genuine, said, "You cannot fight." This is an important part of Dr. Niyitunga's account; it is the moment when the former classmate, in his own words, acknowledges Dr. Niyitunga's stance. He continues, quoting the convoy leader's instructions to the captors, "You must respect his faith. So, I'm ordering him to be taken back to where you took him from. And this, we must do it in the next two weeks. Just prepare the soldiers. Take him back and make sure his father and mother see him and he sees you as well. Then come back."

As we said, this extraordinary life-saving reversal of fortune is only a small part of Dr. Niyitunga's narrative of many years as a refugee in which he encountered both benefactors and tormenters. Today, Dr. Niyitunga frames the narrative in two ways, first, from his present position in which he actively helps local refugees in South Africa by providing them with clothing and other resources; and second, in his more general observations about the subjectivity of the refugee. He says that as a refugee, "you feel not welcomed as a person in the community," referring to the long process of officials requesting documentation that refugees do not have and thus being denied food and other fundamental needs for survival. These two frames are, of course, connected. Dr. Niyitunga explains that he helps refugees currently *as* someone who has suffered as a refugee. The obligation, whether to tell a story, to recognize others, to help others, is not only personal; it also represents an adjacency, *as* someone positioned by events in a particular way, responding to a solicitation. Refugees more often experience refusal, rather than connection and recognition; moments like the exchange between Dr. Niyitunga and his former classmate are rare. The frames of connection/recognition and refusal also reference the crucial different possibilities for adjacency in refugee experience.

Dr. Niyitunga's encounter with his former classmate exhibits Bamberg's three kinds of positioning, described above but these three positions are not sufficient to account for adjacency and we must additional consider 4) the multiple encounters Dr. Niyitunga experienced as a child to his parent, captive to his captors, object of suspicion in refugee camps, benefactor to other refugees, etc.; and 5) larger frameworks of dependency and privilege in which we recognize our mutual responsibility and the ways we become implicated in each other's lives.

Many scholars have focused on the final category and the ways that we are implicated in each other's lives. Positioning is not only a matter of a choice to promote particular identities or relationships. The positions are both created by our narratives and also exist prior to them. Judith Butler writes that "we are acted on, and solicited, ethically, prior to any clear sense of choice."[22] Co-witnessing, as described by Irene Kacandes and others is a response to a solicitation.[23]

In his story, after encountering his classmate who recognized him and appointed soldiers to return him to his home, Dr. Niyitunga says, "I became now a person because he was so respected." He went from being a captive, a nonperson, to a person. Personhood is both deprived by and conferred by others. In Dr. Niyitunga's narrative, the commander makes an ethical decision based on

---

**22** Judith Butler, "Precarious Life, Vulnerability, and the Ethics of Cohabitation," *The Journal of Speculative Philosophy* 26, no. 2 (2012): 139.
**23** Irene Kacandes, "Narrative Witnessing as Memory Work: Reading Gertrud Kolmar's *A Jewish Mother*," in *Acts of Memory: Cultural Recall in the Present* (Hanover, NH: University Press of New England, 1999), 55–74, and Kacandes, *Daddy's War*.

recognition. For Dr. Niyitunga, the story is not only about being released from captivity; it's also about being recognized as a person who could not join the militia because of his faith and because he would not take up arms against his mother's tribe.

Many years ago, Amy Shuman published an essay based on her research on stories about coincidental meetings, small world stories, in which, at the end, the tellers often remark, "small world, isn't it?" Dr. Niyitunga's story is not such a story. As Dr. Shuman observes in her essay, small world stories are a luxury; although they often include near misses or lucky rescues, they are told in a world in which life and death do not depend on the possibility of meeting a friend in an unlikely place or discovering that one has a connection to a seeming stranger. Being a refugee, by definition, means that the assumed categories of strangers and friends have been disrupted. In this disrupted state, as Hannah Arendt famously said, refugees are pariahs.[24] Dr. Niyitunga similarly observes that in addition to being "persecuted at home and marginalized abroad," refugees are, in some cases, categorized as terrorists.[25] The categories of stranger and friend, ally and perpetrator, are overwhelmed by questions of security. Thus, in Dr. Niyitunga's story, the moment of recognition, the fortune-changing, life-saving event, is not a coincidence but an occasion in which two people recognized their relationships to each other, putting aside their overdetermined roles defined by the conflict.

Dr. Niyitunga's scholarship focuses on diplomacy, state sovereignty, and the state of refugees, not his personal struggles. However, his personal story reproduces some of the same issues he addresses in his scholarship, especially the non-personhood of the refugee and the recognition of obligations. In both his personal story and in his scholarship, he observes the fact that the relationships are not reciprocal but instead reveal imbalances of power and injustices that position the refugee as victim, even terrorist, rather than deserving of the refuge dictated by law. It is these imbalances and injustices that are also at the heart of objections to particular narratives.

The narratives political asylum seekers and refugees tell are about complex relationships, whether about neighbors who have become enemies, enemies who become benefactors, or tribal affiliations that suddenly require secrecy and careful, clandestine arrangements. Thus, the simple act of reporting the atrocities people have experienced is embedded in constant assessments of who is safe,

---

24 Hannah Arendt, "We Refugees," in *International Refugee Law*, ed. Hélène Lambert (New York: Routledge, 2017), 3–12.
25 Niyitunga, "Impact of Global War on Terror," 48.

who is under suspicion, and who is dangerous. As many scholars have observed, telling one's own story, then, is not about assessments of accuracy but instead is an assessment of a wide range of considerations, from social status to moral stance to danger/safety.[26] As Coco Fusco argued, the question is not who does this story belong to or who can represent someone else's suffering. It is more important, exigent, crucial, to know that telling stories cannot restore balance or reciprocity in the face of injustices. It is sometimes important for people to tell their own stories.

We need to better understand the positions from which people speak, whether speaking for themselves or for others, or as a collective, or don't speak, either because they refuse to or can't. And when they can't speak, is it because others are speaking for them, supplanting their voices? Amy Shuman wrestles with speaking for and with her son, knowing that when he speaks alone, he is rarely heard. Amy Shuman and Carol Bohmer have spent decades writing about asylum seekers and being careful to protect the identities of the individuals they wrote about. Eric Niyitunga speaks for himself, but the story we report here is about speaking and not being heard by his captors and then speaking and being heard by the leader who recognized him. As he says, he "again became a person." In each of these situations, telling our own and others' stories is a response to a solicitation implicating our mutual interdependencies. We argue for a more complex understanding of what it means to be adjacent to a situation, to own a story as one's own, to object or fulfill an obligation to tell.

# Works Cited

Alcoff, Linda Martin. "The Problem of Speaking for Others." In *Who can Speak: Authority and Critical Identity*, 97–119. Edited by J. Roof and R. Wiegman. Urbana: University of Illinois Press, 1995.

Arendt, Hannah. "We Refugees." In *International Refugee Law*. Edited by Hélène Lambert. New York: Routledge, 2017.

Bamberg, Michael. "Positioning Between Structure and Performance." *Journal of Narrative and Life History* 7, nos. 1–4 (1997): 335–342.

Bérubé, Michael. *Life as We Know It: A Father, a Family, and an Exceptional Child*. New York: Vintage, 1998.

Bérubé, Michael. *The Secret Life of Stories: From Don Quixote to Harry Potter, How Understanding Intellectual Disability Transforms the Way We Read*. New York: New York University Press, 2018.

---

**26** Kacandes, *Daddy's War*.

Bohmer, Carol, and Amy Shuman. *Rejecting Refugees: Political Asylum in the 21st Century*. New York: Routledge, 2007.

Bohmer, Carol, and Amy Shuman. *Political Asylum Deceptions: The Culture of Suspicion*. New York: Springer, 2017.

Butler, Judith. "Precarious Life, Vulnerability, and The Ethics of Cohabitation." *The Journal of Speculative Philosophy* 26, no. 2 (2012): 134–151.

Charlton, James I. *Nothing About Us Without Us: Disability Oppression and Empowerment*. Berkeley: University of California Press, 2000.

De Fina, Anna. "Discourse and Identity." *Discourse Studies: A Multidisciplinary Introduction* 263 (2011): 263–282.

Fusco, Coco. "Censorship, Not the Painting, Must Go: On Dana Schutz's Image of Emmett Till." *HyperAllergic*. Accessed 27 March 2017. https://hyperallergic.com/368290/censorship-not-the-painting-must-go-on-dana-schutzs-image-of-emmett-till. (2017).

Goodwin, Charles. "A Competent Speaker Who Can't Speak: The Social Life of Aphasia." *Journal of Linguistic Anthropology* 14, no. 2 (2004): 151–170.

Greenberger, Alex. "Controversial Philip Guston Show Postponement Met with Shock and Anger from Art Community." *ARTnews*, 28 September 2020. https://www.artnews.com/art-news/news/philip-guston-retrospective-postponement-reactions-1234571844/.

Grue, Jan. "The Problem With Inspiration Porn: A Tentative Definition and a Provisional Critique." *Disability & Society* 31, no. 6 (2016): 838–849.

Hirsch, Marianne, and Leo Spitzer. "The witness in the archive: Holocaust studies/*memory studies*." *Memory Studies* 2.2 (2009): 151–170.

Ionescu, Arleen, and Anne-Marie Callus. "Encounters Between Disability Studies and Critical Trauma Studies: Introduction." *Word and Text: A Journal of Literary Studies and Linguistics* 8 (2018): 5–34.

Jackson, Michael. *The Politics of Storytelling: Variations on a Theme by Hannah Arendt*. Copenhagen: Museum Tusculanum Press, 2013.

Jacquemet, Marco. "Crosstalk 2.0: Asylum and Communicative Breakdowns." *Text & Talk* 31, no. 4 (2011): 475–497.

Kacandes, Irene. "Narrative Witnessing as Memory Work: Reading Gertrud Kolmar's *A Jewish Mother*." In *Acts of Memory: Cultural Recall in the Present*. Edited by Mieke Bal, Jonathan Crewe, Leo Spitzer, 55–74. Hanover, NH: University Press of New England, 1999.

Kacandes, Irene. *Daddy's War: Greek American Stories. A Paramemoir*. Lincoln: University of Nebraska Press, 2009.

Mahon, Derek. "A Disused Shed in Co. Wexford." *The Poetry Ireland Review* 101 (2010): 38–39.

Mintz, Susannah B. "Strangers Within: On Reading Disability Memoir." *Life Writing* 9, no. 4 (2012): 435–443.

Niyitunga, Eric Blanco. "Impact of Global War on Terror on the Rights of Refugees and Asylum Seekers in the Horn of Africa – An Analysis of the Somali Refugees and Asylum Seekers in Kenya." *Africa Insight* 45, no. 3 (2015): 46–62.

Piepmeier, Alison. "Saints, Sages, and Victims: Endorsement Of and Resistance to Cultural Stereotypes in Memoirs By Parents of Children With Disabilities." *Disability Studies Quarterly* 32, no. 1 (2012).

Rapp, Rayna, and Faye D. Ginsburg. "Enabling Disability: Rewriting Kinship, Reimagining Citizenship." *Public Culture* 13, no. 3 (2001): 533–556.

Shuman, Amy. "Entitlement and Empathy in Personal Narrative." *Narrative Inquiry* 16, no. 1 (2006): 148–155.

Shuman, Amy, and Katharine Young. "The Body as Medium: A Phenomenological Approach to the Production of Affect in Narrative." In *The Edinburgh Companion to Contemporary Narrative Theories*. Edited by Z. Dinnen and R. Warholz, 399–416. Edinburgh: Edinburgh University Press, 2018.

Williams, Val. *Disability and Discourse: Analyzing Inclusive Conversation with People with Intellectual Disabilities*. Hoboken, NJ: John Wiley & Sons, 2011.

Young, Stella. "Inspiration Porn and the Objectification of Disability."TEDxSydney. Retrieved from http://tedxsydney.com/site/item.Cfm (2014).

Priscilla Layne
# Suspicious: On Being Policed in an Anti-Black World

" . . . the evidence we have of racism and sexism is deemed insufficient *because* of racism and sexism. Indeed, racism and sexism work by disregarding evidence or by rendering evidence unreliable or suspicious – often by rendering those who have direct experience of racism and sexism unreliable and suspicious."

Sara Ahmed, "Suspicious"

the equation is the same worldwide
the amount of melanin produces your
racial profile, your race is profiled and
your profile is suspicious . . .

Excerpt taken from Trisha P. Schultz's poem "Suspicious"

I begin with these two excerpts because of the many ways in which they reso-nate with my experience moving through the world as a Black woman, both in the United States and in Germany. Trisha P. Schultz is a Black German artist from Bremen who studied American Studies and Cultural Studies and writes in German and English. Her transnational perspective is reflected in her poem, where she addresses sociopolitical realities that crisscross the Atlantic, lament-ing the people who have been "killed for wearing a hoodie or hijab." Schultz's use of the term "suspicious" might lead one to believe she is only speaking of those suspected of having committed a criminal act. But the truth is, the kind of policing identified by Schultz is not just committed by policing agencies and it's not just about policing a person's behavior as it relates to a code of law. Schultz is naming a policing that is directed towards *all* Black bodies, not to determine whether they have transgressed some law, but because their very presence is understood to be an act of transgression. In this essay I want to reflect on the ways that my Black, femme body has been policed in Germany. In particular, I want to link different kinds of policing to variations of implicit violence directed at Black people. This is not to say that the types of violence I have experienced are more devastating than others. I realize, that as a straight, cisgender, mid-dle-class African American woman, there are types of violence that my sexual-ity, gender, class and citizenship protect me from. These are just snapshots of my experience that help reveal the ways in which anti-Blackness permeates

https://doi.org/10.1515/9783110753295-004

German society. In a way, these examples serve as the kind of evidence that is often deemed "insufficient," as Sara Ahmed puts it, because not believing People of Color is such a central part of maintaining a racist society.

I've chosen to focus on my experiences in Germany, because white Germans often ask me where I'd rather like to live, hoping that I will say Germany because it is allegedly less racist than the United States. And I used to give in to their desire for exceptionalism, because my experience with American police had been so much more traumatic. In the United States, when I encountered police, it was always only as a Black person. Regardless of my age, whether I was ten, eighteen or twenty-four, each and every encounter I had with police positioned me as a criminal. In Germany my experiences had been different. Police weren't necessarily hostile towards me; they often ignored me. For a long time, I thought that was because Germans acknowledged more layers of my identity than Americans did. Strolling around Tübingen, I wasn't just a Black woman, I was a visiting scholar; someone teaching a course at the university and staying in university housing. But looking back on these experiences now, I realize that any neutral experiences I had with German police had to do with mitigating factors. For example, I remember sitting in the S-Bahn (a train system that's part of city public transportation), with a bunch of German punks returning to Berlin from a show in Potsdam. We were all drinking, there were beer bottles scattered throughout the car and people slouched in the seats. When two police officers walked by, I thought we were done for. I was familiar with police harassment at punk shows in my native Chicago. I was baffled when they just kept walking, barely noticing us. At the time, I thought how progressive it was that they didn't single me out as the only Black person and chastise me for my behavior. Now I realize that in that moment I was protected by the white privilege of the people I happened to be traveling with. If I had been a lone Black passenger behaving in the same way, things probably would have gone much differently. The same process of demystification happened during my stay in Tübingen. At first, I reveled in being in the quaint town. I camped out in the library, took my son to play at the Botanic Garden and regularly ate ice cream with him in the Altstadt. I never felt like my Blackness was a problem. Until I met Black people who actually lived in Tübingen and weren't, like me, visiting scholars there for a few months. They told me about the racism they dealt with, the feeling of exclusion. And then I realized that what I interpreted as the absence of racism, was really the privilege of being an African American scholar there temporarily. Like W. E. B. DuBois and Audre Lorde before me, I was a special guest whose presence could be functionalized to demonstrate how *not racist* Germany was. But this experience could never match that of the Black people who grew up there and lived there permanently. It is because of these realizations that I no longer answer the question, "Where would you rather live, Germany or the US?" I simply respond, "The entire world is anti-Black, so it doesn't really matter where I live."

When I refer to racism in Germany, white Germans usually try to discredit my claims. They say, I must be talking about xenophobia, because "there is no racism in Germany." And this stance probably has a lot to do with the tendency in Germany to only ever understand racism in its extreme cases. Racism is a group of neo-Nazis beating a Black man unprovoked. Racism is an arson attack on an asylum seekers' home. But this understanding of racism as spectacle ignores all of the other, non-violent ways in which racism presents itself in Germany. Thus, in order to counter the push back I receive when I describe Germany as anti-Black, in this essay I want to focus on the different kinds of "quotidian violence" I have faced there. Kara Keeling describes "quotidian violence" as:

> . . . the violence that the reproduction, reinscription, and survival of what exists relies upon and enacts in order to manage . . . 'the only lasting truth.' In other words, quotidian violence names the violence that maintains a temporality and a spatial logic hostile to the change and chance immanent in each now . . .[1]

In short, quotidian violence maintains the white supremacy of German society.

## Linguistic Policing/ Linguistic Violence

Linguistic violence can refer to "excitable speech" or derogative terms others have used to address me in Germany, most notably the N-word. I can't recall all the times I have had this slur directed at me. But there are a few particular moments that stand out. For example, the time a neo-Nazi yelled at me on the train platform at Berlin-Köpenick because I was eating an ice cream cone: "Ich hoffe, Dir schmeckt das Eis, Du N****" (I hope the ice cream tastes good, you n*****). One would think that would be the time I felt most threatened. But I was standing with my then boyfriend at the time, and I felt confident that if someone physically attacked me, he would defend me. It's hard to say what offended the neo-Nazi most. Was it my presence as a Black woman in Berlin; in the "no-go-area" of the East? Or was it the presence of my white boyfriend standing next to me? Or perhaps it was my unapologetic Black joy – that I staked a claim to absent-mindedly enjoying an ice cream on a summer day along with the white people in the neighborhood. Who knows? But I knew my silent refusal to acknowledge his comment

---

1 Kara Keeling, *Queer Times, Black Futures* (New York: New York University Press, 2019), 17.

would hurt him more than anything else I could do. I've had several run-ins with skinheads when I was a student in Germany. These all remained verbal confrontations, but I was worried they'd escalate. However, the time when linguistic violence hurt most was when someone yelled N**** at me from a car while I was walking around in Kreuzberg. Out of the three years I lived in Berlin, though I've lived all over the city, the very first year was in Kreuzberg. It was my happy place: a place of punk locales and bars, record shops and bookstores, my favorite falafel and beautiful, shady parks. I felt comforted by the number of Black and brown faces I saw on its streets and by the graffitied buildings and posters advertising concerts. That's what made hearing the N-word there feel so unsettling. It was as if the speaker wanted to remind me that I couldn't have a safe space in Berlin, not even among People of Color.

Some linguistic violence can appear quite banal to bystanders. Like when a bus driver told me "Wir sind nicht hier in Afrika" (We're not in Africa) because I mistakenly tried to board the bus at the wrong stop. Linguistic policing can also take the form of the persistent questioning that so many BIPOC in Germany must contend with. "Wo kommst Du her?" "Wieso kannst Du so gut Deutsch?" (Where are you from? Why do you speak such good German?). These are questions that not only interrogate my right to be in a space, but also interrogate my right to speak the language. Germany's distinction between formal and informal speech also opens up new possibilities for policing myself. When I first went to Germany in 2001, it was still common practice that adults used the formal version of "you" (Sie) when speaking with other adults they didn't know and the informal "you" (Du) to refer to friends and family and to children. Occasionally you would also here strangers use "Du" in less formal milieus, like the punk bars and clubs I frequented in Kreuzberg. The unspoken rules of this practice made navigating spaces more difficult in German. I made sure to use "Sie" in the supermarket and at the bank. But what should I use at a punk concert, when ordering a drink or checking out the merchandise table? In any case, when a stranger did address me with "Du," I often had to search a catalogue of reasons for why they were being so informal. Because, if we clearly did not share some subcultural affinity, it may have been because I looked young, it may have been because I was a Black woman and didn't deserve the same respect or because I was a Black woman, they assumed I would be more laidback about these kinds of rules. In any case, one instance stands out in my mind in particular. I had just finished a tutoring job in an apartment building and returned to the building's foyer to retrieve my bike in order to head home. I had left my bike leaning against the wall, under the mailboxes with a lock around the wheel and the frame. When I reached the bottom of the stairs, a tall, white man turned to me and said "Ach, Du warst das!" (Oh, it was you!). I was immediately taken aback. Why was this stranger referring to me as "Du"? And what did he mean by "It was you"? What did I do? I quickly learned he was referring to my bike in the hallway – that this wasn't an appropriate spot for a bike. And though I had seen people throughout Berlin leave their bicycles under the mailboxes, I didn't question his right to police me. I had already learned that the banal, everyday minor transgressions white people commit that go uncommented on can become major offenses when committed by Black bodies. I didn't even bother to express displeasure with his addressing me as "Du." I simply grabbed my bike and slouched away. I soon learned I wouldn't have to find a better place to park my bike for that tutoring job when the pupil's father called me later that day and fired me for speaking "American" with his daughter, rather than "English." This incident is of course not only about race, but about both race and gender. Even if the person only based their word choice off of my youth, as a Black woman I cannot separate these parts of my identity.

That experience happened when I was still an undergraduate – an exchange student at the Free University for the year. But even now, as a tenured professor with graying hair in the front, a few more wrinkles in my face and a child by my side, people may not assume I am a child myself, but I still have to grapple with linguistic violence, just a more sophisticated form – like when people refuse to refer to me as Dr., even after being corrected. This happened when I did an interview with *Der Spiegel* (the approximate German equivalent of *Time* magazine) in the summer of 2020. When they sent me the transcript, I objected to them referring to me as "Frau Layne" (Mrs. Layne) and not "Dr. Layne." I asked them to correct it. My request was ignored. And this happens in the United States too. If I am on a panel with several white men, the audience always addresses them as "Dr." or "Professor," while I am simply called "Miss."

# Policing Space/ Policing Wealth and Resources

In fall 2018, I was spending several months in Berlin as a fellow at the American Academy. I was living on-site, in the privileged district of Wannsee, on the far outskirts of the city. But, as always, on the weekends I ventured to the city center to socialize with friends. And this stay in Berlin had been especially fulfilling, because I had met many more Black people than I'd ever met in Germany – an entire rainbow of the diaspora. Thanks to this network of Black folx, I had been invited to celebrate a Black woman's birthday – a fellow American ex-pat had rented a boat to take us on a tour of the Spree. It was the perfect day for such an outing, and I was eager to get there; I had a backpack full of goodies I was bringing to share. But on my way to the dock, a stranger walked past me and mentioned that my backpack was unzipped. I quickly stopped and set it down on the sidewalk to check to see if anything had fallen out before I could rearrange everything to fit comfortably. I just happened to be next to a restaurant that had several tables lining the sidewalk. Suddenly, a waiter approached me. I can't recall exactly what he said, but he insinuated that I was trying to steal a bottle of wine. He felt that was why my backpack was open on the ground; it must be an intricate ruse to slip the bottle of wine on the table next to me into my bag. For a brief minute, I contemplated the irony of the situation. Here I was, during my umpteenth stay in Berlin – my first time in Berlin as a tenured professor, with a stable income and not only that, on a fellowship from a prestigious organization. I was literally living in a villa in Wannsee by the lake. As someone who grew up the daughter of working-class Caribbean immigrants in Chicago and paid for my year abroad with student loans, I had never been in such a privileged position in my life. And yet, here was a waiter accusing me of stealing what was at most a 10-Euro bottle of wine simply because I was a Black woman. What made the feeling even worse, was that the waiter appeared to be of South Asian descent. It felt as if he was targeting me in an attempt to increase his proximity to whiteness. I felt betrayed by someone who I felt should understand what it's like to be judged based on your race or ethnicity in a predominantly white space. I was incredulous. And though I couldn't think of a good response to him in that moment, other than an angry look, when he returned to the restaurant, I followed him inside and yelled to his face that I could afford to buy several bottles of wine if I chose to. And I called him a racist asshole. Unfortunately, this is just one example of the times I've been policed by other marginalized individuals who feel they can at least attack me, as a Black woman. That has ranged from other People of Color who are trying to increase their proximity to whiteness, to the homeless man who told me to "Go back to Africa" when I didn't respond to his request that I buy a copy of the *Motz*.[2] On that night I was pretty distraught, because my boyfriend had just broken up with me. This is what made it impossible for me to respond to his request for a donation. I merely walked by in silence, not because I didn't care about his suffering, but I was just too engrossed in mine. But he responded to my pain with incredulousness. Because a Black person isn't allowed to have

---

**2** The *Motz* is a Berlin Street magazine that was started in 1995. It was the fusion of two previous magazines: *Mob-Magazin* and *Haz* (Hunnis Allgemeine Zeitung, Hunnis General Newspaper). The name *Motz* reflects the fusion of the two: *Mo*b + Ha(*t*)*z* = Motz. Proceeds from the magazine are used to help the homeless.

feelings or different moods. His "Go back to Africa" remark was intended to remind me that I am just a guest in Germany and as a white man, his needs come before mine. Thus, if I can't open up my purse when he requests a donation, I should best just leave the country.

# Policing/Profiling Sexuality

The final form of policing I want to consider is the act of policing one's sexuality. This can be repressive, such as when there are laws that determine with whom you can have sex. But this form of policing, or perhaps profiling, can also lead people to be overly permissive when it relates to the assumptions that men make about Black, femme bodies. On several occasions in Germany, I became acutely aware that just by being a Black woman in a pre-dominantly white space, the white men around me assumed I was a sex worker, or at least there for the taking. Sometimes this came out in quick remarks, like when a man would whisper "Schokolade" (chocolate) when I walked past him in a bar. Not all men tried to be as discrete, which reminds me of the elderly white German man in San Francisco at a Goethe Institute event who straight-up asked me if he could call me "Schoko-Mädchen" (chocolate girl). But those incidents, though disturbing, did not feel as threatening as the moments when I felt a white man had no problem with imposing his projection onto me. The most traumatic experience was in Dortmund on a very late night. I had been living in Bonn for the summer, doing an internship as a translator at the Deutsche Welle. Bonn was small and quaint, and usually there was little to do on the weekend. Thus, when I heard about the German Skateboarding Championship being held in Dortmund, with a lineup of punk bands playing after the competition, I was determined to make the trip. I had grown accustomed to travelling alone, going to concerts and movies alone. I'm an only child and an introvert. I can entertain myself. What I hadn't thought about was the logistics of trying to get back to Bonn, from Dortmund, with trains running infrequently late at night. As a result, I had to wait for what felt like hours at Dortmund's main train station. While I sat there, at some point an inebriated, white German man approached me. He was eating a sausage. I don't recall what he said. All I remember is him insisting on sitting next to me, attempting to put an arm around me and mockingly trying to feed me the sausage he was eating. I was repulsed by the smell of alcohol and mustard on his breath (I notoriously hate condiments). And I was grateful when he finally stopped badgering me and moved on. But what had frightened me were not only the liberties he took invading my space, but also that no one around me reacted or intervened. I had had similar experiences before. There was a time when a homeless man walking through the subway car I was on in Berlin sat next to me and tried to "cozy up" to me. From my perspective, it should have been clear to the other passengers that I didn't know him and that he was making me uncomfortable. I probably asked him to leave me alone. But no one said anything. They did their best to look away or look through me. Their reactions could just be part of the culture of a big city – everyone minds their own business. But it also felt like no one intervened in these incidents of sexual harassment, because as a Black woman, it wasn't possible to sexually harass me. My body was read as always available and lacking any boundaries.[3]

And yet despite all of these experiences, I know that being an academic and having an American passport also shields me from certain types of violence – such as when some German police asked to see my papers while I was waiting on the S-Bahn at Zoo Station – ironically this was not only during the World Cup when the world was theoretically "Zu Gast bei Freunden" (A Guest at Friends') but also when I had just returned from a seminar on border crossing at Viadrina University in Frankfurt/Oder. When the police saw I had an American passport, they immediately became much friendlier. And I know if I had been an African immigrant or a refugee, things would have gone much differently.

## Policing Childhood

Regardless of how these incidents of violence have made me feel, they cannot compare to the pain I've felt when having to witness my son face similar patterns of racialization. I have traveled with him to Germany since he was eight months old, but it was not until he was around four or five when I began to notice the ways in which people would police his behavior. Germany is known for being a place hostile to children. It is a stereotype, but it is a stereotype that rests on the fact that Germany is very much a country of rules and failing to conform to the rules can have consequences regarding how people perceive you. This is true no matter who you are. However, if you are a racialized *child* it

---

**3** Sharon Dodua Otoo also addresses similar experiences of sexual harassment from aggressive white men in the essay "Liebe": "I don't know anymore, how often I have been approached, hit on or sexually harassed by drunk, white men." Sharon Dodua Otoo, "Liebe," in *Eure Heimat is unser Albtraum*, eds. Fatma Aydemir and Hengameh Yaghoobifarah (Berlin: Ullstein fünf, 2020), 63. All German to English translations by Priscilla Layne.

feels like even more is at stake. When my son happens to block the entrance to an elevator or when he is eagerly inspecting some toys for sale, these might be typical occurrences in the life of a small child, but white Germans seem to respond in a much more exaggerated manner.[4] They scoff at him for always being in the way. They warn him to not dare touch any toys in the store. His cries of joy that are music to my ears are noise that they must drown out with their earbuds. But one beacon of light is when I think about how differently my son behaves in the world, in part, because he doesn't totally understand the ins and outs of racism. He doesn't yet understand all of the negative things people project onto Black bodies. I'm an introvert; he's an extrovert and has no problem being expressive and taking up space. I've realized that over the years, as a Black girl and now a woman, I have learned to make myself small and take up as little space as possible. I tried to blend into the background, hoping people wouldn't notice me and therefore I wouldn't become a target. This was especially how I felt when I lived in Germany, not only because of racism and sexism, but also because there were so many rules in German society about how one should behave. But my son is unabashedly himself wherever we go. It made me giggle to see him zip through crowds of appalled white Germans on his little scooter when we'd walk around Berlin. On the subway, while I would avoid eye contact with people, he would just randomly ask people questions. He is too young to understand that you "just don't do x or y." He is just himself. And that brings me so much joy. I hope we're able to change this world for the better before that light in him gets extinguished. Because if one day I notice that he too has learned how to make himself small in white spaces, it will be heartbreaking for me.

## Being Black in the Field of German Studies

Making myself small is ultimately a strategy I developed to deal with hostile spaces. And I often wonder how it has impacted my professional life, as academia is certainly often a hostile place for Black women. I have been able to make a home for myself within this space, by focusing on issues of race in my work,

---

4 Otoo also discusses this phenomenon in the essay "Liebe." There, she not only reflects on what kinds of tools a Black parent must equip their child within a majority white society, but also how hostile white people often respond to her Black children in otherwise mundane situations. See Otoo, "Liebe," 62.

including working in Black German Studies where I have found many dear friends and fellow scholars. I first learned of the existence of a Black German community when I was an undergraduate. I was working in a bookstore a few blocks from campus, and one day I happened to find Hans Jürgen Massaquoi's book *Destined to Witness* on one of our display shelves. I was drawn to its eerie cover: a black and white photo of a young Black boy, wearing a sweater vest with a handmade swastika patch on the front. I knew little to nothing about the presence of Black people in Germany and I certainly didn't know there were Black survivors of the Holocaust. I bought the book right away and read it fairly quickly.

I didn't think much about Black German history after that, until I spent my junior year in Berlin and happened to meet a Black German woman who gave me a flyer to an ADEFRA (Afrodeutsche Frauen, Afro German Women) meeting while I was at the FU. I went to the meeting but being there made me suddenly very aware of my privilege vis à vis the other women at the meeting. I was a guest, an exchange student – an American citizen who just happened to be living in Germany for a year. I could not relate to what it was like to grow up in Germany as a Black person, to do all of my schooling there, deal with puberty and dating, and try to find my first job there. I understood to keep my mouth shut and listen closely to the women's stories. It was an important lesson in learning when to be quiet and let other people's voices be heard.

Black German experience wasn't on my radar again until I started graduate school. My first semester, I took a class in stylistics. One of our assignments was to choose a poem, recite it for the class and write a short analysis of it. I don't know any more how I found it, but my first choice was to recite May Ayim's poem "Deutschland im Herbst" (Germany in Autumn). It draws a comparison between the violence of the Nazis, to the nationalist violence directed against minorities following German reunification. I was unsure whether or not to recite the poem. It spoke to me – to my soul. But reciting it in front of the white graduate students and my white professor felt like I was revealing something, outing myself. It was one thing that I was the only Black grad student in our program, it was another thing to draw attention to it by reciting a poem about racism by a Black German poet. I felt especially discouraged by the way my professor talked about diversity – to her multiculturalism was "a fad." Another professor warned me to choose something else. "If you're a Black scholar working on a Black poet, people will put you in a box and think this is all you can do." She wasn't trying to hurt me. In her own way, she was trying to protect me. She had dealt with a lot of sexism in academia; she was probably trying to shield me from racism. But it rubbed me the wrong way, that because I was Black certain topics should be off limits because my analysis would seem too subjective, while all of my white peers were free to research anything they wanted because a white perspective is always "objective." I decided to recite a poem by Hermann Hesse instead. My professor still wasn't very pleased with me.

I did not have many of those experiences in graduate school. Usually, the cause of my anxiety were class issues, rather than race. These days I am typically made most aware of my Blackness standing at the front of the classroom. I feel exposed again; as a Black woman, as a child of working-class immigrants, as a "Northerner" teaching in the South. I fear my students will judge me. I know that many of them have not had a Black teacher or professor; they may not be used to seeing a Black woman in a position of authority. But despite my anxieties, my job is to educate them, to get them to think critically. So, I push back when they say, "immigrants are taking over." I push back when they say, "I don't understand why blackface is offensive." I push back when they say, "that Black kid should have listened to the cops, then he wouldn't have been shot." And with each conversation, I only hope that they are able to rethink their assumptions, just as I had to rethink my assumptions about the South.

Nowadays, besides being attuned to the systemic racist violence that is occurring in cities around the United States, I also try to think more about another kind of violence: the violence of representation: who gets to speak for whom? This question was made very real to me in June of 2020, when I received a phone call from a German TV show looking to get some input from an African American about the Black Lives Matters protests that summer. A good friend of mine, another Black woman in German Studies, had passed on my number after checking with me. I thought they'd just want a sound bite for a story. I thought, sure I'm Black and I know American racism very well. I can describe what it's like to be Black in America right now. What I hadn't realized was the show, Maischberger, had been criticized recently for planning an all-white panel to discuss racism. Reaching out to me was their Hail Mary. They hoped to mitigate the backlash by having an African American join the panel. The call came Tuesday night inviting me to actually appear on television Wednesday afternoon. By Wednesday morning, I awoke to the fallout on Twitter. People commended them for having invited *a* Black person to speak; but why was this Black person an American and not a German? I agreed. Why hadn't they invited a Black German on to speak about this issue? I knew plenty of Black German scholars who had lived and even taught in the United States and could offer an effective comparison between racism in the United States and Germany: Fatima El-Tayeb, Peggy Piesche, and Natasha Kelly readily came to mind. Then it occurred to me, the reason they *did not* reach out to these Black German experts on this topic was because they hadn't done their research. If they had spent some time speaking with Black German activists and organizations, they would have been given a host of names that were not mine. But because my name was readily available, and because I was an American, they felt it was sufficient to stop looking. It is convenient to ask an American to speak about this topic because it further allows Germans to claim that racism is an *American* problem. This incident made me realize how much Black *American* scholars have to be vigilant not to let white German institutions misuse them to further silence Black German communities.

The question of representation and power, as it pertains to Black German Studies, was first made clear to me when I saw Black German author Noah Sow speak at the first conference of the Black German Cultural Heritage and Research Association. Sow questioned why African Americans seemed to be dominating the field of Black German Studies. She later took this up again in a keynote address at the 31st African Literature Conference in Bayreuth, though there she addressed the issue of privilege in the Black Diaspora more widely. One of the things Sow implores Black authors to do going forward is to demand authority over their own stories. To clarify, she recalls an experience she once had with " . . . an American editor with a British publisher" who sought to "'inform the audience' – whatever audience they had in mind – 'about Black German writings,' doing so by translating and republishing works of contemporary Black German authors, without having asked the authors' permissions."[5] Sow warns others "Being published, being studied, being-well received does not hold any intrinsic value when one's intellectual agency is seized."[6]

What Sow is describing, is what she later refers to as "geopolitical privilege" – when scholars with more privilege in publishing and the academy (usually both white and Black Americans) objectify Black Germans for research in order to further their own careers. As an African American of Caribbean descent who was trained as a Germanist and comes to Black German Studies out of sheer interest and solidarity, this is a charge I take seriously and one which I try to reconcile with the kind of work I do and how I do it. This geopolitical privilege can rear its head in a variety of ways. It is present when African American scholars are invited to German conferences as experts, while Black German scholars are overlooked. It is present in the reality that African Americans can take courses on Black German Studies, write their Masters and PhD theses about Black German Studies and get hired and promoted based on their work in

**5** Noah Sow, "Diaspora dynamics: shaping the future of literature," *Journal of the African Literature Association* 11, no. 1 (2017): 31.
**6** Sow, "Diaspora dynamics," 31.

Black German Studies, while it is much more difficult for Black Germans to get accepted to do a PhD in Germany, let alone be hired as a professor there. It should not go unscrutinized that so many publications and conference papers about Black German studies are produced in the United States.

While my students at the University of North Carolina can take a freshman seminar with me on Germany and the Black Diaspora or a graduate seminar on Black German writing, one would have a hard time finding such a course in German curricula. And for that reason, it is often much more difficult for Black Germans to encounter the very texts that have been written *by* them, *for* them and *about* them. In *Was weiße Menschen nicht über Rassismus hören wollen aber wissen sollte* (What White People Don't Want to Hear about Racism, but Should Know), Alice Hasters recounts what a revelation it was for her to find the text *Farbe bekennen*, edited by May Ayim, Katharina Oguntoye and Dagmar Schultz in 1986. If Black Germans never encounter such a text in their formal schooling, on bookshelves, or at their universities; if they rarely, if at all, have a Black German teacher or professor; if they cannot take a class in, let alone get a degree in, Black German Studies, then the Black German community ultimately has less access to higher education, and it is a structural problem and a deliberate problem. African Americans actually benefit from this structural inequality, even if not consciously, because we are more prevalent in the academy and there is more of a chance that those in power (tenured professors, administrators, editors and publishers) will listen to us. Reflecting on my positionality as a Black American academic vis a vis Black Germans has made me realize the ways in which I might be suspicious to them, because of the privilege that comes along with my citizenship, my academic training and my university.

To address this in my work, I try to be an engaged scholar and be as transparent as possible about who I am and what my intentions are – actions that deliberately contradict how white scholars have often conducted their research. I have usually met all of the Black German artists about whom I write. We have shared coffee, meals, and babysitters. And for that reason, I am beholden to them and their communities. I seek to position myself not as an "expert" on their lives, but as someone who is intrigued by their work and hoping to create a dialogue between Germany and the United States that can generate even more discussions and collaborations. As we try to change our field and address questions of diversity and equity, it is important that we not only change the content of what is researched, but also change *how* we research. This is the way I attempt to act ethically as a Black woman in my field, having learned from all the experiences I have described here.

# Works Cited

Ahmed, Sarah. "Evidence." feministkilljoys. Accessed 1 October 2020. https://feministkill joys.com/2016/07/12/evidence/. 12 July 2016.

Ayim, May, Katharina Oguntoye, and Dagmar Schultz. *Farbe bekennen: Frauen auf den Spuren ihrer Geschichte*. Berlin: Orlanda Verlag, 1986.

Hasters, Alice. *Was weisse Menschen nicht über Rassismus hören wollen aber wissen sollten*. Munich: Hanserblau, 2019.

Keeling, Kara. *Queer Times, Black Futures*. New York: New York University Press, 2019.

Massaquoi, Hans-Jürgen. *Destined to Witness: Growing Up Black in Nazi Germany*. New York: W. Morrow, 1999.

Otoo, Sharon Dodua. "Liebe." In *Eure Heimat ist unser Albtraum*. Edited by Fatma Aydemir and Hengameh Yaghoobifarah, 56–58. Berlin: Ullstein fünf, 2020.

Schultz, Trisha P. "Suspicious." In *Arriving in the Future*: *Stories of Home and Exile: Poetry and Creative Writing*. Edited by Asoka Esuruoso and Philip Khabo Köpsell, 97. Berlin: epub., 2014.

Sow, Noah. "Diaspora dynamics: shaping the future of literature." *Journal of the African Literature Association* 11, no.1 (2017): 28–33.

Ann Cvetkovich

# Hidden Places: The Indigenous Presence in My Affective Turn

## An Archive of Feelings

One way the Indigenous presence in my work makes itself felt is through the footnotes. In *An Archive of Feelings*, my interest in bringing trauma theory to queer and sexual experience led me to a practical experiment in testimony through interviews with lesbian AIDS activists.[1] In addition to being inspired by the emerging role of testimony and trauma in Holocaust studies, I was thinking through the theoretical training that had taught me to be critical of the presumption that the subject could speak her experience, which complicated understandings of testimony and added ballast to the notion of trauma as unspeakable or as "unclaimed experience." My key sources were Gayatri Spivak's "Can the Subaltern Speak?" and Joan Scott's "The Evidence of Experience," two classic essays of feminist and postcolonial theory.[2] But standing alongside those citations was also a book called *Resistance and Renewal: Surviving the Indian Residential School*, which is based on interviews with survivors of the Kamloops Residential School in the interior of British Columbia.[3] I knew about this book because it was written by my aunt, Celia Haig-Brown, based on the MA thesis that she was completing at the same time I was reading high theory in graduate school. It was part of my intellectual landscape but in an adjacent rather than direct way.

Although there are now many books about the residential school system, and Canada has had a Truth and Reconciliation Commission (inspired by South Africa's TRC, which is in turn a response to trials for Nazi war crimes) that has produced Calls to Action across the country, there was not much work of its kind when *Resistance and Renewal* was published, and, even as the work of a

1 Ann Cvetkovich, *An Archive of Feelings* (Durham, NC: Duke University Press, 2003).
2 Gayatri Chakravorty Spivak, "Can the Subaltern Speak?," in *Marxism and the Interpretation of Culture*, eds. Cary Nelson and Lawrence Grossberg (Urbana: University of Illinois Press, 1987); Joan Scott, "The Evidence of Experience," in *The Lesbian and Gay Studies Reader*, eds. Henry Abelove, Michele Aina Barale, and David M. Halperin (New York: Routledge, 1993).
3 Celia Haig-Brown, *Resistance and Renewal: Surviving the Indian Residential School* (Vancouver: Arsenal Pulp, 1988).

https://doi.org/10.1515/9783110753295-005

non-Indigenous scholar, it remains often cited.[4] My aunt came to her project in part because she had established friendships with former students of the Kamloops Indian Residential School through her involvement in rodeo production and high school teaching. Her work with the Native Indian Teacher Education Program (NITEP), a program founded in 1974 for Indigenous students seeking university degrees in teacher education, provided further impetus, especially because this UBC-based program was located on the site of the former Kamloops Indian Residential School.[5] Her work resembled the gathering of Holocaust testimony as a way to remember and give voice to those silenced and erased by genocidal practices. Drawing from British cultural studies (as well as Paolo Freire) to frame its ethical relation to the unheard, my aunt's theoretical framework overlapped with my own training. Like me, she was informed by critiques of anthropology, the crisis of representation, and the power relations between researchers and their participants that were present in Spivak and Scott (as well as Visweswaran), but with the additional urgency of Indigenous push back about being the subjects of settler research.[6] In her attention to Indigenous critique, she was a model to me for the complex methodological and ethical contradictions in the testimonial project of listening and giving witness to the undocumented.

Another distinctive hallmark of my aunt's focus on Indigenous cultures was her insistence on categories of "resistance and renewal," which she drew from British cultural studies. She wanted to emphasize that Indigenous youth in residential schools were not irrevocably damaged by their experience but instead found ways to resist and circumvent violence in order to survive – thus providing foundations for the Indigenous resurgence that future generations are now bringing into being. Looking back at work that has been many years and generations in the making and has expanded to include the Truth and Reconciliation Commission of Canada, an archive for those testimonies in Winnipeg and Vancouver, many many more books, and perhaps most importantly Calls to Action that might produce real forms of sovereignty and not just memorial, it is easy to forget how hard it was to fight for this public acknowledgement through the power of testimony.

---

**4** For more information on Canada's TRC, see www.trc.ca. For more on the history of truth commissions as a geopolitical format rooted in testimonial practice, see Priscilla B. Hayner, *Unspeakable Truths: Transitional Justice and the Challenge of Truth Commissions* (New York: Routledge, 2001, 2010).

**5** For more information on NITEP, see https://nitep.educ.ubc.ca.

**6** See Kamala Visweswaran, *Fictions of Feminist Ethnography* (Minneapolis: University of Minnesota Press, 1994).

Celia Haig-Brown's book on the Kamloops Residential School was just a tiny footnote late in *An Archive of Feelings,* and not nearly as prominent as the influence of a critical trauma studies rooted in Holocaust studies. Although one of the book's main goals was to make the leap from the trauma of genocide and war to sexual trauma, especially queer trauma, and from there to ordinary forms of violence, the available resources were based firmly in Holocaust studies – especially work by Cathy Caruth, and Shoshana Felman and Dori Laub, among others, that was informed by critical theory.[7] Of particular value to me in being able to connect different sites of trauma were scholars in this collection, such as Irene Kacandes and Marianne Hirsch, who were using the resources of feminism to talk about the experiences of the children of survivors as a form of adjacency to historical violence.[8] As the field of trauma studies has grown, it has been applied in many different locations and geopolitical contexts, but I find it important to acknowledge its base in European and American warfare – and to be alert to discrepancies rather than transporting it out of context or creating false analogies or easy comparisons. I faced challenges in using the tools of trauma studies when located too firmly in Holocaust studies in part because it led to problems of incommensurability – of seeming to place lesbian sexual trauma, for example, or HIV/AIDS activism alongside the Holocaust. I tried to draw from a range of sources – and was pulled in the direction of the tools for talking about insidious, ordinary, or everyday trauma rather than the experience of genocide. This is not to say that Holocaust and Jewish studies was not helpful – it was and it remains so, and this collection is one way to acknowledge that debt, especially to the "adjacent to historical violence" work on next generations and attenuated forms of transmission across generations, as well as the methodological work on practices of testimony (and its links to TRCs of many kinds).

---

**7** See Cathy Caruth, *Unclaimed Experience: Trauma, Narrative, and History* (Baltimore: Johns Hopkins University Press, 1996) and *Trauma: Explorations in Memory* (Baltimore: Johns Hopkins University Press, 1995); Shoshana Felman and Dori Laub, *Testimony: Crises of Witnessing in Literature, Psychoanalysis, and History* (New York: Routledge, 1992); Marianne Hirsch, *Family Frames: Photography, Narrative, and Postmemory* (Cambridge, MA: Harvard University Press, 1997). More recently, Bessel Van der Kolk's *The Body Keeps the Score: Brain, Mind, and Body in the Healing of Trauma* (New York: Penguin, 2014) has achieved broad visibility for work that originated with Holocaust survivors.
**8** Marianne Hirsch, *Family Frames and The Generation of Postmemory: Writing and Visual Culture After the Holocaust* (New York: Columbia University Press, 2012); Irene Kacandes, *Daddy's War: Greek American Stories. A Paramemoir* (Lincoln: University of Nebraska Press, 2009); as well as Marianne Hirsch and Nancy K. Miller, eds., *Rites of Return: Diasporic Poetics and the Politics of Memory* (New York: Columbia University Press, 2011).

But Jewish and Holocaust studies located predominantly in Europe, albeit one where distinctions between East and West were complex, couldn't be everything. In *An Archive of Feelings*, I was also grappling with my sense that African diaspora slavery – or what has come to be called, following Saidiya Hartman "the afterlife of slavery" – was as important as the Holocaust in creating multiple "archives of feelings," especially in the Americas. Here, too, I was working with available resources that have now expanded considerably. I would cite Toni Morrison's *Beloved* and other post-slavery work on the impossible and irretrievable archive of slavery, especially the emotional and intimate lives of the enslaved. I was reading Hortense Spillers, Patricia Williams, Avery Gordon, and Saidiya Hartman, among others.[9] Just trying to think the Holocaust and slavery together – never mind Indigenous genocide or queer and sexual trauma – was a task in itself with respect to issues of incommensurability and comparative sites of historical trauma. As Hartman points out in *Rites of Return*, the institutions of slavery often extinguished the familial transmission of post-memory and its histories necessitate attention to an ongoing present of systemic racism and violence.[10]

Although the sources I was using in American Studies weren't very directly focused on Indigenous genocide, I always tried to get it in there, even if it was just a footnote. Because I was already struggling to insert queer experience into the frame of trauma, I was used to the sense that something was missing. I would often find myself on panels where I had to discuss lesbian HIV/AIDS activism alongside – or adjacent to – the Holocaust, slavery, and genocide; the incommensurability of these sites of violence often made my cases seem insignificant and/or presumptuous. But in my version of a queer method, all of these things are connected – Holocaust testimony with sexual trauma, with AIDS activism, with the afterlife of slavery, with diasporic migrations, with Indigenous

---

**9** Toni Morrison, *Beloved* (New York: Knopf, 1987) and *The Source of Self-Regard: Selected Essays, Speeches and Meditations* (New York: Knopf, 2019); Hortense Spillers, "Mama's Baby, Papa's Maybe: An American Grammar Book," *Diacritics* 17 (Summer 1987): 65–81; Patricia Williams, *The Alchemy of Race and Rights* (Cambridge, MA: Harvard University Press, 1991); and Avery Gordon, *Ghostly Matters: Haunting and the Sociological Imagination* (Minneapolis: University of Minnesota Press, 1997) and *The Hawthorn Archive: Letters from the Utopian Margins* (New York: Fordham University Press, 2017); and Saidiya Hartman, *Scenes of Subjection: Terror, Slavery, and Self-Making in Nineteenth-Century America* (New York: Oxford University Press, 1997) and *Lose Your Mother: A Journey Along the Atlantic Slave Route* (New York: Farrar, Straus and Giroux, 2007).
**10** Panel discussion Hartman, Daniel Mendelsohn, and Eva Hoffman in Hirsch and Miller, *Rites of Return*.

survival – even if it can seem like a huge stretch to link them.[11] In writing the introduction for *An Archive of Feelings*, I tried to make the bridge by turning to queer solo performance and personal narrative rather than theory – placing Lisa Kron's *2.5 Minute Ride*, a queer take on being the child of a Holocaust survivor, alongside Carmelita Tropicana's *Milk of Amnesia*, about queer Cuban exile.[12] Their use of images such as the roller coaster at Cedar Point amusement park in Sandusky, Ohio, which is compared with Auschwitz as a site of tourism, and a Spanish conquistador's horse, whose memories of arrival in the Americas are channeled by Tropicana, provided a queer approach to testimony and made it possible for me to link Indigenous genocide in the Americas with the camps of Auschwitz. The Indigenous presence was there – in the form of a horse's memory! And a footnote.

## *Depression* and Return to the River

My work on trauma as ordinary led me to depression, and in my efforts to show that what gets called clinical depression is caused by capitalism not biochemical imbalance, the Indigenous presence in *Depression: A Public Feeling* became more explicit than it was in *An Archive of Feelings*. One of the book's slogans, as I sought not only to diagnose what causes depression but to figure out how to address it, is "just saying that capitalism or colonialism is the problem doesn't help me get up in the morning."[13] In putting colonialism alongside capitalism, I was thinking about the long arc of racialized capitalism that includes the settlement of the Americas and Indigenous genocide and displacement, as well as the enslavement and displacement of peoples of African descent. I sought to make what Lisa Lowe calls "the intimacies of four continents" my frame for depression – to

---

**11** On queer method, see Eve Kosofsky Sedgwick's foundational notion of the queer as "an open mesh of possibilities, gaps, overlaps, dissonances and resonances, lapses and excesses of meaning" in "Queer and Now," *Tendencies* (Durham, NC: Duke University Press, 1993), 8, or Kathryn Bond Stockton's notion of the "sideways" in *The Queer Child, or Growing Sideways in the Twentieth Century* (Durham, NC: Duke University Press, 2009). These and many other texts in queer theory inform my practice of adjacency – of setting seemingly disparate things alongside one another – as a form of queer method.

**12** See Lisa Kron, *2.5 Minute Ride* (New York: Theatre Communications Group, 2001) and Carmelita Tropicana, *Milk of Amnesia* in Alina Troyano, *I, Carmelita Tropicana: Performing Between Cultures* (Boston: Beacon, 2000), as well as the discussion of them in Cvetkovich, *An Archive of Feelings*, 20–29, 30–41.

**13** Ann Cvetkovich, *Depression: A Public Feeling* (Durham, NC: Duke University Press, 2012).

somehow include all of the intersecting histories that converge in ordinary feelings of malaise.[14] I was also interested in how the affective structures of feeling that emerge from those histories are racialized. What might be the differences among Black sadness and Brown sadness and Indigenous sadness, and white sadness – and how might it be important not to conflate them or to let the predominant focus on slavery and African American culture in the United States erase Indigenous experience and the legacies of settler colonialism? I wanted to explore depression's whiteness and its entanglement with racialized histories that produce white sadness, exhaustion, isolation, and other affective conditions or structures of feeling that are the result of the occluded exploitation of others that maintains white supremacy.[15]

An Indigenous presence can be most felt in my turn in *Depression* to memoir, as another genre for testimony about trauma, to write about my own experience of depression as a political and cultural problem. At certain key moments in the narrative, I return to the Campbell River, the site of my maternal grandparents' house and the place that I most feel to be "home." The notion of "return to the river" is borrowed from my grandfather Roderick Haig-Brown's writing about salmon and their miraculous efforts to return to their home rivers to spawn and die. My grandfather, a regionally renowned fly fisherman and environmentalist, wrote a novel about the chinook runs of the Columbia River called *Return to the River*, and his first book was called *Silver: The Life Story of An Atlantic Salmon*.[16] Across many books, both fiction and nonfiction, much of his writing about fish was drawn from his experience living in a house called Above Tide on the banks of the Campbell River, a place far enough upstream from the opening of

**14** Lisa Lowe, *The Intimacies of Four Continents* (Durham, NC: Duke University Press, 2015). On Black/Indigenous relations, see Tiffany Lethabo King, *The Black Shoals: Offshore Formations of Black and Native Studies* (Durham, NC: Duke University Press, 2019). There is also a growing practice of land acknowledgement, which has been common in Canada for much longer, in the United States – catalyzed in part by BLM politics.
**15** See, for example, José Esteban Muñoz in *The Sense of Brown* (Durham, NC: Duke University Press, 2020) on the specificity of Latinx structures of feeling, and Billy-Ray Belcourt's use of Muñoz's work and mine in *A History of My Brief Body* (2020) and *This Wound Is a World* (2017) to forge connections between colonialism and forms of negative affect, such as sadness and loneliness. For more on this, see my essay, "Billy-Ray Belcourt's Loneliness as the Affective Life of Settler Colonialism," *Feminist Theory*, forthcoming.
**16** Roderick L. Haig-Brown, *Return to the River: A Story of the Chinook Run* (New York: William Morrow, 1941) and *Silver: The Life Story of An Atlantic Salmon* (London: A. & C. Black, 1930), as well as *A River Never Sleeps* (Toronto: Collins, 1944). Like many Canadian writers, Haig-Brown is not well-known outside Canada, especially since his writing on fish is often considered a minor genre, known only to specialists. But he was often referred to as the Izaak Walton of twentieth-century fly-fishing in North America.

its mouth onto the Discovery Passage that the water is fresh and runs rapid. Although Roderick Haig-Brown fished all over the world, his primary source of knowledge was his adopted home territory, that of the Kwakwaka'wakh of the coastal islands of what is now called British Columbia, some of whose reserve lands were just down the road and on whose unceded land his house sat. In the 1950s, he fought unsuccessfully against the damming of the Campbell River at Buttle Lake for power to run the pulp mill, and before his death in the 1970s, he sat on the International Boundary Commissions that sought to control fishing in the borderlands between the United States and Canada. A half century later, his efforts are all the more urgent as salmon runs are being destroyed by overfishing, pollution, and corporate monopolies.[17] Both land and water are in jeopardy, although there are also efforts to restore fish habitat for spawning, including the restoration of Kingfisher Creek on my grandparents' property, which is now an environmental center owned by the provincial government and run by the local museum.[18]

My own work of "return to the river" for the histories of both my grandparents' presence and that of Indigenous peoples is ongoing. My efforts to do some of it for this essay brought more questions than answers. And more footnotes. Going down the rabbit hole of my grandfather's "return to the river" yielded not only his novel by that name, as well as his many books of essays about fishing the Campbell and other rivers, but also an early novel named *Pool and Rapid* drawn from his logging years on the Nimpkish River north of Campbell River, in which a fictional version of the river is the central character.[19] In thinking about how my grandfather (and my grandmother too since the matriarchal influence is strong) fostered my relation to place through their home on the river, I also turned to *What We Learned*, a collection of essays by my mother and her siblings about the lessons in environmental

---

**17** For more on the destruction of fishing as a small worker-owned business, see the work of another Haig-Brown, my uncle Alan, in Alan Haig-Brown, *Still Fishin': The BC Fishing Industry Revisited* (Madeira Park, BC: Harbour Publishing, 2010), which also uses interviews and oral history to collect stories.

**18** See website for Haig-Brown House, http://www.haig-brown.bc.ca/.

**19** See Roderick Haig-Brown, *Pool and Rapid* (London: A. & C. Black, 1931). There is much more to say about whether *Pool and Rapid*'s depiction of the river as a living presence borrows from Indigenous ways of thinking, especially given the novel's representation of Indigenous people as peripheral to the story. That Haig-Brown chose not to reprint it suggests that he felt that the novel was not a success. Yet his experiments with genre, and the interplay between fiction and nonfiction across his work, provide an important historical context for contemporary practices of writing about the entanglements of environment and the violence of colonial settlement.

and social justice they received from their parents. My aunt's observations about the "uncanny resonance" between when she learned from her father and what she has learned from her work in Indigenous thought has been especially provocative as I seek to understand my own experience of this connection.[20] How is it that I love this place – in part because of the deep respect for the land that my grandparents and their children had – when my presence on it is the product of historical violence? What does it mean to know and love a place well as a settler, while also acknowledging the theft of Indigenous land? I certainly grew up with an awareness of the Kwakwahk'wahh people on the reserve just down the road, but I still don't know the exact history of how the land on which Above Tide sits, which my grandfather bought from a friend who owned it before him, became private property. That's another future research project.[21] "Return to the river" has meant peeling back the layers of multiple histories – situating my own memories of a place by the river within histories of settlement, that of my family and those before them, histories of Indigenous displacement and dispossession, and ongoing Indigenous relations to place. What would it mean to sense all of that when making one of my returns to the spot by the river that was once an Indigenous canoe bay?[22]

The image of the salmon returning to the river was a powerful guide for questions about displacement and home that are central to the *Depression* book. Although my critical training taught me to be suspicious of "nostalgia for lost origins" or notions of "return home," I was also interested in how what we call "depression" might be a function of displacement and forced diaspora, whether for Jews, people of African heritage, and Indigenous people – or for white people whose colonizing trajectories and tendencies to "follow

---

**20** See Celia Haig-Brown in Haig-Brown Family: Mary, Valerie, Alan, Celia, *What We Learned.* Third Haig-Brown Memorial Lecture (Campbell River, BC: Haig-Brown Institute, 2012), as well as her essay "Decolonizing Diaspora: Whose Traditional Land Are We On?" *Cultural and Pedagogical Inquiry* 1:1 (2009): 4–21, which inspires my own efforts to write decolonial autobiography that begins with a relation to land and Indigenous presence on it. For an example of Indigenous thought about learning from the land, see Leanne Betasamosake Simpson, *As We Have Always Done: Indigenous Freedom Through Radical Resistance* (Minneapolis: University of Minnesota Press, 2017).

**21** I asked my Uncle Alan, and he replied, "do you want to open a can of worms?" Still, he did proceed to share information about the use of the land and the river by local Indigenous people for a fishing camp, and he ended with the story of an Indigenous elder saying to my grandfather, "You're on Indian land, Haig-Brown."

**22** I'm thinking here alongside Dylan Robinson's use of the term "sensate sovereignty" to describe Indigenous relations to place that are sensory and embodied – and wondering whether there is a counterpart for settler experiences of land that include feeling or sensing both Indigenous presence and the violence of dispossession. See *Hungry Listening: Resonant Theory for Indigenous Sound Studies* (Minneapolis: University of Minnesota Press, 2020).

the money" have sustained cultures of rootlessness and violent settlement. My own trajectory from British Columbia in the west to Toronto in the east and then south across the border to the United States was central to my story about the academic structures of feeling that lead to depression. The dislocations of academia have people moving all the time, with the mixed blessing of a fellowship or a job or admission to graduate school meaning yet another move, often with only a temporary post that prevents a sustained relation to place (that could include a relation to Indigenous people's fight for sovereignty).

The Indigenous presence in a return to the river is also a significant part of *Depression*'s more scholarly inquiry, especially the chapter about race and depression, which focuses on accounts of the psychic effects of the "afterlife of slavery" in Saidiya Hartman's *Lose Your Mother* and Jacqui Alexander's *Pedagogies of Crossing*. I track a geographic story of displacement in Hartman's journey to the slave routes and dungeons of Africa, and in Alexander's embrace of the multiple and overlapping histories of Black and Indigenous knowledges in the Caribbean. Following their methodological lead, I wanted to think about how whiteness, displacement, and depression – or race, place, and affect – are connected. In the case of Sharon Cameron's *The Family Silver*, Irish American identities rooted in the migrations of the potato famine are important, and in the case of Jeffery Smith's *When the Roots Reach for Water*, white Appalachian cultures are traced to migrations and settlements that are filled with loss and trauma, and in Smith's later move to Montana his relation to land is informed by his contact with the local Blackfeet cultures. Through these stories, I argue that, far more so than serotonin and pharmaceuticals, locations affect our moods. As Indigenous ways of knowing suggest, we are all co-constituted by the lands in which we live, whether because they are home, or because, for those of us who are settlers, we know at some level that we are living on stolen land or living comfortably in segregated neighborhoods because of racist structures of economic inequality.[23]

Although it is increasingly informed by Indigenous scholarship, my work on location and displacement, both in memoir form and in scholarship, is also indebted to and enabled by feminist scholarship on second-generation Holocaust survivors, including concepts such as Marianne Hirsch's postmemory and Irene Kacandes's co-witnessing.[24] It is significant that this work on transgenerational transmission of trauma has been accompanied by a turn to the personal, made

---

23 In addition to Simpson's *As We Have Always Done*, see also Kim TallBear on being co-constituted by land and on critical relationality in "Caretaking Relations, Not American Dreaming," and her website: www.kimtallbear.com.
24 See Hirsch, *Family Frames* and *The Generation of Postmemory*, and Kacandes, *Daddy's War*.

more legitimate by feminist theory and method, and also necessitated by silences and absent archives that often accompany locations that are "adjacent to historical violence." Their feminist research practice has often entailed personal journeys of many kinds, whose comparative dynamics are explored, for example, in Hirsch and Miller's *Rites of Return*.[25] Their work, which is centered in or originates from Jewish diaspora but also reaches for comparative or connective relations to other histories, especially African diaspora, has been formative for me, allowing me to unearth, for example, uncanny connections with my Serbo-Croatian heritage in the former Yugoslavia.[26] But I wish to stress here the way in which this body of work, and the invitation of this collection, have allowed me to bring to the fore an Indigenous presence always running (like river water) below the surface of my work on trauma and affect, along with questions about my relation to a "home" in Campbell River, even though I haven't lived in Canada for many years.

## Another Return to the River

The metaphor of a return to the river inspired by the salmon going upstream to spawn has again become literal for me because, after spending most of my adult life in the United States, including 30 years in Texas, I have returned to Canada and to a homeplace by a river. It is not the Campbell River or the Fraser River of my British Columbia childhood – instead, I find myself by the Ottawa River, where its confluence with the Rideau and the Gatineau have made this a good spot to gather for the Indigenous Algonquin people on whose unceded territory I sit. And also for the white settlers who made this the capital city of Canada. I don't know this river at all but I am learning more about it – about the logging and other industries that settled here just below the Chaudière Rapids, about its importance for the fur trade that went north via the Ottawa River across Lake Nipissing to Georgian Bay and Lake Huron and south to the mighty St. Lawrence that connects the Great Lakes to the Atlantic Ocean. About the Rideau Canal that was cut through the natural landscape so as to have an alternate waterway to serve the capital city in case the contested border terrain of the St. Lawrence were cut off. Although Ottawa's location is vague to many USers, who know it only in abstract terms as the capital city of Canada or the site of the iconic Victorian Gothic Parliament buildings, I have been

**25** Hirsch and Miller, *Rites of Return*.
**26** See Ann Cvetkovich, "The Balkan Notebooks," in *Eastern Europe Unmapped*, eds. Irene Kacandes and Yuliya Komska (New York: Berghahn, 2017), 229–247.

seeking to tune in to the land – and to the waterways that have been even more important than roads as a means of passage. I explore the section of the Rideau River that is only a mile or so from where I live, just upstream from the Hog's Back Dam that forms Mooney's Bay on one side and the start of the canal on the other. From here, I have a chance to come to know a territory by its waters as a way to learn about Indigenous cultures and histories of colonization.

Although I am not by my home waters in British Columbia, I recognize that I am a settler there too – and that the familiar places of my childhood were also covered over by colonial practices and erasures that made it difficult for me to know the land and water in an Indigenous way. We are all learning new names and ways now, in some cases returning to Indigenous names and undoing the practice of colonizing a place by giving it a settler name, usually male. Thus, the park on the Adams River in British Columbia that was named Roderick Haig-Brown Provincial Park in 1978 after his death was renamed in 2018 as Tsútswecw Park, from the Secwepemc word for "many rivers." Renaming or un-naming is also a practice of re-seeing the land and waterways before settlement.[27] I have been learning to pay attention to the traces of settlement and displacement at the river's mouth, since that was often the most lucrative place for industry or settlement and is now the hardest to recover or return. White people took over the river mouths to make their cities often to the point of covering over estuaries, deltas, and forms of cultivation. I recently walked the path along the Campbell that has been restored along the estuary so that you can walk from my grandparents' house to the opening on to salt water past the reservation. As I child, I had no idea how they were connected.[28]

I don't yet know how what I learn here in Ottawa will inform my writing and perhaps make the Indigenous presence more explicit. It often seems like a fool's errand to have moved after so long – and, under conditions of pandemic, I am now a bit of an exile from the United States, which is also my home, and even British Columbia. After many years in the United States, I am acclimatized or co-constituted by those lands, including the bit I know near the Aquarena Springs

---

**27** See Robin Wall Kimmerer, *Braiding Sweetgrass: Indigenous Wisdom, Scientific Knowledge, and the Teachings of Plants* (Minneapolis, MN: Milkweed Editions, 2013) on botany's practice of nomenclature and classification as colonial epistemology – that prevents actually seeing and knowing plants.

**28** Other rivers near which I have lived that I am coming to know better through an Indigenous frame include the Fraser River delta of Vancouver (including the recovery of the Indigenous village of Cesna?em at its mouth), and the Don and Humber Rivers in Toronto/Tkaronto, as well as the Nimpkish River on Vancouver Island, where Roderick Haig-Brown first lived.

on the San Marcos River that is one of the oldest inhabited places in the Americas. Especially as a white settler, I remain mindful of the admonitions of Gayatri Spivak and others about "nostalgia for lost origins" – particularly the feelings of white people whose origins were lost long ago. What can home ever mean for white people in the Americas? It starts I think with coming to know and respect the places where we settle – and coming to know them as land and water not just city and settlement. My settler colonial kin have been on these lands for only two generations and less than a hundred years before me – and issues of displacement and settlement are very present in their histories. As I come to know the river and the Indigenous cultures where I live now, the Indigenous presence in my affective turn will continue to make itself felt.

# Works Cited

Belcourt, Billy-Ray. *A History of My Brief Body*. Toronto: Penguin Random House Canada, 2020.

Caruth, Cathy. *Unclaimed Experience: Trauma, Narrative, and History*. Baltimore: Johns Hopkins University Press, 1996.

Caruth, Cathy, ed. *Trauma: Explorations in Memory*. Baltimore: Johns Hopkins University Press, 1995.

Cvetkovich, Ann. *An Archive of Feelings: Trauma, Sexuality and Lesbian Public Culture*. Durham: Duke University Press, 2003.

Cvetkovich, Ann. *Depression: A Public Feeling*. Durham: Duke University Press, 2012.

Cvetkovich, Ann. "The Balkan Notebooks." In *Eastern Europe Unmapped: Beyond Borders and Peripheries*. Edited by Irene Kacandes and Yuliya Komska, 229–247. New York: Berghahn Books, 2017.

Cvetkovich, Ann. "Billy-Ray Belcourt's Loneliness as the Affective Life of Settler Colonialism," *Feminist Theory*, forthcoming.

Felman, Shoshana, and Dori Laub. *Testimony: Crises of Witnessing in Literature, Psychoanalysis, and History*. New York: Routledge, 1992.

Gordon, Avery. *Ghostly Matters: Haunting and the Sociological Imagination*. Minneapolis: University of Minnesota Press, 1997.

Gordon, Avery. *The Hawthorn Archive: Letters from the Utopian Margins*. New York: Fordham University Press, 2017.

Haig-Brown Family: Mary, Valerie, Alan, Celia. *What We Learned*. Third Haig-Brown Memorial Lecture. Campbell River: Haig-Brown Institute, 2012.

Haig-Brown, Alan. *Still Fishin': The BC Fishing Industry Revisited*. Madeira Park, BC: Harbour Publishing, 2010.

Haig-Brown, Celia. *Resistance and Renewal: Surviving the Indian Residential School*. Vancouver: Arsenal Pulp, 1988.

Haig-Brown, Celia. "Decolonizing Diaspora: Whose Traditional Land Are We On?" *Cultural and Pedagogical Inquiry* 1, no. 1 (2009): 4–21.

Haig-Brown, Roderick. *Pool and Rapid*. London: A. & C. Black, 1931.

Haig-Brown, Roderick. *Return to the River: A Story of the Chinook Run*. New York: William Morrow, 1941.

Haig-Brown, Roderick. *Silver: The Life Story of An Atlantic Salmon*. London: A. & C. Black, 1930.

Haig-Brown, Roderick. *A River Never Sleeps*. Toronto: Collins, 1944.

Hartman, Saidiya. *Scenes of Subjection: Terror, Slavery, and Self-Making in Nineteenth-Century America*. New York: Oxford University Press, 1997.

Hartman, Saidiya. *Lose Your Mother: A Journey Along the Atlantic Slave Route*. New York: Farrar, Straus and Giroux, 2007.

Hayner, Priscilla B. *Unspeakable Truths: Transitional Justice and the Challenge of Truth Commissions*. New York: Routledge, 2001, 2010.

Hirsch, Marianne. *Family Frames: Photography, Narrative, and Postmemory*. Cambridge, MA: Harvard University Press, 1997.

Hirsch, Marianne. *The Generation of Postmemory: Writing and Visual Culture After the Holocaust*. New York: Columbia University Press, 2012.

Hirsch, Marianne, and Nancy K. Miller, eds. *Rites of Return: Diasporic Poetics and the Politics of Memory*. New York: Columbia University Press, 2011.

Kacandes, Irene. *Daddy's War: Greek American Stories. A Paramemoir*. Lincoln: University of Nebraska Press, 2009.

Kacandes, Irene, and Yuliya Komska, eds. *Eastern Europe Unmapped: Beyond Borders and Peripheries*. New York: Berghahn Books, 2017.

King, Tiffany Lethabo. *The Black Shoals: Offshore Formations of Black and Native Studies*. Durham: Duke University Press, 2019.

Kimmerer, Robin Wall. *Braiding Sweetgrass: Indigenous Wisdom, Scientific Knowledge, and the Teachings of Plants*. Minneapolis: Milkweed Editions, 2013.

Kron, Lisa. *2.5 Minute Ride*. New York: Theatre Communications Group, 2001.

Lowe, Lisa. *The Intimacies of Four Continents*. Durham, NC: Duke University Press, 2015.

Morrison, Toni. *Beloved*. New York: Knopf, 1987.

Morrison, Toni. *The Source of Self-Regard: Selected Essays, Speeches and Meditations*. New York: Knopf, 2019.

Muñoz, José Esteban. *The Sense of Brown*. Durham, NC: Duke University Press, 2020.

Robinson, Dylan. *Hungry Listening: Resonant Theory for Indigenous Sound Studies*. Minneapolis: University of Minnesota Press, 2020.

Scott, Joan. "The Evidence of Experience." In *The Lesbian and Gay Studies Reader*. Edited by Henry Abelove, Michele Aina Barale, and David M. Halperin. New York: Routledge, 1993.

Simpson, Leanne Betasamosake. *As We Have Always Done: Indigenous Freedom Through Radical Resistance*. Minneapolis: University of Minnesota Press, 2017.

Spillers, Hortense. "Mama's Baby, Papa's Maybe: An American Grammar Book." *Diacritics* 17 (Summer 1987): 65–81.

Spivak, Gayatri Chakravorty. "Can the Subaltern Speak?" In *Marxism and the Interpretation of Culture*. Edited by Cary Nelson and Lawrence Grossberg, 217–313. Urbana: University of Illinois Press, 1987.

Stockton, Kathryn Bond. *The Queer Child, or Growing Sideways in the Twentieth Century*. Durham: Duke University Press, 2009.

TallBear, Kim. "Caretaking Relations, Not American Dreaming." *Kalfou: A Journal of Comparative and Relational Ethnic Studies* 6, no. 1 (2019): 24–41.

Troyana, Alina. *I, Carmelita Tropicana: Performing Between Cultures*. Boston: Beacon, 2000.

Van der Kolk, Bessel. *The Body Keeps the Score: Brain, Mind, and Body in the Healing of Trauma*. New York: Penguin, 2014.

Viseweswaran, Kamala. *Fictions of Feminist Ethnography*. Minneapolis: University of Minnesota Press, 1994.

Williams, Patricia. 1991. *The Alchemy of Race and Rights*. Cambridge, MA: Harvard University Press, 1991.

Dagmar Herzog

# On Being Adjacent to the Nazi Disability Murder Project

This essay finds me just at the beginning of a larger and already intellectually and emotionally challenging investigation, focused on the theology and politics of disability in twentieth-century Germany. I was initially wanting to explore the evolving contests between orthodox-conservative theological ideas of disability as willed and caused by God (whether as punishment or as test), more moderately conservative variants which assert that God's ways are mysterious but the task of providing loving care for those who are dependent brings out unique goodness in those doing the caring (or even that God is especially present among the weak and vulnerable), and a more radical liberation theology notion that God is disabled. The idea was to reconstruct arguments made on behalf of the value of disabled life, starting from a recreation of the devoted-paternalistic paradigm of Christian charity care I had thought was well ensconced around 1900 and then tracing the ensuing developments through the dark chasm of the hundred-thousand-fold National Socialist "euthanasia" killings of 1939-1945 – including attention to the handful of legendary efforts to resist and protest those killings – but then also charting the long and winding climb out of that abyss, through various stages of recursive memory-work, into the defiant and impressive disability rights movement of the present.

I believed that I was bringing at least a few familial resources to this investigation, as two great-aunts (one on my father's side and one on my mother's) as well as one grandmother (and a great-grandmother I didn't know but about whom I'd heard pertinent details), all bore direct witness to some of the events I am researching. Nor did it seem irrelevant that my concern with these issues extends back to my teenage years when I recurrently spoke with (or perhaps one might say with the benefit of hindsight, *interviewed*) one of those great-aunts about what she had experienced during the Third Reich at the fabled Protestant charity institution known as Bethel, where she had worked as a deaconess for many decades, dating back to the 1920s, and where I too worked one long-ago summer (1980) as a volunteer, and where, under ambiguous but ultimately marvelous circumstances, several thousand individuals with intellectual disabilities had managed in the first half of the 1940s to be saved from the Nazi murder machinery intent on slaying them. In a manner of speaking, then, in one fashion or another I had been reflecting on this subject and its various dimensions for my entire adult life. In addition, more obliquely but arguably

https://doi.org/10.1515/9783110753295-006

even more significantly, I had been raised on chronicles of faith-fueled courage and stalwart heroism, many of them, notably, crisscrossing in some way through Bethel – albeit also on stories of conflicts between different branches of my family, as choices in love and friendship ended up shaping the particular constellations of theological and political positions taken during the Third Reich and the at times incompatible lessons drawn from its horrors in the aftermath.

Although this essay will not be probing these personal stories further, they have inevitably, and rather profoundly, inflected my preoccupations, reflexes, curiosities, and expectations, and so it is this background that may explain something about my considerable shock at what the archives – in the year when I began to research the topic academically in earnest – started to reveal to me. But I suspect that my initial archival finds might also challenge suppositions of many German studies and history colleagues. That is because – for whatever complex combination of reasons (or: investments, misrecognitions, taboos) which, I see now, it will eventually also be my task to untangle – the standard narratives available to us about disability in German history do not regularly include the historical episode I focus on here.

This essay looks back to *before* the Third Reich, and to the *pre*history of a particular – and at least in retrospect deeply dismaying – moment when twenty-two leaders in the Protestant Inner Mission, the umbrella organization for all Protestant welfare work, gathered north of Frankfurt/Main in the small town of Treysa in 1931 to issue a kind of Christian charity policy platform which proposed that "differential care" (*differenzierte Fürsorge*) would henceforth be provided to those among their disabled charges who were, due to severe cognitive or psychiatric impairment, unable to engage in productive labor.[1] In short, these Protestant leaders would at Treysa formally endorse a hierarchization of human value.

For decades now, scholars have shown us the apparent breadth of Protestant and Catholic reluctance to defend or protect the Jewish victims of the Nazi Third Reich. What had caused me to assume that Christians had been steadfast on behalf of the disabled? Or to put it bluntly: How had I so completely missed the unpalatable truth that Christians, and not least precisely those whose life's work was care work, had supreme difficulty – and this already before the Nazis came to power – in articulating a vigorous defense of the value of disabled lives? Here are some preliminary discoveries.

---

1 "Die Treysaer Resolution des Central-Ausschusses für Innere Mission (1931)," in *Eugenik, Sterilisation, Euthanasie: Politische Biologie in Deutschland 1895–1945 – Eine Dokumentation*, eds. Jochen-Christoph Kaiser et al. (Berlin: Buchverlag Union, 1992), 106–110, here 106.

## "Work-Capacity"

The Treysa Resolution was quite evidently, at least in part, a *financial* decision, promulgated at a time of renewed crisis for the Weimar Republic and reductions in overall per-person subsidies instigated by the budget-strapped Prussian state. Protestant leaders were factoring in multiple economic exigencies. An institution's debt burden; its tax-exempt status; the size of its endowment; the annual level of charitable donations; the proportion of private-pay to publicly funded residents; how low the salaries for staff could be kept: all of these matters were existential for care institutions. And all through the first decades of the twentieth century in Germany they had reached into the corporeal experiences of the residents as well as the quality of attentiveness they received. Yet the Treysa Resolution was also a response to a shifting *ideological* climate, as there was widespread – and avidly media-fueled – public resentment around the supposedly lavish conditions in which the most severely disabled lived. Critically, however, these much-promoted fantasies around the "welfare-luxury" (*Wohlfahrtsluxus*) enjoyed by the nonworking disabled covered a far more complicated story.[2] For as it happens, a considerable number of larger institutions augmented the small daily per-resident allotments they received from the state with the labor-power of residents. The most economically valuable were the contributions of institutionally affiliated "worker-colonies" (*Arbeiterkolonien*) in which homeless men received housing in exchange for agricultural or other land-work. But a vital additional source was the work of the "moderately" disabled residents, even as the labor they provided was presented (and no doubt experienced by many) as therapeutic – whether in fields and garden, in kitchen, laundry, sheltered workshops or around the grounds. Finally, the Treysa Resolution marked the culmination of profound alterations in *theological* conceptions of disability that had been underway since the end of World War I. And those theological conceptions are my real subject here.

But first some basic facts: For following the thread of the issue of "work-capacity" (*Arbeitsfähigkeit*) leads into the intensely delicate but also, as it turns out, enormously consequential issue of gradations of disability and divisions among the disabled – with convoluted effects ramifying through the Weimar era and ultimately becoming of the most ominous significance during the years of mass murder. Not only was there through the first decades of the twentieth century a stark pecking order among the disabled, with individuals with sensory or physical impairments often insisting, adamantly, on their superiority to

---

2 Friedrich Lensch, "Dennoch!" *Briefe und Bilder aus Alsterdorf* 55–56 (1931–32): 2–5, here 3.

individuals with cognitive challenges – including with regard to the ideal of "hereditary" health – but there were strongly perceived and much-discussed divisions identified *among* those with intellectual disabilities as well. Although the lines were always both blurry and unstable, and the nomenclature varied greatly, three broadly distinguishable groups emerged. To the one side were those with more minor cognitive challenges or only temporary psychiatric crises, able to benefit from remedial education and be self-supporting in the outside world, for the most part functioning perfectly well especially in supervised agricultural or mechanical jobs. To the other side were the institutionalized. But within this group, too, there was that split to which the Treysa Resolution, along with its apparently unembarrassed call for "differential care," had alluded: the divide between those able to provide labor in and for the institution in some way, and those utterly unable to work (even on something as simple as tearing rags) and who instead required above all the time investment and the labor of care from others.

As scholars have established, there would end up being only partial overlap between the two major disabled victim groups of Nazism. Of the four hundred thousand individuals subjected to the coercive "eugenic" sterilizations performed under the auspices of the Nazis' 1933 Law for the Prevention of Hereditarily Diseased Offspring, and although the highest percentage were labeled "feeble-minded" and the next most common diagnosis was "schizophrenic," the majority of all these would actually be best described as the *not-so-disabled*. And although there would be individuals who were sterilized among the approximately three hundred thousand victims of the Nazis' disability murder project, for the most part sterilization acted as a kind of protection against being killed (not causally, but as a correlation, because it was a sign that the individual was in some way economically exploitable by the regime). While fervently promoted albeit often incoherent or poorly documented ideas about "heredity" and "racial hygiene" offered the justifying frameworks for both the sterilization and the killing programs, in practice "work-capacity" became the major determinant of outcomes.[3]

The Nazi disability murder project, as it developed momentum from 1939 on, had many components: In "special children's wards" (*Kinderfachabteilungen*), killing of children by poison injection – sometimes after torturous premortem experiments; in Poland early in the war and later in the Soviet Union, killing of psychiatric patients by mass shootings and gas vans, at times for the

---

**3** Maike Rotzoll et al., *Die nationalsozialistische "Euthanasie"-Aktion "T4" und ihre Opfer: Geschichte und ethische Konsequenzen für die Gegenwart* (Paderborn: Schöningh, 2010).

sheer purpose of clearing hospital beds for German soldiers; and, within the German Reich, from January 1940 to August 1941, first six "T4" carbon monoxide gas chambers – with T4 being a reference to Tiergartenstrasse 4 in Berlin where these seventy thousand killings were centrally coordinated. And although it had begun earlier in local initiatives, a second, "decentralized" phase of murder, primarily by medication overdose or deliberate starvation, claimed twice as many victims as had been killed by T4. Both of these methods continued even a few weeks past the war's end in May 1945. At every stage, it was evident that the "euthanasia" murders' main targets were primarily drawn from those who were long-term institutionalized and especially among those with the most severe cognitive or psychiatric impairments. Indeed, over and over, the killings that were most brazenly pursued were directed at those requiring labor of care from others, rather than being able to provide labor themselves, even as the bar for what was needed to be allowed to survive was rather high. (Thus, for instance, the ability to peel potatoes or mop floors, or to weave mats or glue paper bags, was counted as "mechanical" rather than "productive" work, and deemed insufficient for keeping a person alive.)

The intricacies surrounding these core economic issues of work-capacity and labor-power never stopped mattering. This led to countless hallucinatorily grotesque scenes. An administrator at the Protestant charity institution of Stetten in Württemberg arguing with deportation personnel at the very door of the gray bus meant to take victims to the gas chamber that he absolutely needed to keep his faithful cook Lina L., otherwise he would be obliged to replace her with "a person of fully sound mind" (*eine vollsinnige Person*); he failed.[4] Moderately (as opposed to profoundly) psychiatrically ill and/or cognitively disabled residents at the Wiesloch institution near Heidelberg doing all the labor in kitchen and laundry to sustain care for wounded Wehrmacht soldiers who by that point in the war inhabited the beds freed by an earlier round of killings.[5] A disabled man helping with the corpse-burning in the crematorium at the Kaufbeuren institution in Bavaria – while psychiatrically ill patients at the Pomeranian disability murder center of Meseritz-Obrawalde were tasked each morning with collecting the past night's dead on a little wagon to bring them to the morgue.[6] A disabled woman surviving Meseritz-Obrawalde (as Protestant

---

4 Jakob Rupp quoted in Ernst Klee, *"Euthanasie" im NS-Staat: Die "Vernichtung lebensunwerten Lebens"* (Frankfurt am Main: Fischer, 1983), 237.
5 Frank Janzowski, *Die NS-Vergangenheit in der Heil- und Pflegeanstalt Wiesloch: . . . so intensiv wenden wir unsere Arbeitskraft der Ausschaltung der Erbkranken zu* (Ubstadt-Weiher: Verlag Regionalkultur, 2015), 292–293.
6 Klee, *"Euthanasie,"* 445, 410.

deaconesses from Bad Kreuznach who had been obliged to accompany their charges in locked and guarded trains to that site, and then return without them, later discovered) by providing housecleaning services for one of the perpetrator-physicians stationed nearby.[7] Or a disabled man who would recall forty years after 1945 how he had assisted with the digging of mass graves on the hill above the killing center at Hadamar in Hessen during the decentralized phase of murders there, and was still able to describe exactly the space needed to make the graves the correct size. In the first postwar decades, no one had wanted to hear this man disclose his experience; he was best known in and around the institution's grounds for playing the harmonica.[8]

# Permission to Annihilate

To understand what Protestant charity leaders were thinking when they formulated the Treysa Resolution – and thereby endorsed a hierarchy of human value – requires a return to lawyer Karl Binding and psychiatrist Alfred Hoche's brochure of 1920, *Die Freigabe der Vernichtung lebensunwerten Lebens* (Permission to Annihilate Life Unworthy of Life). This was the text which was to supply such ample *avant-la-lettre* legitimation for the National Socialists' subsequent disability murder project. Coming as the text did in the wake of the mass carnage of World War I, in which so many had died on the battlefield and the majority of the civilian population too had been exposed to extreme hunger and deprivation, and the economy remained devastated, Binding and Hoche's insistence that for both emotional and economic reasons the profoundly disabled should be disposed of gained great public attention and acclaim.

A combination of emotions and economics was crucial, as Binding and Hoche both played on feelings of revulsion at disability and emphasized its financial cost. Although the bulk of their text blurred pain relief eventuating in death with deliberate killing and extemporized at length about assisted suicide in the case of terminal illness, the main purpose of the brochure lay in a small

---

**7** Karl-Adolf Bauer, "Aus der Geschichte der Diakonie-Anstalten Bad Kreuznach," in Evangelische Akademie Mülheim/Ruhr, ed., "Diakonie im Dritten Reich: Tagung in Zusammenarbeit mit dem Diakonischen Werk der Evangelischen Kirche im Rheinland und den Diakonie-Anstalten Bad Kreuznach, 16–17 Mai 1987, Außentagung in Maria Laach," *Begegnungen* 5 (Mülheim/Ruhr, 1987), 61–70, here 68.
**8** Leo Waltermann, "Diakonie im Dritten Reich: Aus der Geschichte Konsequenzen für heute," in ibid., 85–117, here 87.

subplot. Binding was candid in his view with respect to the group he labeled "incurable idiots." Indignant that "years and decades" of the valuable time and labor of caregivers were being spent on these persons, Binding averred that "I can find neither from the legal, nor the social, nor the moral, nor the religious standpoint any grounds whatsoever not to permit the killing of these humans . . . who arouse horror in almost everyone who encounters them." Hoche, for his part, reinforced this position with mathematical calculations as to the money expended every year on food, clothing, and heating for "the total idiots" (whose numbers he estimated nationally in the tens of thousands, of which he thought "the mentally completely dead" constituted approximately three thousand to four thousand). In Hoche's view, these people were "ballast-existences" (*Ballast-existenzen*) – nothing but dead weight burdening a struggling society. The text further expounded on a concept of "affection-value" (*Affektionswert*). The disabled, the text argued, had no such value. Binding and Hoche distinguished here between long-beloved persons who in old age had lost their mental faculties to dementia and those who were from the start "on an intellectual level that we only find again very low down in the ranks of animals." The former group deserved to live until a natural death took them; the latter group most definitely had to be helped into an earlier grave.[9]

*Die Freigabe der Vernichtung lebensunwerten Lebens* provoked a redirection of the German national discussion. Binding had opened the pamphlet with a direct attack on Christianity for what he deemed to be *its* cruelty in rejecting assisted suicide. Binding took offense at what he called "the impure idea that the God of love could wish that a person would only be permitted to die after enduring unending bodily or spiritual torments."[10] Religious leaders took notice. This was not just because the rhetorical jab at Christianity was unequivocally directed at them, but for the double additional reasons that their institutions were among the primary homes for the class of humans whose right to stay alive was now under attack and that their institutions were indeed, recurrently, in financial stress.

Pressure came from all sides, as eugenic theories increasingly became the cultural commonsense. Religious worldviews were eroding more generally – due both to the ascent of biological-"scientific" frameworks and to the expansion of popular secularization, whether socialist or liberal or based in sheer everyday indifference. Eugenics, with its interest not just in uplifting the health of populations but also with curtailing (whether through marriage prohibition,

---

**9** Karl Binding and Alfred Hoche, *Die Freigabe der Vernichtung lebensunwerten Lebens: Ihr Maß und ihre Form* (Leipzig: Felix Meinen, 1920), 31–32, 53–55, 57.
**10** Binding and Hoche, *Die Freigabe*, 6.

birth control, abortion, or sterilization) the reproduction of people believed to be the carriers of inheritable deficiencies, was a movement attracting adherents from across the ideological spectrum (including Jews, leftists, feminists, homosexual rights activists – and on the other end, although the Catholic church was officially opposed, also Catholics). And when we look internationally, an interest in eugenics by no means implied an endorsement of the – only seemingly twinned – phenomenon of euthanasia. However, in the terms of conversation in Germany at the time, and especially when in the vicinity of the severe handicaps that were the specialty of the Protestant Inner Mission, keeping those boundaries distinct turned out to require continuous effort. This was particularly because Christian charity caregiving itself was increasingly being represented as responsible for the looming death of the German Volk, and not just because it diverted labor-power and financial resources, but literally because its tendency to engage in "artificial nurturance" (*künstlich großzupäppeln*) of the less-than-healthy was deemed to be "contra-selective" in a Social Darwinian sense.[11]

As one pastor – the director of a large institution for the cognitively disabled in Hamburg – would summarize the situation, "the objections [to keeping the severely disabled alive] can be subdivided into three groups. The first are economic, the second the eugenic type, the third are generally humanitarian." The economic argument – he said in an explicit gesture to Binding and Hoche's concept of "ballast-existences" – could be refuted fairly easily, since Protestant charities like his, he proudly explained, were extremely cost-efficient. (In fact, he went on to say that "the suggestion always again heard, to get rid of these 'valueless existences' with a sufficient dose of morphine, can under current economic conditions hardly be justified" – pointing out that the disabled were at least, in a time of agricultural overproduction and reduced sales volumes, helping the economy in their function as "consumers" (*Konsumenten*).) The eugenic argument, he averred, had always already been a concern of the Inner Mission, since the mid-nineteenth century, and that was precisely why institutionalization was so important, as it kept the disabled out of sexual circulation and provided a source base of subjects for medical research into "feeblemindedness" (*Schwachsinn*), in expectation that its prevalence could be reduced in the future. And then there were the "humanitarian" objections – the notion that the lives

---

11 Hermann Büchsel, "Euthanasie," *Aufwärts* 280 (30 November 1926), reissued in 1927 as an offprint (and warmly recommended for education of deaconesses), Hauptarchiv der v. Bodelschwinghsche Stiftungen Bethel Kl. Erwerb. 189, 1–4, here 1. Also at the Alsterdorfer Anstalten in Hamburg (in 1927 under the director of Pastor Paul Stritter), this essay had been reprinted and commended to the staff.

of the disabled must be unbearable. Ultimately, this pastor-director argued that his residents loved life, and that it would be "cowardly" to kill them. But not before he had with rather alarming candor confessed that "Those who tour our institution and perhaps for the first time in their lives stand before this abyss of misery . . . always again, shaken, ask us if one cannot, if only for the sake of the ill ones themselves, liberate them from this life that can after all hardly be called life. I freely admit that I felt no differently when I first came in, and these thoughts also creep up on us again and again, like a temptation [*wie eine Versuchung*]."[12]

His was hardly an anomalous voice. The 1920s saw an outpouring of commentary from theologians, pastors, and Protestant welfare institution directors as they all reacted to Binding and Hoche. The possibility of murdering the most severely disabled individuals hovered around the edges of all these conversations. What evolved over the course of the decade was a new way of speaking, a kind of *theo-biopolitics*, one in which religious and moral considerations got confused with financial ones, while issues of biological bodily health or disease got thoroughly re-muddled with questions of guilt and sin.

Some of the responses to Binding and Hoche took place in private correspondence, although they were to end up published in *Das Problem der Abkürzung "lebensunwerten" Lebens* (The Problem of Abbreviating Life "Unworthy" of Life, 1925) by Ewald Meltzer, a Protestant doctor who was director of a major institution in Saxony for individuals with intellectual disabilities. Meltzer was intent on compiling all possible arguments, both religious and secular, *against* murdering the disabled, but he also catalogued the breadth of arguments in favor. Here we learn of Protestant religious leaders endorsing murder of the most seriously disabled as morally acceptable. We read that a theology professor from Magdeburg frankly believed that caring for "idiots" and keeping them alive was pointless in the way that pouring water on sticks stuck into the ground and expecting something to grow was pointless. We find that a noted religious educator in the Saxon town of Auerbach criticized the "palaces built" for the disabled – in contrast to the hovels in which too many healthy poor people had to live, and that in his opinion one could and should not extrapolate from Jesus' views about mercy for the handicapped as expressed in the Bible to the then-present of the 1920s; he averred that as long as "abbreviating" life was done out of "benevolence" (*Wohlwollen*) it was "not immoral." A theology professor in Heidelberg (known as a conservative biblical Pietist) opined that

---

12 Lensch, "Dennoch!" 2–5.

although a *Christian* may not kill, a *state* may, especially when it came to the more severely disabled individuals whose "soul-life had not been raised above the plant or animal level." And a prominent Protestant professor in Göttingen felt that "the elimination" of such "nuclei" or "sprouts" (*Keime* – he meant severely disabled individuals) "for which the force of human care will not be able to bring them to their destiny-appropriate development" could in fact be "compatible with full love of God and genuine humanitarianism." Christians, he contended, believed that unlike nature, which sometimes produced mistakes, God was directed toward perfection. Thus, he wrote that if caregiving staff had tried, but "had the experience of being unable to awaken and nurture the personal life of an idiot, then one cannot in this case speak of a crime against a person."[13] Although none of these men had close continuous contact with individuals with disabilities, their arguments provided religious legitimation for Binding and Hoche's position.

Commentators who were directly involved in care work, by contrast, were at pains in their texts to declare themselves opposed to killing, or to remind readers that Christianity strictly forbade it, and yet their texts often, and peculiarly, amplified rather than challenged a sense of revulsion toward the most severely disabled. Simultaneously, a number of these commentators took the opportunity of public interest in the theme of disability and its supposed high costs to drag in a welter of unrelated subjects, and above all to elaborate a variety of connecting links between sex and sin and disability. Rather than functioning as opposition to Binding and Hoche, then, these responses created an accrual of further negative perspectives on the disabled. Moreover, these commentators expanded the conversation beyond the group of severely disabled targeted by Binding and Hoche to bring in the more mildly or moderately disabled. This had everything to do with conservative Christian distaste for the liberalization of sexual mores that had been going on already since the beginning of the twentieth century, but to them seemed to have reached its apotheosis in the Weimar Republic.

Much mid-to-late 1920s response to Binding and Hoche pseudo-scientifically sourced disability as being a *result* of the sexual disorder of Weimar. Binding and Hoche had not mentioned sex; it took their self-defined Christian *opponents* to introduce, and then keep harping on, the idea that there was – as one author put it – a "connection," indeed "a tight one," "between idiocy and sin."[14] A welfare

---

**13** Karl Weidel, Karl Ernst Thrändorf, Ludwig Lemme, and Arthur Titius quoted in Ewald Meltzer, *Das Problem der Abkürzung "lebensunwerten" Lebens* (Halle: Carl Marhold, 1925), 76, 78, 81, 82.
**14** Büchsel, "Euthanasie," 2.

institution director in Magdeburg (with a doctorate in theology) in 1925 offered as evidence for his own expertise and familiarity with thousands of "abnormals" that "they cannot talk or think . . . they resemble vegetating clumps of flesh whose expressions of life consist in nothing more than in eating, drinking, elimination, and sleeping." (In other words, he confirmed in his own words that the severely disabled were indeed "lesser" beings, inferior, gross and disgusting, and also time- and labor-consuming and expensive.) But – raging against couplings and reproduction resulting from "animal rutting" (*tierische Brunst*) – he additionally took the occasion to argue that "the worst sources of lesser-value progeny are those vices of the Volk": "above all alcoholism and sexual aberrations [*die sexuellen Verirrungen*]." Meanwhile, his rationale for why these "lesser-value progeny" should not be "eliminated" was because they would serve as living reminders to their parents to feel "somehow guilty . . . knowing as they do that so much wretchedness is the result of heredity." If the disabled were to be killed, so his argument went, consciences would be numbed, and then people would just return to producing more disability.[15]

Also, those who made a point of incorporating some sentimental defense of the severely disabled joined in the chorus of lament about the only moderately disabled and their purportedly errant ways.[16] As one Berlin-based theologian involved in welfare work observed, even the most severely disabled nonetheless could have a transformative – indeed redemptive – effect on those around them. Yet for this author, too, the number of "feeble-minded and psychopaths" was "expanding frighteningly," as their sexual "drive-life is stronger." And "the death of the Volk" would soon be "inescapable," for the very existence of welfare-care, he further affirmed, was allowing "everything withered, crippled,

---

**15** Martin Ulbrich, *Dürfen wir minderwertiges Leben vernichten? Ein Wort an die Anhänger und Verteidiger der Euthanasie* (Berlin-Dahlem: Wichern-Verlag, 1925), 3, 9–10, 12–14. Eighty percent of all "abnormality," Ulbrich averred, was caused by one's "own or others' guilt."

**16** Meltzer too (and even as he understood himself as an adamant opponent of Binding and Hoche's proposition, acknowledging that he would be deemed "backwards and sentimental" for being so), in the midst of an otherwise compelling anticapitalist excursus pointing out that the real "useless parasites of society" were not the disabled poor but the greedy rich, argued that if one really wanted to talk about the "waste of physical and psychic energy expended in the care of idiots," one should realize that "it is still nearly nothing against the energy that is spent these days on dirty literature and filthy art which so poison the life of the Volk that it only follows its carnal drives." Meltzer, *Das Problem*, 125, 51. Already in earlier writing, Meltzer had acknowledged that there were plenty of illnesses and unhappy circumstances that could cause progeny to be disabled, whether severely or moderately; "often enough," however, he continued, there were "relationships between idiocy and immoral fornication." Ewald Meltzer, *Zum Kampf gegen Unzucht und Unsittlichkeit!* (Dresden: C. Ludwig Ungelenk, 1917), 5.

ailing to keep growing" and then it "combines itself with the healthy, produces sick progeny and causes immeasurable damage."[17] In a Bethel-based publication, a pastor-director from Saxony-Anhalt offered a yet more intricate version of this associative constellation: "If there was no addiction to alcohol . . ., no brothel and no venereal diseases, no overabundance of lust and no submersion in vice – then most of the institutions of the Inner Mission could close their doors" (in other words, they wouldn't have any clients). Again, this was an argument *against* Binding and Hoche, as the author continued: "Because the world is the way it is and because our Volk simply is the way it has become – that is why there are idiots and epileptics and cretins and people ill with venereal diseases and tuberculosis and the cripples and the insane." This author was particularly indignant that "the media-Jews, the dirty-joke-writers and the real estate speculators, the stock exchange jobbers and the whole pleasure-craving populace that knows nothing more than insipid amusements and hot fashions and tawdry wit, all of them should live" – while "the poor newborn children, that have been put into the world by this adulterous generation and which carry the sins of the parents for the rest of their lives" would (if Binding and Hoche's recommendations were to be followed) be put to death.[18] In this defense of the moral innocence of the disabled, the causal contention could not be more explicit: The purported pervasiveness of sexual dissoluteness and sin was here blamed on Jews among others, and that dissoluteness and sin was proffered as explanation for *how disabled people are produced*. While there was certainly a kernel of sense and logic in the evidence that venereal diseases could cause birth defects, leaders in the Inner Mission evinced a far more global and diffuse dislike for what they saw as proliferating sexual profligacy.

## Compromise Formation

The reconfiguration of theology underway over the course of the 1920s and the growing consensus about the degraded value of disabled lives was to find yet fuller expression in the Treysa Resolution of 1931. Developed at that gathering of the Central Board of the Inner Mission which brought together twenty-two charity institution-affiliated professionals (ten of them theologians and pastors, the others physicians, a social worker, a teacher, and a lawyer) from all over

---

**17** Helmuth Schreiner, *Vom Recht zur Vernichtung unterwertigen Menschenlebens* (Schwerin: F. Bahn, 1928), 3, cf. 6.
**18** Büchsel, "Euthanasie," 2–3.

Germany to update the Inner Mission's positions on eugenics and euthanasia, its promulgation was to mark a decisive transition from a version of Protestant charity committed to provision of services for everyone in need to an explicitly eugenic orientation. And the short version of the resolution's argument turned out to be a kind of theological compromise formation: *Murder no, sterilization yes*. Sterilization had previously been rejected on grounds that an intellectually challenged sterilized female would either no longer have a reason *not* to become promiscuous and/or would be vulnerable to male exploitation and rapaciousness, as sexual conservatism had provided the foundational grid of intelligibility for churchmen's decision-making.[19]

Under growing pressure to consider murder, however, resistance to sterilizations dissolved. Or as the Treysa Resolution put it: "We do not want to eliminate the victims of guilt and sin, but to attempt to prevent them." (Worth noting once more here is the blatant implication that disabled progeny was the result of "guilt and sin.") Then again, however, the causation went in the other direction, as one fresh reason given for the importance of sterilizing was that a body's "God-given functions" might lead to "evil," and hence there was "not only a right, but a moral duty to sterilize out of neighborly love" and – here was a new emphasis on the Volk-collectivity and the future – "out of the responsibility which has been imposed on us not just for the generation that exists already, but also for the coming generation." What Christian "love" meant, in short, had been substantially redefined at Treysa. And once again, there were the side swipes at the sexual culture of Weimar. "Brothel-keepers," it was stressed, were "far worse vermin than the physical and mental invalids." Yet although Binding and Hoche (cited by name) were repudiated for assuming that there was "no soul-life" among the severely disabled, nonetheless the aversion that Binding and Hoche had articulated against that category of human beings was confirmed and re-stirred: "Shattering like almost no other thing, the sight of these

---

**19** A typical summary view is provided by Pastor Friedrich Lensch, director of the Alsterdorfer Anstalten in Hamburg (participating in the conversations among pastors and doctors, 18–20 May 1931, that culminated in the Treysa Resolution): "After short holidays 50 percent of the girls return to us morally degenerated and abused. Is that the residents' libidinal urges, or the irresponsibility of their kin? . . . One vacation day is enough for them to go off the rails." Quoted in Harald Jenner, "Friedrich Lensch als Leiter der Alsterdorfer Anstalten 1930 bis 1945," in Michael Wunder et al., *Auf dieser schiefen Ebene gibt es kein Halten mehr: Die Alsterdorfer Anstalten im Nationalsozialismus*, 3rd ed. (Stuttgart: Kohlhammer, 2016), 185–245, here 210–11. On the pervasiveness of the cultural assumption that sexual exploitation of sterilized females was normative/expectable male behavior, see Gisela Bock, *Zwangssterilisation im Nationalsozialismus: Studien zur Rassenpolitik und Geschlechterpolitik* (1986, reprint Münster: MV Wissenschaft, 2010), 437–439.

ailing and miserable beings admonishes the healthy person to keep his body unscathed and pure and to become conscious of his deep responsibility when he founds a family."[20] Keeping the disabled alive was once more construed as a valuable lesson in chastity for the nondisabled.

By 1933, when the Nazis assumed the reins of government and passed their overtly coercive and expansive sterilization law, the Protestant leaders of the Inner Mission signed off on it – or even declared (as Meltzer for instance did) that they had been in favor of sterilization all along.[21] They also began enthusiastically to

---

**20** "Die Treysaer Resolution," 107, 109.
**21** Ewald Meltzer, "Gesetz zur Verhütung Erbkranken Nachwuchses," *Zeitschrift für die Behandlung Anomaler* 53 (1933), 113–119, here 115. The overall trend was not completely inevitable. Not every Protestant – also not every pro-Nazi one – got on the pro-sterilization bandwagon. One noteworthy exception to the newfound zeal to curtail the reproductive lives of others – demonstrating, further, that it was at least *possible* to imagine and sustain alternate understandings of disability (and indeed of the science of heredity) – was the prominent Lutheran theologian Paul Althaus at the University of Erlangen. More generally, his case scrambles contemporary assumptions in instructive if perturbing ways, as he had in the late 1920s denounced "the Jewish menace to our Volk-character," in early 1933 placed great hopes in Hitler, and simultaneously was a sensitive and eloquent defender of the value of disabled life. Paul Althaus, *Kirche und Volkstum* (Gütersloh: Bertelsmann, 1928), 33. In April 1933, Althaus delivered a lecture at a conference in Bremen in which he not only expressly rejected Binding and Hoche's proposals to kill individuals with severe cognitive disabilities, but also urged great caution with sterilization. Althaus was repelled by his contemporaries' fantasies of broad-based sterilization ("eugenic orgies," he called these), and stressed the lack of self-evidence and numerous ongoing enigmas in patterns of hereditary transmission of "feeblemindedness." He further emphasized that also the most severely disabled individuals – here he explicitly invoked the residents of Bethel – were evidently capable of joy and of receiving human love. And was this not the very point of life? Disability, Althaus moreover averred, provided an incomparable opportunity also to encounter *God's* love. Althaus distanced himself utterly from the notion of "unworthy" life by putting the term in quotes in his lecture title, arguing that it was imperative to distinguish between "worth" (*Wert*) and "dignity" (*Würde*), and insisting that "dignity . . . has no gradations" (*Würde . . . verträgt keine Abstufung*). He decried the ascendant obsessions with health and strength and the notion of life as above all a "struggle for survival" (*Daseinskampf*), observing that illness could enter any life in an instant, and that "perfectibility" (*Vervollkommnung*) was unattainable and in any event undesirable. Paul Althaus, "'Unwertes' Leben im Lichte christlichen Glaubens," in *"Von der Verhütung* unwerten Lebens": *Ein Zyklus von 5 Vorträgen* (*Bremer Beiträge zur Naturwissenschaft*, Sonderband 1933), 79–97, here 79, 82, 84, 86, 88, 91–92. Although the 1933 lecture had slipped into print in a scientific journal sandwiched between the lectures of his (uniformly pro-sterilization) co-panelists, Althaus was denied the right to republish it as he had planned, on grounds that it might discredit public health and state policy in the eyes of the public. Gotthard Jasper, *Paul Althaus (1888–1966): Professor, Prediger und Patriot in seiner Zeit* (Göttingen: Vandenhoeck & Ruprecht, 2015), 217–222. In no way, however, was his professorship endangered.

sterilize. At Bethel and numerous other charitable institutions for the disabled, a regular routine was established, with literally a weekly "sterilization day" – and the doctors could hardly keep up.[22] Charity institutions were encouraged to develop card catalogues and clan charts identifying individual and familial vulnerabilities.[23]

More generally, disability care institutions warmly welcomed the Third Reich. In fulsome terms, directors embraced the rise of Nazism, expressing appreciation that not only "Marxist socialism" had been pushed out of its dominant role in public life, but also that the new government had effectively "broken . . . the influence of that Jewish liberalism that is totally hostile to the Gospels."[24] "Hitler's strong hand has intervened" and brought to an end the "liberal-Marxist era" which had been "leading to Bolshevism, the Jewish dictatorship."[25] Photographs of gatherings of individuals with disabilities, whether accompanied by their caregiving deaconesses amid swastika flags at a sunny festive outing, or with a swastika woven into a handmade basket in an in-house sheltered workshop, conveyed the climate with special poignancy.[26]

Scholars have long known that National Socialists tactically invoked "religious" language as they worked to dispel potential church or popular hesitation about eugenics and euthanasia. But as it turns out, Christians were way ahead of them in preemptively adapting Christian teachings already in the 1920s and first years of the 1930s. Moreover, once the sterilization law passed, churchmen and charity doctors speaking in the name of Christianity developed yet further *religious* justifications for sterilization – as pastors continued to mix up supposed hereditary deficiencies with "evil" and purportedly higher-value people with "goodness," and extended the received Lutheran doctrine of two kingdoms

---

**22** Uwe Kaminsky, "Paternalistische Verschwiegenheit – Bethel, die Zwangssterilisation und NS-'Euthanasie,'" *Lippische Mitteilungen aus Geschichte und Landeskunde* (2020), 69–87, here 72–77; Ernst Klee, *"Die SA Jesu Christi": Die Kirche im Banne Hitlers* (Frankfurt am Main: Fischer, 1989), 92.

**23** Michael Wunder, "Die Karriere des Dr. Gerhard Kreyenberg – Heilen und Vernichten in Alsterdorf," in Wunder et al., *Auf dieser schiefen Ebene*, 137–183, esp. 155–166; cf. "Die Treysaer Resolution," 108.

**24** Fritz von Bodelschwingh quoted in Matthias Benad, "Bethels Verhältnis zum Nationalsozialismus," in *Zwangsverpflichtet: Kriegsgefangene und zivile Zwangsarbeiter(-innen) in Bethel und Lobetal, 1939–1945*, eds. Matthias Benad and Regina Mentner (Bielefeld: Verlag für Regionalgeschichte, 2002), 27–66, here 28.

**25** Meltzer, "Gesetz," 115–116.

**26** Photo from the Alsterdorfer Anstalten in Hamburg on the cover of Wunder et al., *Auf dieser schiefen Ebene*; photo with basket woven at Bruckberg, Neuendettelsau in *50 Jahre Lebenshilfe Fürth – Jubiläumsdokumentation* (Nürnberg: NOVA-Druck, 2011), 17.

(the heavenly and the earthly) into a vigorous defense of the new governing authorities and their policies.[27] The director of the Hamburg charity was particularly exhilarated that "the much-discussed question of sterilization of the hereditarily ill . . . has been torn away from the materialist motivations of the Marxist era, which mainly defended the idea because the sterilization of patients facilitated an unrestricted pursuit of pleasure (*uneingeschränktes Sichausleben*) . . . . This . . . is why we as Germans and Christians resisted the idea. Now . . . we can honestly welcome the new regulation."[28] And a pastor affiliated with one of Bethel's auxiliary institutions asserted in 1934: If there was any past "guilt" for which the Protestant church might self-castigate, it was that it had permitted *too many* hereditarily ill people to marry (and thus reproduce). Joining in the chorus of voices declaring that "the continued existence of our Volk is lethally threatened by the strong expansion of the biologically inferior," cheering that "the liberalistic slogan of the right of the human being to his own body [another reference to Weimar cultural values] has been thoroughly undone by this law," the pastor validated "all enthusiasm for the Third Reich and its Führer . . . God Himself has given this government the power."[29]

The history of disability quite evidently moves with different rhythms, in syncopation rather than in synchrony with standard markers of political change. And the history of Protestant theology in Germany in the first decades of the twentieth century looks different when we examine it from the point of view of ideas about disability. In Germany around 1900, labor of care for the most vulnerable among the vulnerable had been understood as God's work and as a source of reciprocal joy and meaning. By 1930 at the latest, such confidence had unraveled. In their efforts to come up with reasons not to condone murder, Protestant spokesmen had settled on the ardent endorsement of sterilizations – and so ended up not just eugenicizing theology, but theologizing eugenics. Christian receptiveness to eugenic argumentation was not solely a consequence of the at once economic and political crisis of the Weimar welfare state; it

---

27 Gerhard Kunze, pastor in Hannover, and similar examples quoted in Kurt Nowak, *"Euthanasie" und Sterilisierung im "Dritten Reich": Die Konfrontation der evangelischen und katholischen Kirche mit dem Gesetz zur Verhütung erbkranken Nachwuchses und die "Euthanasie"-Aktion* (Göttingen: Vandenhoeck & Ruprecht, 1978), 98.

28 Friedrich Lensch, "Die Alsterdorfer Anstalten im Dritten Reich," *Briefe und Bilder aus Alsterdorf* 58 (1934), 1–4, here 2.

29 Ernst Klessmann, "Auswirkungen des Gesetzes zur Verhütung erbkranken Nachwuchses für den seelsorgerlichen Dienst," in *Dokumente zur "Euthanasie,"* ed. Ernst Klee (Frankfurt am Main: Fischer, 1985), 55–57.

additionally had a great deal to do with sexuality – and especially with distress at the purportedly Jewish-exacerbated liberalization of sexual mores in Weimar. Because they were on the defensive, because they hoped to maintain their cultural status and keep their institutions afloat, but perhaps also because they all along had unresolved ambivalences about their own work, the very people who had the disabled in their care cast aspersions on them and formulated – in the name of Christianity – a novel means of denigrating the disabled. Far from providing an energetic, robust alternative to Binding and Hoche's cost-benefit arguments and emotional revulsion arguments that demanded killing, Protestant leaders effectively intensified antidisability animus.

# Coda

Why is this episode so little known also among scholars of German culture and history? Several dynamics converge. Getting to an answer will certainly require studying critically the self-exculpatory and self-celebratory legends that were carefully cultivated by Protestant church spokesmen after 1945. But no less consequential was the deep shame still accruing to disability for many years after the war. Nazi-exacerbated contempt toward individuals with disabilities, especially cognitive disabilities, whether moderate or severe, had a lasting hold on the populace, and the West German government obstinately refused even to grant recognition as victims of Nazi persecution, to say nothing of monetary reparations, to survivors of the coercive sterilizations or to family members of individuals killed in the euthanasia program. A third factor involves the enduring difficulty of getting the disability murders taken seriously in their intricate interrelationships with the Holocaust of European Jewry – a phenomenon which remains palpably, painfully evident to this day in the United States Holocaust Memorial Museum's permanent exhibit.[30] This is so despite the fact that the murder in carbon monoxide gas chambers of Jewish individuals with disabilities was the first installment of the Holocaust.

---

**30** "The murder of the handicapped" is placed in a side room – and this regardless of the fact that 121 men who got their training and practice in mass murder by killing the disabled in the T4 program followed up by moving their expertise and technology to the Operation Reinhard death factories of Belzec, Sobibor, and Treblinka, where *fully one quarter* of the Holocaust took place. The Museum could well, in short, have led visitors through, rather than around or away from, the euthanasia murders.

It is symptomatic that certain details only came to public attention decades after the war. In the case of Bethel, seven Jewish patients were turned over upon demand to the Nazi authorities in September 1940, although Bethel's director knew at that point that the destination was death. Eight other Jewish residents, however, were temporarily rescued as the Bethel administration notified their kin to retrieve them or found other Jewish families to take them in. The lives of the seven who were given up ended in the T4 gas chamber of Brandenburg. Brief mention in passing of these seven, though without explanation or analysis, had to wait until 1970.[31] It was the twenty-first century before an essay sensitively reconstructed all fifteen of the Jewish victims' experiences.[32] Also the eight that had initially been rescued ended up murdered – in Minsk, in Theresienstadt, and in Auschwitz.

But on what grounds had the seven been forfeited – just as Bethel's director was engaged in delicate behind-the-scenes negotiations with high-ranking Nazi officials in (ultimately successful – or above all simply lucky) desperate efforts to save the approximately 440 of his 3100 residents that would have been the most likely targets of the disability murder program? In 2007, the searing particulars came to light. Heartrendingly, the answer was: "work-capacity." All seven were individuals who required intensive labor of care from others, and who – beyond such simple tasks as sorting aluminum foil out of trash or shredding strips of cloth – could not contribute labor themselves. One of them, Olga Laubheim, had lived in Bethel for forty-five years.

## Works Cited

Althaus, Paul. *Kirche und Volkstum*. Gütersloh: Bertelsmann, 1928.
Althaus, Paul. "'Unwertes' Leben im Lichte Christlichen Glaubens." In *"Von der Verhütung unwerten Lebens": Ein Zyklus von 5 Vorträgen. Bremer Beiträge zur Naturwissenschaft.* Sonderband, 1933: 79–97.
Bauer, Karl-Adolf. "Aus der Geschichte der Diakonie-Anstalten Bad Kreuznach." In "Diakonie im Dritten Reich: Tagung in Zusammenarbeit mit dem Diakonischen Werk der Evangelischen Kirche im Rheinland und den Diakonie-Anstalten Bad Kreuznach, 16–17 Mai 1987, Außentagung in Maria Laach," *Begegnungen* 5. Edited by Evangelische Akademie Mülheim/Ruhr. Mülheim/ Ruhr, 1987: 61–70.

---

**31** Klee, *"Euthanasie,"* 320.
**32** Kerstin Stockhecke, "September 1940: Die 'Euthanasie' und die jüdischen Patienten in den v. Bodelschwingschen Anstalten Bethel," in *Kirchenarchiv mit Zukunft,* eds. Claudia Brack et al., (Bielefeld: Verlag für Regionalgeschichte, 2007), 131–142.

Benad, Matthias. "Bethels Verhältnis zum Nationalsozialismus." In *Zwangsverpflichtet: Kriegsgefangene und zivile Zwangsarbeiter(-innen) in Bethel und Lobetal, 1939–1945.* Edited by Matthias Benad and Regina Mentner, 27–66. Bielefeld: Verlag für Regionalgeschichte, 2002.

Binding, Karl, and Alfred Hoche. *Die Freigabe der Vernichtung lebensunwerten Lebens: Ihr Maß und ihre Form.* Leipzig: Felix Meinen, 1920.

Bock, Gisela. *Zwangssterilisation im Nationalsozialismus: Studien zur Rassenpolitik und Geschlechterpolitik.* 1986. Reprint Münster: MV Wissenschaft, 2010.

Büchsel, Hermann. "Euthanasie." *Aufwärts* 280 (30 November 1926): 1–4.

"Die Treysaer Resolution des Central-Ausschusses für Innere Mission (1931)." In *Eugenik, Sterilisation, Euthanasie: Politische Biologie in Deutschland 1895–1945 – Eine Dokumentation.* Edited by Jochen-Christoph Kaiser et al., 106–110. Berlin: Buchverlag Union, 1992.

Janzowski, Frank. *Die NS-Vergangenheit in der Heil- und Pflegeanstalt* Wiesloch: . . . *so intensiv wenden wir unsere Arbeitskraft der Ausschaltung der Erbkranken zu.* Ubstadt-Weiher: Verlag Regionalkultur, 2015.

Jasper, Gotthard. *Paul Althaus (1888–1966): Professor, Prediger und Patriot in seiner Zeit.* Göttingen: Vandenhoeck & Ruprecht, 2015.

Jenner, Harald. "Friedrich Lensch als Leiter der Alsterdorfer Anstalten 1930 bis 1945." In *Auf dieser schiefen Ebene gibt es kein Halten mehr: Die Alsterdorfer Anstalten im Nationalsozialismus*, 3rd. Edited by Michael Wunder et al., 185–245. Stuttgart: Kohlhammer, 2016.

Kaminsky, Uwe. "Paternalistische Verschwiegenheit – Bethel, die Zwangssterilisation und NS-'Euthanasie.'" *Lippische Mitteilungen aus Geschichte und Landeskunde* 89 (2020): 69–87.

Klee, Ernst. *"Euthanasie" im NS-Staat: Die "Vernichtung lebensunwerten Lebens."* Frankfurt am Main: Fischer, 1983.

Klee, Ernst. *"Die SA Jesu Christi": Die Kirche im Banne Hitlers.* Frankfurt am Main: Fischer, 1989.

Klessmann, Ernst. "Auswirkungen des Gesetzes zur Verhütung erbkranken Nachwuchses für den seelsorgerlichen Dienst." In *Dokumente zur "Euthanasie."* Edited by Ernst Klee, 55–57. Frankfurt am Main: Fischer, 1985.

Lensch, Friedrich. "Dennoch!" *Briefe und Bilder aus Alsterdorf* 55–56 (1931–32): 2–5.

Lensch, Friedrich. "Die Alsterdorfer Anstalten im Dritten Reich," *Briefe und Bilder aus Alsterdorf* 58 (1934), 1–4.

Mayer, Alexander. *50 Jahre Lebenshilfe Fürth – Jubiläumsdokumentation.* Nürnberg: NOVA-Druck, 2011.

Meltzer, Ewald. *Zum Kampf gegen Unzucht und Unsittlichkeit!* Dresden: C. Ludwig Ungelenk, 1917.

Meltzer, Ewald. *Das Problem der Abkürzung "lebensunwerten" Lebens.* Halle: Carl Marhold, 1925.

Meltzer, Ewald. "Gesetz zur Verhütung Erbkranken Nachwuchses," *Zeitschrift für die Behandlung Anomaler* 53 (1933): 113–19.

Nowak, Kurt. *"Euthanasie" und Sterilisierung im "Dritten Reich": Die Konfrontation der evangelischen und katholischen Kirche mit dem Gesetz zur Verhütung erbkranken Nachwuchses und die "Euthanasie"-Aktion.* Göttingen: Vandenhoeck & Ruprecht, 1978.

Rotzoll, Maike, et al. *Die nationalsozialistische "Euthanasie"-Aktion "T4" und ihre Opfer: Geschichte und ethische Konsequenzen für die Gegenwart.* Paderborn: Schöningh, 2010.

Schreiner, Helmuth. *Vom Recht zur Vernichtung unterwertigen Menschenlebens*. Schwerin: F. Bahn, 1928.

Stockhecke, Kerstin. "September 1940: Die 'Euthanasie' und die jüdischen Patienten in den v. Bodelschwingschen Anstalten Bethel." In *Kirchenarchiv mit Zukunft*. Edited by Claudia Brack et al., 131–142. Bielefeld: Verlag für Regionalgeschichte, 2007.

Ulbrich, Martin. *Dürfen wir minderwertiges Leben vernichten? Ein Wort an die Anhänger und Verteidiger der Euthanasie*. Berlin-Dahlem: Wichern-Verlag, 1925.

Waltermann, Leo. "Diakonie im Dritten Reich: Aus der Geschichte Konsequenzen für heute." In "Diakonie im Dritten Reich: Tagung in Zusammenarbeit mit dem Diakonischen Werk der Evangelischen Kirche im Rheinland und den Diakonie-Anstalten Bad Kreuznach, 16–17 Mai 1987, Außentagung in Maria Laach," *Begegnungen* 5. Edited by Evangelische Akademie Mülheim/ Ruhr, 85–117. Mülheim/ Ruhr, 1987.

Wunder, Michael. "Die Karriere des Dr. Gerhard Kreyenberg – Heilen und Vernichten in Alsterdorf." In *Auf dieser schiefen Ebene gibt es kein Halten mehr: Die Alsterdorfer Anstalten im Nationalsozialismus*, 3rd ed. Edited by Michael Wunder et al., 137–183. Stuttgart: Kohlhammer, 2016.

Susannah Heschel

# Sacrament versus Racism: Converted Jews in Nazi Germany

Erna Becker-Kohen, a German Jew who was baptized Roman Catholic in 1936, was deeply pious and found great sustenance in prayer. In addition to singing in the church choir, she attended Mass regularly. All that changed after September 1941. "For a while I was a member of the church choir in our little parish. Singing has always given me much joy, but now I had to give it up because a few singers did not like the idea of a Jew participating. I always remained modestly, even shyly, in the background. Still, I am not wanted."[1]

In the autumn of 1941 in Nazi Germany, Christians suddenly became Jews.[2] Not through a formal conversion process nor out of their own faith decision but rendered Jewish Christians by the race laws, Jews aged seven and older were required by the Nazi regime to wear a yellow Star of David (*Judenstern*) on their clothing. Even Jews who had converted to Christianity now had to wear the star, including in church, Catholic and Protestant.[3] Race versus faith: the Nazi regime mandated that race transcends all other signifiers of identity, including baptism, a sacrament of the church that provides "an indelible spiritual mark" of "belonging to Christ," in the words of the Catholic catechism.[4] Prior to September 1941, those attending church services would likely not have known which parishioners were Jews who had been baptized by their parents at birth or Jews who converted as adults from Judaism to Christianity. The yellow Star of David identified these Christians in public as racial Jews. How parishioners, congregants, pastors, priests, and bishops reacted to the yellow star and how Jews who had converted to Christianity negotiated their marked status is my inquiry; what the reactions reveal about

---

1 Kevin P. Spicer and Martina Cucchiara, ed. and trans., *The Evil that Surrounds Us: The Memoir of Erna Becker-Kohen* (Bloomington: Indiana University Press, 2017), 46.
2 Hartmut Ludwig, *Suddenly Jews: The Story of Christians whom the Nazi Racial Laws Classified as Jews, and of the Good Samaritans who came to their Aid (the Bureau Grüber)*, trans. Martin Nicolaus (Berkeley: Duplex Press, 2015).
3 The 18,000 German Protestant pastors were divided between those who joined the German Christian Movement (Deutsche Christen) and gave Christian theological support to the Nazi regime, including its race laws, and the 7,000 who joined the Confessing Church (Bekennende Kirche) and opposed alterations of Christian doctrine but did not always oppose the Nazi regime.
4 1272, *Catechism of the Catholic Church* (US Catholic Conference/Libreria Editrice Vaticana, 1994), 324.

https://doi.org/10.1515/9783110753295-007

Christian views of Judaism is my larger question; how religions should respond to racism is my deeper concern.

Baptism did not allow Jews to escape the Nuremberg Laws nor guarantee protection from deportation and death, but it helped. Some degree of shelter was available to Jews married to Aryans and to converted Jews, but those regulations, like the Nazi definitions of "Jew," were complex and not always clear or enforced. In the first years of the Nazi Reich, baptism and marriage to an Aryan offered Jews hope for asylum from the race laws. There were 20,454 Jews in Germany married to an Aryan in 1939, but precise numbers of Christians classified by the Nazis as Jews who were arrested deported and murdered is unclear. While marriage to Aryans offered some degree of safety from Nazi oppression, baptism was less effective. The Catholic and Protestant churches may have offered hope and spiritual comfort, and individual priests and pastors could be kind, but ultimately, as institutions, they could not guarantee Christians of Jewish heritage a safe haven within the Third Reich. Even more troubling, some did not want to. As Becker-Kohen's memoir demonstrates, although some Catholics turned away from her after September 1941, some priests tried to help and console her. Mixed attitudes are found among Protestants, though the pro-Nazi German Christian movement within the Protestant church sought to expel what they called "non-Aryan Christians," while their opponents within the church, known as the Confessing Church, offered assistance. The reaction of the public to the star decree was described initially by Nazi opinion surveyors as being "greeted" by the population. In Catholic and middleclass areas, however, there were expressions of pity and some people spoke of "methods out of the middle ages."[5] While we do not know the degree to which baptism, like marriage to an Aryan, offered a measure of security from the Nazi government program to murder all Jews, we do know that German Jews found no guarantee of church asylum from Nazi antisemitism through baptism.

Initially after Hitler came to power, Jews thought baptism might offer protection; 900 converted in the year 1933, but baptism provided no guarantee of safety from Nazi race laws. Controversies raged within the Protestant church about offering assistance to Christians of Jewish background and became a major factor in the split within the Protestant church, with some bishops and pastors advocating assistance for converted Jews, but not for Jews who had not converted, and others insisting that converted Jews, as well as anything "Jewish," should be

---

5 Robert Gellately, *Backing Hitler: Consent and Coercion in Nazi Germany* (New York: Oxford University Press, 2001), 131. Gellately is here citing Heinz Boberach, ed., *Meldungen aus dem Reich 1938–1945. Die geheimen Lageberichte des Sicherheitsdienstes der SS* (Herrsching: Pawlak Verlag, 1984), (9 October 1941), 2849.

banned from the church. Safety concerns did not motivate Becker-Kohen, whose conversion to Roman Catholicism in 1936, a few years after her marriage to a Catholic engineer, Gustav Becker, was for religious sustenance. As she writes, she "saw a light and yearned to be one of them."[6] To see a light and yearn: prayer is a hope for transformation. She sought closeness to God and experienced a conversion of the heart. In March 1938, her baby son, Silvan, was baptized.

Becker-Kohen was shocked that wearing a yellow star brought her social isolation and rejection by her parish community and by some of the Catholic priests whose help she sought as she struggled to survive, with her husband and son, during World War II. She was one of countless so-called "non-Aryan Christians" in the Reich; their number is unclear, but some have estimated it to be around 300,000 in June 1933:[7] about 50,000 German Christians with four Jewish grandparents; 210,000 Germans with two Jewish grandparents; and 80,000 with one Jewish grandparent. Not all were baptized, but few children of intermarriages were raised as Jews, and the intermarriage rate was high. In 1927, 25.8 percent of Jewish men and 16.1 percent of Jewish women had married outside the community.[8] In total, 44 percent of German Jews were married to non-Jews, and 200–400 Jews converted to Protestantism every year between 1900 and 1939, except in 1933, when more than 900 converted, presumably hoping that baptism might offer some protection from Nazi antisemitism.[9] By October 1941, when emigration was no longer possible, only about 164,000 Jews were left in Germany, about half living in Berlin, but the number of remaining baptized Jews in the Reich is not known. By 1 April 1943, 31,910 Jews were left who wore a yellow Star of David and 17,375 Jews were in "privileged" marriages to Aryans and did not have to wear a star.[10] Half the intermarried couples lived in the city of Berlin, but we have no number for the many baptized Jews who remained.

---

**6** Spicer and Cucchiara: *Becker-Kohen*, 152.
**7** Ursula Büttner, "The Persecution of Christian-Jewish Families in the Third Reich," *Leo Baeck Institute Yearbook* 34:1 (1989): 267–289; 271.
**8** Franklin Oberlaender, "The Family Dynamics of German Protestants of Jewish Descent Stigmatized in Nazi Germany and of Their Offspring Born in Postwar West Germany," *Holocaust and Genocide Studies* 9, no. 3 (Winter 1995): 360–377, 361.
**9** Eberhard Röhm and Jörg Thierfelder, *Juden, Christen, Deutsche 1933–1945* 4, no. 1 (Stuttgart: Calwer Verlag 2004), 192–199.
**10** Population statistics can be found here: http://www.statistik-des-holocaust.de/stat_ger_pop.html; For Jews who were "privileged," see Wolfgang Gerlach, *And the Witnesses Were Silent: The Confessing Church and the Persecution of the Jews*, trans. Victoria J. Barnett (Lincoln: University of Nebraska Press, 2000), 175.

# The Theological Conflict between Baptism and Race Prior to 1941

As a sacrament of the church, whether Protestant or Catholic, baptism transforms a person into a Christian; it is a fundamental doctrine. Racist ideology, however, challenged the efficacy of baptism: for racists, a Jew remains a Jew despite baptism. Within the Protestant church, the pro-Nazi German Christian faction affirmed race above doctrine, insisting that baptism does not alter race any more than it alters gender; a Jew does not lose racial identity through baptism any more than men or women lose their gender. The debate over race versus baptism had a long history, reaching a climax with the mass conversion of Jews to Catholicism in fifteenth-century Spain that led to suspicion over the sincerity of the converts, labeled "new Christians" or "Marranos" (slang for swine), and again with the Christian missionaries who joined European colonization efforts but were then reluctant to allow baptized Africans and Asians into European churches.

Jews who converted to Christianity, whether out of religious faith or at the request of a spouse or to assimilate more fully into German society, were fulfilling the longstanding wish of the churches. Racism, however, upended that wish. A pastor at a February 1934 meeting of pro-Nazi Protestants in a Berlin auditorium adorned with swastikas, declared, "They say that everyone is equal before God. But baptism never made a Jew into a German, nor did it ever straighten a crooked, hooknose."[11] His words echoed Nazi rhetoric: "just as a pig remains a pig, even if you put it in a horse's stall, so a Jew still remains a Jew, even if he is baptized."[12] A newspaper commented on the baptism of a Jewish family by declaring: "from a dog you are never going to fry bacon."[13]

"Proving" Aryan status required presenting baptismal certificates for oneself, one's parents, and one's grandparents, requiring churches to hire scribes to copy those certificates, millions of them, by hand. Some churches investigated the racial identity of their staff and their congregants. The regional church of Eutin, for example, conscientiously carried out the relevant administrative regulations of

---

11 Doris Bergen, *Twisted Cross: The German Christian Movement in the Third Reich* (Chapel Hill: University of North Carolina Press, 1996), 22.

12 Anon., "Schluss mit den Judentaufen," *Arische Rundschau* 1933, EZA 1/C3/170 vol. 2, fo. 19; cited by Christopher Clark, *The Politics of Conversion: Missionary Protestantism and the Jews in Prussia, 1728–1941* (Oxford: Clarendon, 1995), 291.

13 Interview with Stephen Nasser USC-Shoah Foundation Visual History Archive; Segment #9, 9:25: He reports that his family's conversion to Catholicism was written up in the local newspaper. My thanks to Professor Doris Bergen, University of Toronto, for generously sending me her student's notes on the interview.

the state concerning identification of Christians who were "racial Jews" in their church records by providing or creating certificates of descent and other genealogical documentation.[14]

After the promulgation of the Nuremberg Laws in September 1935, baptism of Jews was forbidden by some Protestants, and pastors who continued to baptize Jews, old and young, did so at some risk. Baptism did not save Jews from deportation, but it did offer some protection. Some pastors and parishioners of the Confessing Church organized assistance to baptized Jews, but few in the Confessing Church helped Jews who had not been baptized. Horst Gessner, born in 1928 to a Jewish mother and a Christian father who had converted to Judaism, was baptized at age ten, an act he describes as "resistance" by the pastor at the Sophienkirche in Berlin. Gessner reports that as a young man he could walk in public without a yellow star and, when stopped, show his baptismal certificate instead of an identity card.[15]

The conflict over baptism and race led to hefty debates among Protestants and ultimately to the split within the church between the German Christian Movement that wanted Nazi race laws implemented within the church and their opponents, the Confessing Church, that insisted on church doctrine regarding baptism. Considering the issue, the Protestant theological faculty of the conservative University of Erlangen, published a resolution by Paul Althaus and Werner Elert restricting the ordination of baptized Jews. The Protestant theologians at the University of Marburg, by contrast, issued a declaration that a "racial Jew" who had accepted Christianity and was a full-fledged member of the Christian church and should not suffer any discrimination.[16] Yet even those who sided with the Marburg opinion could be ambivalent. The Berlin pastor Martin Niemöller, who became a major figure in the Confessing Church, wrote in November 1933 that Christians found converted Jews disagreeable, despite the demands of Scrip-

---

**14** Lawrence Stokes, "Die Eutiner Landeskirche zwischen November-revolution und Nationalsozialismus," in *Kirche und Nationalsozialismus: Beiträge zur Geschichte des Kirchenkampfes in den evangelischen Landeskirchen Schleswig-Holsteins*, ed. Klauspeter Reumann (Neumuenster: Karl Wachholtz Verlag, 1988), 141.

**15** *Evangelisch getauft – als Juden verfolgt: Spurensuche Berliner Kirchengemeinden*, ed. Hildegard Frisius et al. (Berlin: Wichern 2008), 268–269.

**16** There are numerous discussions of the debate; see Marshall Johnson, "Power Politics and New Testament Scholarship in the National Socialist Period," *Journal of Ecumenical Studies* 23, no. 1 (Winter, 1986): 1–24.

ture, and that "pastors of Jewish descent" should work in parishes where they might not "offend the congregation."[17]

In 1935 the Confessing Church's Steglitz Synod took a stronger stance, declaring it a "sin" to forbid the baptism of Jews. The synod further condemned those who reject baptism of Jews as "a betrayal of Christ," but did not urge Christians to care for all Jews, baptized or not.[18] The theologian Karl Barth had already spoken out more strongly in 1933, warning that "belonging to the Church" is determined not by blood and race, but by the Holy Spirit and baptism: "If the German Evangelical Church would exclude the Jewish Christians or treat them as second-class Christians, it would have ceased to be a Christian church."[19]

For Barth, there was no question that baptism overrode Nazi theories of blood and race, and during the course of my research on this topic in recent years, his views brought me a measure of comfort. Still, his hostility to Nazi racism – one of the greatest theologians of the century – was not widely or vocally shared by other Protestant theologians. In 1935 Barth had to leave Germany and return to his birth country, Switzerland. His opposition on religious grounds to the Nazi race laws echoed for me the strong religious voices against racism that had filled my childhood in the United States during the 1960s. That religious opposition to racism was vital I learned from my father's involvement in the Civil Rights movement and my opportunities to hear and meet leaders of that movement. I grew up with the assumption that religion and racism were as diametrically opposed as life and death, and the role of Jews in Civil Rights, as well as the importance of the Civil Rights movement for Judaism's recalibration after the Holocaust has been an important topic of my own scholarship.[20]

Leaders of the Civil Rights movement in the 1950s and 60s included pastors and rabbis who deployed Scripture to argue that racism is anathema; my father called it "satanism, unmitigated evil." Inspired by Dr. King's ringing evocation of the Hebrew prophets, particularly Amos, I majored in Religion in college, focusing on Hebrew Bible, which I saw as a powerful force in bringing an end to racism. Little did I know . . . When I began studying the Nazi era, I was shocked to find theologians reading the Bible through a racial lens, referring to Hitler as

---

**17** Martin Niemöller, "Sätze zur Arierfrage in der Kirche," *Junge Kirche* (2 November 1933): 269–271; cited by Matthew Hockenos, *Then They Came for Me: Martin Niemöller, the Pastor who Defied the Nazis* (New York: Basic Books, 2018), 100.
**18** Gerlach, *And the Witnesses*, 151.
**19** Karl Barth, *Theological Existence Today*, Heft 1 (June 1933), 24f.; cited by Gerlach, *And the Witnesses*, 32.
**20** Susannah Heschel, "A Friendship in the Prophetic Tradition: Abraham Joshua Heschel and Martin Luther King, Jr.," *Telos* (April 2018): 67–84.

Christ, rewriting the Sermon on the Mount to make it manly and militaristic, and abandoning Christian doctrine in favor of Nazi race laws.

What resources do Christians and Jews have to resist racism? For leaders of the Civil Rights Movement in the United States, the Exodus and the prophets of the Hebrew Bible provided powerful voices and images of liberation from racism and for forging alliances with Jews, a "kinship of strangers who have both know 'Egypt.'"[21] The Black church focused the Old Testament and viewed Martin Luther King, Jr., as a Moses leading his people out of the Egypt of segregation in America. For my father, acknowledging race was antithetical to religion: "Race is a denial of the existence of God."

In giving religious meaning to the destructive experience of racism, the African American theologian James Cone writes that God's choice of enslaved Israel to be his chosen people spoke to God's own character. Cone writes, "If God had chosen as his holy nation the Egyptian slave masters instead of the Israelite slaves, a completely different kind of God would have been revealed."[22] An interpretation of the Bible to justify slavery would violate what we know of God's character.[23]

Protestant theologians in Germany had long focused primarily on the New Testament, and the pro-Nazi faction within the Protestant church banished the Old Testament from the Christian Scriptures. German Biblical scholarship on the prophets did not highlight their warnings against injustice and cruelty but depicted them as ecstatics who lost consciousness and made proclamations without really understanding what they were saying.[24] Without the prophets, they lost one of the precious biblical resources to resist and protest Nazi racism.

My scholarly engagement in Germany has centered in the theological realm, a dimension of German society that many contemporary scholars of German history and literature tend to ignore. Social histories rarely tell us what was transpiring in the churches during the Third Reich; they ignore what the pastors and priests were preaching that contributed to the course of political and cultural

---

**21** Keith Magee, *Prophetic Justice: Essays on Race, Religion, and Politics* (London: New Generation, forthcoming), 5.

**22** James Cone, "Biblical Revelation and Social Existence," *Interpretation* 28, no. 4 (1974): 422–440; 425.

**23** Esau McCaulley, *Reading While Black: African American Biblical Interpretation as an Exercise in Hope* (IVP Academic 2002), 183.

**24** Susannah Heschel, "Ecstasy versus Ethics: The Impact of World War I on German Biblical Scholarship on the Hebrew Prophets," in *The First World War and the Mobilization of Biblical Scholarship*, eds. Andrew Mein, Nathan MacDonald, and Matthew A. Collins (London: T&T Clark, 2019), 187–206.

change, especially regarding Jews and Judaism. Germans have long faced enormous moral dilemmas regarding antisemitism, as Americans face a legacy of slavery, Jim Crow, and systemic racism. We should ask what we, Christians and Jews, learn from the Bible about conscience and religious resources for opposing racism and antisemitism. Can religious communities not only work to combat racism, but also become places of refuge and security from threats of violence? And while law and doctrine can be upheld in the face of racism, the insidious and cloying disgust generated by racism requires the kind of introspective moral self-scrutiny that religious communities are best suited to provide.

## German Reactions to the Yellow Star

It was Josef Goebbels, Nazi Minister of Propaganda, who instigated the mandate that all Jews age six and over living in the Reich wear a yellow badge affixed to their clothing, starting on 15 September 1941. Published in the official gazette, the *Reichsverordnungsblatt*, the ruling also applied to baptized Jews, whether registered as members of the Protestant or Catholic church.[25] The regulations for wearing a yellow star were complicated. The star had to be sewn properly or one risked arrest.[26] For Jews, wearing a yellow star in public brought anguish, as described by Viktor Klemperer in his *Diary*, "Yesterday as Eva was sewing on the Jew's star, I had a raving fit of despair." The star had to be fully displayed, including when opening the door of one's home. Klemperer also reports that a neighbor went out wearing the star, "walking with her umbrella up, even when it has stopped raining – because then her arm covers the star. Or a package or a bag pressed against it. A circular from the Jewish community has warned against it, it is severely punished."[27]

The historian Edward Timms writes, "Being forced to display the Star amounted to a death threat."[28] Who had to wear a star was complex because Nazi race laws were heavily intertwined with religion: Aryan status depended upon baptismal certificates for parents and grandparents, so churches were swamped with requests for copies of baptismal certificates, each of which had

25 http://alex.onb.ac.at/cgi-content/alex?aid=dra&datum=1941&size=45&page=575.
26 *Anna Haag and Her Secret Diary of the Second World War*, ed. Edward Timms (Bern: Peter Lang, 2015), 9.
27 Viktor Klemperer, *The Diaries of Victor Klemperer 1942–45*, vol. II, To the Bitter End (London: Orion, 2000), 410–420.
28 Klemperer, *The Diaries*, 138.

to be copied by hand. Half-Jews were Jews, quarter-Jews had greater leeway. In 1942 forced labor camps were established for those determined by the Nazis to be "Mischlinge," having both Aryan and Jewish ancestry, and some of their Aryan spouses were ultimately sent there as well. Jews married to Aryans and raising their children as Christian were not required to wear a star.[29] By contrast, Jews married to Aryans without a child had to wear the star. Luise Solmitz writes of her family's relief that her husband, Ferdy, would not have to wear a star; he was "privileged" because he was married to an Aryan woman and had a child.[30] Thus, on the one hand, baptism determined Aryan racial identity, and on the other hand, Nazi race laws overrode baptism and could deny Christian identity to Mischlinge, Christians with Jewish parents or grandparents.[31]

The increasing social isolation of Jews after 1938 made the 1941 appearance of Jews on the streets marked with the yellow Star of David all the more startling. Jews who had been forced by Nazi regulations to disappear from public life were now suddenly conspicuous, appearing in the streets, marked yellow and looking wan; most remaining Jews were elderly, poor, socially isolated and struggling for food. Jews were only allowed to shop from 4 to 5 pm, forbidden entry to public places, such as theatres and parks, expelled from schools and universities and civil service jobs, and no one was allowed to converse with them in public. Churches and trams may have been the only enclosed spaces where "Aryans" and Jews mingled.

Riding the trams, Jews encountered a mixed response. Inge Deutschkron reports that one day, riding the tram in Berlin, a man offered her his seat: "'Please, sit down,' he said to me in a loud voice, pointing to his vacant seat. Most of the other passengers pretended not to have heard. It was the morning rush hour, and I was not the only one standing. I was convinced that if it were not for my Jewish star the man would not have offered me his seat."[32] Yet in Stuttgart, Anna Hoag reports in her diary entry of 7 October 1941, a newspaper denounced the sympathy shown to Jews riding the trams in town. The newspaper claimed to have received numerous letters from readers, angry at the sympathy shown to Jews. One reader described someone shaking hands with a Jewish woman in public and concluded, "Here I think there is only one solution: Put her up against the

---

**29** Avram Barkai, "The Final Chapter," in *German-Jewish History in Modern Times*, ed. Michael A. Meyer (New York: Columbia University Press, 1998), vol. 4, 360–388, here 381.
**30** *Anna Haag*, 135–136.
**31** Bergen, *Twisted Cross*, 85.
**32** Inge Deutschkron, *Outcast: A Jewish Girl in Wartime Berlin* (New York: Fromm International Press, 1989), 66.

wall!"[33] Note the prescience: in that very month, Germans began to hear reports about Jews being murdered in Poland, the Baltic States, and the Ukraine.[34]

The goal of racism is to produce a state of revulsion that shatters regulatory practices, whether of law, social etiquette, religious doctrine, or economic status. Nazi antisemitism, such as associating Jews with rats in the propaganda film, *Ewige Jude*, went beyond hate and fear to arouse horror and disgust, and a desire to exterminate Jews as one would exterminate vermin. For Christians in Nazi Germany, as the conflict between race and religion intensified from 1933 to 1941, antisemitic propaganda was trumpeted not only by the Nazi regime, but also by some church publications and sermons, and actions were taken in some of the Protestant regional churches to remove Jewish words and concepts from the hymnal and even from the Bible. Analyzing 910 sermons of 95 Confessing Church pastors delivered in the years prior to the war, William Skiles found 40 that spoke positively about Judaism as the foundation of Christianity, and another 40 that directly denigrated Jews and Judaism, reiterating theological motifs such as Judaism as a religion of law and works, not faith, and asserting that because the Jews, no longer the people of God, had killed Jesus, they were now being punished by God.[35]

By 1941, however, antisemitism in the Church was not simply a theoretical or theological matter for discussion; baptized Jews now produced visceral anxieties over the efficacy of baptism and the presence of Jewishness in Christian space. For German Jews, conversion to Christianity had long carried multiple meanings – a culmination of assimilation efforts, a ticket into European society, an achievement of personal piety – but what did baptism of Jews mean to Christians, especially in the midst of Nazi racial politics? Some parishioners complained that a baptized Jew wearing a yellow Star of David knelt next to them at the Communion rail or sat in the same pew or sang in the choir; such complaints were not doctrinal, but expressions of visceral anxieties. For example, a Protestant pastor in Berlin was told by a congregant that he should not place his hand on her child's head if that hand had baptized a Jew.[36] Baptism could

---

**33** October 7, 1941 report of Anna Haag: "'Hier gibt es, glaube ich, nur eine Lösung: An die Wand stellen!'" (HA 5, 62; TS 135–136); *Anna Haag*, 138.

**34** Robert P. Ericksen, *Complicity in the Holocaust: Churches and Universities in Nazi Germany* (New York: Cambridge University Press, 2012), 135.

**35** On sermons, see William Skiles, "The Bearers of Unholy Potential: Confessing Church Sermons on the Jews and Judaism," *Studies in Christian-Jewish Relations* 11:1 (2016), 1–29. The article is drawn from his doctoral dissertation, *Preaching to Nazi Germany: The Confessing Church on National Socialism, the Jews, and the Question of Opposition*.

**36** Auszüge aus den Gemeindebriefen, 374.

not withstand Nazi racism toward Jews as repugnant, one of the primitive emotions that stimulates autonomic physiologic responses such as nausea and haptic revulsion; disgust cannot be placated by church doctrine.

## Church Responses to the Yellow Star

In September 1941, weeks after hearing of hostile reactions to Christians wearing yellow stars in Catholic churches, the Roman Catholic Cardinal Adolf Bertram, archbishop of Breslau, responded with recommendations. He urged priests to avoid a separate seating section for baptized Jews and instead advised baptized Jews to attend an early morning mass (which presumably would be sparsely attended). Only if parishioners stopped attending mass or walked out in protest of the presence of Christians with yellow stars should a segregated service for baptized Jews take place, he wrote. "Reminders concerning brotherly love and avoiding every form of denigrating treatment of any Catholic non-Aryan marked by the star are to be issued [to members of the congregation attending mass] only when and if disturbances arise." He concluded by citing Galatians 3:28, "In Christ there is neither Greek nor Jew."[37] A similar response was published by Cardinal Innitzer of Vienna.

By contrast, many Protestant church officials wanted baptized Jews excluded. Already in 1933, in his inaugural speech as Protestant Bishop of Saxony, Friedrich Coch urged Reich Church Minister, Hans Kerrl, to establish separate churches for baptized Jews, and he demanded that pastors prove not only their own Aryan ancestry but also that of their wives. A few months later, in his eulogy for the notorious antisemite Theodor Fritsch, Coch praised Fritsch for understanding that the "real significance of world history was conflict between Christianity and the Jews."[38]

According to Becker-Kohen's diary, not all fellow Catholics expressed the generosity that Bertram advised. She reports that in April 1942, "A Jewish convert, who has to wear the Jewish star, was asked to think of the other parishioners, and, if he had to attend Mass, at the very least take a seat in the choir loft. Because he is such a deeply pious man, he now takes a seat in the choir loft where no one can see

----

**37** Anson Rabinbach and Sander L. Gilman, eds., *The Third Reich Sourcebook* (Berkeley: University of California Press, 2013), 427–429.
**38** Richard Gutteridge, *Open thy Mouth for the Dumb! The German Evangelical Church and the Jews, 1879–1950* (Oxford: Basil Blackwell, 1976), 155.

him."[39] Ostracized and isolated, Becker-Kohen was not invited to the homes of her fellow parishioners, not even at Christmas, though at times she was assisted by sympathetic priests as she traveled within the Reich, seeking refuge from imprisonment. Even in Berlin, as she writes in April 1942, "I hardly can venture outside in my neighborhood anymore because people will not leave me in peace. For the same reason, I can also no longer go to the local church in our suburb. I therefore travel every morning to St. Hedwig Cathedral where Father Erwin [a friend] celebrates Mass daily at 10:00 a.m."[40] Nechama Tec, born in 1931 and hidden by Catholics for three years in Poland, later wrote, "I had come to look forward eagerly to being in a church . . . because the quiet and peace were soothing . . . . In the stillness I could whisper my secrets without fear."[41]

Protestant pastors debated a "guest status" for baptized Jews and the possibility of establishing separate church services for them or banning them altogether. Some churches had separate seating for baptized Jews: "at the beginning of the service Jewish Christians in the Berlin-Dahlem church scurried to take a seat on the special benches in the half dark in order to participate in the service."[42] Some pastors suggested placing a sign on church doors, "No Admittance for Jews" or "Admittance for Wearers of the Jewish Star Prohibited." Protestant officialdom in Berlin was "ambivalent at best," writes the historian Wolfgang Gerlach, refusing either to permit or forbid such signage.[43] No known discussion took place among church authorities about making a church a "yellow star free" zone, a place where, like a private home, the badge would not have to be worn at all.

## Consequences for Clergy

The Confessing Church, which arose in opposition to the pro-Nazi German Christian movement, was divided over the so-called "Jewish Question." Generally supportive of Jews who had converted to Christianity, most Confessing Church pastors opposed taking a stance in support of Jews who had not been baptized. Helping even baptized Jews became an act of protest that could bring dire consequences. Pastor Heinrich Grüber ran a "Relief Center for Protestant

---

**39** Spicer and Cucchiara: *Becker-Kohen*, 49.
**40** Spicer and Cucchiara: *Becker-Kohen*, 49.
**41** Nechama Tec, *Dry Tears: The Story of a Lost Childhood* (New York: Oxford University Press, 1984), 107–108; cited in Frederick S. Roden, *Recovering Jewishness: Modern Identities Reclaimed* (Santa Barbara, CA: Praeger, 2016), 119.
**42** Frisius et al., *Evangelisch getauft, als Juden verfolgt*, 392–393.
**43** Gerlach, *And the Witnesses*, 175.

Non-Aryans" who were in danger from the Nazi regime. The Center worked to secure visas to other countries, forged passports, and provided food and medical aid. Grüber was arrested in December 1940 and sent to Sachsenhausen, then to Dachau; he survived. Working with him was Katharina Staritz, a Protestant vicar in Breslau and member of the Confessing church, who circulated a protest in the fall of 1941 against the regulation that Jews wear a yellow star and against Christians who objected to baptized Jews in church. Baptized Jews had had nothing to do with Judaism, she wrote, and have been loyal members of their congregations. She warned against "un-Christian behavior." An attack against her appeared in the SS paper, *Das Schwarze Korps*, in March 1942, leading to her arrest and a year in Ravensbrück.[44] After that, she was relieved of her post by the church consistory of Silesia and moved to Marburg, where she resumed theological studies, including the study of Hebrew.[45]

In some cases, pastors not only offered no welcome to baptized Jews but actually denounced them as Jews. Starting in 1934, Berlin pastor Karl Themel searched for non-Aryan Christians within the Berlin church, and out of 255,469 Aryan certificates, he identified 2,612 Christians of Jewish heritage whom he denounced to the Gestapo.[46] Their dates of deportation and death are not known, but in 2002, Bishop Wolfgang Huber of Berlin suggested that memorials be placed in churches for those baptized Jews who were murdered.[47] At Berlin's Messiaskapelle, in the Prenzlauerberg neighborhood, congregants have gathered the names and fate of more than 700 Jews who were baptized between 1933 and 1940; of those, eighty-six were murdered by the Nazis.[48] Baptizing Jews could also bring Nazi contempt, as in this ditty:

> If a Jew should come walking in
> And wants the pastors to baptize him,
> Do not trust him and never do,
> A Jew will always be a Jew.[49]

Not only church members, but some non-Aryan pastors and priests were also arrested, imprisoned, and even murdered. Bruno Benfey of Göttingen, Paul Leo

---

**44** Gerlach, *And the Witnesses*, 169–173.
**45** Gerlach, *And the Witnesses*, 169–173.
**46** Manfred Gailus, "Karl Themel: Ein Pfarrer als Sippenforscher im Dritten Reich," in *Täter und Komplizen in Theologie und Kirchen 1933–1945*, ed. Manfred Gailus (Göttingen: Wallstein, 2015), 197–215.
**47** Gailus, "Karl Themel," 215.
**48** https://www.landeskirchenarchivberlin.de/wp-content/uploads/2009/12/mk-deportationsliste.pdf.
**49** Cited in Frisius et al., *Evangelisch getauft – als Juden verfolgt*, 374.

of Osnabrück, Gustav Oehlert of Rinteln, and Rudolf Gurland of Mainz, were among those non-Aryan Protestant clergy who were persecuted. Some of the regional Protestant church leadership fired non-Aryan pastors, organists, and religion teachers, while others defended their pastors against state authorities. Some pastors were not protected. Pastor Werner Sylten, a converted Jew, was arrested on 27 February 1941, for his work with Pastor Heinrich Gruber in Berlin assisting the persecuted, and was sent to Dachau, where he was killed on 26 September 1942. Pastor Bruno Benfey, also a converted Jew, who had arrived in Göttingen in 1927 as pastor of the St. Marien church, began being harassed in 1933 by his parishioners. By 1936 church members were protesting with signs outside the church on days he preached. Others got up and walked out, and some "shouted antisemitic slurs at Benfey and his wife as they walked home"; he finally resigned in 1937.[50] Benfey was arrested in November 1938 and taken to Buchenwald; he was released and able to emigrate and spend the war in Holland, returning in 1946 to the same parish in Germany; he died in 1962.[51]

Both Catholic priests and Protestant pastors were affected. In a case described by the psychologist Franklin Oberländer, a non-Aryan Catholic priest, son of a Jewish father and Catholic mother who was baptized as a child, was dispatched after his ordination to the priesthood in 1936 to a remote village as an effort to conceal his non-Aryan status. In 1940, he volunteered as chaplain for the Wehrmacht, where he observed Jews being deported. He was expelled a year later from the Wehrmacht as a so-called "first degree Mischling" and imprisoned in one of the Emsland concentration camps; the Catholic hierarchy did not intervene on his behalf. His father was murdered at Auschwitz and a brother survived in hiding, yet no one in the Catholic church after the war spoke of it to him. Asked if the church hierarchy had tried to help him, he replied, "They did nothing. The Bishop back then was the kind who could do nothing but throw his hands in the air and say, 'I knew it, I knew it! What will we do now, there's not a thing we can do.'"[52] A similar situation faced the Protestant pastor Heinz Wolf (1912–1997), who also had a Jewish father and volunteered as a chaplain in the Wehrmacht. He was tried and convicted of falsifying his background as a Mischling and was imprisoned in the labor camp Börgermoor from December 1943

---

**50** Gerhard Lindemann, ‚Typisch Jüdisch': Die Stellung der Ev.-luth. Landeskirche Hannovers zu Antijudaismus, Judenfeindschaft und Antisemitismus 1919–1949 (Berlin, 1998). Cited by Robert P. Ericksen, Complicity in the Holocaust: Churches and Universities in Nazi Germany (New York: Cambridge University Press, 2012), 120.
**51** Gerlach, And the Witnesses, 127–129.
**52** Franklin A. Oberlaender, "My God, They Just Have Other Interests," Oral History Review 24, no. 2 (Summer 1997): 23–53; 40.

to March 1945. After the war, he served as a pastor of a church in Zeilsheim, outside Frankfurt am Main.[53]

Yet there were efforts at Christian kindness and assistance as well. Pastor Oehlert's church leadership in Rinteln defended him after the SS propaganda journal, *Das Schwarze Korps*, attacked him as a Jew teaching Confirmation classes. A Jewish convert named Ernst H. Steiner reported after the war that at a Berlin church on Good Friday 1942 "a large number of faithful Christians who were present came up to us after the conclusion of the Holy Supper and offered us their hand in a brotherly and sisterly way, thereby demonstrating their sympathy with us in our fate."[54]

In Germany, the final months of 1941 saw not only imposition of wearing the Star of David, but also significant shifts in the public rhetoric regarding Jews: "The restraint of the previous months gave way to an explosion of the vilest anti-Jewish invectives and threats," the historian Saul Friedländer writes.[55] In a speech on 2 October 1941, Hitler declared that the enemy that was poised to "annihilate not only Germany but the whole of Europe" was "Jews and only Jews!"[56]

Nazi regulations limiting Jews' presence in public did not include the churches. Instead, several Protestant regional churches led by pro-Nazi clergy announced on 17 December 1941 that Jews were to blame for the war and approved all measures taken by the Nazi state against them. They denied that Christian baptism could change anything "about the racial essence of Jews" and stressed that "racially Jewish Christians" were not to receive pastoral care from clergy and were unwelcome in churches, which were devoted to the German Volk. In conclusion, the signatories pledged to tolerate "absolutely no influence of the Jewish spirit on German religious and ecclesiastical life" and to "suspend all relations with Jewish Christians."[57] The announcement was based on a theological opinion written by Heinz Eisenhuth, professor of systematic theology at the University of Jena, who declared: "Jewish Christians are to be excluded from religious congregations as enemies of the Reich; German pastors may not extend their official services to Jewish Christians; church taxes may not be collected from Jewish Christians." Eisenhuth justified the exclusion by reference to Martin Luther: "Luther saw the Jews above all as murderers of Christ. We see in them destroyers of God's creation, whose

---

**53** Stolpersteine Gütersloh; Heinz Wolf was pastor of St. Bartholomäus church in Zeitlsheim from 1959 to 1984, when he retired.
**54** Gerlach, *And the Witnesses*, 173.
**55** Saul Friedländer, *The Years of Extermination* (New York: Harper Collins, 2007), 272.
**56** Cited in Friedländer, *The Years*, 272.
**57** Bergen, *Twisted Cross*, 25.

defense is the duty of all Germans and also of the German Protestant church."[58] The rulings were upheld by the national governing board of the entire Protestant church, published three days before Christmas 1941.

# Interpretation One: Border as Baptism and Asylum

Jews who converted to Christianity had to withdraw their registration from the Jewish community and register with the church. Those baptized Jews were no longer eligible for social welfare assistance from the Jewish community in Germany, and, as we have seen, not everyone accepted baptized Jews as Christians. During the Nazi period, and especially after imposition of the wearing of the yellow Star of David in 1941, they stood in a religious wasteland, not fully in one community nor the other. Their very lives were uncertain, as baptism at times brought mitigation, but no full exemption from Nazi racial laws. Marriage to an Aryan brought greater privilege, though also no guarantee of safety.

Attitudes from the Nazi era have affected the historiography. The fate of baptized Jews is rarely mentioned in the thousands of books and articles published on the history of the Protestant and Catholic churches during the Third Reich. Historians of the Jews generally ignore the fate of the many German Jews who had converted as they are viewed as Christian. One of the most assured forms of security for Jews in Nazi Germany was to marry an Aryan, even without converting to Christianity, and historians have not yet determined how many of those marriages ended in divorce or death, resulting in the deportation and death of the Jewish spouse.

A baptized Jew in Nazi Europe lived in a state of limbo, desperate for safety and protection, to say nothing of kindness and compassion. Who is our neighbor? Kenneth Reinhard writes that "the political is the condition of the ethical, the only ground by which we approach ethics, and not vice versa."[59] Within the Nazi realm that had launched its genocide, Jews were desperate to save their lives. For Christians, who also faced threats of imprisonment, though not genocide, the political had reconfigured the ethical. The churches worked on both sides of the equation:

**58** Susannah Heschel, *The Aryan Jesus: Christian Theologians and the Bible in Nazi Germany* (Princeton, NJ: Princeton University Press, 2008), 140.
**59** Kenneth Reinhard, "Toward a Political Theology of the Neighbor," in Slavoj Žižek, Eric Santner, Kenneth Reinhard, *The Neighbor: Three Inquiries in Political Theology* (Chicago: University of Chicago Press, 2005), 49.

copying and distributing millions of baptismal certificates made possible the Nazi determination of Aryan identity; offering baptism and marrying Jews could save numerous individuals from Nazi murder. Christians throughout Europe were divided over baptizing Jews. Baptism in Bosnia under Croatian control offered genuine security to Jews,[60] whereas in Romania, baptism offered no protection from the state or church authorities,[61] and in Slovakia some priests refused baptism to Jews because their motivation was not piety, but "pragmatic."[62]

## Interpretation Two: The Uncanny Intruder

The strong emotions evoked in Christians by the presence of Jews marked with a yellow Star of David reflect the anxieties of living under Nazi rule and the underlying anxiety within Christianity regarding its own roots within Judaism. Despite the wish of the church for the conversion of the Jews, the mass baptism of Jews in fifteenth-century Spain gave rise to a "crisis of classification."[63] The *unheimlich* (uncanny) Jew was now inside the *Heim*, the church. The satisfaction of the Christian wish aroused suspicion, anxiety, and disgust. The Inquisition was called to Spain to investigate these "new Christians" to determine their sincerity and whether they were Judaizing other Christians, infecting them with their Jewishness. The Catholic church then developed its insistence on the "purity of blood," and Spain decided to expel its Jews in 1492. As Jacques Lacan writes, there is a gap "between the articulation of a wish and what occurs when its desire sets out on the path of its realization."[64]

As Christians entered German churches wearing a yellow star, a feeling arose of the uncanny: were these Christians or were they Jews in the wrong place, threatening the church's pure, Aryan Christianness? In his essay on "The Uncanny," Freud links the experience of the uncanny to intrusion. That which

---

**60** Emily Greble, *Sarajevo, 1941–1945. Muslims, Christians and Jews in Hitler's Europe* (Ithaca, NY: Cornell University Press, 2011).

**61** Ion Popa, "Reaction of Converted Jews and their Families to Antisemitic Laws: Romania 1940–1942," *The Historical Journal of the Romanian Jewish Community*, 1, nos. 16–17 (2016): 250–270; 267.

**62** Nina Paulovicova, "Rescue of Jews in the Slovak State (1939–1945)," PhD Dissertation, University of Edmonton, Alberta, Spring 2012, 263–264.

**63** Theodor Dunkelgrün and Paweł Maciejko, "Introduction," *Bastards and Believers: Jewish Converts and Conversion from the Bible to the Present* (Philadelphia: University of Pennsylvania Press, 2020), 1–25, here 2.

**64** Jacques Lacan, *Ethics of Psychoanalysis, 1959–60* (New York: Norton, 1992), 41.

he felt as uncanny appeared to him at first as an intruder, but he then recognized the intruder as a mirror image of himself. He writes:

> I was sitting alone in my wagon-lit compartment when a more than usually violent jolt of the train swung back the door of the adjoining washing-cabinet, and an elderly gentleman in a dressing-gown and a travelling cap came in. I assumed that in leaving the washing-cabinet, which lay between the two compartments, he had taken the wrong direction and come into my compartment by mistake. Jumping up with the intention of putting him right, I at once realized to my dismay that the intruder was nothing but my own reflection in the looking glass on the open door.[65]

The putative "intruder" was Freud himself or rather a reflection of Freud in the mirror, his mirror image. A baptized Jew – that is, a Christian – appears in church wearing a yellow Star of David, a marker of Jewishness. Has the Jew become an intruder into Christian space? Or has a Christian entered, beneath that yellow star, as a replica of the Christian self, Christianity as a baptized Judaism, now misplaced, Christianity suddenly looking at the yellow star and experiencing itself as an intruder, an *Unheimliche*?

Growing German nationalism in the nineteenth century viewed Jews with suspicion, as intruders, and Jewish assimilation only furthered the growing suspicion that Jews were invisible, unmarked, and unrecognizable. The outsider became insider, in Peter Gay's famous formulation about the Weimar period, the Jew intruding into the heart of German culture.[66] However, Jews as intruders into German culture shifted with the rise of racial ideology, which transformed anti-Jewish hostility into a visceral experience. Germany became a body, and the Jew was not simply an intruder, but a medical infection or a sexual penetration, even rape. A baptized Jew was not merely sitting in the wrong pew and asked to move to the choir loft; a baptized Jew now evoked disgust, so that a pastor who baptized a Jew was asked not to place his hand on the head of a Christian child. Sacraments became sites of anxiety: How could a Jewish body become part of a parish or congregation and take Communion? How could the body of Christ be taken into the body of a Jew, even if baptized? A baptized Jew came to be feared as a transplant, not a lack, but an excess, Jewishness transplanted onto the healthy body of Christianness.

Nazi theologians were desperate to protect against charges of Jewish contamination of Christianity and a hefty debate over the Jewishness of the Old Testament and Jesus that began already in the nineteenth century heightened

---

**65** Sigmund Freud, "The Uncanny," trans. James Strachey, *The Standard Edition of the Complete Psychological Works of Sigmund Freud*, Volume XVII, 248.
**66** Peter Gay, *Weimar Culture: The Outsider as Insider* (New York: Harper & Row, 1968).

after 1933. The pro-Nazi German Christian movement, in a 1937 pamphlet, declared: "Is National Socialism Marxist because it emerged in the struggle against Marxism? Just as little does genuine Christianity have anything to do with the spirit of the Jews."[67] That Jesus was not a Jew, but of Aryan ancestry, born in Galilee, a region, many Protestant theologians argued, that was filled with Aryans, was a theme popular with academic theologians as well as demagogues.[68] They faced Nazi leaders who wanted to demote, if not abolish, church power in Germany, and the small neo-pagan movement that mocked Christianity as a Jewish religion.

The problem that faced the churches recapitulates the larger problem of Christianity emerging out of Judaism: Jesus the Jew becoming the first and greatest Christian; the New Testament writing proleptically from the Old Testament; Paul declaring that Gentiles are grafted onto the olive tree of Judaism; Christians as the New Israel of the spirit. Baptized Jews coming to churcharoused the anxieties: is Christianity a baptized Judaism? Wearing a yellow star, converted Jews were not simply intruders, but threatened a deeper Christian fear: that of Romans 11, that gentiles have been grafted onto Judaism.

If the church is a body, Jewishness enters through the womb. Gustav Volkmar, a nineteenth-century Protestant theologian, wrote, "Christianity is born of the virgin womb of the God of Judaism," expressing the wish that God's womb is virginal, preserved from Jewish penetration.[69] If Christianity was born out of Judaism, or birthed by the Jew Jesus, it has already been infiltrated by Jewish presence. That is the anxiety felt in 1941: baptized Jews entering the church with a yellow star constituted more than a Jewish intrusion into Christian sanctified space; they served as a symbolic reminder that Judaism is already part of the Christian body, that the Aryan Christian cannot be protected from contamination with Jewish blood.

---

**67** *Das Ringen der Deutschen Christen um die Kirche* (Verlag Deutsche Christen, 1937), 7. Niedersächsisches Staatsarchiv Oldenburg, Bestand: 134/57: Bund für Deutsches Christentum Mitteilungen.

**68** The academic theologians include Emanuel Hirsch (Göttingen), Gerhard Kittel (Tübingen), and Walter Grundmann (Jena), among others. The demagogues include Hans Hauptmann, *Jesus der Arier/Jesus der Galiläer* (Munich, 1930); Edmund Kutschera, *Nordischer Jesus* (Vienna, 1929), among others.

**69** Gustav Volkmar, *Die Religion Jesu und ihre erste Entwicklung nach dem gegenwaertigen Stande der Wissenschaft* (Leipzig: F.A. Brockhaus, 1857), 37.

# Interpretation Three: Christianity as the Transplanted Judaism

The 1924 German expressionist film, *Orlacs Hände*, by the famed director Robert Wiene, offers a metaphorical expression of the Christian anxiety. The film narrates the surgical transplantation of hands to the body of a pianist, Orlac, after his are crushed in a train wreck. His new hands, the pianist believes, are those of Vasseur, a man executed for stabbing to death a moneylender. Haunted by Vasseur and by the belief that his new hands are those of a murderer, Orlac can neither play the piano nor touch his wife, though he responds to the seductions of the housemaid and murders his father. Begging his surgeon to remove the hands, he ultimately learns that Vasseur was wrongly executed for a crime committed by someone else, a man who had conspired with the housemaid to drive Orlac to madness.

The film shifts from fantasy to reality, from Expressionism to the Neue Sachlichkeit (the New Objectivity), in order to resolve the horror and depression experienced by Orlac. His wife is transformed in the film from an object of her husband's erotic and haptic desire into a forceful woman who takes charge of the family finances. The film turns from Orlac's inner state to the detectives discovering that Vasseur was not the killer after all, so that the audience is given a happy ending.[70]

There is nothing explicitly theological about the film, nor any reference to Christians or Jews, though Vasseur, with a non-German name, is clearly not German and the man he is wrongly believed to have killed was a money lender, a code name for Jew. Yet the image of transplanted hands that arouse profound disgust and rob Orlac of his ability to play the piano and be a husband points to the problem of transplantation in the racial imaginary. A suggestive link is the film's focus on the haptic, echoing the disgust directed at the pastor who places his hand on the head of a baptized Jew. As a metaphor for Christian anxiety, the film suggests Christian horror at having Jews, the murderers of Christ, transplanted onto the Christian body, and Christianity's wish to kill the father, Judaism, fulfilled by Orlac himself.

The horror for Christianity is that the transplantation can be understood as reciprocal and bilateral. Is Christianity grafted onto Judaism as a branch grafted onto the olive tree, as Paul writes in Romans 11, or is Judaism transplanted onto Christianity, with Jesus and Paul as foundational figures of the church, along with Jewish apostles, concepts, and even Jewish Scripture, the Old Testament?

---

**70** Anjeana K. Hans, *Gender and the Uncanny in Films of the Weimar Republic* (Detroit: Wayne State University Press, 2014), 182.

The uncertainty of the bidirectional transplantation of the two religions points to the sense of the uncanny, the theological ambiguity that shapes the relationship between the two religions: Christianity's sense of shame at having originated within Judaism and Judaism's sense of shame at having been superseded by Christianity.

# Conclusion

Church is the place of prayer, a sanctified time for the soul, yet Nazi racism intruded. If prayer brings focus to the soul, race demands attention to the physical body; prayer and race are, like time and space, distinct classifications. Baptism and racism concern themselves with entirely different categorization of human beings, but they were brought together in 1941, arousing anxieties, anger and even disgust. Why disgust? Jews were no longer mere intruders into German or Christian space but were now penetrating the Christian body. German nationalism of the nineteenth century had been transformed by racial theory so that the German nation was not simply a political entity, but a body whose blood could allegedly be poisoned by Jews. For the church, too, Judaism was not simply a different religion, but was viewed as posing a dangerous threat of pollution. Baptized Jews were the present-day incarnation of a problem since the birth of Christianity, which had taken place within the social and theological context of Jews and Judaism. Thus, the crisis of 1941 was not only whether baptized Jews became Christians, but whether Christianity itself was a baptized form of Judaism. The uncertainty lay at the heart of the anxiety that exacerbated the churches' failures in caring for both Jews and baptized Jews.

Baptism forms the borderline of the church; those who are baptized may enter and receive Communion and other sacraments as well as pastoral care and membership in a community. Baptism offers a kind of religious nationality, an immigration visa to the Christian world. In most cases in Europe, baptism offered Jews some degree of refuge, but was not a protected haven for Jews. Clergy, too, were imprisoned in concentration camps and even murdered for their humanitarian beliefs and actions. Even after the war, the remaining fifty thousand to sixty thousand Jews in Germany who had converted to Christianity were trapped between communities.[71] Most had Jewish relatives who had been murdered, but because of their conversion, they did not receive assistance from Jewish aid

---

71 Matthew Hockenos, *A Church Divided: German Protestants Confront the Nazi Past*. Kindle Edition, Location 2108.

organizations. Christian organizations and churches did not understand the complexity of their needs nor their own responsibility for the catastrophe. Nazi hatred of Jews that had been cultivated for the past twelve years could not be overcome quickly but rather lingered. They were Jews and they were Christians, and in certain ways they were treated – like refugees seeking entry to a national border – as neither.

When we in the United States, a country of immigrants, view the refugees seeking asylum at our border, do we see intruders or do we recognize that we are looking at ourselves in a mirror, like Freud, and seeing the uncanny reflection of our own immigrant past? Even without formal race laws, we remain infected with a racism that fails to recognize that the humanity of the refugees is our own humanity as well. Today's refugees are not shipped by the United States to death camps, and they are not executed by firing squads or in gas chambers. Rather, despite legal claims to asylum in the United States under the 1980 United States Refugee Act, they remain, often endlessly, waiting and hoping – or expelled to the horrors they fled. Over 500 children, torn from their parents, have been orphaned because the government has lost track of their parents, an unfathomable trauma. Although immigrants and refugees bring economic prosperity and cultural vitality, too many Americans view them through a racist lens, as if they were Orlac's hands, dangerous and even murderous transplants. Without the proper visa or passport, some are sent back to countries where they face threats of violence and murder. Baptism did not necessarily guarantee Jews asylum from Nazi deportation, though it offered (often false) meagre hope in a time of desperation.

Why have I chosen this topic? As a Jew from a family of holocaust victims and survivors with no interest in conversion, it is precisely the "otherness" of the topic that feels so urgent today. This is not "my topic" with relevance to my own family experience. How, then, am I adjacent? I am living in a country that is building fences, walls, barriers that keep out refugees, people starving, fleeing, asylum seekers, migrants who want to live in democratic countries with thriving economies and, most of all, without war: a guarantee of food, shelter, education and medical care without threat of prison, guns, death. What, then, is our adjacency, if not our moral responsibility? Christian failure to protect converted Jews during World War II, let alone speak out on behalf of all Jews and all those persecuted by the Nazis is a lesson teaching me that the integrity of my own Jewish faith, family, and country decays with my failure to recognize my adjacency and my responsibility for those both outside and inside whatever gruesome barriers my government establishes.

# Works Cited

Barkai, Avram. "The Final Chapter," *German-Jewish History in Modern Times*. Edited by Michael
 A. Meyer, 360–388. New York: Columbia University Press, 1998, vol. 4.
Bergen, Doris. *Twisted Cross: The German Christian Movement in the Third Reich*. Chapel Hill:
 University of North Carolina Press, 1996.
Büttner, Ursula. "The Persecution of Christian-Jewish Families in the Third Reich." *Leo Baeck
 Institute Yearbook* 34, no. 1 (1989): 267–289.
Cone, James. "Biblical Revelation and Social Existence," *Interpretation* 28, no. 4 (1974):
 422–440.
Deutschkron, Inge. *Outcast: A Jewish Girl in Wartime Berlin*. New York: Fromm International
 Press, 1989.
Dunkelgrün, Theodor, and Paweł Maciejko. "Introduction." *Bastards and Believers: Jewish
 Converts and Conversion from the Bible to the Present*, 1–25. Philadelphia: University of
 Pennsylvania Press, 2020.
Ericksen, Robert P. *Complicity in the Holocaust: Churches and Universities in Nazi Germany*.
 New York: Cambridge University Press, 2012.
Freud, Sigmund. "The 'Uncanny,'" *The Standard Edition of the Complete Psychological Works
 of Sigmund Freud*, Volume XVII (1917–1919): An Infantile Neurosis and Other Works,
 217–256.
Friedländer, Saul. *The Years of Extermination*. New York: Harper Collins, 2007.
Hildegard Frisius et al., eds. *Evangelisch getauft – als Juden verfolgt: Spurensuche Berliner
 Kirchengemeinden*. Berlin: Wichern 2008.
Gailus, Manfred. "Karl Themel: Ein Pfarrer als Sippenforscher im Dritten Reich." In *Täter und
 Komplizen in Theologie und Kirchen 1933–1945*. Edited by Manfred Gailus, 197–215.
 Göttingen: Wallstein, 2015.
Gay, Peter. *Weimar Culture: The Outsider as Insider*. New York: Harper & Row, 1968.
Gellately, Robert. *Backing Hitler: Consent and Coercion in Nazi Germany*. New York: Oxford
 University Press, 2001.
Gerlach, Wolfgang. *And the Witnesses Were Silent: The Confessing Church and the
 Persecution of the Jews*. Translated by Victoria J. Barnett. Lincoln: University of
 Nebraska Press, 2000.
Gutteridge, Richard. *Open thy Mouth for the Dumb! The German Evangelical Church and the
 Jews, 1879–1950*. Oxford: Basil Blackwell, 1976.
Hans, Anjeana K. *Gender and the Uncanny in Films of the Weimar Republic*. Detroit, MI: Wayne
 State University Press, 2014.
Heschel, Susannah. "A Friendship in the Prophetic Tradition: Abraham Joshua Heschel and
 Martin Luther King, Jr." *Telos*. April 2018: 67–84.
Heschel, Susannah. *The Aryan Jesus: Christian Theologians and the Bible in Nazi Germany*.
 Princeton, NJ: Princeton University Press, 2008.
Heschel, Susannah. "Ecstasy versus Ethics: The Impact of World War I on German Biblical
 Scholarship on the Hebrew Prophets." In *The First World War and the Mobilization of
 Biblical Scholarship*. Edited by Andrew Mein, Nathan MacDonald and Matthew A. Collins,
 187–206. London: T&T Clark, 2019.
Hockenos, Matthew. *A Church Divided: German Protestants Confront the Nazi Past*. Kindle
 Edition, Location 2108.

Hockenos, Matthew. *Then They Came for Me: Martin Niemöller, the Pastor who Defied the Nazis*. New York: Basic Books, 2018.

Johnson, Marshall. "Power Politics and New Testament Scholarship in the National Socialist Period." *Journal of Ecumenical Studies* 23:1 (Winter, 1986): 1–24.

Klemperer, Viktor. *The Diaries of Victor Klemperer 1942–45*. Vol. II, To the Bitter End (London: Orion, 2000).

Lacan, Jacques. *Ethics of Psychoanalysis, 1959–60*. New York: Norton, 1992.

Ludwig, Hartmut. *Suddenly Jews: The Story of Christians whom the Nazi Racial Laws Classified as Jews, and of the Good Samaritans who came to their Aid (the Bureau Grüber)*. Translated by Martin Nicolaus. Berkeley: Duplex Press, 2015.

Magee, Keith. *Prophetic Justice: Essays on Race, Religion, and Politics*. London: New Generation, forthcoming.

McCaulley, Esau. *Reading While Black: African American Biblical Interpretation as an Exercise in Hope*. Westmont, IL: IVP Academic 2002.

Oberlaender, Franklin. "The Family Dynamics of German Protestants of Jewish Descent Stigmatized in Nazi Germany and of Their Offspring Born in Postwar West Germany." *Holocaust and Genocide Studies* 9:3 (Winter 1995): 360–377.

Oberlaender, Franklin A. "My God, They Just Have Other Interests." *Oral History Review* 24:2 (Summer 1997): 23–53.

Paulovicova, Nina. *Rescue of Jews in the Slovak State (1939 – 1945)*. Ph.D. Dissertation, University of Edmonton, Alberta, Spring 2012.

Ion Popa. "Reaction of Converted Jews and their Families to Antisemitic Laws: Romania 1940–1942." *The Historical Journal of the Romanian Jewish Community* 1, nos. 16–17 (2016): 250–270.

Rabinbach, Anson, and Sander L. Gilman, eds. *The Third Reich Sourcebook*. Berkeley: University of California Press, 2013.

Roden, Frederick S. *Recovering Jewishness: Modern Identities Reclaimed*. Santa Barbara, CA: Praeger, 2016.

Röhm, Eberhard, and Thierfelder Jörg. *Juden, Christen, Deutsche 1933–1945* 4:1. Stuttgart: Calwer Verlag 2004, 192–199.

Spicer, Kevin P. and Martina Cucchiara, trans. and ed. *The Evil that Surrounds Us: The Memoir of Erna Becker-Kohen*. Bloomington: Indiana University Press, 2017.

Skiles, William. "The Bearers of Unholy Potential: Confessing Church Sermons on Jews and Judaism." *Studies in Christian-Jewish Relations* . 11:1 (2016), 1–29.

Timms, Edward, ed. *Anna Haag and Her Secret Diary of the Second World War*. Bern: Peter Lang, 2015.

Volkmar, Gustav. *Die Religion Jesu und ihre erste Entwicklung nach dem gegenwaertigen Stande der Wissenschaft*. Leipzig: F.A. Brockhaus, 1857.

Žižek, Slavoj, Eric Santner, Kenneth Reinhard. *The Neighbor: Three Inquiries in Political Theology*. Chicago: University of Chicago Press, 2005.

Doris L. Bergen
# Buried Words, Exposed Connections

> We sewed a dress for Bacha, four panties, a blouse, an apron and socks. It is a pleasure
> for me. My finger was peeling and hurting. And on April 20 Kitten was suffering from
> weakness of the heart, so I convinced her to invite Ciuruniu over to strengthen it. And she
> wrote the following note: "If you want to, come over for a while in the evening, because I
> am lonely and sad." And he did. And there was some [sex]. And her heart became stronger
> right away because she needed it as medicine.

These are the words of Molly Applebaum, from a diary entry in April 1944. Thirteen at the time, Applebaum, together with her older cousin Helen ("Kitten"),
was confined to a wooden box in a barn, at the mercy of the Polish brother and
sister who had agreed to hide them. Applebaum's extraordinary diary, together
with a memoir she wrote decades later and an introductory essay by historian
Jan Grabowski, is published as *Buried Words: The Diary of Molly Applebaum*.[1]
This book – and encounters with its author, her family, my students, and myself – is at the heart of this essay.

I opened with the words of Molly Applebaum because in all my work, I try to
put people at the center. This may seem like an obvious thing to do, but it is surprisingly rare in historical scholarship. Jews who went through the Holocaust created a wealth of personal accounts, from letters, diaries, poems, and songs, to
memoirs, documentaries, and interviews, in dozens of languages. Living in Toronto, I have been fortunate to get to know many Holocaust survivors and to become close to some of them. These connections have been transformative for me.
With respect to my scholarship, they have helped me see and understand events
and developments I would never have noticed on my own. Applebaum is a case in
point. Reading her book has illuminated aspects of hiding and rescue, agency,
family dynamics in the midst of violence, and the importance and multiple meanings of sex during the Holocaust that would never have occurred to me. Hearing
her talk about her experiences and memories in interviews and reflecting on her
life and family now complicates and deepens those insights.

What does it mean to be adjacent to historical violence? This question brought
me to consider the historian's role – my role – in an approach to the Holocaust

---

1 Molly Applebaum, *Buried Words: The Diary of Molly Applebaum* (Toronto: Azrieli Foundation,
2017). The opening quotation is on page 25. My thanks to Anna Shternshis and Mita Choudhury
for their help with this essay. I would also like to express my appreciation to Sharon Wrock,
Neal Applebaum, and Jason Chalmers, to Stephanie Corazza, Marilyn Campeau, students in
my class on the Holocaust in spring 2020, and above all to Molly Applebaum.

https://doi.org/10.1515/9783110753295-008

that centers the people on the receiving end of violence. As a scholar, I wield a significant amount of power and to pretend otherwise is disingenuous and impedes understanding for me and my audience. I have always been uncomfortable with how easily we as scholars of the Holocaust can make it about ourselves. At the same time, I am not merely a channel for the words and experiences of others. Influenced by my own needs and biases, I decide whose voices and experiences to present and then craft the narratives in which to frame them. Participating in this project *On Being Adjacent* is a challenge to lower my defenses and see myself as present in my work, an active member of a community of inquiry and witness.

When I speak of putting people at the center, what I am really talking about is their words, recorded at the time of given events in diaries and letters, or afterward in memoirs and testimonies, and preserved, collected, and in most cases published. People who produce such texts want to be heard; they are not voiceless victims. Scholars, however, can amplify their impact. This is the case with Molly Applebaum, whom I first encountered literally as a book. Stephanie Corazza at the Azrieli Institute introduced me to *Buried Words*, and I also heard a talk by Jan Grabowski on it. The author was described as a Polish Jewish teenage girl in hiding, who had sex with her rescuer, a middle-aged Polish Catholic farmer. This part of the story was what I heard in clips from interviews with Applebaum,[2] and it was also the focus of the "Buried Words" Workshop on Sexuality, Violence, and the Holocaust, held in Toronto in October 2018.[3] The sexual encounter is what features in publications that address Applebaum,[4] and it is how I introduced her in a series of talks I gave in 2019 on sexual violence in the Holocaust. I realize now that profile was inadequate, because it failed to convey Applebaum's depth of experience and complex personality, and inaccurate, because it reduced the sex to abuse, which Applebaum does not accept.

The book – Applebaum's diary and memoir and Grabowski's introduction – provides an outline of the author's biography, which I will summarize here. Molly

---

2 Interviews with Molly Applebaum: USHMM (1997), Azrieli Foundation (2018).
3 "Buried Words: A Workshop on Sexuality, Violence and Holocaust Testimony," sponsored by the Azrieli Foundation, Toronto, 11 October 2018. Details available here: https://memoirs. azrielifoundation.org/conferences/buried-words-a-workshop-on-sexuality-violence-and-holocaust-testimonies/.
4 See Doris L. Bergen, "Ordinary Men and the Women in their Shadows: Gender Issues in the Holocaust Scholarship of Christopher R. Browning," in *Beyond "Ordinary Men": Christopher R. Browning and Holocaust Historiography,* eds. Thomas Pegelow Kaplan, Jürgen Matthäus, and Mark W. Hornburg (Paderborn: Schöningh, 2019), 15–29; and Sara R. Horowitz, "What We Learn, at Last: Recounting Sexuality in Women's Deferred Autobiographies and Testimonies," in *The Palgrave Handbook of Holocaust Literature and Culture,* eds. Victoria Aarons and Phyllis Lassner (Cham, Switzerland: Palgrave Macmillan, 2020), 45–63.

Applebaum was born Melania Weissenberg in 1930 in Krakow. Her parents had a dry goods store. After her father died in 1938, her mother remarried, and she gained a stepfather and soon also a younger brother, Zyga. Months after the German invasion, the family fled Krakow and ended up in the ghetto in Dąmbrowa Tarnowska. Melania's mother arranged a hiding place on a farm, with Victor (Wiktor) Wójcik and his sister Emilia (Eugenia) Kułaga. When it became too complicated to keep Zyga concealed, Molly's mother returned to the ghetto with him. Molly never saw them again. Together with her cousin Helen, Molly remained hidden on the farm. The two of them were the only members of their extended family to survive the war. Later Molly came to Canada on a program for Jewish orphans. In Toronto she met and eventually married Rubin Applebaum, a survivor of Auschwitz. They had three children. By the time I read her book, Molly Applebaum, a widow since the 1980s, also had grandchildren and great-grandchildren.

In 2018, Anna Shternshis, Yu Wang and I taught the diary portion of *Buried Words* in a small class on the Holocaust in literature, and I subsequently used the book in several seminars. For the spring term of 2020, I decided to assign it to my large class on the Holocaust (75 students). My idea was to pair *Buried Words* with the Azrieli Foundation's 2018 interview with Molly Applebaum, which I had already done to powerful effect in a graduate seminar the previous summer in Jena, Germany.

As a historian, I am comfortable with books. People can be more challenging. Like everyone who works in this field, I have seen Holocaust survivors publicly excoriate scholars for getting it wrong, and I have been that scholar myself. Probably some discomfort with the unpredictability and demands of actual people contributes to a preference for books. In addition, academics share a tendency, widespread in our society, to write off elderly people. Ever since I taught my first class on the Holocaust thirty years ago, I have heard the refrain, "the survivors are leaving us," as if they were already gone, if not literally, then practically. All of this is to say that, although I had heard Molly Applebaum lived in Toronto, somehow, I must have assumed she was too old or infirm to appear in person. Perhaps I reached this conclusion because she was not present at the workshop in October 2018, even though I saw a clip from an interview conducted with her just months beforehand. Later I learned she only found out about the workshop after it happened.

Anna Shternshis wrote a column in the *Canadian Jewish News* about the symposium[5] and Applebaum read it. Her daughter, Sharon Wrock, contacted

---

5 Anna Shternshis, "Should we talk about sexual violence during the Holocaust?" *Canadian Jewish News* (24 October 2018).

Shternshis to convey her mother's appreciation. She also communicated that Applebaum had been disappointed not to win a Canadian Jewish Book Award for *Buried Words*. Soon Dan Rosenberg, a producer I know, called to tell me he was planning to make a radio program about Molly Applebaum. He had interviewed her and learned she was keen to talk about her book. She had also expressed to him her disappointment that the book had not won a prize.

Thinking about that elusive prize made me see Applebaum as a person, not simply as a metonym for her book. Suddenly I got the idea, which should have been obvious all along, to invite her to speak to my class. Rosenberg asked her daughter, whom he knew, to relay the invitation, and to my delight, Applebaum agreed right away. She was pleased to hear students were reading her book. I was surprised that she did not realize what a sensation her book was, perhaps no one had told her.

Applebaum is in fact frail, so we came up with a plan for her to Skype with us. We would send questions ahead of time, and either Rosenberg or Applebaum's daughter would film her responding. Dan warned me that based on his experience, she would answer in a few words, so we should collect a lot of questions. I scheduled discussions of the book in advance of Applebaum's visit, and students started reading.

We were still figuring out exactly how we were going to do it all when COVID struck. The university moved classes online, and in-person events were canceled. I was devastated and could not bear to tell my students, who had expressed so much excitement about meeting a woman they already referred to fondly as "Molly." Much about the pandemic was still unclear, and particularly the devastating impact it would have on patients and staff in long-term care facilities was not yet evident. More than 80 percent of the total deaths due to COVID-19 in Canada to date have been people in retirement and nursing homes.[6] The scandalous neglect of the lives and the physical, mental, and social wellbeing of elderly people that produced this outcome highlighted for me the often-casual dismissal of older people as having something valuable or even coherent to contribute, even to conversations about themselves. I see my own assumptions about Applebaum and her ability to speak for herself as part of that disregard.

By the end of March, when we had arranged for Rosenberg to interview Applebaum, residents of long-term care facilities were limited to one designated visitor each. Soon they would not be allowed any visitors at all.

---

6 The rate of deaths in long-term care homes in Canada is twice that of other countries in the Organization for Economic Cooperation and Development (OECD). Samuel Freeman and Alan Freeman, "Canada's first wave: a pandemic report card," *iPolitics*, 26 June 2020 (online).

There was an additional problem. I have a hearing disability and was anxious about the turn to online classes. I already knew from experience that meetings on Skype are almost a complete waste of time for me. So, we had to come up with creative solutions that would enable Applebaum to speak to us while everyone was in a different location. With the help of my teaching assistant, Marilyn Campeau, we developed a multi-part plan. Our students read the book, watched the interview, and discussed it all online in written posts that included questions they wanted to ask Applebaum. Even when I edited out the repeats, those questions filled eight pages.

We sent the questions to Sharon Wrock, now the only person allowed to visit her mother in the nursing home. Wrock read the questions to Applebaum and recorded her answers on her phone. The result was about a forty-minute audio interview. At the end of it, Applebaum offered to answer more questions, and she called her daughter later to add details about the food they ate in the ghetto. Wrock generously agreed to answer questions, too, and she also said she would relay questions to other members of the family.

The online format allowed students to articulate their questions thoughtfully and build on one another's ideas. We had a wonderful exchange, first with the audio recordings of Wrock reading the questions and Applebaum responding, and then with further rounds of questions and answers via email with Wrock, her brother, and an adult child of their sister. My hope that students would connect with Molly Applebaum was more than achieved. Instead of the image of a survivor as someone old and obsolete or frozen in time, students came to see Applebaum as a complex person with a past, present, and future. In short, COVID-19 restrictions, so often isolating, in this case brought unexpected but welcome pedagogical benefits.

"I was very touched to listen to the audio interviews by Mrs. Applebaum and her daughter," one student wrote: "I enjoyed hearing Molly's answers and have to add that hearing her voice and the age that comes with it makes the experiences more real, as I'm almost able to hear the journey and experience that life has brought to her." The same student commented on the organization of the questions, which was a combination of my handiwork and Wrock's: "I enjoyed how her daughter structured the interview in a way that was very freeing and didn't force her to answer the question a certain way. Often interviews have a goal in how the questions are to be answered – often in a way that is expected by the audience. With the way this interview was conducted we were able to hear raw responses to the questions asked."

Another student analyzed Applebaum's depiction of belonging: "A sentence from Molly that stuck with me: 'Home is somewhere that someone wants me to stay.' Molly had been a part of a lot of families but was never one of them.

I cannot even imagine what she felt like being an add-on when in hiding, looking for a country that would take her in, going through different foster families." And yet, the student observed, even if imagination was limited, some understanding was possible: "It is intimate personal experiences like this that allow us to really know Molly, more than just a survivor, but a human being."

One student found it particularly important that Applebaum and members of her family had expressed their appreciation of our interest: "I'm glad that the feelings were mutual . . . because it helped me emotionally connect with her, even though we weren't communicating in person." The audio component also had an impact: "Hearing Molly's voice put her story in a new light, it made it more real for me, and hearing her daughter's voice made me think about the impact of her survival and that it allowed her to continue a family and a legacy." Themes of family and community resonated in many students' responses: "I also think it's so great that we have heard a lot from her family members. What her children and grandchildren think about her story is a really interesting perspective to hear."

One student put into words what I was experiencing in the weeks that we were immersed in this project: "This is feeling like more of a community, and hearing and being able to engage in Molly's story has made it bigger than just that – especially during this time of self-isolation."

The responses from Applebaum, her children and grandchild, and my students helped me understand the multiple ways that family shapes experiences and memories of violence. For Applebaum as a child and teenager, family was both a refuge from violence and a site of violence. As an adult survivor, her family continued to function in both ways, as memory tangled together with the ongoing experience of abuse at home. The German Jewish diarist Victor Klemperer observed that, "All private suffering is multiplied and poisoned a thousand times over by the political circumstances."[7] Home and family are where those vectors meet.

Family constitutes a particular form of adjacency. As one student observed, the topic of family also allows people outside the circle of Holocaust survivors and their close relatives to empathize: "The fact that my own sense of identity and security come greatly from my family again reminds me how the destruction of the Holocaust carries on long after the war was over." Reflecting on family, in other words, opens the possibility of seeing the Holocaust not only as a

7 Victor Klemperer, entry from 18 October 1936, *I Will Bear Witness: A Diary of the Nazi Years*, vol. 1, *1933–1941*, trans. Martin Chalmers (New York: Modern Library, 1999), 197.

rupture but as a site of continuity and the quotidian, including everyday domestic violence.

The comments from Applebaum's family members were generous and engaged. Here is how Molly Applebaum's son, Neal Applebaum, answered a student's question: "Which parts of the diary and memoir impacted you the most?"

> I guess I was most surprised to hear about any sexual related recounting of her days in hiding. I guess I didn't even think about it, I mean who thinks of their mother as a teenager experiencing regular teenage feelings. Reading the thoughts of a young girl pondering life (hers and everyone's) and belief/faith in God was extremely emotional. While it may have lost something in the translation, I found it was still overpowering in its impact. If I were to try to write a poem or in a poetic style I don't think I could have achieved the same power as her simple words, simple words of a teenage girl. That's what impacted me the most.

Applebaum's daughter Sharon Wrock shared her process of discovering her mother's story, which started with a brush-off from her father: "Our father who had a number tattooed on his arm told us as children that it was his phone number, and this way he'd never forget it." Only as an adult did she begin to ask "more pressing questions." They were always answered, she said, "but without details." Wrock pointed out that it was she who convinced her mother to publish the diary:

> When I found out about the existence of the diary, only about 10 years ago, and asked my mother to translate it, she said it was just the ramblings of a young girl. It wasn't until I persisted and eventually had it translated that I realized the significance of the diary. It was around this time that my mother began speaking up freely about her experiences and encouraged us to ask her any questions. And she answered them freely, just as she has for your class.

A comment from Applebaum's grandchild hit especially close to home for students and for me, because it demonstrated how a university professor connected Applebaum, her daughter, and the potential audience for her book:

> My aunt asked, "Did you know Baba kept a diary when in hiding? Do you know anyone who could translate it?" This was the first I heard about the diary, but I said that I'd try to find someone. I ended up giving a copy to Dr. Grabowski . . . After my next class he asked if we could meet . . . He told me, "Her diary confirms everything that's in her memoirs – although it also has some content that she left out." This was of course the content on sex and sexuality. I was a little shocked, but it mostly didn't faze me. It didn't disrupt or change anything that I already knew – it just added new information. After getting a preliminary translation, what shocked me most was the author's voice: it was so vividly the voice of my grandmother. It was written seventy years earlier, and translated from Polish, but it was nevertheless my grandmother's voice, perspective, response to events.

Through our correspondence, I learned that Molly Applebaum's grandchild, Jason Chalmers, had just defended their dissertation in Sociology at the University of Alberta,[8] where I completed my master's degree. Chalmers's scholarship on settler colonialism, genocide, and public memory in Canada speaks directly to my work on the Holocaust. I was a member of the design team for the National Holocaust Monument in Ottawa, the subject of an article by Chalmers that begins with this challenge: "This paper considers how Holocaust memory in Canada reproduces settler colonialism and Indigenous genocide."[9]

Reading Chalmers' work in the midst of a summer of Black Lives Matter protests against systemic racism and its deadly manifestation in police brutality sparked a shock of recognition. In May and June 2020, I was teaching the class on the Holocaust in Literature with Anna Shternshis, and I felt acutely the need to articulate connections all of us were feeling. In Canada too, COVID cases and deaths were disproportionately among Black and Indigenous people and people of color. Here, too, people from these same groups are killed by police far more frequently than their white counterparts.[10] During the two months of our class, Abraham Natanine, Regis Korchinski-Paquet, D'Andre Campbell, Everett Patrick, Rodney Levi, Chantel Moore, and Ejaz Choudry all were killed by police or died in the presence of police in Canada. Reflecting on Applebaum's book and my encounter with her and her family in the context of a global pandemic and a reckoning with systemic racism in the United States and Canada, reveals multiple layers of relationships, the individual as connected to others, immediately and distantly, in imagined and unacknowledged communities.

Applebaum continues to write and connect the chapters of her life to the lives of others. According to Wrock, "She has since put all of her thoughts and memories into a journal from before the war, during and after, so that her next generations can read them when they're ready. And we've sent a copy of this journal to Helen's children so that they can know more about their mother, who died when they were just little boys." Many descriptions of the book speak of "two girls," but Helen, fifteen years older than her cousin, was an adult woman at the time. She died in the 1950s, and it appears the diary was in her husband's possession for some time, unread.

---

**8** Jason Chalmers, "National Myth in Canada: Reproducing and Resisting Settler Colonialism at Memorial Sites," PhD Dissertation, Sociology, University of Alberta, 2019.
**9** Jason Chalmers, "Settled Memories on Stolen Land: Settler Mythology at Canada's National Holocaust Monument," *American Indian Quarterly* 43, no. 4 (Fall 2019): 379–407.
**10** Pamela Palmater, "Yes, Canada has a racism crisis and it's killing Black and Indigenous peoples," *Canadian Dimension* (3 June 2020), online.

Applebaum's son and grandchild both touched on the subject of sex in *Buried Words,* but in a way that regards it as just part of the author's life, not its essence. The situation in hiding, Applebaum shows, was intense, vulnerable, and crushingly boring. The cousins spent time talking, sewing, writing, and having sex. In an interview Applebaum insisted that "for us it was entertainment." This statement is jarring to a feminist scholar, attuned to the silences and distortions surrounding sexual abuse. Yet it is a vital assertion of agency, of Applebaum's desire to create and control her narrative. Already in her diary, as evident in the passage I quoted at the beginning of this essay, Applebaum presented herself as a force in shaping events. She suggested that Helen write to Victor and invite him to have sex, with the result that Helen felt stronger and healthier. She willingly did the sewing work Emilia assigned to the girls, and she enjoyed it, although her finger hurt. Maybe that physical pain itself was comforting, even pleasurable, like people who cut themselves report. Cutting, it is said, can help temporarily to control emotional pain.

In early 2019, after the symposium on *Buried Words* but before our class's encounter with Molly Applebaum and her family, my mother died. Only while trying to write this essay did I start to understand the entanglements between my mother, me, and my connections to Applebaum and her book. My mother was born in 1923 in Soviet Ukraine and died in a nursing home in a small town in Western Canada. Her name, like Applebaum's cousin, was Helen. She was neither a Holocaust survivor nor a refugee. The big sadness at the center of her life was the death of her first child, a daughter, Judy, who was hit by a drunk driver while getting off the school bus. Judy had been named for my mother's younger sister, who also died as a child, shortly after the family immigrated to Canada. My mother had six more children, and she loved us and cared for us, but my sisters and I felt that, in her gentle way, she kept us emotionally at a distance. I found it satisfying to cooperate with Sharon Wrock to bring the interview with Molly Applebaum to fruition. I felt like I was helping a daughter get closer to her mother.

My mother had a number of things in common with Applebaum, including their intelligence, beautiful handwriting, and interest in literature. I was taken aback to see the epigraph that Applebaum chose for her book, from Robert Frost's "Stopping by Woods on a Snowy Evening":

The woods are lovely, dark and deep,
But I have promises to keep,
And miles to go before I sleep,
And miles to go before I sleep.

My mother loved poetry, and "Stopping by Woods on a Snowy Evening" was one of her favorite poems. I remember arguing with her after I studied it in a class in my first year of university. I told her it was about death, but she insisted it should be read more literally, as about introspection and the beauty of nature.

When I listened to the audio recording of Sharon Wrock reading our questions to her mother, I realized something she and I have in common. Neither of us speak our mothers' mother tongue. Wrock's evident discomfort with the Polish place names that came up in the questions reminded me of my inability as a child to understand my mother speaking "Plautdietsch" with her own mother. This linguistic gap is both a metaphor and a manifestation of the space between my mother and me. Learning German was probably an attempt to bridge it, and the relationships I have formed with certain Holocaust survivors – women and men the same age as my parents – as mentors and friends could also be connected.

Molly Applebaum had her own issues with her mother. Her father died when she was eight, and her mother never even told her. She found out from other children at school. In an interview, Applebaum said she could not trust her mother after that, and that lack of trust is why she chose to stay on the farm with Helen rather than returning to the ghetto with her mother and Zyga. This aspect of Applebaum's story resonated with me, although I am still not sure if it is because of the resentment toward her mother or the longing for the absent mother. Most likely it is both: feelings of betrayal and loss can be difficult to pull apart.

At the same time, I am so struck by the fact that Applebaum never acknowledges her mother's role in saving her life. It is always Victor, and to a lesser extent his sister Emilia, whom Applebaum lauds, and she worked hard and successfully to see them recognized as Righteous among the Nations for their role in rescuing her and Helen. Yet her mother, Sara (Salomea) Weissenberg, had the resourcefulness, quick thinking, skillful negotiating, and perfect Polish to get the family out of Krakow. She cultivated the relationship with Victor and Emilia, arranged the hiding place on the farm, organized the transfer of furniture and other supplies to compensate the Polish brother and sister, and made it possible for Molly and Helen to remain there by taking the little brother back to the ghetto. In the process, she gave up her own best chance of surviving.

An important chapter at the end of Applebaum's book is titled "Awakening." It opens with a sentence that appears in many survivors' accounts of postwar life: "Life went on."[11] But the awakening Applebaum describes is not the usual triumphant account of gaining children, grandchildren, friends, and financial security.

---

11 Applebaum, *Buried Words*, 105.

Instead, she talks about her husband's volatile temper and the verbal and emotional abuse she endured from him. "He had a hairline temper and flew off the handle at the drop of a hat," she writes, "but he could control himself if certain people who mattered to him were around. I do not cry easily, but I often cried from frustration. I felt belittled."[12]

Whenever I have taught *Buried Words*, I notice that students steer clear of this chapter. Only one perceptive graduate student observed that Applebaum's account of her abusive relationship in fact deepens her depiction of the loss and isolation she suffered during the Holocaust and after. She knew how to protect herself emotionally in the abusive relationship with her husband because the pattern was familiar. In Applebaum's words, "I was walking on eggshells and sweeping hurts under the rug. I seldom expressed my feelings and said only what I felt he wanted to hear. I had no guts to stand up for myself; I didn't know I had any rights. I was a wimp, letting myself be dominated. To this day, the memories of these incidents, and so many others, prey on me. I do not feel able to erase them."[13]

I also neglected this part of the book. Although I have read *Buried Words* many times and discussed it with students and colleagues in detail, I realized when I returned to this section that I misremembered something Applebaum had said. I remembered correctly that she expressed regret for not showing more affection to her children. "Neither Rubin nor I hugged or kissed the kids when they were not babies anymore," she wrote: "We didn't tell them that we loved them, and we didn't tell them they were cute or good-looking or any other things that parents usually tell their children. We assumed that taking physical care of them was sufficient, or an occasional outing or vacations taken together." Applebaum explained that, "We had no models from our own childhood."[14] Until now, writing this essay, I had assumed that sentence referred to the devastation of their families in the Holocaust. But Applebaum was a teenager when she and her mother parted, so her lack of a model of affectionate parenting was not the result of a collective, historical trauma but a private, domestic one.

In my mind, I distinctly recalled Applebaum worrying that she had been a bad mother. I recognized a similarity to my own mother who, in her last years, often asked me if I thought she had been a good mother. Those questions embarrassed me, and I never knew what to say or even what I thought. Mostly I tried to

---

12 Applebaum, *Buried Words*, 111.
13 Applebaum, *Buried Words*, 111–112.
14 Applebaum, *Buried Words*, 112–113.

**124** —— Doris L. Bergen

reassure her she had been "good enough" and then change the subject.[15] Applebaum, however, did not raise that same question. Instead, she wondered "how much of the tension the children sensed." They knew about their father's "terrible temper," and each of them responded in their own way, she observed: "Marlene removed herself from the home situation and went away to school, never to return for good. [. . .] Sharon tried to live with it, as she had no other place to go; I think she was relieved to get married. Neal learned how to grin and bear it and cooperate wherever possible."[16]

I wish Applebaum's book had won a prize, and I hope it will still do so and receive the recognition it deserves as a masterpiece of writing about the Holocaust. Given its departure from what have become the conventions of survivor memoirs, however, it may be unlikely. Rather than ending with photographs of children, grandchildren, and great-grandchildren, Applebaum concludes her book by returning to what she describes as the most meaningful relationship of her life, her bond to Victor, Emilia, and their descendants: "I'd be remiss if I did not write about my connection with the family that saved my life," she writes. "My original saviours, Victor and Emilia, brother and sister, are no longer alive, but their progeny lives on [. . .] We exchange photos and I also periodically help them out financially."

The first few times I read the book, I found it strange that Applebaum decided to end this way. But on reflection, I feel like I am beginning to understand what she is saying: "Their parents risked their lives for me for two and a half years, and I wonder who would do it now even for a day. I translated most of this memoir into Polish and sent it to them, and they found it fascinating. [. . .] I did not want to have regrets that I could have accomplished it and didn't. I got it done before it was too late."[17]

Applebaum's words illuminate for me both the general importance of feeling valued and doing something worthwhile and my own need to feel I am making a significant contribution. I understand the feeling of satisfaction she describes, because I have felt something similar when succeeding in communicating about the Holocaust to students or public audiences, or in connecting receptive listeners with a survivor – Gerhard Weinberg, Livia Prince, Sara Ginaite, Eli Pfefferkorn, Marijke Brown, N. N. Shneidman, Molly Applebaum. When I take pride in putting people at the center, I am also reinforcing my own sense of doing something

**15** The concept of the "good enough mother" originated with D. W. Winnicott, "Transitional Objects – Transitional Phenomena: A Study of the First Not-Me-Possession," *International Journal of Psychoanalysis* 34 (1953): 89–97.
**16** Applebaum, *Buried Words*, 111.
**17** Applebaum, Epilogue in *Buried Words*, 116–117.

valuable, bringing people and ideas into view, drawing strength and a sense of purpose from being adjacent to violence. To quote Applebaum, "I did not want to have regrets that I could have accomplished it and didn't."

# Works Cited

Applebaum, Molly. *Buried Words: The Diary of Molly Applebaum*. Toronto: Azrieli Foundation, 2017.

Applebaum, Molly. Oral history interview 57177. Interviewed by Liam Romalis, 26 February. The Azrieli Foundation Collection, Visual History Archive; USC Shoah Foundation, 2018.

Applebaum, Molly. Oral history interview with Molly Applebaum. Interviewed by Marilyn Potash, 22 January 1997. United States Holocaust Memorial Museum RG-50.431.1017, 1997.

Chalmers, Jason. "National Myth in Canada: Reproducing and Resisting Settler Colonialism at Memorial Sites." PhD dissertation, Sociology, University of Alberta, 2019.

Chalmers, Jason. "Settled Memories on Stolen Land: Settler Mythology at Canada's National Holocaust Monument." *American Indian Quarterly* 43, no. 4 (Fall 2019): 379–407.

Freeman, Samuel and Alan Freeman. "Canada's first wave: a pandemic report card," *iPolitics*, 26 June 2020, https://ipolitics.ca/2020/06/26/canadas-first-wave-a-pandemic-report-card/.

Horowitz, Sara R. "What We Learn, at last: Recounting Sexuality in Women's Deferred Autobiographies and Testimonies." In *The Palgrave Handbook of Holocaust Literature and Culture*. Edited by Victoria Aarons and Phyllis Lassner, 45–63. Cham, Switzerland: Palgrave Macmillan, 2020.

Klemperer, Victor. *I Will Bear Witness: A Diary of the Nazi Years*, vol. 1, *1933–1941*, translated by Martin Chalmers. New York: Modern Library, 1999.

Palmater, Pamela. "Yes, Canada has a racism crisis and it's killing Black and Indigenous peoples." *Canadian Dimension*, 3 June 2020, https://canadiandimension.com/articles/view/yes-canada-has-a-racism-crisis-and-its-killing-black-and-indigenous-peoples.

Shternshis, Anna. "Should we talk about sexual violence during the Holocaust?" *Canadian Jewish News*, 24 October 2018.

Rosenberg, Dan. "Cafe International - Buried Words - Molly Applebaum radio documentary," Oct 28, 2020. https://audiomack.com/cafeinternational/song/buried-words-molly-applebaum-radio-documentary

Winnicott, D. W. "Transitional Objects – Transitional Phenomena: A Study of the First Not-Me-Possession." *International Journal of Psychoanalysis* 34 (1953): 89–97.

Erin McGlothlin
# Affiliative Adjacency and Generational Grief: Ruth Klüger, Ursula Mahlendorf, and the Passing of a Generation

For some of the contributors to this volume, "being adjacent to historical violence" denotes their encounter with violence and its aftermath in direct, personal, and often visceral ways, either through their own lived experience of historical oppression, marginalization, or assault on selfhood or through their mediated connection to such violence via familial connections, generational inheritance, or community membership. For others, including me, "adjacency" refers not to personal or familial involvement in or experience of violent historical events, but rather to an elective affinity with the legacy of such violence, a self-established commitment that is rooted not in the biographies of ourselves or our families, but rather in our professional, intellectual, and pedagogical lives, which we devote, at least in part, to the study of violence and its ramifications. (In my case, this means the study of the Holocaust and its aftermath in narratives of both survivors and perpetrators.) In other words, although we have no personal or familial connection to particular manifestations of mass violence or oppression, those of us in the latter group forge our own ethical relationship to such historical events through our attempts to understand, commemorate, and – since most of us are researchers and teachers – write and teach about their impact. Marianne Hirsch describes such forms of adjacency as "affiliative," an adjective that implies that such connections are intentionally chosen and deliberately maintained.[1] Affiliative relationships to legacies of violence, often derived from a conscious sense of solidarity with victims and survivors of violence, can be ethically constructive and can contribute in positively impactful ways to the scholarship and teaching of such topics. At the same time, however, they also bear the capacity for problematic repercussions, an issue suggested by the term "affiliation," which can also signify "adoption" or "association," concepts that can come uncomfortably close to outright appropriation. To engage in affiliative adjacency thus obliges one to walk a narrow line between ethically committed and self-aware exploration of the impact of violence on unrelated persons who have suffered (or perpetrated) it and uncritical personal alignment with their experience

---

__FOOTNOTE__
1 Marianne Hirsch, "The Generation of Postmemory," *Poetics Today* 29, no. 1 (2008): 103–128, 114.

https://doi.org/10.1515/9783110753295-009

or even dubious subsumption of it under ethically questionable intellectual frameworks. Moreover, it also requires one to remain aware of and manage one's own intellectual and affective investment in the narratives of those who have suffered and survived violence, which invariably impact – to some degree, at least – one's own sense of self and one's place in the world. For a person like me, who chooses to enter into relationships of affiliative adjacency through my scholarly work on the Holocaust, the challenge is to acknowledge the personal and emotional ramifications of my work with narratives of survivors (and perpetrators) of the Holocaust without allowing myself to imagine that my feelings are somehow constitutive of those narratives. In other words, while I at times experience a range of sometimes powerful feelings that emerge as a result of my affiliative work, my ethical code entails that I recognize that such narratives are, above all, not about *me*.

And yet, at moments, my carefully managed practice of ethical distance breaks down, at least in part. Such an occurrence is happening to me as I sit down to write this essay. It is Friday, 9 October 2020. I have just lived through one of those weeks that occur rarely but are experienced intensely, during which I am suddenly confronted with a current event that seems to crystallize the dynamics that drive much of my life and my work. I am not talking about the most prominent political event of the week (in an extremely politically chaotic time), in which an extremely narcissistic and autocrat-wannabe American president attempted to rewrite his contraction of COVID-19, which almost certainly had been caused by his own willful negligence and months-long pathological minimization of the dangers of the virus to the vulnerable population whom he is charged to protect, as a kind of personal heroic test that he passed with flying colors. (At the same time, he also downplayed the severity of the virus, touted an unapproved "cure," and directed his attorney general to arrest his opponent in the upcoming election. It was a very busy week, but not an atypical one for the Trump administration.) After all, that episode is itself just one in a long series of moral offenses and feats of incompetence that said president has been committing on a daily basis for nigh on four years now. Seen from that perspective, it is barely a noteworthy event at all. (And who knows what will happen between now, as I write these words, and the moment in which they appear in print? Fresh and ever more extreme outrages, perpetrated on a daily and sometimes hourly basis by the president, the morally bankrupt political party he represents, and white supremacist and nationalist militants who eagerly follow him, threaten to turn each line written into a hopelessly outdated record of seemingly naïve indignation and impotent fury. At this point I am no longer convinced we have seen the worst, even if the election proceeds smoothly and the US electorate rejects Trumpism.) No, the event I am talking about was a much smaller quake, one that barely registered among the intense seismic activity of a worldwide pandemic and a turbulent presidential election. Nevertheless, it has rocked me to my core, inducing sharp feelings of loss. It was the news that Ruth Klüger, the eminent German Studies scholar, avowed feminist, and child survivor of the Holocaust, had passed away at the age of eighty-eight.

Klüger was a trenchant and at times unapologetically pugnacious figure in discussions of Holocaust memory. Through her incisive and occasionally contentious memoir *weiter leben: eine Jugend* (1992), which played a major role in German discourse on the Holocaust in the 1990s, and its English-language counterpart *Still Alive: A Holocaust Girlhood Remembered* (2001), which she called "neither a translation nor a new book," but rather "another version, a parallel book, if you will, for my children and my American students," Klüger played a crucial role in the transnational scholarly discourse on the Holocaust.[2] As their titles indicate, Klüger's memoirs reconstruct her experience as a child and young teen in Nazi Vienna, in the Theresienstadt ghetto, in Auschwitz-Birkenau, and in the labor camp Christianstadt, and they additionally explore her life in early occupied Germany and as a new immigrant to the United States. Klüger embeds her exploration of her traumatic childhood and posttraumatic young adulthood in an intricate, gender-sensitive examination of how the effects of these experiences resound into the present both in her interactions and relationships with others and in public narratives of the Holocaust, the latter of which, she finds, tend to reduce the events to bromidic platitudes. She accomplishes this dual focus on past and present through a complex narrative design that often concurrently focalizes both the younger experiencing self and the older evaluating and narrating self. Klüger's nuanced, critically aware perspective insists on a rigorous evaluation of both the historical events of the Holocaust and contemporary memory of them and passionately rejects the cathartic sentimentality, simplistic binaries, and trite narratives that often characterize public discourse on survivors' experience. Her memoirs, which readers uniformly find both moving and challenging, reflect Klüger's own reputation in the fields of German Studies and Holocaust Studies; she was notorious for her acerbic wit and her readiness to take offense, but she was also known for her generosity with others, her impressive knowledge of a wide range of literatures and intellectual traditions, her prodigious talent for reading literary texts, and her gift for writing, which she accomplished in an inimitable voice that was unremittingly honest and that refused to conform to the expectations of others.

---

**2** Ruth Klüger, *Still Alive: A Holocaust Girlhood Remembered* (New York: The Feminist Press of the City University of New York, 2001), 210.

I knew Ruth Klüger only very slightly; I met her at the 2004 Lessons and Legacies of the Holocaust conference, where I chatted with her briefly about a recent article I had published on the two versions of her memoir (and where she gently – and quite characteristically – corrected me with regard to a minor factual mistake I had made in it[3]). Shortly after that encounter, she and I corresponded briefly over e-mail about her introduction to a then-new edition of the masterful novel *Der siebente Brunnen* by the Austrian-Jewish Holocaust survivor Fred Wander. (Klüger saw in Wander's work "an optimism which I don't share but which I admire," but she also found it "a loveable work" with "so much (moral) goodness in it," an ambivalence that was fairly typical of Klüger's simultaneously critical and generous spirit.[4]) I saw her again briefly at the 2013 German Studies Association conference, where she gave a keynote address and where Irene Kacandes had organized a panel on her work. That was the totality of my actual relationship with Klüger, and had that represented the extent of my encounter with her, the news of her passing would likely have not had such a strong effect on me. But Klüger's memoirs have loomed large in my scholarship and teaching on the Holocaust, as they have done for many scholars of Holocaust literature of my generation, who attended graduate school in the 1990s and began their careers in the early 2000s. I first read her 1992 German memoir as a graduate student, and an essay I wrote on it eventually became my first scholarly publication.[5] In 2001, when her English-language memoir appeared, I, a new assistant professor, immediately felt compelled to write about how the two iterations of her autobiography challenge our ideas about the coherence of autobiographical writing, and I published my second article (the one with the slight mistake identified by Klüger) on her work in 2003.[6] Further, I have taught *weiter leben* and *Still Alive* a total of seventeen times together in graduate and undergraduate courses on Holocaust literature, contemporary German-Jewish literature, German-language autobiography, and children's experience in the Third Reich, the Holocaust, and World War II; indeed, I have found her texts – along with Primo Levi's *Survival in Auschwitz* and Art Spiegelman's *Maus* – to be indispensable for modeling to students how to think in more critically

---

3 Oddly, although I remember the encounter with Klüger well and can visualize where it happened, I no longer recall the mistake that she identified in my article.
4 E-mail from Ruth Klüger to author, 14 November 2004.
5 Erin McGlothlin, "'Im eigenen Hause . . . vom eigenen Ich': Holocaust Autobiography and the Quest for 'Heimat' and Self," in *Erinnerte Shoah: Die Literatur des Überlebenden*, ed. Walter Schmitz (Dresden: Thelem Verlag, 2003), 120–134.
6 Erin McGlothlin, "Autobiographical Re-vision: Ruth Klüger's *weiter leben* and *Still Alive*," *Gegenwartsliteratur* 3 (2004): 46–70.

aware ways about received narratives of the Holocaust. And, as recently as a few weeks ago, I finished writing an essay for inclusion in a volume entitled *The Ethics of Survival* (edited by Gerd Bayer) that takes Klüger's autobiographical texts as a point of departure for my consideration of readers' affective consumption of survivor memoirs and their representation of the experience of liberation or escape. At this point, I can quote passages almost verbatim from both Klüger's German-language text and her English-language memoir; unparalleled in its critical force, its conceptual precision, and its ethical clarity, her distinctive voice perennially resounds in my thinking about the Holocaust, bearing a lasting influence on how I teach and how I write about this topic, the abiding commitment of my scholarly life. For this reason, although I barely knew Klüger the flesh-and-blood person, it not an exaggeration for me to say that, over the course of the last quarter-century, Klüger, as a literary voice, has served as a sagacious guide and intimate textual companion on my intellectual journey through the memory and representation of the Holocaust. At the same time, however, while Klüger's voice certainly endures in her writing, the loss of Klüger the person engenders in me – and, I imagine, for many others like me, who felt a strong connection to her work – a poignancy that borders on real grief.

The news of Klüger's passing this week came at the exact moment at which I have been processing the death of another scholar-memoirist from the same generation as Klüger: Ursula Mahlendorf, also a foundational feminist critic in the field of German Studies (and, not incidentally, a friend of Klüger's). Like Klüger, Mahlendorf penned a rigorously honest autobiographical account of her childhood and youth in the Third Reich and World War II; however, whereas Klüger experienced that time as an intended victim of the Nazi regime and its project of genocidal destruction of the European Jews, Mahlendorf, born in 1929, a year before Klüger, lived through the period on the other side of the National Socialist racial divide, as an "Aryan" and an ostensibly favored member of the National Socialist *Volksgemeinschaft*.[7] Her remarkable memoir, *The Shame of Survival: Working Through a Nazi Childhood* (2009), is a sensitive but self-critical reckoning with that experience, providing a perspective that has rarely been offered in English (and furthermore has rarely been offered so critically in German): that of a

---

7 More than fifteen years before Mahlendorf published *The Shame of Survival*, she wrote an astute review of Klüger's *weiter leben*, praising it above all for its "relentless honesty," "unflinching honesty," "unflinching confrontation" and "relentless truthfulness." Although, to my knowledge, Mahlendorf did not mention Klüger's work as a model for her own, her emphasis on scrupulous honesty in the review provides insight into the qualities she valued in autobiographical writing and worked to achieve in her own memoir. Ursula Mahlendorf, "World Literature Review: Germany," *World Literature Today* 67, no. 3 (1993): 607.

former BDM (League of German Girls) leader who was an avid follower of Hitler and a loyal supporter of the Nazi regime until its demise, when she endured a series of traumas and hardships at the end and in the aftermath of the war, most notably her flight from the Soviet army, her difficulties under the Soviet and the Polish occupations, and her expulsion from her hometown by the Polish government. Adamantly intent on not portraying herself as a victim, Mahlendorf contextualizes her experience of expulsion within the larger frame of German military aggression and the Holocaust, not only demonstrating the brutal consequences of her unquestioning childhood loyalty to the Nazi regime, but also describing how her own traumatic experiences caused her to reevaluate everything she had learned as a child and young person. Eloquently written and lacking in sentimentality or unreflective nostalgia, *The Shame of Survival*, like Klüger's two Holocaust memoirs, alternates between narration of the historical events in which Mahlendorf was a participant and commentary on those events from a contemporary perspective informed by a lifetime of hard-won reflection and self-examination, a structure that emerges from what she terms "a head-on conflict between memory and history."[8] Moreover, Mahlendorf's perspective, like that of Klüger, purposely employs the lens of gender in its analysis of personal and historical dynamics, providing a first-hand look at how women and girls were cynically co-opted by the Nazi regime. *The Shame of Survival* is deeply affective in its humility, in its honesty, and in the courage Mahlendorf demonstrates through her readiness to lay bare the painful dynamics of her young life in order to "show today's students, through an early life history like my own, how the right leaders in an ordinary small town could produce potential perpetrators."[9] Mahlendorf took her self-imposed pedagogical duty extremely seriously, and in the latter years of her life, she spent a great deal of time and energy speaking in public about her first-hand experience growing up and initially identifying with the Nazi state.[10]

As with Klüger, I knew Ursula Mahlendorf only fleetingly; I was a peer reviewer for *The Shame of Survival*, and I met her briefly at the 2007 German Studies Association conference not long after I had read the manuscript. After her memoir was published, I conducted a short correspondence with her in Fall 2010 about my experience teaching it in the classroom. My students were completely

---

**8** Ursula Mahlendorf, *The Shame of Survival: Working Through a Nazi Childhood* (University Park: The Pennsylvania State University Press, 2009), 5.
**9** Mahlendorf, *The Shame of Survival*, 4.
**10** Mahlendorf's public educational work, which she found necessary and rewarding, sometimes took its toll on her. In an e-mail to me from 30 September 2010, she confessed that she had "used my resources in strength a bit too much (the last one was 6 campuses)" and was experiencing health problems.

engrossed in it and had lots of questions about her experience; I gathered their questions together and e-mailed them to Mahlendorf, who, though recovering from surgery, generously wrote thoughtful replies to each question, which I then presented to my class. That was the extent of my personal engagement with Mahlendorf the person, although I have gone on to teach her memoir a total of six times, most recently this very week. Each time I read her text, I am gripped by Mahlendorf's radical candor, by her attempt to reconstruct her childhood in all its beauty and horror, and by her intrepid willingness to probe both her shame for her own misguided entanglement – however limited it was – in the murderous movement of the Third Reich and her grief for the many personal losses she endured as a result of Germany's defeat. The fortitude, openness, and humbleness with which she faces her past not only make her memoir a fascinating model of autobiographical practice, they are also exemplary for the lessons they convey as part of one woman's quest for meaningful self-knowledge. Mahlendorf's profoundly moral probity with regard to her former actions and beliefs, her anguish at her actual and potential complicity in the Nazi regime, and her sorrow for the losses she endured during its dissolution and aftermath establish with the reader a relationship of intimacy and trust in which the latter becomes aware that she is engaging with something important, elemental, and real. Like Klüger, Mahlendorf actualizes in her memoir a singular voice through which pain, remorse, and a desire to understand and perhaps even make peace with the past are articulated in unforgettable ways.

On Monday of this past week, two days before I learned of Klüger's death, I performed a quick internet search in order to compile some information on Mahlendorf's biography to present to my students. I was astonished to learn that she had passed away in October 2018, just a month after I had previously taught her text. I had not heard of her passing at the time, or at least I do not remember now to have heard of it; I know that sometimes I do not always immediately absorb the news of the death of a person whom I do not know well and integrate it into my thinking about that person, so it could be that I heard about it but simply did not retain that information. In any case, when I learned (or re-learned) of her death on Monday, I felt somewhat distraught and unnerved; I had lost track of the years since I had been in touch with her, so in my mind she remained a woman in her late seventies, the age she had been when I met her in person. This impression had of course been fostered rather than dispelled by my frequent reading and teaching of her text in the interim; for me, the voice that emerged in her memoir was as fresh as it had been when I first read her manuscript in the summer of 2007. As with Klüger, who I knew had been aging but whose voice in her memoirs was, in a sense, ageless, with Mahlendorf I had implicitly assumed that the actual woman would remain as enduringly present as the narrator in her text. The knowledge of her death on Monday, together with the news of Klüger's recent death on Wednesday, has triggered in me a bout of emotional turmoil, whereby my workaday reserve and insistence on a measure of ethical distance to my work have momentarily collapsed, leaving me somewhat bereft. Klüger's and Mahlendorf's deaths have done something to me, something I cannot quite describe but nevertheless can discern.

Of course, I wasn't fully unprepared for the reality that these women, whose work I so admire and who have helped me to understand, at least in part, events and dynamics that I find fascinating on account of their very resistance to easy comprehension, would at some point die. Both women were in their late eighties and had lived past the life expectancy of their demographic. At this, I perceive with regard to both Klüger and Mahlendorf something of what Klüger experienced when her own aged mother, also a survivor of the Holocaust, died: "I felt a sense of triumph, because this had been a human death, because she had survived and outlived the evil times and had died in her own good time, almost a hundred years after she was born."[11] Certainly one has cause to celebrate the fact that Klüger, one of the youngest people to have survived Auschwitz, died not in the Holocaust but in her old age, but the same can be said for Mahlendorf, who could have easily fallen victim of the deadly violence that she witnessed at the end of the war. Their deaths weren't entirely unexpected, and the fact that they occurred after both women had enjoyed long, even fulfilling, lives, assuages the sense of sorrow they engender. Moreover, while some members of their generation continue to write and speak publicly about their experience during the Nazi period, a large number of their cohort, meaning both the child and teenage Jewish survivors of the Holocaust whom Susan Rubin Suleiman terms the "1.5" generation and their non-Jewish counterparts – German men and women who were children and adolescents during the Third Reich and World War II – had already passed in recent years (or earlier).[12] Among the writers whose work has been important to me (many of whom I have written and/or taught about), this group includes Jurek Becker (1937–1997), Jakov Lind (1927–2007), Raymond Federman (1928–2009), Harry Mulisch (1927–2010), Arnošt Lustig (1926–2011), Christa Wolf (1929–2011), Cordelia Edvardson (1929–2012), Günter Grass (1927–2015), Elie Wiesel (1928–2016), Geoffrey Hartmann (1929–2016), Imre Kertész (1929–2016), Edgar Hilsenrath (1926–2018), and Aharon Appelfeld (1932–2018). As I have progressed through the profession, from graduate student, to assistant professor, to associate professor, and most recently to full professor, I have seen this generation

---

**11** Klüger, *Still Alive*, 211.
**12** Susan Rubin Suleiman, "The 1.5 Generation: Thinking About Child Survivors and the Holocaust," *American Imago* 59, no. 3 (2002): 277–295.

of writers and thinkers, who were so profoundly affected by the events of 1933–1945 in Europe but were too young to have contributed to their making or to have responded to them as full agents, slowly diminish from year to year. The progressive attrition of this group of writers takes the opposite trajectory of my encounter with them and their writing; the longer I read and write about them, the smaller the group becomes. To engage in such affiliatively adjacent work in this context means to attend to the gradual passing of an entire generation, some of the last witnesses of the historical events of the Third Reich and the Holocaust who are able to remember and mediate them. The grief I have been experiencing this week stems in part from my sudden, acute recognition of a larger, unalterable, accelerating phenomenon, the reality of which Klüger's and Mahlendorf's deaths accentuate.

But I think my response goes deeper than that. Throughout the week, I have asked myself why I feel such desolation in response to the news of Mahlendorf's and Klüger's passing and why this generation has had such a profound effect on me. To a large extent, I think it has do with the ways in which the writers of this cohort work to comprehend their childhood experience of violence, which occurred before they had developed secure frameworks for understanding it and even, as Suleiman argues with regard to the younger members of this generation, "before the formation of stable identity that we associate with adulthood, and in some cases before any conscious sense of self."[13] As writers from this generation – in particular, Mahlendorf and Klüger – seek to understand their own involvement and implication in the larger forces of history that governed their childhood, they endeavor to untangle knotted and broken threads of memory, to establish causal linkages between events, to fathom the motivations and behaviors of the adults responsible for them, and to excavate their own emotional responses to the chaos that enveloped them. In doing so, they draw the reader into their efforts to reconstruct their childhood selves and even grant her a role within their process. In short, they invite the reader to engage in a relationship of affiliative adjacency.

At the same time, however, I know that my present sorrow stems from a greater source. For not only do Klüger and Mahlendorf represent poignant examples of the long shadow that fascism, war, and genocide can cast on adults who were reared in such violent circumstances, they also represent for me something more personal, something against which my own biography rubs up rather than remaining affiliatively adjacent to it. Both women belong to the same generational cohort as my own parents. My father, who passed away in

---

13 Suleiman, "The 1.5 Generation," 277.

1996, was born in 1929, one month before Mahlendorf, while my mother was born in 1930, a year after Mahlendorf and a year before Klüger. (My parents were somewhat older than average when they had me; my mother had just turned thirty-eight and my father was almost thirty-nine.) Moreover, my mother's death occurred in between those of Mahlendorf and Klüger at analogous temporal distances; she died in 2019, one year after Mahlendorf and one year before Klüger. In fact, like both Mahlendorf and Klüger, my mother died within a month of her 89th birthday; Klüger and my mother were just shy of eighty-nine, while Mahlendorf had just reached that age when she passed. Hearing of Mahlendorf's and Klüger's deaths this week reactivated my grief at my own mother's death, which had ebbed somewhat amidst the chaos of pandemic time. The loss of these two women, whom I mostly knew from their intimate memoirs, suddenly also seemed related to the loss of my own mother, and beyond that to that of my father as well.

To be sure, apart from the issue of generational belonging, there are few overt connections between the biographies of my parents and those of Mahlendorf and Klüger. Most pertinently, my parents never experienced the extreme violence, persecution, political upheaval, physical danger, and existential insecurity suffered by Mahlendorf and especially Klüger. They were born on this side of the Atlantic, and they remained in the United States during the years of hardship, suffering, and precarity that affected Europeans so devastatingly. They both experienced World War II from the American home front, which means that, in comparison to Mahlendorf and Klüger, they barely experienced the war at all. (My father, desperate to fight in the war, lied to military recruiters on two separate occasions about his age, but both times they discovered he was too young to enlist; for her part, my mother, along with her entire German-Texan community, was suddenly forced to cease speaking German in public). Moreover, my parents also never experienced the United States as refugees or immigrants; they were "at home" here their entire lives, unlike Klüger and Mahlendorf, whose lives were both characterized by significant displacement. At the same time, however, from my perspective, as someone whose childhood in the 1970s and 1980s was characterized by TV sitcoms, Hamburger Helper, and the Cold War, the lives of my parents had more in common with Mahlendorf's and Klüger's than they had with my own life. (My parents, whose frame of reference was the Great Depression, Benny Goodman, and Glen Miller, also had little in common with the significantly younger parents of my contemporaries, who were reared on postwar prosperity, Elvis, and The Beatles.)

In particular, my father, whose childhood was marked by the extreme poverty and upheaval caused by the Dust Bowl storms of the 1930s in Oklahoma and Texas, endured conditions that were unthinkable to me in the security of my own childhood. Having lost their farm, his family migrated often during his early years. His father died of tuberculosis when he was ten, and his mother left him and his sisters with her parents, who were tenant farmers, while she went to Dallas to find work. He suffered at times from hunger, often went without shoes or other basic necessities, and was forced as a young teen to permanently leave school, which he had loved and in which he had excelled, so that he could help support his family. While his volatile childhood experience is certainly not comparable to that of Klüger, who was persecuted by a genocidal regime that murdered her father and brother, or to that of Mahlendorf, who was caught up in the maelstrom of the final days of the war and its brutal aftermath and then forced to flee her hometown in Lower Silesia, it was clear to me even as a child that the physical insecurity and emotional instability he underwent when he was young affected him throughout his life, whether in his fraught relationship to food, his difficult interactions with others, his frequent bouts of melancholia, or his generally pessimistic outlook on his own life and the lives of his children. I can detect an echo, however faint, of the impoverished conditions and emotional responses that Klüger and Mahlendorf describe in their memoirs in my own father's story, and this resonance encourages me to consider his life with a greater degree of compassion than I once was able to muster.

My mother, on the other hand, experienced a generally stable, if spartan and religiously intolerant, childhood on a family farm in central Texas, despite the general austerity and indigence that characterized Depression-era rural life. For that reason, I don't perceive any particular resonance between Mahlendorf's and Klüger's childhoods and that of my mother. However, the feminist lens with which the two memoirists view their lives connects them, in my mind, to my mother, who, like Mahlendorf and Klüger, struggled to build a professional career in the 1950s and 1960s, when gender bias was rampant (even if my mother did not have the vocabulary to identify it as such) and there were few opportunities for women to advance in either the academic or the corporate world. My mother experienced on a daily basis smaller and greater indignities that were not unlike the overt sexism that Klüger describes in *weiter leben*, in *Still Alive*, and in a third autobiographical text about her postwar life, *Unterwegs verloren: Erinnerungen* (2008). In the mid-1960s, my mother was the highest-ranking woman in her company of thousands, but she still was paid less than the men who were her subordinates; when she challenged her supervisor about the disparity, he justified it, remarking that the male employees all had families to support, while she had a husband who supported her and could consider herself lucky to even keep the job she had. (Of course, she refrained from telling this man that her husband suffered at times from undiagnosed depression and had trouble earning a steady wage, and that she was therefore obliged to assume the responsibility of chief breadwinner in the family.) Even as a child, I was able to discern the humiliation she was sometimes forced to endure and the despair she felt at continually having to prove herself in an often hostile workplace while working almost singlehandedly to care for a family and manage a household. The decidedly feminist frameworks through which Mahlendorf and Klüger evaluate their experiences of misogyny and personal diminishment not only encourage me to view their experiences in light of those of my mother, but also evoke in me the same admiration and even wonder that I feel for my mother, whose inner strength and fortitude, I am certain, were equal to that of the two memoirists, who survived much greater travails. In fact, over the course of the past week, during which I have thought intently about the convergences between these three women, I have come to realize that, although I met both Klüger and Mahlendorf in person and certainly know what they looked like, in my mind's eye, they very much resemble my beloved mother, despite there being no commonality in appearance. It is no wonder that the news of the two women's passing dislodges painful nuggets of grief within me; I feel as if I am reexperiencing my own mother's death from last summer as I acknowledge and mourn their deaths today.

Above all, this recent concatenation of affect and sorrow has made clear to me what a critical role the generation of Ruth Klüger, Ursula Mahlendorf, Charles McGlothlin, and Velma Weiser McGlothlin has played in my life. Sandwiched between what in the United States has misleadingly been called the "Greatest Generation" of adults who fought in the war (the cohort my father longed to join) and the later generation of Baby Boomers, who more forcefully made their mark on the postwar and contemporary cultural and political landscape, the generation that included Klüger, Mahlendorf, and my parents was born into a world that was already fundamentally out of kilter. As children, as youths, and as young adults, they were forced to negotiate the shifting sands of their respective environments, to assume responsibilities beyond their years and capabilities, and to adapt to new and sometimes quite harsh realities. That they were able to survive – and perhaps even eventually to thrive – under such conditions is a testament to their perseverance and resilience. At the same time, they all carried with them their difficult pasts, which continued to challenge them throughout their lives and which at times threatened the peace and prosperity of the existence that they had created for themselves. To be sure, Mahlendorf and especially Klüger survived physical privations, political calamities, and existential threats that dwarfed the challenges faced by my parents. In particular, both women witnessed and experienced varied forms and degrees of violence which my parents were altogether spared. However, through their autobiographical writing, Mahlendorf and Klüger also worked to understand and reconcile themselves to their pasts, committing themselves publicly to a sober examination of the challenges they had endured in their young lives and how they had responded to them. In doing so, they have bequeathed perspicacious accounts of their struggle to make amends with their childhoods to which I find myself turning again and again in order to understand their generation, my parents included. My parents, on the other hand, left no such public record of their lives; apart from the diaries my mother wrote in middle age (which she specifically requested that I not read) and a few letters, little in the way of written narrative exists from which I might glean insight into the intimate details of their lives and how they felt about them. I feel deep sorrow about this, an emotion exacerbated by the nagging sense that I could have done more to rectify this situation while they were (or at least my mother was) still alive. (Perhaps, part of me thinks, had I been less interested in the traumatic narratives of other people, I might have had a little extra energy to devote to family history. It is an uncharitable way to think about one's scholarship, but there it is.) At the same time, however, I have come to the conclusion that, in my continued reading, teaching, and writing about Mahlendorf and Klüger, I am somehow also forging a connection to my parents, however contrived, tenuous, or meaningful only to me.

Moreover, I have learned from my week of mourning Mahlendorf and Klüger – and, by extension, my own parents – that I am more personally implicated in the narratives of violence about which I read, teach, and write than I had previously thought (even if such implication is associative and mediated rather than historically contiguous or direct). I still maintain, as I noted at the beginning of this essay, that the experiences of the Holocaust that so engross me require that I take ethical care not to make them about me. At the same time, however, I have come to understand that, as a potentially fruitful byproduct of my choice to engage intellectually and emotionally with such narratives and to take on a relationship of affiliative adjacency to the historical violence that they convey, I have the opportunity to learn more about myself and my own past. I hope I am not engaging entirely in wishful speculation by thinking that both Mahlendorf and Klüger would approve.

# Works Cited

Hirsch, Marianne. "The Generation of Postmemory." *Poetics Today* 29, no. 1 (2008): 103–128, 114.

Klüger, Ruth. E-mail to author. 14 November 2004.

Klüger, Ruth. *Still Alive: A Holocaust Girlhood Remembered*. New York: The Feminist Press of the City University of New York, 2001.

Klüger, Ruth. *Unterwegs verloren: Erinnerungen*. Vienna: Paul Zsolnay Verlag, 2008.

Klüger, Ruth. *weiter leben: eine Jugend*. Göttingen: Wallstein Verlag, 1992.

Mahlendorf, Ursula. E-mail to author. 30 September 2010.

Mahlendorf, Ursula. *The Shame of Survival: Working Through a Nazi Childhood*. University Park: The Pennsylvania State University Press, 2009.

Mahlendorf, Ursula. "World Literature Review: Germany." *World Literature Today* 67, no. 3 (1993): 607.

Suleiman, Susan Rubin. 2002. "The 1.5 Generation: Thinking About Child Survivors and the Holocaust." *American Imago* 59, no. 3 (2002): 277–295.

Chunjie Zhang
# Identity Freedom or On Choosing Who We Are

In the Spring quarter in 2020, I taught a virtual seminar "What is Enlightenment." I considered it meaningful to start the course with the historian David
Hollinger's essay "The Enlightenment and the Genealogy of Cultural Conflict in
the United States."[1] Hollinger's text discusses multiculturalism and identity
politics and argues that one of the key Enlightenment legacies for twentieth-
century America is: "Physical characteristics such as skin color and shape of
the face should not be allowed to determine the cultural tastes and social associations of individuals."[2] Indeed, one can't choose one's physical appearance
determined by the genes of one's parents. Yet cultural tastes such as aesthetic
styles, manners, or consumer behavior and social associations such as friendships, relationships, religions, ideological convictions, sexual orientations, or
political alliances are things one can choose. Physical characteristics, usually
associated with race, ethnicity, or country of origin, do not speak for one's personality, intelligence, moral stature, spirituality, work ethics, education, or social status. One has the ultimate freedom to not be associated with the group to
which one is customarily ascribed. Hollinger's essay makes clear that freedom
in choosing one's sociocultural or gender identity is a legacy of the Enlightenment that we should fiercely protect and insistently continue practicing.

The ideological Enlightenment identity freedom finds its positivist support
in recent genetic research. The geneticist Bryan Sykes tells us in *DNA USA: A
Genetic Portrait of America* (2012) that Americans have converged genes from
Asia, Africa, and Europe.[3] Once a client of Sykes's could not believe that he had
a Viking ancestor because, the man said, "I am dark haired and short." Sykes
comments: "At that point I gave up, realizing once again that DNA always struggles to reverse the deepest of psychological perceptions or identity associations."[4] A very recent piece of research, published on 17 September 2020 in
*Nature*, reveals that many Vikings had brown hair and were not merely from

---

1 David Hollinger, "The Enlightenment and the Genealogy of Cultural Conflict in the United
States," in *What's Left of Enlightenment? A Postmodern Question*, eds. Keith Michael Baker and
Peter Hanns Reill (Stanford, CA: Stanford Unviersity Press, 2001) 7–18.
2 Hollinger, "The Enlightenment," 16.
3 Bryan Sykes, *DNA USA: A Genetic Portrait of America* (New York and London: W. W. Norton
and Company, 2012).
4 Sykes, *DNA USA,* 153.

https://doi.org/10.1515/9783110753295-010

Scandinavia. They also had genes from Asia and Southern Europe.[5] The finding changes the misconception about this medieval people being only blond and Nordic, something that could be traced back to biological racial conservatism around 1900. An individual of "pure" race or ethnicity rarely exists in human genetic information. There is rather a constant process of combination and integration at the microbiological level. Sykes reports that even though he could be "easily classified as a white Caucasian," he in fact has African genes that run vital organs in his body next his European heritage.[6]

Skin color, a common signifier used to racialize people, is the result of a random selection of only one basic substance – melanin. "There will be Americans whose DNA is almost all African in origin, yet if the pigmentation genes are not included in these segments and instead come from European ancestors, then their skin color will be white. Similarly, it would be entirely possible for a European American with only a small overall component of African DNA to be very dark skinned if these ancestral segments were to include the pigmentation genes."[7] Hence racial categories are "the artificial divisions we have created for ourselves."[8] Genetic information reveals inherent diversity, which we have inherited over millions of years from countless ancestors. Sykes mentions that his Hispanic clients in Mexico and the southwestern United States are surprised to find out about their Jewish genes. Early Jewish immigrants in America were afraid of the Spanish Inquisition and thus hid their identity. Even a colonial governor in Mexico couldn't save his own family from the Inquisition. In 1579, the Portuguese colonizer Luis de Carabajal consulted with his priest one day because his wife spoke a strange language to their children. Carabajal was concerned about her sanity. Then the priest realized that they were speaking Hebrew and reported the case to the Inquisition. The mother and the children were consequently burned in Mexico City, and the governor died in prison.

A similar report also occurs in the eighteenth-century drama *Nathan der Weise* (Nathan the Wise, 1779) by the German Enlightenment thinker Gotthold Ephraim Lessing. Set in the era of the Crusades in the twelfth century, a templar, pardoned by the Islamic Sultan Saladin, tells a Christian Patriarch that a girl, with whom he has fallen in love, was born to Christian parents but has been raised in a Jew's house. The Patriarch immediately judges that the Jew, Nathan, shall be burnt despite the fact that he raises his foster child with love

---

5 Accessed 17 September 2020, https://edition.cnn.com/2020/09/16/europe/vikings-blond-scandinavian-study-scn-scli-intl-gbr/index.html.
6 Sykes, *DNA USA*, 296.
7 Sykes, *DNA USA*, 298.
8 Sykes, *DNA USA*, 299.

and care but without religion.[9] Fortunately, Nathan is not burned. The drama ends with a harmonious union of people of three conflicting faiths: Christianity, Judaism, and Islam. While *Nathan the Wise* has been routinely interpreted in German scholarship after the Second World War as a classical plea for Christian tolerance toward the Jews, it also manifests an idea of universal humanity that moves beyond the identity boundaries set by religion, culture, and geography. It is still topical for today's discourse on identity politics based on reified racial, cultural, and religious categories. Lessing's drama is a literary illustration of the Enlightenment identity freedom in moving beyond one's given physical or religious confinement and choosing one's belonging as a non-partisan cosmopolitan.

The contemporary German Jewish writer Max Czollek's provocative book *Desintegriert euch!* (*Disintegrate Yourselves!* 2018) calls for Jews and Muslims to disintegrate themselves from mainstream German culture and thus free themselves from the roles ascribed to them by German society and politics where they are expected to either play the victims of the World War II Judeocide or the absolute other to the German Christian identity. Even though Czollek's message of disintegration seems at odds with Lessing's imaginative plea for integration and universal humanity, I argue here that identity freedom is the driving force behind both works across more than two hundred years. This freedom, whose foundation is human diversity since millennia, as evidenced by genetic science, proves a principle that we need to diligently practice, cultivate, share, and protect. Reading Lessing's drama and Czollek's prose together, I aim to show that identity freedom is not only a legacy from the Enlightenment but also an urgent issue still demanding our attention today.

## Enlightenment Identity Freedom

I once wanted to write my doctoral dissertation on Lessing's *Nathan the Wise* and the British-Indian writer Salman Rushdie's *Midnight Children* (1981) because I was fascinated by the diversity and hybridity in the different historical and cultural contexts of these literary works. The project, however, involved too many languages and crossed too diverse historical periods to be feasible for a beginner. Yet my interest in Lessing's drama and its entanglement with contemporary issues of identity has not vanished. As a *New York Times*'s theater review of an off-Broadway production of *Nathan* in 2016 comments: "Ultimately it

---

**9** See Gotthold Ephraim Lessing, *Nathan der Weise* (Frankfurt am Main: Suhrkamp, 2003), 105–107.

proves to be a moving story that speaks, as you might guess, to conflicts that roil the world today."[10] Most striking in Lessing's drama is the entwinement between diverse religions and cultures. Set in twelfth-century Jerusalem, the Jew Nathan has a foster child Recha, born to Christian parents. At the point a Christian knight brought the baby girl to Nathan for adoption, Nathan's wife and seven sons have just been killed by Crusaders.[11] Nathan raises Recha in Jerusalem under Islamic reign. When the Sultan Saladin befriends Nathan, they find out that Saladin's long-lost brother had followed a Christian woman to Germany and was the father of Recha. He was also a friend of Nathan's. Recha was begotten by an Islamic father, who most probably had converted to Christianity, and a Christian mother, and she grows up in the house of a Jew. The Christian templar, with whom Recha has fallen in love, turns out to be her brother. The drama ends in embraces of family members. The drama shows that, while religious faiths separate people, they are all part of a bigger family of the human species. If religious institutions use violence and wage war like the Inquisition and the Crusades, they cause damage to all humans including themselves. Lessing's literary imagination breaks with the immediacy between religion and identity and emphasizes rather a deeply seated human connection and peaceful love as fundamentally vital. When the templar discovers that Saladin is his uncle, he exclaims: "I of your blood! – Then those dreams, with which they cradled my childhood, are, now, more than dreams, much more!"[12] If we substitute the word "blood" here with genes, then Lessing's drama predicts the genetic entanglement that Bryan Sykes's research proves.

In a conversation between Nathan and the templar, Nathan contends that there are good people in every country. The templar responds that they have differences. Nathan agrees that they are different in "color, clothing, and body shape" (*an Farb', an Kleidung, an Gestalt verschieden*).[13] The templar responds that things are not as peaceful as Nathan's relativism describes. He reflects that the Jews first declare themselves as the only chosen people by god; then came the dark and blind Crusades, in which different religions claim their deity as

---

**10** Accessed 7 August 2020, https://www.nytimes.com/2016/04/14/theater/review-nathan-the-wise-brings-a-morality-tale-to-today.html.
**11** Nathan's narration of his past in Act 4, Scene 7, is reminiscent of Job and his story in the Bible. See Ingrid Strohschneider-Kohrs, "›Nathan‹ – poetische Chiffre der religio-Erfahrung," in *Gotthold Ephraim Lessing. Neue Wege der Forschung*, ed. Markus Fauser (Darmstadt: Wissenschaftliche Buchgesellschaft, 2008).
**12** Lessing, *Nathan*, 159. "Ich deines Bluts! – So waren jene Träume, womit man meine Kindheit wiegte, doch – doch mehr als Träume!" All translations from *Nathan der Weise* are mine.
**13** Lessing, *Nathan*, 57.

the sole true god and force that idea onto the entire world. Nathan enthusiastically avows to the templar:

> We must, we must be friends. Despise my people so much you like. We both did not get to choose a people for ourselves. Are we our peoples? What's a people then? Are Christians and Jews more Christian and Jewish than they are human beings? Ah! if I found in you one more, for whom it is enough to be a human![14]

Christians and Jews are not merely religious identities because the word *Volk* is used in the German original, which in the eighteenth century meant ethnic and cultural communities.[15] Nathan's comment insists on the fundamental importance of universal humanity over all other categories that separate people. When accepting the Lessing Prize in 1959, the Jewish thinker Hannah Arendt praised Lessing as an exemplar of humanity in dark times because his works are not about the identity of being Jewish or German but about our shared humanity.[16] Lessing's vision is one of identity freedom, free from religious, ethnic, racial, or gender constraints. It provides a space for the choice of belonging and imagination.

In Lessing scholarship, however, this identity freedom so important to my analysis does not take center stage. Critics mostly agree that the idea of religious tolerance (*Toleranz*) is the key message in Lessing's drama.[17] The ring parable, which Nathan uses to deconstruct Saladin's question about the sole true religion, has been considered the essence of the drama. While Monika Fick, in her comprehensive book summarizing Lessing scholarship from 1945 to date, mentions that tolerance doesn't mean an arrogant lenience (*herablassende Duldung*) but rather an acceptance of non-Christian religions as equal and comparable, the interpretive perspective that critics use after 1945 still reveals a Christian dominance. The word *Toleranz*, along with the verb *tolerieren* and the adjective *tolerant*,

---

**14** Lessing, *Nathan*, 58–59. "Wir müssen, müssen Freunde sein! – Verachtet / Mein Volk so sehr Ihr wollt. Wir haben beide / Uns unser Volk nicht auserlesen. Sind / Wir unser Volk? Was heißt denn Volk? / Sind Christ und Jude eher Christ und Jude, / Als Mensch? Ah! wenn ich einen mehr in Euch / Gefunden hätte, dem es genügt, ein Mensch / Zu heißen!".
**15** See the definition of *Volk* in the eighteenth-century dictionary *Adelung – Grammatisch-kritisches Wörterbuch der hochdeutschen Mundart*: https://lexika.digitale-sammlungen.de/adelung/lemma/bsb00009134_5_1_1401.
**16** Hannah Arendt, *Von der Menschlichkeit in finsteren Zeiten* (Munich: R. Piper & Co Verlag, 1960).
**17** See Monika Fick, *Lessing-Handbuch: Leben – Werk – Wirkung* (Stuttgart: J. B. Metzler Verlag, 2016), 449.

lexically means to accept a different view, but it does not exclude that there is a dominating "normality."[18] In German legal language, it is usually formulated that illegal immigrants or drug addicts could be tolerated (*toleriert*) to a certain extent. *Toleranz* reflects a passive attitude of allowing others to exist on the margin and still betrays the hierarchy between the majority of Christianity and the minority of Judaism in German society.

In Fick's analysis of Lessing's drama, the word "tolerance" concerns a privileged Christian position toward the Jews.[19] It is not understood as a possible position that all religions could take toward other religions. Fick mentions that several scholars consider a historical anecdote important for Lessing's drama, in which the King of Aragon, Don Pedro I, should be persuaded to "tolerate" the Jewish minority with the help of a ring story. The hierarchy, in which the tolerance comes from the Christians to the Jews, is made clear in this story. Remarkably, the word *Toleranz* is not used in Fick's discussions of Islam and natural religions. Even if it is not meant to be a *herablassende Duldung*, as Fick states, the tendency reveals that Lessing scholarship has, in fact, taken a normative Christian position (toward the Jews). Moreover, the word tolerance is not used in Lessing's drama; it is rather imposed on Lessing's drama by postwar critics. Nathan also does not use the ring parable to ask Saladin for permission to live, as in the story of the King of Aragon. It is actually Saladin's financial crisis that coerces him to ask Nathan a difficult or unanswerable question so that he has an excuse to get the rich man's money. Indeed, Saladin's question resembles the Gretchen question (*Gretchensfrage*) about Faust's position on religion in Johann Wolfgang Goethe's *Faust I* (1808): "Now tell me, what do you think about religion? You are a hearty good man, but I think, you don't take it seriously."[20] A cultural icon in the German tradition, the Gretchen question is both inconvenient and critical in that it contains a prevalent anxiety about the loss of Christianity's social dominance and predicts the irreversible trend of secularization at the dawn of global modernity. Fick rightly

---

**18** *Tolerant* is thus defined: (in *Fragen der religiösen, politischen o. a. Überzeugung, der Lebensführung anderer*) "bereit, eine andere Anschauung, Einstellung, andere Sitten, Gewohnheiten u. a. gelten zu lassen; Toleranz means: 1. das Tolerantsein; Duldsamkeit; 2. zulässige Differenz zwischen der angestrebten Norm und den tatsächlichen Maßen, Größen, Mengen a. Ä." See Duden at https://www.duden.de/rechtschreibung/tolerant#Bedeutung-1.
**19** Fick, *Lessing-Handbuch*, 448.
**20** Johann Wolfgang Goethe, "Faust I und II; die Wahlverwandtschaften," in *Goethe Werke*, eds. Albrecht Schöne and Waltraud Wiethölter (Frankfurt am Main: Insel Verlag, 1998), vol. 3, 122. "Nun sag', wie hast du's mit der Religion? Du bist ein herzlich guter Mann, allein ich glaub', du hält'st nicht viel davon."

points out that the question about the truth is epistemologically unsolved in the ring parable.[21] Indeed, the ring parable, which Lessing adapted from Boccaccio's *Decamerone* (1470), effectively deconstructs Saladin's question about the true religion and reveals it as invalid. The drama is thus not about the true religion or tolerance. It is about universal humanity and the identity freedom of choosing one's faith or even atheism, as the Gretchen question audaciously suggests.

While Lessing's drama is now celebrated as a piece of canonical literature, he did not write it in a supportive environment. Rather the drama was a product of desperate adversity in Lessing's life. In 1777, Lessing published fragments out of the deceased Hamburg professor Hermann Samuel Reimarus's treatise *Apologie oder Schutzschrift für die vernünftigen Verehrer Gottes* (*Apologia or Defence of the Rational Worshippers of God*, written between 1735 and 1768), questioning the authority of the Lutheran Church and the Christian faith and promoting pantheistic natural religion and universal reason. The publication enraged orthodox theologians, in particular the Hamburg pastor Johann Melchior Goez. Lessing was attacked and made responsible for Reimarus's anti-Christian position. The scandalous theological conflict, known as the *Fragmentenstreit*, ended in 1778 when Duke Karl of Brunswig rescinded Lessing's exemption from censorship, forbidding him to publish nearly anything on religion. Lessing was officially ostracized from the public discussion and sank into financial difficulty. *Nathan the Wise* emerged both as a literary manifestation of Lessing's ideas and a means to lessen his financial burden.

Interestingly enough, Lessing never publicly announced his own religious position, despite repeated demands by his adversary Goeze. Lessing's biographer H. B. Nisbet comments: "Not only Goeze, but even the Berlin rationalists were unsure where Lessing stood, and Nicolai declared: 'The theologians think you are a freethinker, and the freethinkers think you have become a theologian.' In more recent times, different commentators have presented him as a Christian fideist, a secularist or Spinozist bent on undermining Christianity from within, as simply inconsistent, or as caught in a cognitive crisis between rational and empirical paradigms."[22] Alas, Lessing's death at the age of fifty-two in 1781, less than two years after the publication of *Nathan*, didn't give him enough time to declare his own position in a time of severe censorship and

---

**21** Fick, *Lessing-Handbuch*, 449.
**22** H. B. Nisbet, *Gotthold Ephraim Lessing: His Life, Works, and Thought* (Oxford: Oxford University Press, 2013), 565–566.

hostility. Yet, I venture to argue, that Lessing was indeed a Spinozist. The historian Jonathan Israel argues that the Dutch thinker Baruch de Spinoza's pantheism and universalism are essential for some eighteenth-century thinkers to formulate their secular and democratic ideas of equality and freedom, which have hugely influenced later sociopolitical and cultural movements in the West.[23] Lessing discussed Spinoza already in 1755 with the Jewish Enlightenment thinker Moses Mendelssohn. After Lessing's death, Friedrich Heinrich Jacobi published in 1785 a controversial book *Über die Lehre des Spinoza in Briefen an den Herrn Moses Mendelssohn* (*On the Doctrine of Spinoza in Letters to Herr Moses Mendelssohn*) and claimed it to be in Lessing's papers. Jacobi also unwaveringly insists on Lessing being a Spinozist, which he more or less equals to an atheist and blasphemer. Even though Lessing's contemporaries and today's critics question the authenticity of Jacobi's claim and the authorship of the work, there are traces in *Nathan the Wise* that fit the paradigm of a secular universalist discourse of freedom and equality. Lessing may not be a blasphemer but he could well be a Spinozist pantheist or a radical Enlightenment thinker in Israel's interpretation.[24]

The intellectual historical background supports my argument about identity freedom in Lessing's drama. Saladin, the templar, and Nathan are three equally openminded persons who already carry the idea of identity freedom in them. Saladin and the templar open up themselves when they encounter Nathan. The final scene, in which all the family relations are revealed and confirmed through the Persian document of Saladin's brother, Nathan's friend, the father of the templar and Recha, is probably even more meaningful than the ring parable. The reliance on the document discloses the preference of rationality and scientific evidence over religious revelation. Saladin's invisible brother connects all three religions and has both a German name Wolf von Filneck and a Muslim name Assad. He symbolizes the countless invisible connections among humans since ancient times. This person functions as a DNA carrier and transmitter that unnoticeably ties humans deep within and subtly whispers that they should not hate and fight against one another.

Despite *Nathan*'s market success with more than three thousand sold copies, the drama was not publicly appreciated by the authorities and prominent

---

**23** See Jonathan I. Israel, *Radical Enlightenment: Philosophy and the Making of Modernity 1650–1750* (Oxford: Oxford University Press, 2001); *Enlightenment Contested: Philosophy, Modernity, and the Emancipation of Man, 1670–1752* (Oxford and New York: Oxford University Press, 2006); *Democratic Enlightenment: Philosophy, Revolution, and Human Rights 1750–1790* (Oxford: Oxford University Press, 2011).
**24** See more discussion in Nisbet, *Gotthold Ephraim Lessing*, 625–641.

intellectuals of the time. In private, while Lessing's drama was met with enthusiasm among some of his contemporaries and especially in the German Jewish community, followers of orthodox theology and some of Lessing's friends fiercely rejected it. During the Nazi period, *Nathan* was officially banned on stage and in school textbooks. Yet after 1945, it has become one of the most publicly acclaimed literary works in Germany.[25] Scenes in the drama such as that of Nathan's indictment of the murder of his wife and sons or that of the patriarch's vow to burn the Jew echo the reality of the twentieth century. When the *Deutsches Theater* performed *Nathan* for its first reopening after the war on 7 September 1945, "numbers of people broke down and had to leave the auditorium on hearing Nathan's narrative of how his entire family was massacred."[26] Nisbet mentions that *Nathan* has gained new significance after 11 September 2001, and was performed twenty-four times in Germany shortly thereafter; its English-language productions powerfully impacted the post-9/11 audience in Washington D.C. in 2011 and in New York in 2012. Of course, the latest performance of *Nathan* is not merely about the German-Jewish relationship but also about the West's relationship with Islam, an aspect that was downplayed in earlier reception of the drama.

The topicality of *Nathan* in the twenty-first century resonates with the importance of identity freedom today. It is not necessarily the integration of all religions in one family that serves as a utopian social model. Rather it is the Enlightenment legacy of practicing the freedom to choose and determine one's identity that still keeps *Nathan* remarkable throughout centuries. The German-Jewish writer Max Czollek's book *Disintegrate Yourselves!* continues in the spirit of Lessing's sense of identity freedom. Even though Czollek calls for Jews and Muslims to disintegrate from mainstream German culture, something seemingly contrary to *Nathan*'s message of integration, the intellectual incentive behind the strategies of disintegration also draws its energy from identity freedom. In particular, the message of not playing the roles that are ascribed to Jews and Muslims by German politics and society is prominent in the book. Czollek promotes the freedom to determine one's own role in the society through personal choice.

**25** More on the reception of Nathan see Barbara Fischer, *Nathans Ende? Von Lessing bis Tabori. Zur deutsch-jüdischen Rezeption von ›Nathan der Weise‹* (Göttingen: Wallstein Verlag, 2000).

**26** Nisbet, *Gotthold Ephraim Lessing*, 621. Atina Grossmann also discusses this performance described in Marcel Reich-Ranicki's autobiography. Yet the performance is remembered quite differently from that in Nisbet's book. See Atina Grossmann, *Jews, Germans, and Allies: Close Encounters in Occupied Germany* (Princeton: Princeton University Press, 2009), 23.

# The Strategy of Disintegration

Czollek claims that he wrote this highly critical and angry interrogation of German identity politics after 1945 from the perspective of a lyric poet, a Berliner, and a Jew (*Lyriker, Berliner und Jude*).[27] Czollek intends to articulate his anger about the wounds inflicted on him and his family with a work of art.[28] The target of Czollek's polemic is the German immigration politics of integration. This reminds me that, during my student time in Germany, the chancellor candidate Edmund Stoiber emphatically claimed in 2002 in a TV presentation that Islam does not fit in the occidental Christian tradition of Germany. Debates about "integration" and lamentations about its difficulties among Muslims in Germany permeated media reports and private conversations then. Czollek's book, obviously, writes against integration policy. The imperative of the verb *desintegriert* manifests a resistance that attracts attention and promises controversy.

Officially, according to the Federal Office for Migration and Refugees (*Bundesamt für Migration und Flüchtling*),

> Integration is a long-lasting process. Its goal is to include in society all people who legally live in Germany as long-term residents. Immigrants should be able to have a comprehensive and equal share in all areas of society. They have the duty to learn German and to know, to respect, and to follow the constitution and laws.[29]

Czollek points out that this integration policy, however, does not expressly state that it acknowledges and accepts the diversity that immigrants and refugees contribute to the German society. On the contrary, the policy aims to coerce the new arrivals to accept the dominance and homogeneity of German culture and to play certain roles to confirm and conform to an imagined German identity. Integration doesn't address everyone in the German society; rather it only targets certain population groups, alternately labeled as "Eastern European Jews, asylum seekers, Turks, migrants, Muslims, economic refugees, refugees, Nafris."[30] The policy insinuates that the standard for being a good citizen is to become as German as possible. This demand not

---

**27** Max Czollek, *Desintegriert euch!* (Munich: btb Verlag, 2020), 11.
**28** Czollek, *Desintegriert euch!*, 174–175.
**29** Last accessed 3 September 2020, https://www.bamf.de/DE/Service/ServiceCenter/Glossar/_functions/glossar.html?nn=282918&cms_lv2=282958. "Integration ist ein langfristiger Prozess. Sein Ziel ist es, alle Menschen, die dauerhaft und rechtmäßig in Deutschland leben, in die Gesellschaft einzubeziehen. Zugewanderten soll eine umfassende und gleichberechtigte Teilhabe in allen gesellschaftlichen Bereichen ermöglicht werden. Sie stehen dafür in der Pflicht, Deutsch zu lernen sowie die Verfassung und die Gesetze zu kennen, zu respektieren und zu befolgen."
**30** Czollek, *Desintegriert euch!*, 64. "Ostjuden, Asylanten, Türken, Migranten, Muslime, Wirtschaftsflüchtlinge, Flüchtlinge, Nafris [. . .]".

only disrespects the backgrounds of immigrants but also ignores the internal diversity in German society and history. Jews or Muslims who were born and lived in Germany for centuries and decades are not properly recognized by this homogenizing and narrow policy, either. Czollek uses his own experience to indignantly question the role of victim that the Jews have to play in a memory culture of the Holocaust choreographed by the Germans. Czollek points out that Jews still are not accepted as an integral part of the German society after a centuries-long history of integration. They are labeled and treated as Jewish in order to play the role of victims to confirm the invention of a German identity that is imagined to be monocultural, traditional, Christian, and exclusionary. Czollek considers it high time to break with the attempt to integrate by using a strategy of disintegration. He understands disintegration as a Jewish contribution to the project of post-migrant society. Disintegration aims to introduce radical diversity into German society through aesthetic and political means.[31] Czollek's strategy of disintegration also mocks the Nazi exclusion and persecution of Jews. Now without being passively ostracized by the political authorities, Jews and Muslims should actively step out of the policy of integration and voluntarily distance themselves from German expectations.

Czollek uses the notion of *Gedächtnistheater* (the theater of memory), a concept developed in Y. Michal Bodemann's book *Memory Theater: the Jewish Community and Its German Invention* (*Gedächtnistheater. Die jüdische Gemeinschaft und ihre deutsche Erfindung* 1996), to critique German memory culture of the Holocaust and World War II as an insincere performance.[32] The German word *Theater* has an ironic connotation that something is insincere, untrue, fake, and deceiving, a quality that prominent Enlightenment thinkers such as Jean-Jacques Rousseau ascribe to the theatrical stage in the eighteenth century.[33] A theater is a place where fiction is performed. It is thus separate from the reality and is deceptive. Lessing only decided to write *Nathan* after being banned from publishing non-fictional essays on religion. The term "memory theater" implies that German memory culture is staged and insincere.

Czollek contends that the German authorities have staged a new normality after World War II with the aid of memory theater. He considers it inappropriate to call the end of the war a "liberation" of Germany from the Nazis, which Bundespräsident Richard von Weizäcker once announced. It is equally unfitting to count the resistance against fascism, racism, and antisemitism as part of being

---

31 Czollek, *Desintegriert euch!*, 133.
32 Y. Michal Bodemann, *Gedächtnistheater. Die jüdische Gemeinschaft und ihre deutsche Erfindung* (Hamburg: Rotbuch, 1996).
33 See Rousseau's Letter to M. D'Alembert on Spectacles (1758).

German (*Deutsch-Sein*), as Frank-Walter Steinmeier proclaims.[34] These claims create the image of a good German who should have suffered under the Nazis and have had little to do with the present rise of the rightwing populist political party AfD (*Alternative für Deutschland*, Alternative for Germany). Czollek argues, however, the Germans were not liberated (*befreit*) but defeated (*besiegt*) on 8 May 1945.[35] The majority of the German population relentlessly supported the Nazi regime, which was a mass movement (*Volksbewegung*). Thus, the image of a good German hardly existed in reality. This twisted staging of memory renders the Germans as victims of the war instead of its perpetrators.[36] This staged positive self-image posits a liberated and purified German identity (*befreite und geläuterte Deutsche*), which heavily depends on the Jews as the victim in the theater. The Jews now function as saviors for an imagined German identity that is liberated from the Nazis. Either before or after 1945, the Jews are not accepted as Germans and by the Germans. Even after 1945, they are still kept outside of the German identity and have to play the victims of the Nazis. The purpose of the memory theater, according to Czollek, lies in enthralling people to believe that the past belongs to the past, and a new normality has arrived. Yet the success of AfD being elected to the federal parliament, serves as evidence for Czollek that the fascist ideas of antisemitism, racism, exclusion, discrimination, exploitation, and extermination have not vanished. It is thus high time to wake up from the memory theater and reform the immigration politics to introduce more diversity into social consciousness.[37]

Czollek observes that there are three repeated themes reserved for the Jews as victims: antisemitism, the Shoah, and Israel.[38] As soon as Jewish citizens respond to these three areas, they find themselves on the stage of the German memory theater and have become the "Jews for the Germans." German authorities have developed strategies to patronize the Jews for their own needs and expectations while ignoring the complexity of Jewish experiences after the War. Czollek criticizes the German writer Magnus Enzensberger for portraying the Jewish poet Nelly Sachs not primarily as a writer but as a Jewish victim because Enzensberger highlights that Sachs's writings show neither hatred nor revenge, a German wish for the Jews after the war. Czollek argues that the Germans expect the Jews to commemorate the Holocaust as heavy-hearted victims. He sarcastically calls this

---

**34** Czollek, *Desintegriert euch!*, 20, 107.
**35** See Czollek, *Desintegriert euch!*, 20.
**36** Czollek, *Desintegriert euch!*, 22.
**37** Czollek, *Desintegriert euch!*, 13.
**38** Czollek, *Desintegriert euch!*, 27.

the Jewish service (*Dienstleistung*) for the salvation (*Erlösung*) of the Germans.[39] It is *Holocaustkitsch*.[40] Jews currently living in Germany have acquiesced to play the role of the victims, though not without reluctance. Czollek ironically comments: "Let's talk about antisemitism, brothers and sisters, let's become Jews."[41] This type of expectations of the Germans from other ethnic and religious groups is not only confined to the Jews. When I once discussed with a German professor my idea of writing a master's thesis in comparative literature on a French ethnologist and writer, I was immediately told that I should write about a Chinese topic because I am from China. I also heard that non-German students are constantly asked to write about their "home country" or their "native culture." They feel categorized and narrowly defined by their German professors.

Czollek proposes that Jews change this compliant acceptance of the German perspective and develop a more independent identity. He argues that the effort to strive for a German-Jewish symbiosis had already failed before the end of World War II when six million Jews were murdered. It is time to break free from the illusion of a successful German-Jewish integration in the German society.[42] Czollek explains that disintegration does not mean to avoid or forbid discussion about antisemitism, Shoah, and Israel; rather it means to cultivate a different type of reflection on these themes. Czollek provocatively challenges the heavy-hearted solemnity associated with Auschwitz and the Holocaust, which he considers imposed by the Germans. He suggests that one could also laugh and joke about Auschwitz. Heavy and gray Holocaust memorials serve German political needs to create a unified identity for the Germans more than the recognition of Jewish participation and the inherent diversity within German culture. Despite the establishment of such monuments, the social reality is quite different. Sociological findings in *Grandpa was not a Nazi* (*Opa war kein Nazi*, 2002) and *MEMO Germany* (*MEMO Deutschland*, 2018) reveal that third-generation Germans after the war have come to glorify their Nazi grandfathers as heroes and victims of the war; they have even begun to cultivate a memory of victimhood in the family. Two-third of the Germans participating in the survey deny that there were any Nazi perpetrators in their families.[43] Czollek considers this result a

---

**39** See Czollek, *Desintegriert euch!*, 79–80.
**40** Czollek, *Desintegriert euch!*, 84.
**41** Czollek, *Desintegriert euch!*, 87. "Reden wir über Antisemitismus, Brüder und Schwestern, lasst uns Juden werden!".
**42** Czollek, *Desintegriert euch!*, 91.
**43** Czollek, *Desintegriert euch!*, 96–97. Omer Bartov also argues that the German society has failed to recognize Jews as a constitutive element of German culture. In literary, cinematic, and scholarly representations, a significant absence of the Jews is symptomatic for non-Jewish

sheer impossibility because the majority of the German society was involved in the Nazi movement. The social change to deviate from the Nazi past is not a sudden development but a gradual transformation of the German identity that has been nourished by the memory theater.[44]

Czollek considers it indispensable to change the German style of "heavy and gray remembering" and counter it with humor and satire. "There are of course alternatives for dealing with the Shoah different from the current trend. Everybody, who has once heard an Israeli joke about the Holocaust or watched an American Jewish TV series, knows that."[45] Projects such as *yolocaust.de* by the Jewish satirist Shahak Shapira or *Dancing Auschwitz* by the Australian artist Jane Korman are good examples.[46] A joke causes confusion and irritation through its peculiar uncommonness that breaks with the mainstream style of representation. Czollek also considers hiphop an effective artistic expression of xenophobic experience in Germany and a powerful political resistance against such hostilities. Czollek argues that humor demonstrates the triumph of the survivors over their persecutors. He also calls out to people who are not directly involved in the German-Jewish constellation to break the taboo and secularize the holy

---

Germans who portray themselves as victims. Despite high-volume media attention to the Holocaust and the Third Reich, the representation of the Jews still remains insubstantial. See Omer Bartov, "'Seit die Juden weg sind . . .': Germany, History, and Representations of Absence," in *A User's Guide to German Cultural Studies*, eds. Scott D. Denham, Irene Kacandes, and Jonathan Petropoulos (Ann Arbor: University of Michigan Press, 1997). Atina Grossmann also observes that postwar Germans consider themselves victims of the war primarily because of the mass rape of German women. The Judeocide was rarely discussed in the public sphere after 1945. Grossmann, *Jews, Germans, and Allies*, 7 and 276.

**44** Czollek, *Desintegriert euch!*, 98.

**45** Czollek, *Desintegriert euch!*, 100. "Natürlich gibt es Alternativen zum hiesigen Umgang mit der Shoah, das wissen alle, die schon mal einen israelischen Holocaust-Witz gehört oder eine US-amerikanische jüdische Serie gesehen haben."

**46** Czollek's point here reminds me of a scene in the movie *Er ist wieder da* (2015). Hitler wakes up in 2014 and becomes a very popular TV star in comedy shows by propagating his fascist ideas. Yet before his first show, the TV station director warns him that the question of the Jews is not funny. Even though the film shows that the audience laughs at Hitler's ideas, which should indicate their absurdity, the huge popularity of Hitler and his ethnically homogenizing statements contains an undercurrent of the acceptance or even revival of the Nazi ideology in the form of a comedy. The name of the film maker who introduces Hitler to the media and makes him a star would not be accidentally Sawatzki, which contains all letters to form the word swastika. Hitler's claim, toward the end of the movie, that he can't be killed because he is in every German negatively testifies to Czollek's point about the undercurrent of fascist ideas in the German society. Major German newspapers and weeklies rightly warned in their reviews that the film has a dubious message.

grail of German memory of the Holocaust, because Auschwitz means different things to different people. Czollek tries to introduce more diversity and multivalence into the narrowly guarded, heavy, and gray German-Jewish memory theater. Indeed, knowledge and knowledge production could be neutral if we treat them to be so. German language and history do not belong to the ethnic Germans but could be treated as a set of knowledge to be studied by Germans and non-Germans alike. Speaking again from my own experience, after I had to write my master's thesis on a China-related topic, I decided to do something that has nothing to do with China for my doctoral thesis. I chose to disintegrate in Czollek's terms from Western expectations for me to talk about China because I am from China.

If the Germans consider the Jews a homogeneous group, then they err. In Germany today, Czollek counts, ninety percent of the Jewish population in today's Germany are immigrants from the former Soviet Union. The second largest group of Jewish immigrants came from Israel to Berlin in the twenty-first century. In addition to West German Jews, a smaller group of Jews comes from the former GDR. Czollek argues that such an internal diversity in the Jewish community could serve as the incentive for disintegration.[47] "Jewish men and women come from different places, have different sexual preferences, religious or political positions and many more differences. Those who take these differences seriously are a step further in the direction of a new German, European, and international Judaism."[48] Most Jews in today's Germany are people with immigration background (*Migrationshintergrund*). Czollek contends that it is possible to create Jewish art with no relation to the Judeocide because the Jewish experience is richer than the three recurring themes of the Holocaust, Shoah, and Israel, as needed by the Germans.[49]

Czollek makes clear that he respects the efforts by those Germans who are aware of the historical violence and endeavor to make up for it in the present through a meaningful memory culture. He thus condemns even more the right-wing and *völkisch* effort to disavow the Holocaust and reestablish a German identity based on pre-1945 examples. Both groups, however, grapple with the same question about German identity after 1945 and stage together the memory theater. Czollek rejects any empathy with the AfD that interprets their racist policies as political frustration and patriotism.[50] This rhetoric of tenderness

---

**47** Czollek, *Desintegriert euch!*, 150.
**48** Czollek, *Desintegriert euch!*, 152. "Juden und Jüdinnen kommen von unterschiedlichen Orten, haben unterschiedliche sexuelle Präferenzen, religiöse oder politische Haltungen und vieles mehr. Wer diese Unterschiede ernst nimmt [. . .] der ist einen Schritt weiter in Richtung eines neuen deutschen, europäischen, internationalen Judentums."
**49** Czollek, *Desintegriert euch!*, 150.
**50** Czollek, *Desintegriert euch!*, 117.

(*Rhetorik der Zärtlichkeit*) toward the AfD is accompanied with a rhetoric of hardness (*Rhetorik der Härte*) toward immigrants, Muslims, and refugees. Czollek points to the fact that, in 2017, there were on average four personal attacks on refugees every day and all together 251 attacks on refugee residences. Czollek is stunned that the AfD claims that they are friends of Jews and points out that they use the conservative political slogan of a Jewish-Christian (*jüdisch-christliche*) tradition to exclude the Muslims.[51] The resemblance between antisemitism and the current Islamophobia, however, is obvious to him: "Next time, maybe the mosques burn first, and then the synagogues. I don't have any illusions."[52]

Czollek argues: "Every human being consists of many parts, which also always move and change. The unbendable identity is a dangerous illusion."[53] It is indispensable to move away from the imagined unified German identity and recognize the inherent diversity in Germany. Coming back to my story, I consider this diversity in German and European cultural tradition fascinating. I found in eighteenth-century literature and philosophy a transculturality that has not been highlighted in the scholarship framed within the paradigm of national literature. After studying the German eighteenth century and European colonialism, I realized that I do not need to intentionally avoid "Chinese" topics as a negative or disintegrating reaction to the German expectation. Global entanglements and their inherent multivalence in the past centuries offer an intellectually stimulating field. China and the East Asian cultural sphere played and still play a vital role in the global context, and I shouldn't avoid that but rather need to integrate it more into my research agenda. As for the German expectation that Chinese should and can only deal with Chinese-related topics, I can just push it aside and not think about it any longer. I now feel that I have found more internal freedom in this respect.

# Concluding Remarks

Even though Lessing seemingly preaches integration and Czollek argues explicitly for disintegration, they both are invested in the freedom to choose one's

---

**51** Czollek, *Desintegriert euch!*, 77.
**52** Czollek, *Desintegriert euch!*, 191. "Beim nächsten Mal brennen vielleicht zuerst die Moscheen. Aber dann brennen auch wieder die Synagogen. Ich mache mir da keine Illusionen."
**53** Czollek, *Desintegriert euch!*, 192. "Jeder Mensch besteht aus vielen Teilen, die sich immer wieder verschieben. Die ungebrochene Identität ist eine gefährliche Illusion."

identity and to defend universal human dignity. Lessing's drama and Czollek's book both emerged out of a similar polemic. Lessing was, as Czollek is, combatting dominant Christian or German discrimination against other faiths and cultures, in particular Judaism and Islam. While Lessing was put under censorship, Czollek's book has been reviewed in major German newspapers and celebrated internationally.[54] The time of censorship and massacre has passed, but some undercurrents and obvious trends remind us of the gruesome history of discrimination.

Saidiya Hartman proposes that "a history of the present strives to illuminate the intimacy of our experience with the lives of the dead, to write our now as it is interrupted by this past, and to imagine a *free state*, not as the time before captivity or slavery, but rather as the anticipated future of this writing."[55] My writing on Enlightenment identity freedom and the contemporary critique of immigration politics endeavors to show the inherent diversity in history, in the present, in societies, and in our biological genes. I have tried to imagine greater liberty in choosing who we are. Yet I am also aware that time does not seem to have much influence on the inequity and injustice of the present. The repeated interruptions of the past, such as the civil rights movement in the United States, the decolonialization movements worldwide, the anti-imperialist and antifascist campaigns, and the scholarship that deconstructs nationalism and monolithic identity politics in the past half century, seem to have had little effect on the present. The moment of history in which I am writing is rather discouraging. I choose to grapple with the notion of identity which now divides people through violence instead of uniting and enriching humanity – an issue that has not lost any topicality since Lessing's time. Still, with hope, I pledge for the greatest freedom for all in choosing their belonging, taste, style, language, spirituality, and way of life. Our genes are mixed up after all.

---

**54** Most of the book reviews in major newspapers and weeklies including *Die Zeit*, *Frankfurter Allgemeine*, and *Spiegel* are non-emotional and matter-of-fact-ish. They provide a good summary of the book without further comments. Only one review in *Süddeutsche Zeitung* is critical of Czollek's book and comments that it expresses anger but does not contain effective strategies. While the book receives no strong objections, it has no strong supporters, either. Czollek was interviewed about his book in *New York Times* in early 2020. See https://www.nytimes.com/2020/01/16/books/max-czollek-germany-desintegriert-euch.html.
**55** Saidiya Hartman, "Venus in Two Acts," *Small Axe* 12, no. 2 (June 2008): 1–14, here 4.

# Works Cited

Arendt, Hannah. *Von der Menschlichkeit in finsteren Zeiten*. Munich: R. Piper & Co Verlag, 1960.

Bartov, Omer. "'Seit die Juden weg sind . . . ': Germany, History, and Representations of Absence." In *A User's Guide to German Cultural Studies*. Edited by Scott D. Denham, Irene Kacandes, and Jonathan Petropoulos, 209–226. Ann Arbor: University of Michigan Press, 1997.

Czollek, Max. *Desintegriert euch!* Munich: btb Verlag, 2020.

Fick, Monika. *Lessing-Handbuch: Leben – Werk – Wirkung*. Stuttgart: J. B. Metzler Verlag, 2016.

Fischer, Barbara. *Nathans Ende? Von Lessing bis Tabori. Zur deutsch-jüdischen Rezeption von ›Nathan der Weise‹*. Göttingen: Wallstein Verlag, 2000.

Goethe, Johann Wolfgang. "Faust I und II; die Wahlverwandtschaften." In *Goethe Werke*. Edited by Albrecht Schöne and Waltraud Wiethölter, vol. 3. Frankfurt am Main: Insel Verlag, 1998.

Grossmann, Atina. *Jews, Germans, and Allies: Close Encounters in Occupied Germany*. Princeton: Princeton University Press, 2009.

Hartman, Saidiya. "Venus in Two Acts." *Small Axe* 12, no. 2 (June 2008): 1–14.

Hollinger, David. "The Enlightenment and the Genealogy of Cultural Conflict in the United States." In *What's Left of Enlightenment? A Postmodern Question*. Edited by Keith Michael Baker and Peter Hanns Reill, 7–18. Stanford, CA: Stanford University Press, 2001.

Israel, Jonathan I. *Democratic Enlightenment: Philosophy, Revolution, and Human Rights 1750–1790*. Oxford: Oxford University Press, 2011.

Israel, Jonathan I. *Enlightenment Contested: Philosophy, Modernity, and the Emancipation of Man, 1670–1752*. Oxford and New York: Oxford University Press, 2006.

Israel, Jonathan I. *Radical Enlightenment: Philosophy and the Making of Modernity 1650–1750*. Oxford: Oxford University Press, 2001.

Lessing, Gotthold Ephraim. *Nathan der Weise*. Frankfurt am Main: Suhrkamp, 2003.

Nisbet, H. B. *Gotthold Ephraim Lessing: His Life, Works, and Thought*. Oxford: Oxford University Press, 2013.

Strohschneider-Kohrs, Ingrid. "›Nathan‹ – poetische Chiffre der religio-Erfahrung." In *Gotthold Ephraim Lessing. Neue Wege der Forschung*. Edited by Markus Fauser, 161–181. Darmstadt: Wissenschaftliche Buchgesellschaft, 2008.

Sykes, Bryan. *DNA USA: A Genetic Portrait of America*. New York and London: W. W. Norton and Company, 2012.

Christina Matzen
# Prisoner Experiences in Times of Crisis

Before the summer of 2020, people who inquired about my plans once I graduate from the University of Toronto with a PhD in History learned of my desire to become a professor of German history and also engage with United States prisoner-advocacy and re-entry programs. I knew that such a combination might raise eyebrows. As a scholar working on the history of women's prisons in Nazi and postwar Germany, and as a person committed to addressing contemporary issues of mass incarceration in my country, I was aware that most academic departments in the humanities in North America did not particularly encourage such community activism. The events of 2020, in particular the intersection of COVID-19 and criminal-justice reform in the wake of Black Lives Matter protests, reinforced my conviction that my professional career can and should reflect both my historical interests and my will to act in urgent human-rights matters.

Lessons of German history can teach us much about current events, and my scholarly research is well suited for such application because prisoners across time and space share many of the same experiences. I also aim to use my platform as a historian and teacher to give a voice to women prisoners who deserve to be heard and to be treated with dignity. Prison authorities go to great lengths to silence prisoners and keep their inhumane approaches to justice hidden. With the proliferation of smartphones and social media, the United States public has become increasingly aware of the rampant systemic brutality perpetrated against criminalized people who are "free" and imprisoned, and the United States mainstream is more willing than ever to discuss reforming a criminal justice system that many people, especially white people, have ignored comfortably.

The genealogy of my scholarly and activist motivations is directly linked to my family history, which includes proponents of Nazism as well as of the Confederacy. A subject about which I used to keep private, I now see my family history as underscoring the necessity to interrogate the conditions under which state and personal violence proliferate, and my focus has shifted to my immediate surroundings in the United States. In other words, my interest in studying women's prisons in twentieth-century Germany is now tied symbolically with a profound need to confront racism and mass incarceration in the society to which I belong and also re-examine my own experiences and subject position.

My earliest encounter with race and imprisonment in the United States happened when I was in the sixth grade. My first boyfriend invited me to his house one day after school. We walked into his backyard, and he rolled out a grill. "I want to cook you a steak," he said. "That's how my dad showed my mom he

https://doi.org/10.1515/9783110753295-011

loved her." While we were eating, he explained to me that he was nervous to tell me that his dad was in prison. I felt sad that he was separated from his dad, and I was confused as to why he thought that I might reject him once I learned more about his family. The stigma that he knew so well, that he had internalized, was lost on a young, naïve, and privileged white girl. Then it clicked that it was not a coincidence that his dad – a Black man – was the only person I knew who was serving a prison sentence. At that moment, I started to realize that being Black meant having a radically difference experience of the United States, and that was heartbreaking.

I have since learned that for Black people, Hispanic people, Muslims, Native Americans, sex workers, individuals with disabilities, the LGBTQI+ community, and other marginalized groups in the United States, any contact with the criminal justice system can lead to a lifetime of compounding legal troubles. Moreover, incarceration often perpetuates cycles of generational imprisonment, poverty, and trauma, especially for women and families.[1] For many, life after prison is just as difficult as it was before, if not more so. Many struggle with culture shock, mental health issues, addiction, disenfranchisement, and unemployment. The patterns of injustice plaguing the United States legal system today – racialized hierarchies, lack of accountability, the politicization of bodies, transnational eugenics policies, and violent power structures – were also present in the German prisons on which I focus my research. In my dissertation, I argue that prisons have historically been extremely isolated places where officials have highly variable and often low levels of accountability. I show that prisoners, as a criminalized, vulnerable, and disenfranchised population, were among the first groups in Nazi Germany to experience the ruthlessness of the new regime. Against this backdrop, I analyze federal policies that targeted political, religious, and social outsiders, and I examine how ideological and medical programs affected the lives of women prisoners more specifically. The complex dynamics of staff relations are also crucial to any understanding of how power

---

1 Women currently comprise 7% of the federal prison population and similarly, are a smaller percentage of total inmates in state and local facilities. While there are many fewer female than male inmates in the overall population, over the past thirty years the phenomenon of confining more women to federal, state, and local correctional facilities has exploded at an increase of 700%. Although there has been an increase in women convicted of violent crimes, most incarcerated women are serving sentences for property and drug offenses. African American women are twice as likely to be incarcerated as white women: 96 per 10,0000 v. 49 per 10,0000. See Beryl Ann Cowan, "Incarcerated women: Poverty, trauma, and unmet need," *American Psychological Association*, April 2019, https://www.apa.org/pi/ses/resources/indicator/2019/04/incarcerated-women.

operated within prisons. Despite the repressive conditions to which female inmates were subjected, many found ways to resist the isolating and demoralizing effects of incarceration.

Trying to finish my dissertation while living through the monumental events of 2020 has been a challenging experience. As I write and edit, I think of the pain experienced by so many imprisoned people, past and present. Elsa Albrecht, a woman sentenced in Berlin in July 1944 to six months in prison for "associating with a prisoner of war," elucidates many of the struggles faced by prisoners in Hitler's Germany.[2] Since she had given birth to her daughter in May and was still breastfeeding, the judge gave her until 2 October to continue feeding her child before having to surrender herself to the Barnim Street Women's Prison in Berlin's Friedrichshain district. "During my time in the Barnim Street Prison . . . 86 women were brought to Plötzensee Prison for execution, at least four to six women every Friday," Elsa recalled in a 1995 interview. "I never forgot that number."[3] Also traumatic for Elsa were the Allied bombings and the prison administration's failure to protect most inmates. With limited space in the prison's air raid shelter, only privileged prisoners who worked in the administrative office and the guards could seek protection there; the rest of the women were simply locked in their cells upon the sound of an air raid siren. "The attacks were pretty severe," she said. "The prison was not far from the center, from Alexanderplatz."[4] Christmas was especially difficult. Elsa recalls a group of girls singing in the prison chapel while all the inmates sobbed. They were offered a festive meal of vegetables and meatballs – a major improvement over the usual maggot-infested barley soup they were served – but it offered little reprieve from their grim reality. "It will all pass, it will all pass," Elsa would repeat to herself.[5]

Free civilians in Germany could take shelter upon hearing an air-raid siren, but prisoners were subject to the directives of their overseers, who often forced them to remain in place. Taking shelter, it was thought, would disrupt order and leave a prison susceptible to escape attempts. This is likely why women in the Barnim prison were not allowed to take shelter during an air raid in the spring of 1944. The Army Office in Brandenburg, where the men's prison outside Berlin was located, reprimanded the Barnim administration for such negligence:

---

**2** Claudia von Gélieu, *Barnimstraße 10: Das Berliner Frauengefängnis 1868–1974* (Berlin: Metropol Verlag, 2014), 23.
**3** von Gélieu, *Barnimstraße 10*, 236.
**4** von Gélieu, *Barnimstraße 10*, 236.
**5** von Gélieu, *Barnimstraße 10*, 236.

The work was immediately halted [after the air raid siren]. All civilians, such as soldiers and male prisoners, could enter air-raid shelters and ditches in a timely manner. After about 15 minutes our anti-aircraft defense shot at the incoming enemy planes and only after another 15 minutes did the first enemy bombs fall. The circumstances as to why the female prisoners were not led to a ditch at all and were instead forced to remain in their living quarters are both unknown and inexplicable to me. In my opinion, difficulties due to the nature of the women inmates at the time is out of question . . . . I have arranged for the commandant to be expected to act independently and have presence of mind during sudden cases of panic.[6]

Former prisoner Hannelore Thiel remembered being in a cell on the top floor of the Barnim prison during bombings, and along with her fellow inmates, pounded on the locked doors, threw stools, and shouted that they should be let out. "The women felt buried alive," Thiel recalled.[7]

In May 1944, the Reich Ministry of Justice ordered that the women inmates be taken to shelters because their "screams and noise" disturbed air-raid procedures. The Ministry saw these women as less dangerous than male prisoners and believed women could be trusted to take shelter on the lower-level floors of the prison, provided they wear identification tags so they could be accounted for immediately after enemy attacks.[8] Future East German leader Erich Honecker worked at the Barnim prison in early 1945, and he recollected an Anglo-American attack on Berlin at the end of February 1945:

the whole area from Alexanderplatz to Lichtenberg sank to rubble. We were on duty on the roofs of the Barnimstraße women's prison and they soon looked like miners coming out of the shaft. That afternoon the light of the sun darkened, and the day turned to the night, lit by many fires. It was all hell. Again and again, the roofs of the women's prison burned, and we threw them [the women] down. That was not without risk. In the meantime, there were delay-action bombs, and we never knew when they would explode. During this bombing, a cell wing was hit and destroyed. I took charge of rescuing the women trapped in their cells. We took lamps and shovels from the hospital bunker and recovered the severely injured survivors from the rubble. For 23 of them, help came too late.[9]

I reflected on these experiences as I read about COVID-19 ripping through United States prisons.

---

**6** Memo, Heeres-Nebenzeugamt Brandenburg (Havel) an den Vorstand des Frauengefängnisses, 26.4.1944, Landesarchiv Berlin, A Rep. 365 Nr. 50.

**7** von Gélieu, *Barnimstraße 10*, 203.

**8** von Gélieu, *Barnimstraße 10*, 204.

**9** Erich Honecker, *Aus meinem Leben* (Berlin: Dietz Verlag, 1980), 169. Honecker was incarcerated in the Brandenburg-Görden Prison and due to good behavior was sent to the Barnim prison to repair buildings damaged by bombs.

Since early 2017, I have had a pen pal through the SWOP Behind Bars organization.[10] My pen pal and friend Rosemary, a Mexican American, is incarcerated in an Arizona women's prison, and she gave me permission to share her harrowing experience as she has worked hard to survive the pandemic, physically and psychologically. Like Elsa's fear that a bomb could kill her at any moment, Rosemary is aware that COVID-19 is lurking in her prison but does not know if or when contracting the coronavirus will be her death sentence. Even before scientists knew the virus spread primarily through respiratory droplets, inmates were forbidden from using hand sanitizer. Also, they were also not allowed to wear masks until the end of July 2020. Just as Rosemary was on the verge of using her underwear as a mask, given that inmates were prohibited from purchasing them, all inmates were finally given a mask. Many inmates refused to wear them, and they give Rosemary dirty looks for both wearing a mask and remaining at least six feet away from others.[11]

Rosemary's primary approach to coping with the outbreak has been to keep a distance from all the officers and medical staff who go in and out of the prison. She also tries to calm her mind by avoiding the news, watching soap operas, and listening to meditation music. Staying at least six feet away from people and attending to her mental health are her survival mechanisms in a place with poor sanitation. On 29 April, after weeks of trying to convince me that she was safe and taking care of herself, she admitted she was scared because the prison staff was not taking the virus seriously. Missing visits with her family and worrying about their safety (her relatives are all in their fifties and sixties), Rosemary began struggling with depression. She maintained her sense of humor, though, teasing me at the end of a message for being "locked up like her."[12]

In a California women's prison, inmates spent weeks in a sewing factory stitching masks, even as they were banned from wearing them.[13] The *Los Angeles Times* reported that the fabric they used came from a local men's prison, where a coronavirus outbreak had killed twenty-three inmates. Supervisors would raise daily mask quotas and threatened workers with disciplinary action

---

10 According to its website, SWOP Behind Bars "is a national grassroots social justice network dedicated to the fundamental human rights of sex workers and their communities, focusing on ending violence and stigma through education, community building, and advocacy," accessed 9 October 2020, https://www.swopbehindbars.org/about-new/.

11 Rosemary to author, 28 July 2020, JPay App.

12 Rosemary to author, 29 April 2020, JPay App.

13 Kiera Feldman, "California kept prison factories open. Inmates worked for pennies an hour as COVID-19 spread," *Los Angeles Times*, 11 October 2020, https://www.latimes.com/california/story/2020-10-11/california-prison-factories-inmates-covid-19.

if they refused to work because of fears of COVID-19. Eventually, the virus reached the women's prison in Chino, killing Robbie Hall, who worked twelve-hour days for less than a dollar an hour. This story reminded me of wartime Germany. In 1943, about a thousand women prisoners in Aichach, Bavaria worked for sixteen different companies performing duties such as manufacturing gas masks, grenades, bombs, nozzles for aircraft engines, among other jobs with varying levels of peril. Many other inmates engaged in agriculture, washed laundry, and cooked, which was necessary to maintain the facility.[14] The safety of prisoners on worksites appears to have been of little importance to staff. Several records comment on poor air ventilation in workrooms at both the Aichach and Barnim prisons because windows were blocked off on account of security, resulting in dangerously high temperatures and inhaling toxic vapors.[15] Another record tells the story of a woman in charge of unloading vehicles at the Barnim prison. She was run over and killed, and staff members who witnessed the incident did nothing to help her.[16]

Rosemary found herself in a similar predicament to Robbie Hall in the Chino women's prison. On 6 July Rosemary wrote me the following message:

> I am just glad you're keeping safe. Because I have some bad news. Christina, the coronavirus is now here in my prison: – ( So I'm very scared to be outside, or around the officer/staff/white shirts/medical staff . . . too. There was a staff member here on my yard that has it. I'm just now wanting to quit my job and stay the hell away from anyone, everyone. But I need to work because it's hard in here. As I'm sure it is out there. God, why does this have to have happened. I'm kinda freaking out . . . .[17]

In the following weeks, Rosemary struggled with the prison administration's mixed messages about the virus. She wrote,

> As you know, I'm worried about getting the coronavirus now in here. In this wonderful place . . . : – ( Yeah, I'm being a smart ass. It's just that this place is like Oh you're OK, everything is OK, nothing's wrong when they've shut down a whole unit. Because that's how bad it is in here. But that's not what they're telling our family, TV, media, news, the

---

**14** *100 Jahre: Justizvollzugsanstalt Aichach, 1909–2009*, unpublished chronicle of the Aichach prison produced by multiple staff members for the 100th anniversary, in author's possession, 114.

**15** Memo, Der Generalstaatsanwalt an den Herrn Vorstand des Frauenzuchthauses und der Frauenverwahrungsanstalt Aichach, "Betrifft: Dienstaufsicht über die Vollzugsanstalten," 1.11.1943, Staatsarchiv München, General STAANW beim OLG München 51; von Gélieu, *Barnimstraße 10*, 202–203.

**16** von Gélieu, *Barnimstraße 10*, 202–203.

**17** Rosemary to author, 6 July 2020, JPay App.

outside world. Yeah, I'm freaking out. We have God knows how many people here just in Az. That has it now . . . So, I'm just a little going out of my mind.

With love,
Your crazy friend[18]

More than a month after Rosemary knew of a confirmed COVID-19 case in her prison, she heard rumors of coronavirus testing for all prisoners in her institution. After a negative test on 11 August, she wrote to me that the paperwork indicated that this was the second COVID-19 test administered to inmates in her prison. Angered by this lie, as it was her first test, she crossed out the line and corrected the information. She was relieved to be tested but was also forced into lockdown until she received the results. For Rosemary, the psychological ramifications of lockdown were the hardest. Working enabled her to earn money and provided a much-needed distraction.

Prisoners are among those most vulnerable to contracting the coronavirus. They cannot choose to isolate in ways they deem safe for themselves; they are at the whims of authorities at state, county, and institutional levels, many of which have reportedly taken blatantly deficient COVID-19 precautions. My home state of Ohio was one of the first states in the country to test its prisoners for the coronavirus. In May, when results were in, the Marion Correctional Institution just outside of Columbus found that eighty percent of the prison population tested positive. Moreover, the Ohio National Guard had to be called to replace many staff members who were sick or in quarantine.[19] As of late August, at least 5,783 cases of the coronavirus were reported among prisoners in Ohio, resulting in eighty-nine deaths from the virus. That equals 1,251 cases per 10,000 prisoners and 1,156% higher than Ohio overall. There have been 19 deaths per 10,000 prisoners, which is 463% higher than Ohio overall.[20] As of 15 September, there have been 125,730 cases of prisoners confirmed to have COVID-19 nationally, with at least 1,066 prisoner deaths and 77 prison employee deaths.[21] Even as prisoners are susceptible to high rates of exposure, few states prioritized detention centers in their vaccination implementation plans.

---

**18** Rosemary to author, 13 July 2020, JPay App.

**19** Jenny Hamel, "Inside Marion Correctional With COVID-19: 'We Just Passed It Around,'" ideastream, 14 May 2020, https://www.ideastream.org/news/inside-marion-correctional-with-covid-19-we-just-passed-it-around.

**20** *The Marshall Project*, "A State-by-State Look at Coronavirus in Prisons," accessed 23 September 2020, https://www.themarshallproject.org/2020/05/01/a-state-by-state-look-at-coronavirus-in-prisons?utm_source=The+Marshall+Project+Newsletter&utm_campaign=6cac8ab2c3-EMAIL_CAMPAIGN_2020_09_18_10_14&utm_medium=email&utm_term=0_5e02cdad9d-6cac8ab2c3-174571107.

**21** *The Marshall Project*, "A State-by-State Look at Coronavirus in Prisons."

Global and national crises often exacerbate injustices and disproportionately harm systemically marginalized communities. Many Black, Latino, and Native American communities, with multi-generational and multi-family households and limited access to healthcare, have suffered stark inequalities of infection and death. People of color make up just under forty percent of the United States population but accounted for approximately fifty-two percent of all the "excess deaths" through July 2020, according to an analysis by *The Marshall Project* and The Associated Press.[22] In a Ryerson University Yellowhead Institute article, Sefanit Habtom and Megan Scribe write, "since the World Health Organization declared COVID-19 a pandemic in early March, American and Canadian responses to the highly infectious illness have made it plain whose lives matter. As feel-good phrases circulated, claiming that 'we're all in this together' and 'the virus does not discriminate,' Black and Indigenous peoples stressed that our communities would be among the most affected." Indeed, the Navajo Nation COVID-19 infection rates per capita were the highest in the United States as compared with any individual state.[23]

Mass incarceration is a crisis and an extension of slavery. German prisons today are far from perfect but are significantly more humane than those in the US, where punitive measures are favored over rehabilitative programs despite what the Department of Corrections leads many to believe. Germans have a shameful record of weaponizing prisons to subjugate people perceived *The Marshall Project* to be racial, social, and political threats. An important part of this history entailed American GIs rebuilding much of the prison system after WWII. In Germany today, correctional officers are more like therapists than guards.[24] According to Maurice Chammah, journalist for *The Marshall Project*, "There are different expectations [in Germany] for their corrections officers –

---

**22** Anna Flagg, Damini Sharma, Larry Fenn, and Mike Stobbe, "COVID-19's Toll on People of Color Is Worse Than We Knew," *The Marshall Project*, 21 August 2020, https://www.themar shallproject.org/2020/08/21/covid-19-s-toll-on-people-of-color-is-worse-than-we-knew?utm_ source=The+Marshall+Project+Newsletter&utm_campaign=a3687a4e52-EMAIL_CAMPAIGN_ 2020_08_21_08_28&utm_medium=email&utm_term=0_5e02cdad9d-a3687a4e52-174.

**23** Joshua Cheetham, "Navajo Nation: The people battling America's worst coronavirus outbreak," *BBC*, 15 June 2020, https://www.bbc.com/news/world-us-canada-52941984. In McKinley County, New Mexico, which includes part of the Navajo Nation, one of every 277 residents has died from COVID-19. "COVID in the U.S.: Latest Map and Case Count," *The New York Times*, accessed 12 October 2020, https://www.nytimes.com/interactive/2020/us/coronavi rus-us-cases.html.

**24** Maurice Chammah, "How Germany Does Prison: Americans on a mind-boggling incarceration road trip. Day One," *The Marshall Project*, 16 June 2015, https://www.themarshallproject. org/2015/06/16/how-germany-does-prison.

who are drawn primarily from the ranks of lawyers, social workers, and mental health professionals to be part of a 'therapeutic culture' between staff and offenders – and they consequently receive more training and higher pay."[25] Chammah writes further that German prisons also see almost no violence, and inmates can receive no more than eight hours of solitary confinement.[26] The German example shows a way toward a new, more humane approach to helping inmates on the path of rehabilitation. For the United States, this process must begin with a nationwide commitment to decarceration by granting clemency to prisoners with nonviolent drug offenses, banning for-profit prisons to deincentivize mass incarceration, and channeling federal, state, and local funding away from prisons and into social programs. We must acknowledge and protest brutality, violence, and the injustices continually faced by criminalized people at the hands of the police and our carceral system. Hopefully, Rosemary's assessment of racial justice, which she describes in a 4 June letter, will be proven correct:

> Oh, Christina. I'm so glad to hear from you. I've been feeling a little sad. I know it's everything that's happening out in the real world. ALL THE PAIN. Today was George Floyd's funeral. And it's so sad to even have to lose someone. But to lose them in this way. It's even harder. I have a very good friend that lost her brother at the hands of the cops. And she lost it. I mean, she lost her job, lost her kids, started using drugs. She really needed help. And it took her a really long time to come to terms with the loss. But I really feel that all this had to happen. For the people to awake to demand a change.[27]

## Works Cited

*100 Jahre: Justizvollzugsanstalt Aichach, 1909–2009*. Unpublished chronicle of the Aichach prison produced by multiple staff members for the 100th anniversary. In author's possession.

"COVID in the U.S.: Latest Map and Case Count." *The New York Times*. Accessed 12 October 2020. https://www.nytimes.com/interactive/2020/us/coronavirus-us-cases.html.

Chammah, Maurice. "How Germany Does Prison: Americans on a mind-boggling incarceration road trip. Day One." *The Marshall Project*. https://www.themarshallproject.org/2015/06/16/how-germany-does-prison. 16 June 2015.

---

**25** Maurice Chammah, "Germany's Kinder, Gentler, Safer Prisons: Blank stares and culture shock. How Germany does prison, day two," *The Marshall Project*, 17 June 2015, https://www.themarshallproject.org/2015/06/17/germany-s-kinder-gentler-safer-prisons.
**26** Maurice Chammah, "Germany's Kinder, Gentler, Safer Prisons."
**27** Rosemary to author, 4 June 2020, JPay App.

Chammah, Maurice. "Germany's Kinder, gentler, Safer Prisons: Blank stares and culture shock. How Germany does prison, day two." *The Marshall Project*. https://www.themarshallproject.org/2015/06/17/germany-s-kinder-gentler-safer-prisons. 17 June 2015.

Cheetham, Joshua. "Navajo Nation: The people battling America's worst coronavirus outbreak." *BBC*. https://www.bbc.com/news/world-us-canada-52941984. 15 June 2020.

Cowan, Beryl Ann. "Incarcerated women: Poverty, trauma, and unmet need." *American Psychological Association*. https://www.apa.org/pi/ses/resources/indicator/2019/04/incarcerated-women. April 2019.

Der Generalstaatsanwalt an den Herrn Vorstand des Frauenzuchthauses und der Frauenverwahrungsanstalt Aichach. "Betrifft: Dienstaufsicht über die Vollzugsanstalten." 1.11.1943, Staatsarchiv München, General STAANW beim OLG München 51.

Feldman, Kiera. "California kept prison factories open. Inmates worked for pennies an hour as COVID-19 spread." *Los Angeles Times*. https://www.latimes.com/california/story/2020-10-11/california-prison-factories-inmates-covid-19. 11 October 2020.

Flagg, Anna, Sharma, Damini, Fenn, Larry, and Stobbe, Mike. "COVID-19's Toll on People of Color Is Worse Than We Knew." *The Marshall Project*. https://www.themarshallproject.org/2020/08/21/covid-19-s-toll-on-people-of-color-is-worse-than-weknew?utm_source=The+Marshall+Project+Newsletter&utm_campaign=a3687a4e52-EMAIL_CAMPAIGN_2020_08_21_08_28&utm_medium=email&utm_term=0_5e02cdad9d-a3687a4e52-1745. 21 August 2020.

Hamel, Jenny. "Inside Marion Correctional With COVID-19: 'We Just Passed It Around.'" *ideastream*. https://www.ideastream.org/news/inside-marion-correctional-with-covid-19-we-just-passed-it-around. 14 May 2020.

Heeres-Nebenzeugamt Brandenburg (Havel) an den Vorstand des Frauengefängnisses. 26.4.1944, Landesarchiv Berlin, A Rep. 365 Nr. 50.

Honecker, Erich. *Aus meinem Leben*. Berlin: Dietz Verlag, 1980.

Howard, Marc M. "Prisoners May Hold the Key to Releasing Us from Coronavirus Lockdown." *The New York Times*. https://www.nytimes.com/2020/05/29/opinion/coronavirus-prison-outbreak.html. 29 May 2020. *The Marshall Project*. "A State-by-State Look at Coronavirus in Prisons." Accessed 23 September 2020. https://www.themarshallproject.org/2020/05/01/a-state-by-state-look-at-coronavirus-in-prisons.

Rosemary to author. JPay App.

von Gélieu, Claudia. *Barnimstraße 10: Das Berliner Frauengefängnis 1868–1974*. Berlin: Metropol Verlag, 2014.

Viktor Witkowski
# Borderlands

My parents and I crossed the border in the fall of 1983. I was four years old. We had left our village in southwestern Poland early in the morning and reached the inner German border *Helmstedt/Marienborn* late that night. I remember passing watch towers and East German border guards shining flashlights inside our car. I also remember being terrified and pretending to be asleep. If I kept my eyes closed long enough none of this would be able to touch me. I also kept my eyes closed, pretending to sleep, when we arrived at our final destination, at my dad's German cousin who had agreed to take us in for a few months.

But this is not exactly how our escape from Poland to West Germany took place. After we were allowed to pass the *Helmstedt/Marienborn* border crossing, we continued to the refugee and immigrant transit camp *Friedland*. Originally *Friedland* was set up for returning German prisoners of war as well as civilians fleeing Germany's lost eastern territories after World War II. We spent two weeks there before our case seeking asylum in West Germany was processed. In my head, though, my eyes remained closed from the moment we entered the border crossing up to our arrival at my dad's cousin's house. The two weeks between those events have forever disappeared, and the only images I can evoke are a watchtower, a border guard with a shouldered rifle, and a flashlight directed at my face. Other than that, there is just darkness without sound.

It makes perfect sense that I became a painter and eventually a filmmaker. My earliest memories consist of images in isolation from any spoken word or conversation. Thinking back to our border crossing and additional experiences prior to our escape, I can access about a dozen short scenes. They are fairly clear, although they appear remote, as if witnessed from a distance. The only sound bit that is contained within this succession of vignettes from my early childhood is the 1978 song *Rivers of Babylon* by *Boney M* (thanks to my dad's large collection of western music which he had to leave behind when we left Poland). This particular song has become a soundtrack of sorts for the years 1979–1983 that I can access and view as a permanently fixed set of moving images.

In preparation for this volume, the contributors were given a chance to participate in *ZOOM*-meetings organized by Irene Kacandes over the course of a couple of months in the summer of 2020. During one of these sessions, we were discussing our understanding of "violence." In particular, comments made by Ann Cvetkovich and Priscilla Layne resonated with my artistic work. Both of them spoke out in favor of widening our understanding of violence by pointing

https://doi.org/10.1515/9783110753295-012

out the "different kinds of violence" (Layne) and how there is a "need to speak of violence not in terms of atrocity" (Cvetkovich).[1]

The visual representation of violence has been a central subject of my work for the better part of two decades. Even though I have been addressing a range of global conflicts such as the post-9/11 *War on Terror* in the context of Afghanistan and Iraq (both of which involved the United States, Germany, and Poland among other coalition members), the origins of my interest in different forms of violence lie within my family's history.

As is the case in many European families, personal narratives surrounding the events of WWII are abundant and scarce, both detailed and vague, at times exaggerated or partially forgotten and either relevant to a family's identity or no longer of importance. There are also many grey areas when it comes to one's own family history: some members cannot forget quickly enough or do not even try to remember, while others are digging through what is left, in hopes of obtaining a more complete picture and narrative.

**Figure 1:** Hans Sönnke, German soldiers and German custom officials reenact the removal of the Polish border crossing in Sopot on September 1, 1939, Das Bundesarchiv.

---

1 For more information on these concepts, see the essays by Ann Cvetkovich and Priscilla Layne in this volume.

To address historical events by visual means only – as opposed to using written language – comes with its own set of limitations and possibilities. In what follows, I propose that painting about personal experiences and histories can be used as an extension of thinking and writing. When I say "personal experiences," I do not mean that these experiences have to be eyewitness accounts to become worthy subjects of painting. The long history of painting is full of examples in which the personal/historical trauma has been a source for ambitious painting – no matter if artists experienced trauma first- or second-hand. Some prominent examples include Francisco Goya's prints from *The Disasters of War* series which depict scenes of torture by the invading French forces in the early 1800s in Spain. Later on, the horrific magnitude and loss of World War I created a generation of artists, like Otto Dix and John Singer Sargent, who did not shy away from giving their anti-war sentiment a concrete and often explicit visual form. Contemporary artists like Gerhard Richter, Vija Celmins, Luc Tuymans, Kara Walker, and Titus Kaphar have looked at their respective nations' historic sins and their own personal narratives to investigate how to make the unspeakable visible.

**Figure 2:** Viktor Witkowski, *German Troops Crossing the Polish Border (1939)*, oil on raw linen, 16″ x 20″, 2006.

The painting *German Troops Crossing the Polish Border (1939)* (Fig. 2) is one of my earliest attempts at representing a historical event. I based my painting on an iconic news photograph that represents German troops crossing the Polish border right at the start of WWII. It was taken by Hans Sönnke, a professional photographer from the free city of Danzig (Fig. 1). On the afternoon of 1 September 1939 Sönnke had joined several photographers who were gathered at a checkpoint on the northern outskirts of Danzig. Earlier that day, German troops had already started shelling the city's harbor where Sönnke took his first photos. The gathered photographers wanted to use the checkpoint to document the early stages of the German invasion. After a while, the present photographers decided that they needed a photo opportunity which would capture the dramatic and quickly evolving situation on the ground. Photographs of long columns of marching German troops were not exciting enough. When a unit of *Danziger Landespolizisten* (a Danzig police force which was composed of German soldiers who had been secretly transferred to Danzig during the previous summer) passed the checkpoint, the photographers saw their opportunity. While one of the photographers started sawing at the checkpoint's border post, the others directed the passing German unit to pretend to be breaking the border post into two pieces. In addition to the photographers, one army cameraman from an official government propaganda unit filmed the scene. This staged scene was later used in a September 1939 *Wochenschau* reel to present to Germans at home how successfully and swiftly the invasion of Poland had progressed.[2] I was fascinated by the photograph's casual nature (the smiles on some of the soldier's faces) and by its subsequent iconic status as visual proof – albeit faked – of the moment the German army crossed into Poland.[3] But more importantly, this episode of Polish-German history offered a perfect foil for my own conflicted feelings about my German heritage. My dad's mother was German, married to a German Air Force officer during WWII who died toward the end of the war during a training exercise in Poland. After the war, she remained in Poland with their first child, alone and in a desperate situation as a German war widow. This is when she met my Polish grandfather, who was a grocer, whom she married shortly thereafter. She was a survivor and intended to keep her child and herself alive. My Polish grandfather and German grandmother led a loveless marriage. And when I started making this painting, I kept thinking about that gesture – removing a border post – and how it violently

---

**2** Solveig Grothe, "Zwölf Mann und eine Schranke," https://www.spiegel.de/geschichte/ deutscher-ueberfall-auf-polen-1939-zwoelf-mann-und-eine-schranke-a-7c7a74f5-6300-40b9- a6cc-2a6cfad006b2, SPIEGEL Online, 28 September 2020.
**3** Even though Sönnke distributed his photograph as postcard in Danzig, it only gained in significance after the war when it was rediscovered and, at first, falsely attributed.

and irreparably changed so many lives, millions of lives, and the lives of my grandparents. The fact that the photograph was staged, only added to how I perceived the dynamic between my grandparents: their relationship hinged on something essentially untrue.

The personal is political in many of my paintings, but the personal can be universal, too. *German Troops Crossing the Polish Border (1939)* is not about my family; it implicates every perished family and those who survived. If we consider violence as a demonstration of power, then we do not need to see explicit images of mutilated bodies to understand that violence manifests itself in varying degrees. At times, violence is not even recognizable and visible as such. In the original photograph of the Polish border removal, an air of amusement can be detected in the expressions of some of the participating Germans. Yet, their act set the stage for the unimaginable violence that followed soon after (and had already occurred in other forms within the previous decade). By removing the agents of violence from my painting and rendering the border post as unfinished (by revealing the raw

**Figure 3:** Viktor Witkowski, *Border Marker*, 40" x 26", oil on panel, 2016.

linen of the canvas), I turn away from the historic event and the declaration of war it illustrates. My focus is on the border as a site of trauma and transformation: borders that are built, crossed, conquered, torn down, and sometimes forgotten.

•

When I painted *German Troops Crossing the Polish Border (1939)*, I saw it as a duty to create an altered version of the infamous photograph. The act of painting offered me a chance to claim agency over a part of my family history. I did not linger too much on the thought that other painters had already offered their own powerful interpretations based on black and white photographs depicting WWII.[4] My family's crossing of the West German border was one of the many ripples caused across time by the removal of the Polish border post in September of 1939. But I also thought of my painting as a rendition of a distant and perhaps too impersonal episode in history. I needed to make a painting that spoke more directly to my family's experience, and I wanted it to be centered on another object typically associated with borders.

The first time my parents and I were able to visit our family in Poland was in 1987, four years after having fled. That is when we passed through the German-Polish border town of Görlitz/Zgorzelec. Back then, the lines of cars queuing at the border crossing on both sides posed a challenge that demanded patience, a sense of humor, sufficient food and drink on hand and occasionally some bribes (in the form of packs of Western cigarettes that my dad offered to the East German and Polish border guards). Because everyone crossing the border before the Fall of the Wall and through the early 1990s had to wait for hours on end, we had more than enough time to investigate the specifics of the border installation. One of the very few colorful objects near the border crossing were the border markers. I recall the East German border marker with diagonal stripes in black, red, and yellow. At the top, it included an aluminum plaque depicting the GDR's national emblem with a hammer and compass surrounded by a garland of corn. This object is unremarkable, mundane even. It is unassuming and it does not betray its significance and power. When I decided to make a painting that speaks to our border-crossing, I did not want to include any of the obvious features such as barbed wire, fences, guard towers, or concrete walls. Instead, I wanted to bring attention to the border markers and their status as reminder of a nation's divided past along physical and ideological fault lines. Some of these border markers can be found in their original spot to this day. One prominent

---

4 The most prominent examples of contemporary painters include Gerhard Richter, Vija Celmins, Luc Tuymans, and Wilhelm Sasnal.

example is the East German border marker at *Checkpoint Charlie* in Berlin, and there is at least one at the banks of the river Neisse in Görlitz/Zgorzelec.

For my painting *Border Marker* from 2016 (Fig. 3), I wanted this object to appear freshly painted, as if it had been recently manufactured and installed. In contrast to its appearance as a new marker, I surrounded it by tall grass to indicate the passage of time and with it the marker's potentially defunct status. This contrast of a new marker surrounded by tall grass also speaks to the urgency of our contemporary discourse around borders. Border markers in East Germany have become memorials to a dark chapter of post-war German history; they no longer function as demarcation of two opposed political systems. In other parts of the world borders continue to be fortified, extended and contested. The ongoing standoff between North and South Korea, Russia's annexation of Crimea and incursion into eastern Ukraine, China's ambitions toward Hong Kong, Taiwan and the Himalayan region, Turkey's stake in northern Syria, the Trump administration's Southern Border Wall. Similar to the visual juxtaposition of a newly manufactured object overgrown by vegetation, the effectiveness of borders is a function of their physical permanence and political porosity. Borders are, in fact, highly flexible and solid at the same time which makes them prone to abuses of power. With enough political (ill-)will, a dormant border can be reactivated and become permanent again. Under different conditions, when a society has learned from the errors of its past, borders turn porous: they can be visited, climbed, crossed, spray painted and photographed. They have lost their function as divider and with it their permanence.

Once we crossed back into Poland after years in exile, I was under the impression that all color suddenly faded and that I had entered a world of grey. It took several hours to drive one hundred kilometers to my family's village, since the network of highways had fallen into disrepair over the preceding decades. On the narrow, pot-holed mountain roads, speedy driving was impossible. I do not recall what car we owned at that point, but we were driving through small towns and villages bearing West German license plates. At this point, for many of the people we passed on our way, a car from Germany offered a rare sight in this remote part of Poland. We regularly saw children waving at us from the side of the road and cupping their hands to ask for sweets. My parents would stop at times and give them chocolate bars or gummy bears. I remember these stops vividly because these Polish kids were my age. I felt ashamed most of all. Ashamed, because I did not have to stand on the side of the road hoping to receive candy from passing Westerners. In these moments I felt strongly about not identifying as German. At the same time, it was hard to shake off my German identity. Germany was my adopted home, even though Poland remained my homeland. Even at that age, I had a hazy understanding of my role as a Pole in exile and how I was condemned – like many others in my position – to live in two worlds and with multiple identities.

One time, when we were driving through those villages in the early 1990s, I saw a child, a little younger than me, standing on the side of the road and not waving at us. Instead, he had raised his hand to perform the Hitler salute. This sight struck me with pure horror. I knew the salute from World War II documentaries I had seen with my dad, and I understood this gesture as a symbol of hate and death. By directing the salute at us, the anonymous Polish kid associated us with Nazism. It felt wrong and twisted to me, and once again, I wanted to explain and distance myself from my Germanness. Looking back at this moment, I realize that it has become one of the most defining moments of my childhood. What makes this experience so particular and lasting is its iconic value. I would not be surprised if this kid had been waiting for a while to perform this gesture in front of a German car. His choreography and timing were effortless. And I also wonder if he had watched the same documentary footage I had: Hitler riding in an open-top car, passing masses of people who were enthusiastically greeting him with that salute.

It was not long after this episode that I encountered my first Neonazis in Germany. One day, my German grandmother, who had immigrated to Germany with my Polish grandfather in 1991, was out on an afternoon stroll with my mom and me. This was in the city of Espelkamp where we lived. We always spoke Polish with each other and had made a conscious choice to do so in public, even though we knew other Polish and non-Polish immigrants who only spoke in their native tongue at home. Suddenly we were approached by three Skinheads who had all the visual markers that exposed their ideological beliefs right away (unlike their current *Alt-Right* and *New-Right* successors who have learned to blend in with the general public): shaved head, bomber jacket, high ankle leather boots with white laces. Espelkamp is not a big city, and I had never seen these people before. Also, they had entered our neighborhood which was mostly comprised of low-income, subsidized government housing with many immigrant families. Naturally, they stuck out like a sore thumb. The encounter was brief, but fierce. "You are in Germany, speak German!"; one of them jumped at us. My grandmother responded in her native language and asked them if they had nothing better to do than to harass women and children. The gang of three moved on without another word.

After this encounter, I was shaken. But luckily, we were physically unharmed. Other immigrant families in Germany bore the full brunt of racial hate. It does not come as a surprise that in the years 1989–1992 many of Germany's far-right parties had garnered more votes than in previous years. The initial euphoria surrounding the events of the fall of the Berlin Wall had dissipated and fear of immigrants, refugees and foreigners became a permanent state of mind throughout the early 1990s. Violent attacks against asylum seekers, refugee housing (Hoyerswerda in 1991) and even killings (Mölln in 1992) were a sign that reunified Germany was not immune to the ghosts of its past. In fact, those ghosts were very much alive, and they were among us. To the shock of a nation, they had never left.

Figure 4: Viktor Witkowski, *Escheburg (and elsewhere)*, 40" x 26", oil and colored pencil on panel, 2016.

From late 2015 to mid-2016, during the height of the so-called refugee crisis when hundreds of thousands of Syrians (but also Iraqis and Afghanis) fled the intensifying wars in their homelands, I was reminded of the early 1990s in Germany. With the influx of refugees, overwhelmed German municipalities and Chancellor Angela Merkel's promise to take in those who were seeking refuge, a rise in hate crimes followed promptly. Once again, refugee homes went up in flames. In one such instance, a freshly renovated wooden house in the northern town of Escheburg was targeted. Luckily, nobody was injured because the six Iraqi men had not yet moved in. Photographs accompanying the news story depicted the charred remains of the building's interior while the meticulously restored outside was nearly untouched. The white, wooden window frames stood out amidst the destruction. In the aftermath of this attack and the charred remains, I saw the unassuming white window frames as synonymous with whiteness and specifically white supremacy. About a week after this arson attack, a thirty-nine-year-old local financial advisor was arrested. According to his court defense a few months later, he wanted to prevent the Iraqi men from being resettled in "his" town. He was angry at the local government for not informing the Escheburg residents early enough about the planned arrival of refugees. He was also fearful of the danger that these men would pose to his family.[5]

Within painting's long history, windows offer a view onto the world. It is a way to frame our visible world and by doing so we demonstrate how we perceive and understand it. In early modern European portraiture traditions, a sitter is often placed in front of a window while looking out at the viewer. In the background the depicted landscape is idealized, at times revealing aspects of an actual location. In German Romantic paintings of the early 1800s, figures look out through the window into the world that lies beyond. This self-reflective mode (and the attempt to make our interior lives visible) is developed further toward the end of the nineteenth century, when the structure of the window is likened to the structure of the canvas (and its wooden support found in the back of the canvas). Here the view of a window and through a window is compared to the act of painting. With that realization in mind, painting and painters no longer need to rely on the visible world as a way to measure what constitutes the value of painting. In *Escheburg (and elsewhere)* (Fig. 4), the *trompe-l'oeil* window frame upholds and frames a view of destruction. It does not depict a particular object; it does not reveal what has been consumed by fire. Yet, it is not entirely abstract or devoid of language either. The painting's title is part of the work. It situates and contextualizes what has been painted. It speaks to one particular event, and it simultaneously makes a broader claim that goes beyond the Escheburg attack: *Escheburg (and elsewhere)* is emblematic of systematic and structural racism within German society and the violence and lasting trauma it instills.

To conclude this essay, I would like to mention how the three paintings I introduced here are connected. One the one hand, they are still-lives in the sense that their main subjects are inanimate. On the other hand, they address varying degrees of violence in the absence of humans. Violence is not tied to an individual, no matter if they are a victim or perpetrator. In this set of paintings, violence is less about the physical, spectacular act itself and more about the mechanisms and ideologies that lead up to violent acts and abuses of power. The thirty-nine-year-old arsonist who resides in Escheburg is not a member of a

---

5 https://www.shz.de/regionales/luebeck/brandanschlag-in-escheburg-bewaehrungsstrafe-fuer-39-jaehrigen-id9679886.html.

militant far-right group. He is a family man and inconspicuous member of the community he set on fire. It is easy to overlook a person like him. For some, it is easy to make excuses for his behavior and cast it as an anomaly of an otherwise law-abiding individual. If anything, paintings about violence (versus paintings "of" violence) can sharpen our focus and ability to detect some underlying causes of every-day injustices. We have to look past violent acts to identify and alter the systematic deficiencies which cause them in the first place. Whenever violence occurs, we have missed an opportunity to prevent it from happening.

**Figure 5:** Viktor Witkowski, *Lasgin (Espelkamp, Germany)*, 48" x 60", oil on canvas, 2018.

With all of that being said, we must remember and represent the individual human cost. In the episodes from my childhood that I described above, I presented several intimate examples that speak to the experience of and interactions with individuals across places and time. As pivotal as it has been for my artistic practice to be making work about violence, it is equally as crucial to represent those affected by violence in works "of violence." When I was in Germany in early 2016, I was working on my first short experimental documentary about the refugee crisis. One of the communities I visited was in my German hometown of Espelkamp. My former elementary school had been transformed into a temporary refugee housing site, and when I found out about it, I decided to pack my camera and pay a visit. After a conversation with the social workers assigned to this site and its all-male inhabitants from Syria, Iraq and East Africa, I was allowed to speak with the residents. I returned a couple of times over the next few days to have conversations, take photographs, and record some footage. One of the residents was a Syrian man named Lasgin, in his early thirties. He was separated from his wife and daughter and had not seen them in months. Both of them had arrived with him in Germany, but they were housed in a different part of the country. He was supposed to visit them in a month. To make the best of his time while he was waiting to start taking German language classes, he had picked up drawing. He shared some of the drawings with me. Some depicted his wife and daughter, and in other cases his fellow residents had commissioned him to create drawings for them. I told him parts of my family's story and how we had arrived in Espelkamp thirty-three years prior, the difficulties my parents faced, how I had attended elementary school in the building where he was currently housed. I also mentioned that I was a painter. This piqued his interest and after a while I asked him if I could paint a portrait of him. At first, he was not sure and smiled at me. But then he gave the idea some more thought and agreed. I asked him to get in a position he was comfortable with and that I would take a photograph of him which would later serve as a template for the painting. Both of us were squatting outside with the school's glazed brick wall behind us, sharing a cigarette. Lasgin decided that this was exactly how he wanted to be painted, and so I took a few pictures and some additional close-ups: his face and his eyes that are fixed at something beyond the pictorial frame, his left hand with his daughter's name tattooed across four fingers, his drawing hand holding a cigarette, the purple shimmer of his pants. We finished our cigarettes and said goodbye to each other. I wished him all the best, and Lasgin asked me to send him a picture of the painting when it was finished. I promised him that I would do so, shook his hand, and started walking back to my parent's apartment.

Our shared moments, exchanges, and conversations made their way into a life-sized portrait (Fig. 5). Painting is a form of communication by visual means. In the case of Lasgin, his portrait is the result of a collaboration that took shape before the painting did. We are looking at an individual who lived through the experience of being adjacent to violence. Without explicitly touching on the violence which led him to flee his home, his portrait allows for other narratives to enter the frame. Through the portrait, Lasgin challenges his assigned role of refugee-victim. Painting's non-verbal quality emphasizes his claim for agency and autonomy; no words are needed, and no trauma has to be exposed to render a full image of Lasgin. His portrait is not devoid of his past and memories. But the things he carries are to be found deep within, and they are his. Lasgin thought it best to let them be. Instead, he directed me to produce a painting. His portrait is a testimony to a moment in his life with an unresolved past and future. This territory of human experience which language has not been able to enter yet is where painting shines at its fullest. And with it, Lasgin appears at both his strongest and most vulnerable. He has found a place set in distance to his violent past, while inviting us to share a moment with him. His image has the last word, and it speaks of perseverance.

## Works Cited

dpa. "Brandanschlag in Escheburg: Bewährungsstrafe für 39-Jährigen." 11 May 2015. Accessed 1 December 2020. https://www.shz.de/regionales/luebeck/brandanschlag-in-escheburg-bewaehrungsstrafe-fuer-39-jaehrigen-id9679886.html.
Grothe, Solveig. "Zwölf Mann und eine Schranke." Accessed 28 September 2020. https://www.spiegel.de/geschichte/deutscher-ueberfall-auf-polen-1939-zwoelf-mann-und-eine-schranke-a-7c7a74f5-6300-40b9-a6cc-2a6cfad006b2, SPIEGEL Online.

Part Two: **Families**

Susan Rubin Suleiman
# A Postcard to Zircz (Budapest, 1944)

When my mother died, my sister and I inherited her photographs. A few were displayed in rickety frames on top of her dresser, the rest were jumbled in manila envelopes and plastic bags in a drawer. We divided them as best we could; I got to keep the original wrappings.

That was many years ago. I put mine away, still in a jumble, telling myself that one day I would put them in order. I had seen most of the photos while Mother was alive, on the rare occasions when she consented to talk about our life in Budapest, the life we had left behind when I was ten years old. Usually, she avoided the subject: "That's behind us, forget about that," she would say firmly – but she said it in Hungarian, a fact that undercut the meaning of her words. The past is another country, historians have said. Can you forget a country whose language you still speak? My mother and I always spoke Hungarian when we were together.

I visited her in Miami Beach once or twice a year, and she would bring out the photos if I begged her, naming the people who were unfamiliar to me: provincial uncles and their families who had perished in the Holocaust; two of her cousins, beautiful young women who had given her photos inscribed on the back with "Love to Lilly." We would also look at pictures of me as a baby, a toddler, a schoolgirl. ("That was when I was a boy," I'd joke at the photo of me at six months, lying on my stomach, almost completely bald). "How cute you were, see, so blonde!" she would say. Or: "I always liked that blue dress on you – we had it made in 1948, the year Granny left." And then, at last, she would start reminiscing: "Do you remember our Sunday hikes in the Buda hills? Our trips to the bookstore on Andrássy Avenue to buy you the books in that series about Zsuzsika, the little girl who had your name? And the iceskating rink in City Park, where I would watch you skate?" Yes, I remembered. I always shivered from the cold at the iceskating rink and didn't enjoy it, but she wanted me to learn how to skate and take ballet lessons and play the piano, all the things she had not been able to do as a child. I was her only child (my sister was born much later), while she had grown up with three brothers and a sister in a family that had become fatherless when she was still a teenager. She had had no time for luxuries, and she wanted me to have them.

•

One afternoon, about ten years after her death, I finally undid the plastic bags, shook out the manila envelopes. Some of the black and white photos were torn,

**Note:** An earlier version of this essay appeared in *Harvard Review*, No. 24 (Spring 2003), pp. 112–121.

https://doi.org/10.1515/9783110753295-013

stained, of people whose names I no longer knew, if I ever did. She had carried them around since August 1949, when we escaped from Hungary. These photos and a damask tablecloth with twelve napkins, once owned by my grandmother, were the only talismans linking me to that world; now they were also my links to her, my inheritance. They were mixed in with colored snapshots I had sent her over the years: me with my two boys, me and my husband before our divorce, before our marriage. Among the black and white photos is one of me at my college graduation, standing next to her, smiling, looking lost.

A publisher had asked for a childhood photo to include with an autobiographical essay of mine in an anthology. I fished out the two pictures I was looking for: me at about age four, posed in a photographer's studio in Budapest: *A Mosoly Albuma*, The Album of Smiles, its name, address and phone number imprinted on the bottom right-hand corner. In one photo (Fig. 1), my hair is loose around my

**Figure 1:** Feb. 1944, from private collection of the author.

shoulders with a big silk bow on top; I wear a pretty cotton dress and a gold chain around my neck; a straw shopping basket hangs from my left arm, while the right hand holds the leash of a toy poodle on wheels. I am a young lady going out to do her daily shopping with her dog. Facing the camera, I smile, and my eyes sparkle.

In the other photo (Fig. 2), I am a busy housewife. With my hair in braids and two small bows instead of the big silk one, an apron over my pretty dress, I stand in front of a toy stove, my right hand resting on the lid of a pot sitting on the stove. Once again, I'm smiling at the camera, this time even more broadly – I must have been aware of the silliness of the pose.

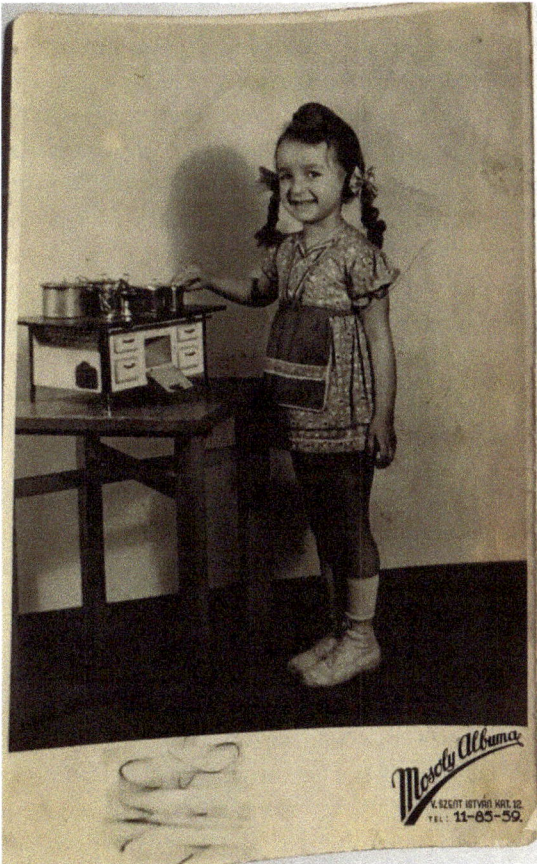

**Figure 2:** Feb. 1944, from private collection of the author.

Mother had made a special appointment with the fancy photographer, and the session probably lasted for several hours. In the end, she chose these two pictures.

I flip them over and receive a shock. I had never noticed before that there was writing on them. Both are in the form of postcards, with a line down the middle: space for the address on the right, the message on the left. The picture with the dog carries no address, but there is a message in Mother's handwriting: "To my sweet good Magduska, with much love and many kisses from little Zsuzsika." My aunt Magda, Mother's sister, had emigrated with her husband and daughter to France in the 1930s. Evidently, this picture was enclosed in a letter: to Magduska from Zsuzsika, the diminutives showing intimacy and familiarity, though in fact she had never seen me. Beneath the message, the date: Budapest, 14 February 1944.

February 1944, St. Valentine's Day: Lilly Rubin, a Jewish woman living in Budapest, sends a picture of her four-year old daughter to her sister Magda, who is hiding somewhere in France. A month later, the German Army will invade Hungary and Adolf Eichmann will set up his headquarters in the capital; the roundup and deportation of provincial Jews will be accomplished with amazing efficiency, eliminating all of Lilly's extended family. But she doesn't know that now. Does she know about the Warsaw ghetto, about Treblinka, about the deportations from France? Does the name "Auschwitz" mean anything to her? I dream about this as I finger the photograph and examine her spidery handwriting. She loved beautiful things, the big silk bow, the gold necklace. Maybe she didn't want to know. Maybe she thought we were safe. (Hungary was Germany's ally, though by then a reluctant one. The Germans invaded to make sure it didn't bolt, and to take care, at last, of the Hungarian Jews).

About forced labor, at any rate, she knew: the other photo, the one with the toy stove, is addressed to her brother Izsó Stern, with a postbox number and the number of a regiment, in the city of Zircz. The Hungarian Army had begun conscripting Jewish men into forced labor service in 1940, and my uncles Laci and Izsó were among them. So was my father. My mother had told me about that, and so had my aunt Rózsi, my father's younger sister – they both claimed that they had been the one to obtain his release after a few weeks, by repeated trips to a government office, badgering the authorities: he was a Rabbi, they pleaded; he worked for the Orthodox Community Bureau; he had a young child to support. I found it amusing that they each wanted sole credit for his release, though of course the main thing was that he had returned safe and sound. But a document I discovered only recently, which my father filled out in 1945, after the war was over, states that he had been in forced labor from December 1942 to February 1944 – that was more than a few weeks. The document also states, in his careful handwriting which I recognize immediately, that he had been in Russia, the worst possible place to be in forced labor. If the dates are correct (I say if, because he may have had reasons to prolong his captivity on paper – documents are not always one hundred percent trustworthy), he had returned from captivity on 9 February, five days before my mother sent the photo of me to her sister. But she must have taken me to the photographer before that, which strikes me as an odd thing to do at the time. I suppose life went on, even if your husband was in a forced labor unit in Russia.

I've read about the forced labor units, seen a few films too. At first, the men were relatively well treated, if one can say that about civilians forced to work long hours clearing woods and carrying rocks, undernourished and ill-clothed. Later, it got much worse. In the winter of 1943, the Hungarian Army was routed by the advancing Russians in Ukraine; Jewish conscripts, now seriously deprived of clothes and shoes and food, were set to clearing minefields ahead of the regular troops. If they weren't blown up, they starved or froze to death by the thousands, or died of exhaustion, or were beaten or shot for sport.

•

My uncle Izsó died in forced labor, though we never knew exactly when or where. His brother Laci, who came back, had lost track of him early on. For a long time he kept repeating, whenever Izsó's name was mentioned: "He was too gentle, he didn't know how to scramble, allowed others to take advantage of him – if only I'd been there to help!" Laci always landed on his two feet, or so he thought. I think he was just lucky.

Now here is this postcard (Fig. 3), which I am looking at again as I write these pages, featuring me as a four-year old housewife, addressed to Izsó in Zircz.

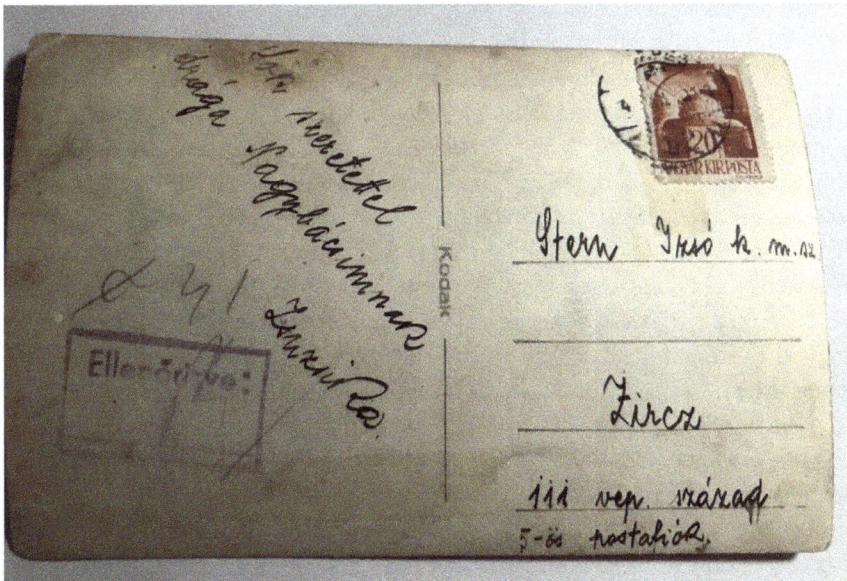

**Figure 3:** Feb. 1944, Verso, from private collection of the author.

My mother's message to him reads: "With much love to my dear uncle," signed with my name, Zsuzsika. No date. There is a stamp with a postmark, indicating that it was actually sent; but the postmark is illegible, even with a magnifying glass. Assuming that she sent this photo around the same time as the one to her sister, the card must have gone out in mid-February 1944. Was he dead by then? I hope he wasn't. I hope he received the card and smiled at his little niece's absurd pose. He was very fond of me, everybody said – I hope the photo made him smile.

But he did die, if not then, then later. How did the picture get back to my mother? Perhaps it was returned, "Addressee deceased"? But there is no return address on the card. Or maybe it came back with a package of my uncle's effects after his death? No, that would have been too civilized – and besides, there was never any official notification of his death. Maybe, when he saw that he was dying, he gave the photo to someone along with other papers and asked him to return it to the family if he survived. I imagine the scene: 1945, summer or early autumn, my grandmother is still waiting for her son. A man knocks on the door, stands in the doorway, looking uncomfortable, shifting from foot to foot: "I brought you this, Izsó asked me to." No, that's melodrama. I don't know how the photo got back to us. Maybe he was given a leave from his unit in 1944 and brought it home with him, before he returned to die.

I look for more signs. The card is addressed to Stern Izsó, Hungarian style, last name first. After his name, the letters *k.m.sz.* In Hungarian, forced labor is called, euphemistically, labor service, *munkaszolgálat*; *m.sz.* is obviously an abbreviaton for that, but what does *k.* stand for? This is the time to consult my bible on these matters, Randolph Braham's big book on the Holocaust in Hungary. Braham gives at least two words beginning with K that went with "munkaszolgálat": *közérdekű,* "public," *különleges,* "special." Public labor service in the Army, special labor service in the Ukraine. It was special all right, unarmed, with no uniform, checking out minefields.

Where is Zircz? Sounds foreign, not Hungarian. Ukraine, the eastern front? I need a detailed map, which today is easy to find. No sooner said than done: Zircz or Zirc – the latter is what the map says – is the site of an ancient Cistercian abbey in northwestern Hungary, toward the Austrian border, not far from the city of Veszprém, which has been renowned as a center of learning since the days of Hungary's first Catholic king St. Stephen, who reigned in the eleventh century. Not Ukraine but authentic, historic Hungary – it was just one of those ironies, to put a labor camp in such a town. But then, Dachau is proud of its medieval churches, and Buchenwald was a stone's throw from Weimar, where Goethe had lived and worked. Culture and barbarism, such close cousins.

At least it wasn't the eastern front. When my mother sent him that picture of me in front of the toy stove, Izsó – if he was still alive, and still in Zircz – was not checking out minefields, merely suffering from cold, hunger, backbreaking labor, the brutality of guards.

He was a gentle man, with an oval face, brown eyes, wearing small, black, owlish glasses that look oddly contemporary on the portrait photo I find of him in the pile. His brown hair, smoothly combed, sits high up on his forehead; he's starting to get bald. He wears a dark grey suit, white shirt with soft collar, and wide striped tie; he is smiling slightly, with closed lips. A gentle man.

On the back of this photo, some very faint writing for which I need the magnifying glass: Stern Izsó, then his address in Budapest (which was also our address and my grandmother's; we all lived in one large apartment) and a date and place: 1904 Nagymihály, evidently his year of birth and birthplace, then a more precise date: 22 March, no doubt his birthday. Beneath that, his mother's maiden name: Lebovits Teréz, then an official stamp with a notation in black ink and an illegible signature. The notation consists of a number, like an ID number, below which is a word that could be "láttam," "I've seen him," followed by a date: 1942, 12/30, and the signature. On 30 December 1942, an official person saw him, or gave him permission to leave, or to return. He was thirty-eight years old. The official stamped the back of his photo, and it too eventually found its way back to my grandmother and from her to my mother.

On page 338 of Braham's book, I find the following information: in November 1942, the Ministry of Defense required all Jewish men aged eighteen to forty-eight to register. "Each affected Jew had to bring along two snapshots, one of which, after it was stamped by both the recruitment center and the local bureau for vital statistics, was to serve as a 'registration certificate.'" What I hold in my hand now is Izsó's registration certificate, stamped and dated and bearing an official signature. Braham reproduces a similar photo, front and back, and I see that it carries the same vital statistics as Izsó's: date and place of birth, permanent address, mother's maiden name. It too has a stamp and a signature; its date is 22 December 1942. Evidently, people were following the Ministry's orders. The unidentified man in the book (it was Braham himself) was born in 1922; he was much younger than my uncle. But Izsó wasn't old. Thirty-eight isn't old, when you lead a normal life.

Hungary has many placenames beginning with *Nagy*: *Nagyyaranypuszta*, great golden plain, *Nagyerdő*, great forest, *Nagylengyel*, great Pole, *Nagyszentjános*, great Saint John. *Nagymihály*, Izsó's birthplace, is Great Michael, named after the Archangel. In 1904, it was a busy industrial town in northeastern Hungary near the Carpathian Mountains, part of the Austro-Hungarian monarchy. World War I put an end to the monarchy and rearranged the map of Central Europe, distributing many border regions to other countries. Today Nagymihály is in far eastern Slovakia and is called Michalovce, the county seat of its region. In 1910, 32% of the population was Jewish, just slightly less than the Catholics

(39%). Today, as far as I can tell, there are no Jews living there at all. The Stern family all came from that part of the world. Izsó's parents, my grandparents, were married and lived for a while in Ungvár, which today is Uzhgorod in Ukraine, less than 20 miles east of Michalovce. Before the war, Ungvár had an even larger Jewish population than Nagymihály; in 1944, many trainloads of Jews were sent to their death from Ungvár.

Izsó was the couple's first child, born before they moved to Budapest; but like his siblings, he was a child of the capital, at home with its pleasures. I rummage for more pictures of him. Here are two small snapshots, clearly from before 1944. In one, Izsó stands with bare chest among a group of smiling men and women, all wearing bathing suits. Maybe it's a Sunday morning at the swimming pool on Margaret Island, or at the Gellért baths. Izsó leans forward, his elbows resting on the back of a wicker armchair in which a young man is reclining; Izsó's shoulder touches that of a blonde woman who is sitting on the arm of the same chair. At the bottom of the picture, two little girls lean against the legs of the man and woman behind them, probably their parents. I have no idea who these people are; only Izsó's face is familiar. (Odd, that I should recognize him instantly on the photo when I have no memory at all of the living man). He is wearing his owlish black glasses. His hair is slightly tousled; his expression, though serious, is relaxed. He never married, but he evidently had good friends. He looks happy.

In the second picture (Fig. 4), he wears his dark grey suit, with a white shirt and tie as in the portrait. He sits on a park bench next to a young couple dressed in summer white.

**Figure 4:** Date, place unknown, from private collection of the author.

The woman's dress has a polka dot pattern; she wears a fashionable straw hat with flowers and a ribbon; the hat is at an angle, low over one eye, giving her a coquettish look. She is smiling. The man's hat is a light grey fedora, almost incongruous with his white suit. They sit very close to each other, and she wears a diamond ring on her left hand. Maybe they've recently gotten engaged. Despite their well-dressed appearance, there is something provincial about the way they look. Izsó is bareheaded; he sits at a slight distance from them. Maybe they're provincial cousins from the large Stern family, on a visit to the capital – the three of them have been walking around City Park and have sat down for a rest. Or maybe the scene is not in Budapest, and it's Izsó who has gone to visit them in their hometown. Wherever it is, it's an occasion worthy of a photo, of celebration on a sunny summer day.

Provincial Jews from Hungary were almost all deported, close to half a million. Very few came back. Like Izsó, this couple will soon be dead, but they don't know it. I feel offended by the obviousness of the tragic irony, the scalding but too easy sense of time's irreversibility these images produce. They are gone while I am here, and I know what happened to them. But please, no philosophizing! Look at the way she leans against her lover's arm, the way she has put her hat on, at an angle, like Marlene Dietrich. The way the shadows play over her neck and face, over her smile. Just look.

The last picture I find of Izsó is a postcard-size photograph of a group of men, like a class picture (Fig. 5): one row sitting, three rows standing, tallest in back. Behind them is a large open door, as of a hangar or barrack. There is a long tear down the middle of the photo, mended with transparent tape on the back. Someone had torn it in half, then changed their mind; or maybe it was folded to fit in a pocket and got torn that way. The men are dressed in winter clothes, jackets and sweaters and overcoats. They wear no uniforms, yet look regimented. Most of them appear to be wearing a tricolor armband on their left arm (some of their arms are hidden), no doubt the colors of the Hungarian flag, red, white and green. Most are bareheaded. A few wear wool work caps. One man, chubby, wears an Army cap; he sits to one side, has a white cook's apron over his wool jacket and no armband.

My uncle Izsó stands in the back row at the extreme left, bareheaded, in a dark overcoat, with a light-colored scarf around his neck. He is unsmiling, looking straight at the camera. They are all unsmiling, except the cook and one other man squatting in the front row. Why is that man smiling? He is one of the conscripts, wearing an armband. It must be the camera – some people can't help smiling whenever a camera appears. But it's not a happy smile; he is in half profile, glancing sideways at the camera.

**Figure 5:** Forced laborers group shot, from private collection of the author.

Nothing is written on the back of the card, nor does any label or marking appear on the photo itself. Yet there is no mistaking it, this is a group of Jewish forced laborers: their pose, their look, their armbands all indicate what they are. I count twenty-nine of them, plus the cook. Why is the cook wearing an Army cap? He has no armband, so maybe he belongs to the regular Army. A dishcloth is slung over his shoulder, touching the strap of his apron.

The men all look in relatively good health; the picture must have been taken in the early days, maybe right after Izsó registered in December 1942. Or maybe even earlier: many Jewish men were called up in '40 or '41, allowed to return home after a few months, then called up again. But why the class picture, and how did it reach my mother? Odd idea, a souvenir photo of a forced labor camp. Was this in Zircz, or did Zircz come only later, when no more photos were taken?

Were there similar pictures, now lost, of the surviving brother Laci? Did he too receive a photo of me in his labor camp, busy at my stove or holding my toy dog by the leash? Was she also planning to send a photo to my father, not knowing he was about to arrive home? I don't know, and there is no one around anymore to ask. My father died decades before my mother, and my aunt Magda and uncle Laci are gone too. While they were alive, we never talked about these things. And it appears I never turned the pictures over to see what was written on the back. Or if I did, the words didn't mean anything to me then. The only people who mourned Izsó after the war were my grandmother and his siblings who survived.

•

I put the pictures back in the drawer, without trying to sort them. Someday I'll do it, for sure. I'll get a nice album, put them all in order, write notes about them for my children.

Meanwhile, I ask myself: was she blind and frivolous, taking me to the photographer like that, in February 1944? Of course, she couldn't know what would happen a few weeks later: all her provincial cousins and uncles and aunts rounded up and murdered, we ourselves in hiding, hunted. But still, how could she continue her bourgeois rituals – the appointment with the photographer, the satin bows and silly poses – so late into the war? Unless it was that very perseverance that showed courage and insight: she would continue to affirm her life, our life, even in the face of destruction.

Sylvia Flescher
# Elsa Lost and Found

## Joachim

My father taught me to play chess when I was eight years old. Whenever we'd finish a game (which he inevitably won), he'd impress upon me how important it was to count all the pieces as I returned them to the box. "You must be very careful, Sylvia," he'd say. "You don't want to lose anyone." I'd pair up black king and black queen, white king and white queen, castles, knights, bishops and pawns. I thought of them as a family that needed to be kept together. I knew that my father had "lost his family" in something called the Holocaust. As a little girl, I thought, *perhaps my daddy had lost his family because he wasn't careful enough*. And I vowed to never, ever be careless, and to try my best not to lose my family, because then I'd be all alone in the world.

I cannot remember a time when I didn't know about the Holocaust. My sister and I simply accepted that we had no grandparents, no aunts or uncles or cousins. It was not that unusual among our classmates whose parents were also refugees from Europe. Looking back, I now understand that we tried very hard not to compound any sense of loss that my parents undoubtedly had, by drawing attention to who was missing at our holiday dinners. But to quote Holocaust educator Alan Berger,[1] there was a "presence of an absence" in our home.

I also carried with me the scar of an interaction with my father that has haunted me for years. We were visiting the annual boat show at the New York Coliseum. I was ten years old and thrilled to have Daddy all to myself. A salesman began to pitch a particular sailboat my father had stopped to admire. But then the salesman switched to a harsh guttural language I didn't understand, and my father abruptly took my hand, saying "Come, Sylvia. It's time to go home." In the taxi, I tried to puzzle out what had just happened, and my father explained through his clenched jaw, "He heard my accent and thought that I was German." To my mind, it was unfair of my father to assume that the salesman had been a Nazi, and so I asked, naively, "But Daddy, weren't there any *good* Germans?" My father, clearly enraged and now at me, stopped talking to me until, later that day, at my mother's urging, I went to apologize to him and ask for his forgiveness. Since then, I've always been a bit scared to ask the wrong question.

---

1 Alan Berger, *Second Generation Voices: Reflections by Children of Holocaust Survivors and Perpetrators* (Syracuse, NY: University of Syracuse Press, 2001).

https://doi.org/10.1515/9783110753295-014

My father was a psychiatrist and psychoanalyst. Many of his patients had, like him, suffered unimaginable trauma and loss in World War II. When I grew older, I learned that their treatment was often paid for by the German government as part of reparations. My father hadn't actually been in a concentration camp. He hadn't experienced first-hand the effects of Nazi brutality; I never heard him wake up screaming from nightmares. But he often spoke of his childhood friend, Josef Bibring, who had emigrated to Palestine before World War II. My father would say "Bibring is the only person alive who knew me before the age of twenty-four." Try as I might, I was unable to wrap my head around this. How was it possible that his entire family, along with all his friends and neighbors in his hometown in Poland had been exterminated, but he alone had survived? As a child, I sensed my father's deep sadness and also his quiet rage. What I could not understand until much later was how he struggled with survivor guilt. (Indeed, the very concept of survivor guilt, now well-established and well-understood, was only coined and described in 1968 by psychiatrist William Niederlander, a colleague of my father's.)[2]

My father was born in 1906 into a moderately observant Jewish family in the town of Buchacz, Poland, then part of the Austro-Hungarian Empire. After World War I, the family moved to the small city of Stanislawow. A very bright and ambitious young man, my father knew at an early age that he wanted to become a doctor, and then a psychiatrist. He was inspired by the writings of a Jewish doctor from Vienna, Sigmund Freud, who was shaking things up with some radical ideas about human sexuality, the importance of dreams, and the existence of the unconscious. Because of the quota limiting how many Jews could study medicine at Polish universities, my father left home for Vienna to pursue his medical degree. Struggling with depression, he also hoped to consult with Freud himself, denying the reality that he had not one extra zloty to pay for treatment. At the university in Vienna, he witnessed Nazi sympathizers throw Jewish students down the stairs while the police stood by. By some miracle, he was spared being attacked himself. But he vividly described to me examples of the blatant antisemitism of many professors. After completing his studies, my father returned to Poland and to his family, only to discover that Jews were not being given any opportunity to practice medicine. He decided to move to Italy for his psychiatric training and eventually hung out his shingle in Rome in 1936 (Fig.1).

---

**2** William Niederlander, "Clinical Observations in the 'Survivor Syndrome'," *International Journal of Psychoanalysis* 49 (1968): 313–315.

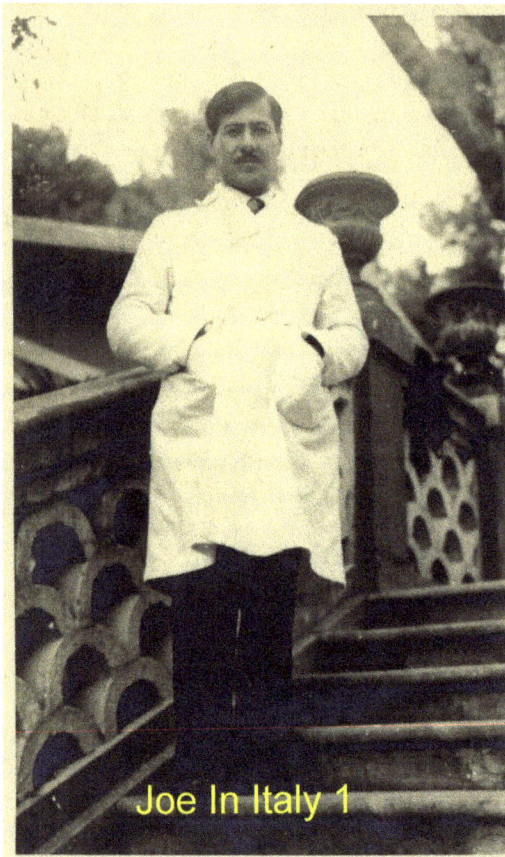

**Figure 1:** Joachim Flescher in Italy. From the private collection of the author.

On 1 September 1939, the Nazis invaded Poland, slaughtering thousands of Poles who fought valiantly but whose fire power was dwarfed by that of Hitler's military. One week earlier, Hitler and Stalin had secretly signed a non-aggression pact, agreeing to split Poland between them. Thus, my relatives, living in the eastern part of Poland, fell under Soviet rule for two years, until in 1941, Hitler broke his promise to Stalin and declared war on Russia. My family in Stanisla-wow then came under Nazi control, in the so-called "General Gouvernement." My father kept sending weekly care packages to his family, not knowing if they ever actually received any of them. With mounting desperation, he used the connec-tions of his high-status patients to wrangle an audience with the Pope, begging him to intervene and get the family out of Poland. But the Pontiff only declared his own helplessness, saying "You know they are killing our priests, too." My fa-ther even managed to get a petition to Mussolini himself, only to get it back with this scrawled comment in *Il Duce*'s handwriting on the top: "This is no time for Jews!"

Eventually, my father would learn that his entire family – over forty relatives – had all perished. Indeed, that there were no Jews at all left in Stanislawow, the first Polish city to be declared *judenfrei* by the Nazis. My grandmother and my two aunts had tried to escape, dressed as peasants, but were discovered and shot in the back at the railway station. I have no idea how the rest of the family met their end.

In September 1943, the Nazis occupied Rome and my father had to go into hiding. My mother, Anna, had been his secretary, and they had become lovers (Fig. 2). They shared an apartment on Via Flaminia. Swiss-born and not Jewish, my mother helped my father pull off a daring scheme to hide him in their apartment by declaring to the authorities that he had decamped and that she had no idea where he was. Hiding a Jew was a crime punishable by death. In 2008, a few months before she died, my mother was honored at Yad Vashem, the Holocaust memorial museum in Jerusalem, as a *Righteous Among the Nations,* for hiding my father and saving his life.

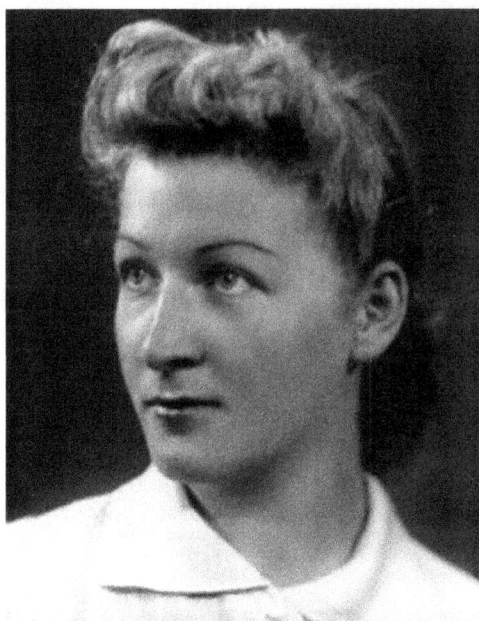

**Figure 2:** Anna Flescher. From the private collection of the author.

A few years after the war ended, my parents emigrated from Italy to the United States and were married. I was born in 1952 and grew up on Manhattan's Upper East Side. My father practiced psychiatry from his office in our apartment. It was the time of the Cold War and the threat of nuclear annihilation hung in the air. My father wrote and self-published a passionate treatise, entitled *Nazi Holocaust and Mankind's Final Solution*,[3] seeking to derive some lesson from the Holocaust that would alert mankind to its potential for total self-destruction. Our kitchen table was covered by books about the "war against the Jews." But for all that he was immersed in the subject, I never saw my father mourn specific relatives, or light a *yahrzeit* candle, or recite the Kaddish prayer for the dead. My father was a passionate Zionist but, following Freud, had a dim view of organized religion and saw no need for ritual. We never constructed a family tree, named the dead, or teased out the web of connections. No doubt that would have made my father's losses too real and too specific, only intensifying his survivor guilt.

I, on the other hand, found myself drawn to ritual and to Judaism. I devoured books on the history of the Jews – reading Leon Uris's *Exodus* four times. I believe I filled the emptiness of having no extended family by identifying with the larger Jewish family. I had complied with my father's dream for me to become a psychiatrist and psychoanalyst, but we differed when it came to the importance of community and the need to affiliate with like-minded souls. When I learned that by Jewish law, I would not be considered a Jew because my mother was a gentile, I decided at the age of eighteen to formally convert.

## The Letters

On a rainy day sometime in 2003, I was rummaging through papers in the storage room of my office. I came upon an aged, unlabeled cardboard box and discovered a treasure: letters from my grandfather Kalman, my grandmother Sala, and my two aunts, Zofka and Gusta, to my father in Vienna and Rome (Fig. 3 & 4). Some of the envelopes were stamped with the Nazi eagle, were marked "General Gouvernement," and had clearly passed through censorship.

Astonishingly, my father had never told us of this trove of letters from his family before he died from pancreatic cancer in 1976. It terrifies me to think how easily they could have been lost in my family's several moves over the years.

---

**3** Joachim Flescher, *Nazi Holocaust and Mankind's Final Solution* (New York: self-published, 1971).

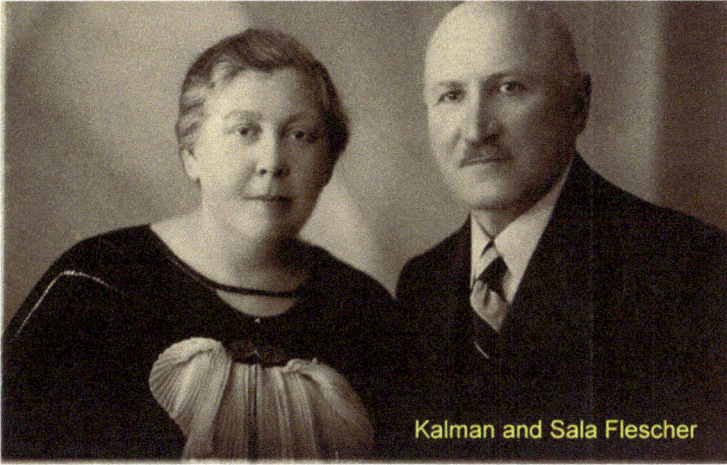

**Figure 3:** Kalman and Sala Flescher. From the private collection of the author.

**Figure 4:** Zofka Flescher with her daughter Emma. From the private collection of the author.

Of course, the letters were in Polish, German, and Italian, and I procrastinated for years about getting them translated. I feared stirring up an even greater sense of loss. Surely the words of my grandparents and my aunts would make them come alive again, and I'd need to mourn them in a new, more painful way. My sister and I did eventually arrange to have some of the letters translated by the Holocaust Memorial Museum in Washington, D.C., where we then donated them for safe-keeping and for future study.

One of the most heartbreaking letters was dated 22 December 1941. My Aunt Zofka in Stanislawow wrote to my father in Rome, asking him not to be angry with them for not telling him sooner how ill their father (my grandfather Kalman Flescher) had been. Kalman had died, and presumably my father had written to his family, expressing his anger at feeling deprived of the chance to come home to see his father one last time before his death or at least to attend a funeral service.

## Roots Trip

In 2010, I attended a panel at The Museum of Jewish Heritage in New York City. The subject was "Writing Our Parents' War." I heard Irene Kacandes read from her book, *Daddy's War: Greek American Stories,*[4] and her concept of the "paramemoir" resonated with me. I reached out to her and soon felt empowered by her encouragement to tell my own story.

My paramemoir-essay, "Googling for Ghosts: A Meditation on Writer's Block, Mourning and the Holocaust," was published in a psychoanalytic journal in 2012.[5] In it, I explored my conflict over feeling that it was my duty to write about my father's losses. I resented the way the Holocaust chronically overshadowed my life. I resonated with the title of Israeli psychologist Ilana Kogan's book *The Cry of Mute Children*[6] and understood my writer's block as a struggle against muteness. Did I have the right to tell my own story? If I wrote about my murdered relatives, it felt to me like they were continuing to take over my life. But if I did not write about them, was I not killing them off again?

In "Googling for Ghosts," I recounted the powerful experience of the Yad Vashem ceremony honoring my mother. Having thus entered my parents' story into

---

4 Irene Kacandes, *Daddy's War: Greek American Stories. A Paramemoir* (Lincoln: University of Nebraska Press, 2009).
5 Sylvia Flescher, "Googling for Ghosts: A Meditation on Writer's Block, Mourning, and the Holocaust," *Psychoanalytic Review* 1.99 (2012): 1–17.
6 Ilana Kogan, *The Cry of Mute Children* (London: Free Association Books, 1995).

the public record, I hoped that I would feel somewhat freed of the burden to memorialize my family's losses. In writing that piece, I realized for the first time that I had always thought of those who died in the Holocaust as my *father's* relatives. But they were *my* grandparents, *my* aunts, *my* cousins.

Irene Kacandes then invited me to speak on a panel at a conference sponsored by the Holocaust Education Foundation, a community of Holocaust scholars I hadn't even known existed. I wrote about discovering the letters and what that meant to me. At the conference, I felt that I'd found my tribe. Here were others immersed in the subject of the Holocaust; I felt less alone, more connected. For many years I'd put off making a so-called roots trip to Poland because I felt it would be too depressing. At the urging of my new colleagues, I decided that I was ready to join the HEF trip to Poland, the Czech Republic and Germany. This trip, offered every summer as part of their mission to train scholars and teachers in the history of the Holocaust, was planned around visits to several death camps, ghettos and memorial museums. I convinced my husband Tom to come with me.

It was time to become more educated about the details of the Holocaust. I needed to witness the geography of the land, to learn not just how my relatives had died, but also how they had lived. As I studied a map of Poland and came upon names that I'd heard all my life (Buchacz, Stanislawow and Yaremcze, the mountain resort town where the Flescher family vacationed), I realized that these places had taken on a mythical quality, like Brigadoon. But here they were, on the map: the actual towns in black and white.

Our first stop, not far from Warsaw, was Treblinka, the death camp where nine hundred thousand people were murdered. The camp buildings were all destroyed by the retreating Nazis, but in their place a huge field of jagged memorial stones now stretches over dozens of acres. Each stone is inscribed with the name of a village, a town, a city. Our guide helped me to find the large stone for Stanislawow, and I placed a small stone on its top, as a Jewish mourning ritual.

The next day we toured the Warsaw ghetto and the one synagogue still standing (there had been three hundred before the war). The Nazis had turned the elegant temple into a stable for their horses. But most impressive to me was the Jewish cemetery. Here was evidence of the Jews living in Warsaw for centuries, of prosperous families erecting monuments to their dead. Here were the graves of over two hundred and fifty thousand Jews who had died of natural causes over more than two hundred years. The day before we'd walked through a place where nearly four times as many people had been killed in less than two years and their bodies reduced to ashes, the victims remaining nameless, deprived of any civilized society's respectful treatment of the dead. The contrast between Treblinka and the Warsaw Jewish cemetery could not have been starker. In the days that followed, we visited several other infamous death camps and memorials.

In the new Museum of the Jews in Poland, all the signs are in three languages: Polish, English and yes, Hebrew! Hitler thought he was going to wipe the Jews from the face of the earth, and yet the ancient Hebrew language continues to thrive, alive and well.

The trip exceeded my expectations. I told people that it had been a transformative experience, but even then I did not realize just how transformative it would turn out to be.

# Elsa

It is summer, 2015, shortly after our return from Europe. My husband and I drive up to our lake house in the Adirondacks, where my family has spent summers since the early 1950s. My father always said this area in upstate New York reminded him of the Carpathian Mountains in Poland where he had vacationed with his family when he was young. Across the lake, a friend of ours, Karen Cohen Holmes, listens closely as we recount the powerful experiences of our recent trip. When she hears we're traveling the following week to Montreal, Karen says excitedly: "Well then, you must meet my dear friend Andrea Axt who lives there. She lived for many years in Poland after the war and has worked on translating diaries written during the Holocaust. I know you'll have a lot to talk about!"

Andrea and I exchange e-mails and we agree to meet for Sunday brunch. She's a petite woman in her late seventies, with reddish curly hair and a vivacious manner, yet I detect an underlying sadness. We've barely sat down when she takes a slim purple volume out of her pocketbook and hands it to me. Its title: *Two Diaries from Witnesses and Victims of Extermination of the Jews in Stanislawow* – the very city my father came from! I look more closely at the cover and read the names of the two diarists at the top. One is Elsa Binder – my world tips slightly – Binder, but that was my grandmother's maiden name! She was, I tell Andrea, Sala Binder Flescher. "Then certainly you are related!" Andrea says excitedly. "Binder is not that common a name." My head spinning, I try to take in some of the details Andrea is providing us: Elsa's notebook was found in July 1943, in a ditch by the side of Batory Street in Stanislawow. This is the street that leads to the Jewish cemetery, which is where the Nazis performed mass executions of the local Jews. A high school teacher was riding by on his bicycle and spied it.

I wait to read the diary until later that day when we return, exhausted from sight-seeing, to our little Airbnb. As I open the book to the first page, I'm aware that I feel anxious. Am I hopeful that this diary, this voice from the tragic past will shed light on my family's fate? What if there is no connection, and I'll be disappointed, sorry I'd gotten my hopes up? On the other hand, what if there *is* a connection and I will learn some awful details about how my relatives met their end? I take a deep breath and begin to read.

First comes Andrea's introduction, telling us that Elsa was twenty-two years old on the date of the last entry to the diary, which is 18 June 1942. This is when Elsa probably perished in a round-up of victims, a so-called *aktia*. She and her parents were likely transported to Belzec and gassed. So, here's a mystery: if the diary was found a full year later, how did it survive the snows of a Polish winter, and the melting that spring would have brought, surely blurring if not erasing the ink? Perhaps Elsa entrusted the diary to someone who placed it there a year later, in a spot where someone would be certain to see it? Or someone who moved into their vacated apartment discovered it, and felt compelled to anonymously ensure it survived, even if its youthful author did not? So many questions with no one to supply the answers.

I turn the page to a map of Stanislawow. I search for street names that correspond to addresses on the envelopes of the letters from my father's family, but the writing is too small. Then comes a sample of Elsa's handwriting in Polish, so much like the many letters from my family hidden for decades in that old card-board box. I turn the page and read the first entry:

> 12/23/41 – Yesterday, my aunt, tears in her eyes, brought over a letter from her son in Italy . . .

Wait. I read the line again: a letter from her son in Italy. Can this really be? I am dumbfounded. The son in Italy that Elsa refers to in these opening lines can only be my father! And the aunt she refers to is my grandmother, Sala! At that moment, I realize that I'm holding in my hands the diary of my father's first cousin. I read on:

> Yesterday, my aunt, tears in her eyes, brought over a letter from her son in Italy. In it he complains that he wasn't told earlier of his father's illness. In his innocence he claims he would have come to save his father (straight into the jaws of the lion.) [Here is the first example of Elsa's characteristic sarcasm.] He asks that flowers be taken to the grave. How little idea he has of our conditions!

Elsa is referring to my father's reaction upon learning of the death of his father Kalman, my grandfather. She must have realized that her cousin in Italy has absolutely no clue as to how dire the circumstances for the Jews in Stanislawow are, otherwise he'd never ask that flowers be put on his father's grave. How could he think, had he known earlier of his father's condition, that he'd have been able to come to see him?

The day after that first entry, all the Jews are forced into a ghetto area of the city and Elsa begins to intensely record, over the following six months, an eyewitness account of the day-to-day conditions of the ghetto. Did my father's letter, with its challenging tone, inspire her to bear witness? Moreover, had Elsa not begun her diary with those lines referring to my father's letter, I could have read the entire notebook never knowing how, or even that we were related.

Here are some excerpts from Elsa Binder's diary.[7]

> Dec. 27 1941 . . . Like thunder from the sky a decree fell on us that Jews must, under penalty of death, give up all furs and pieces of fur. The cheaper ones were taken long ago. To make sure that the order will be followed, a few dozen hostages were taken. I look at my threadbare coat which has the wind for a lining . . . and I wonder whether the Germans, caught off-guard by the winter, intend to dress their troops in Jewish furs. I think that if everyone's furs are as pitiful as mine, I don't envy the Germans.

Elsa often uses sarcasm and refers to literary texts and Latin words. She's intelligent, educated, clear-eyed and matter-of-fact. In between episodes of heartbreaking suffering, she also confides the more mundane preoccupations of family life, musing that her mother loves her younger sister Dora more than her.

The next entry is written exactly eleven years before the day I come into this world, and I realize with a shock how recent my father's losses were when I was born.

> Dec. 31, 1941 . . . Today is New Year's Eve. The last day of 1941. That also means the end of my twenty-first year, still embattled with the gods. They've endowed me with a face that basically makes no good impression and a set of rounded shoulders that are even less exciting . . . As it stands all Jewish girls share the same fate. Neither death nor prison considers one's looks. Many of my peers, so much more naturally endowed than I, have gone to their fate.[8] And I, as we all stand in the face of death, desire life but am not afraid. I'm not sure whether all my dear ones who are dead might not be better off than I am today or will be tomorrow. I see them every night still together, just as when they left, lying together, happy and laughing and above all, free. Truly free for now and forever . . . It's an icy winter. Bony-fingered frost knocks on the window. We have coal for only a week and food to last a month . . . What are you bringing me, long-awaited 1942 . . . I greet you, 1942, may you bring salvation and victory. May you be the first year of grace for this ancient and unlucky race whose fate lies in the hands of the unjust. And one more thing. Whatever you bring for me, be it life or death, bring it quickly.

---

7 Elsa Binder, "Eliszewa's Diary," in *Two Diaries from Witnesses and Victims of Extermination of the Jews of Stanislawow* (Montreal: Polish-Jewish Heritage Foundation, 2015), 17–64.

8 Here Elsa is referring to the massacre known as "Bloody Sunday," when twelve thousand Jews were rounded up and executed in a mass grave.

Three days later she writes:

> The evenings are the worst. At 7:00 pm, following orders to shut off lights, the whole family goes to bed. Only I waste a candle by which to read or write. Later, I go to my unheated room and to a bed warmed by a brick and there I have a chance to think about my life. Life? I doubt that you can even call it life. Vegetating? Not that either. I do think, feel and suffer. Everything in me is just one chord of pain and hope . . . . Yes! I want to live. I want to eat well (butter is a legend to me and milk a distant memory of the past), I want to dress well (I can't remember the last time I did), I want intellectual stimulation, and here comes the last and dearest wish – I want to love and be loved. This is the law of nature. It's sad but it's healthy. My God: At the fresh grave of my peers here I am dreaming of such trivial things.

On January 12, the three-month anniversary of Bloody Sunday, Elsa writes, addressing her murdered friend, Samek:

> I don't even know how you died. Did a sadist's brick, tossed into a crowd, hit you? Were you smothered by masses of people retreating to save themselves? Perhaps a well-aimed shot in the breast and then you were swallowed in the huge common grave? Oh God, what if you were slightly wounded, but lay slowly dying in a pile of corpses?

The next day's entry continues:

> The number of victims ranges between ten and twelve thousand. Mother Earth cried over them, the wind moaned, the tops of trees bent their branches, and the cloudy sky cried rain and snow. The victims were accompanied by cries and moaning and the mighty call of "Shma Yisrael." And God was silent. Mothers were left without children, children without mothers, houses and apartments stood empty. And hearts were left hardened by pain, hate, and helplessness. But in my heart, strangely, there is no hate, only pain, shock, and the universal: "why?" How could sons of mothers and fathers of children drive proud wrinkled elders, whimpering babies, healthy youths and pregnant women to the cemetery where fresh common graves were waiting? Was it in the name of the love that their religion teaches that they forced mothers to smother their children in the terrible crush and children to trample their mothers? Supposedly, international law demands that criminals and spies sentenced to death by shooting be blindfolded so that they need not look death in the face . . . But here little children clinging to their mothers cry, "Mama, make me go first I don't want to see when they kill you" . . . [these children] are not only made to face death directly, but they get to see how one can hold a pistol in one hand and a cigarette in the other . . . . they also see how mothers tell their children about death, and how they prepare them for it. And when they cry "Mama, we want so badly to live," she cries out "children, since this is not given to us, we will die together" . . . She can see an infant's head crushed under a sadist's boot, and rifles pointed at pregnant bellies. The crowd is silent. Perhaps it's praying.

Every time I read these words, I need to fight the urge to look away. I think that my enduring ambivalence about immersing myself in the horrors of the Holocaust results from being saddled too young with the duty to compensate my father for his losses. The Israeli psychologist Dina Wardi said it best when she called children of survivors "memorial candles."[9] I don't want to be anyone's memorial candle. But now, Elsa has helped me to not look away. My sense of being alone with the Holocaust trauma has gradually been mitigated over the past decade by several key events: first, the Yad Vashem ceremony, then writing about it in "Googling for Ghosts," then joining the Holocaust Education Foundation community, then going on the "roots trip" and ultimately discovering Elsa and her diary. This chain of events beautifully exemplifies the Buddhist saying, "when the student is ready, the teacher appears." I feel that the universe has given me a precious gift.

Ironically, while I was mourning not having any evidence of my lost relatives, academics had been teaching the diary for twenty years. Excerpts had been published in Alexandra Zapruder's collection, *Salvaged Pages*.[10] My father died without knowing that his cousin's diary had survived. Would it have been a comfort, I wonder, or instead painful evidence of how his family suffered? Of one thing I can be certain: he would be pleased at the sustained scholarship being devoted to studying and teaching the Holocaust.

Elsa was known as Eliszewa to her Zionist youth group. Only two photographs of her exist that I know of. In them, she is with other girls from that group, all wearing the uniform of *Hashomer Hatzair* (Fig. 5). A photograph of my Aunt Gusta, around age thirteen, shows her in the very same uniform (Fig. 6).

Elsa's last name remained unknown for a long time. When parts of the diary were translated into Hebrew by a member of kibbutz *Sha'ar Haamakim*, another kibbutz member recognized Eliszewa as the name given to Elsa Binder in their shared *Hashomer Hatzair* youth group. I can now google "Elsa Binder" and find several links. Her diary appears in an MTV documentary called *I'm Still Here*,[11] based on the Zapruder anthology. The filmmakers begin their film with a line from Elsa's diary: "There is no hate in my heart . . . Only the universal question, why?"

The psychoanalyst Hans Loewald has written some much-quoted words about the necessary process of turning ghosts into ancestors, in order to live more comfortably with one's personal history. Ghosts haunt us, ancestors connect us.[12] No doubt the fact that my mother was not Jewish had left me yearning to connect with female Jewish ancestors, to counterbalance my father's outsize influence on

**9** Dina Wardi, *Memorial Candles: Children of the Holocaust* (New York: Rutledge, 1992).
**10** Alexandra Zapruder, *Salvaged Pages: Young Writers' Diaries of the Holocaust* (New York: Yale University Press, 2002).
**11** *I'm Still Here: Real Diaries of Young People Who Lived During the Holocaust*, Director Lauren Lazin (MTV Networks, 2005).
**12** Hans Loewald, "On the therapeutic action of psychoanalysis," *International Journal of Psychoanalysis* 41 (1991): 16–33.

**Figure 5:** Eliszewa (Elsa Binder), center, with Zionist youth group. From private collection of Andrea Axt, with her permission.

my development – and Elsa has come into my life to do just that. She's become the representative of all the relatives I never knew. Her eloquent words underscore what I've been cheated of, but they also serve to fill in some small way the emptiness from my past. I can hear her voice and connect with her intelligent and soulful spirit. Now when I sit down to write, I feel that I'm carrying on a family tradition, and that somehow Elsa's spirit is being channeled through me.

There are no graves for my murdered relatives. I cannot place a pebble on their headstones. But the diary of Elsa, my gifted second cousin, lives on.

**Figure 6:** Gusta Flescher. From private collection of the author.

# Works Cited

Berger, Alan. *Second Generation Voices: Reflections by Children of Holocaust Survivors and Perpetrators*. Syracuse, NY: University of Syracuse Press, 2001.

Binder, Elsa. "Eliszewa's Diary." In *Two Diaries from Witnesses and Victims of Extermination of the Jews of Stanislawow*, 17–64. Montreal: Polish-Jewish Heritage Foundation, 2015.

Flescher, Joachim. *Nazi Holocaust and Mankind's Final Solution*. New York: self-published, 1971.

Flescher, Sylvia. "Googling for Ghosts: A Meditation on Writer's Block, Mourning, and the Holocaust." *Psychoanalytic Review* 1, no. 99 (2012): 1–17.

*I'm Still Here: Real Diaries of Young People Who Lived During the Holocaust*. Dir. Lauren Lazin. MTV Networks, 2005.

Kacandes, Irene. *Daddy's War. Greek American Stories. A Paramemoir*. Lincoln: University of Nebraska Press, 2009.

Kogan, Ilana. *The Cry of Mute Children*. London: Free Association Books, 1995.

Loewald, Hans. "On the Therapeutic Action of Psychoanalysis." *International Journal of Psychoanalysis* 41 (1991): 16–33.

Niederlander, William. "Clinical Observations in the 'Survivor Syndrome.'" *International Journal of Psychoanalysis* 49 (1968): 313–315.

Wardi, Dina. *Memorial Candles: Children of the Holocaust.* New York: Routledge, 1992.

Zapruder, Alexandra. *Salvaged Pages: Young Writers' Diaries of the Holocaust.* New York: Yale University Press, 2002.

Leo Spitzer
# "Something Dreadful . . ."
# *[Etwas Schreckliches . . .]*

An early memory:

*My mother, grandparents, and uncles are congregated around my father in the kitchen of the apartment in Miraflores where we all live in La Paz in 1945. I am with them, a six-year-old, present but on the margins. My father is sobbing – something I have never seen him do – and he is holding a letter he retrieved from the central post-office earlier that morning. It's from Frieda, he tells us. She sent it from a Displaced Persons camp in Germany addressed to him and to her brother Ferry at "Poste Restante, La Paz, Bolivia." But the family's joy in receiving this sign of life from a close relative in post-war Europe is almost immediately shattered when my father, reading Frieda's letter out loud, voices her first line:*

"Ich bin die Einzige, die überlebt hat."
*["I am the only one who survived."]*

What ensued is less well-etched in my mind. Probably, Frieda's startling and unambiguous opening declaration of survival and death was the preamble for her account of the Nazi killing of her parents and youngest sister, along with hundreds of others, shortly after they arrived in Riga, Latvia, in February 1942 on a transport train crammed with Jews from Vienna. It framed her determination to handwrite a letter, after years without possibility to do so, in order to tell about horrors she witnessed and terrors she and thousands of others suffered during their lengthy incarceration in the Riga Ghetto and Kaiserwald concentration camp as slave laborers for the Nazi regime.

I have no precise recollection of my childhood response at the time. How did I react? Did I also weep and wail like the adults surrounding me? Did I rush to my mother or one of my grandmothers for solace? I just cannot remember. Yet I do know that the contents of Frieda's letter on that day, and the spectral scene of family sorrow and despondency that immediately enveloped my father's reading of her words in our somber kitchen room, confirmed and fed some of my already dark childhood fears. Implanted and grown within my consciousness during our Bolivian refuge by whispered, yet overheard, adult conversations about war and relatives left behind, about Hitler and Nazis, about cruelty and persecution – and nourished by a nervous threat and invisible dread of a terrifying "over there" – my fears were now validated by the testimonial concreteness of the written

https://doi.org/10.1515/9783110753295-015

document from Frieda that we had come to possess. Over the decades, I have realized how profoundly these fears marked me. They have remained with me like the everlasting scars from an acid burn.

•

Frieda Kohn became my aunt when she married Julius Wolfinger, my mother's younger brother, not long after receiving a visa and being admitted to the United States in 1949. But as a daughter of my father's half-sister, Gisi Spitzer Kohn, she was also my first cousin, and both of us prized the fact that we were doubly related. Leopold Spitzer, my paternal grandfather (after whom I was named) was also her grandfather. We joked and laughed about our augmented kinship, but we also believed it was special. "When I tell you to do something, remember I am your *Tante* Frieda," she would sometimes warn me while smiling and shaking her index finger. "The rest of the time I can just be your *Cousine* [cousin] Fritzi." Certainly, the funny idea that "just your *Cousine* Fritzi" could pull a version of parental rank on me as my senior "*Tante* Frieda," strengthened our close bond even more – the lovingly affectionate friendship that developed after I landed in the United States as a ten-year old in March 1950, almost a year before my parents were granted their visas and admitted to this country. I spent much of that year living with Frieda and Julius, cared for and indulged by them and my maternal grandparents in the small tenement apartment they all shared on Second Avenue in Manhattan. While we each launched into our various immigrant adoptions and adaptations of a new language and what, in the US, was imperiously called "Americanization," it was in that apartment that Frieda and I began to nurture a close, often playful, connection which deepened over the years to the point where we felt we could trust each other to talk about most anything.

I deliberately use the adverb "most" because throughout the 1950s, when I spent much time with Frieda and Julius and my grandparents – and even in the '60s and '70s when I attended university and started to teach – she and I chatted openly and unreservedly about numerous subjects in our lives. Numerous subjects, that is, except those touching on things that happened to her in the nearly four years of captivity during the Holocaust.

Her silence might seem puzzling given that, not long after Germany's surrender and her own post-war liberation, Frieda herself had written the letter to our family in Bolivia that conveyed aspects of her experience during this immensely traumatic phase of her life. But her subsequent lengthy avoidance of talk about these matters to all (except, most probably, her husband Julius) was in fact not unusual. At first, like Frieda, many survivors in the years immediately after the war, often from DP Camps where they were living, also attempted to reconnect by mail with relatives and others from their past – and, in so doing, provided narratives about their existence under Nazi domination. Some, like the one hundred thirty survivors interviewed and wire-recorded in various European shelter houses a year after the war's end by the Chicago-based psychologist David Boder, or those who labored to reconstruct scribbled scraps of notes into more coherent written diaries and memoirs, readily testified to a potentially wider audience about what they and fellow Nazi victims had suffered and borne.[1] And yet, in the following decade of the 1950s, public witnessing by Holocaust survivors declined substantially. "As many people have pointed out," Dominick LaCapra observed about this occurrence,

> right after the events there was a rush of memoirs and diaries, and then it all sort of died down for a long period of time – what is tempting to interpret as a period of latency after a traumatic series of events. One of the reasons is that survivors found – in different countries, for different reasons – that they didn't have an audience; they didn't have people who wanted to listen to them.[2]

I have no knowledge that this was applicable to Frieda at the time. But, like many other surviving contemporaries in the 1950s, she did not speak about her life under Nazi domination during this decade – and, I believe, no one in our family asked her to do so. I surely did not. This, however, again changed appreciably in the aftermath of the 1961 Eichmann trial in Jerusalem, which relied extensively on spoken witness testimonies, and which employed radio broadcasting and documentary film in Israel to disseminate Holocaust accounts publicly to a large national and international audience. Indeed, it was this trial and its widescale reception that influenced the upsurge and proliferation of Holocaust testimonies in the ensuing years – in the "Era of the Witness" [*L'ère du témoin*], as the historian Annette Wieviorka has labelled them.[3] And "witnessing," oral and video recording, and the archiving of Holocaust life history narratives gained further impetus after the October 1985 premiere of Claude Lanzmann's groundbreaking film *Shoah*, the nine-and-a-half-hour documentary consisting of interviews, accounts, and reflections by Holocaust victims, bystanders, and even perpetrators.[4]

**1** David Boder, "Voices of the Holocaust, Oral Histories" – the earliest known recorded oral histories of the Holocaust can be accessed through the Illinois Institute of Technology [https://lit.aviaryplatform.com/collections/23/].
**2** Dominick LaCapra, "Interview for Yad Vashem (9 June 1998)," *Writing History, Writing Trauma* (Baltimore: Johns Hopkins University Press, 2001), 158.
**3** Annette Wieviorka, *The Era of the Witness*, trans. Jared Stark (Ithaca, NY: Cornell University Press, 2006).
**4** Wieviorka, *Era of the Witness*, 56–144.

In April 1987 Frieda watched *Shoah* on television when it was broadcast over the course of four days in the United States by PBS. She did this together with her friend, Lore Dühring Markusfeld, with whom she had been close in the Riga Ghetto and ever since, and with Lore's husband, Kuba Markusfeld, who had fought in the Warsaw Ghetto uprising. Not long afterwards, the three pulled Marianne and me aside at a gathering in New York and began to speak, initially about Lanzmann's *Shoah* but then about their own captivity and experiences during the war years. It was the first time that Frieda, as well as Lore and Kuba, had ever initiated such a talk with us. The television showing of Lanzmann's documentary, the extensive buzz and discussions it engendered, and the brief appearance within *Shoah* of their friend and fellow Riga survivor, Gertrude Schneider, along with Schneider's mother and sister, seems to have validated the fact for them that they also could speak about their traumatic pasts to willing and sympathetic listeners.

•

By this time, I did of course already know quite a lot about Frieda's past. My father, and others in the family, had spoken about the forced displacement of the Kohns (together with the Spitzers) after the Nazi "cleansing" of Jews from the Burgenland following the *Anschluss* of Austria in March 1938. I learned about the dire circumstances in which the Kohns had subsisted in Vienna prior to their train transport "eastward," to Riga, in 1942. Frieda's mother, Gisi, had been accosted, slapped, and spat upon by teenage Nazi-brownshirts on a Vienna street, and she was so traumatized in the aftermath of this incident that she largely stopped speaking. It was from my parents as well that I first found out that in late 1939, when Jews still had some chance to flee Austria, Frieda's father Leopold Kohn had let his son, Ferry, make his own way out of Vienna – across the border to Switzerland and then to Bolivia – but he had stubbornly forbidden his wife and daughters to leave without him. He had a brother near Budapest, and it was safe in Hungary, Leopold had maintained. He expected to be able to join his brother "any day." Both Frieda and Ferry would later confirm this account to me.

There was also a shattering claim, which I had heard my mother make without attribution on several occasions during my teens and early twenties, that "something dreadful" (*etwas Schreckliches*) had happened to Frieda in the Riga Ghetto, an event that would forever prevent her from becoming pregnant and bearing a child. My mother did not elaborate on this. Both she and my father were always quite reticent, seemingly embarrassed, to discuss sexual matters openly with their children. Admittedly, however, I assumed that the "something dreadful" referred to sexual violence – perhaps to rape and forced abortion: the mandated elimination of pregnancies by the Nazi Riga Ghetto authorities.[5] I've always considered Frieda's extraordinarily meticulous home cleaning, her craving to put everything in order and make things neat, as a reaction formation against the harrowing memories of a chaotic time when everything, certainly, was out of her control.

•

I videotaped two interviews with Frieda after she viewed *Shoah* – the first, a very abbreviated one, in November 1989 and the second, nearly three hours long, in July 1992. In these she speaks about her rural childhood and schooling in the town of Hodis, Burgenland, Austria, where she was born, and about her early relationship with her parents and siblings. She recalls her early friends and playmates – many, if not most, non-Jewish children – and believes her experience to have been "a pretty nice childhood, for the most part." She tells about the awkward day visits of our very orthodox Jewish grandfather, Leopold Spitzer, who would travel to Hodis from nearby Rechnitz where he lived, but who would never touch food served in their secular, unkosher, home. And she speaks of her mother, Gisi, "a beautiful, joyful, person both physically and spiritually during those years" – one who "always moved about the house singing and happy," in marked contrast to the "shocked and devastating wreck" she would become after the family was forced to move to Vienna.

The bulk of Frieda's video narrative, however, concentrates on her displacement and Holocaust years: on the grim months she and her family spent in Vienna under Nazi domination, and on her subsequent ordeals as a slave laborer in Latvia and Germany before her escape from Nazi control in the final days before the German surrender. She describes her train journey arrival in Riga in the bitter-cold winter of 1942 and the almost immediate separation from her parents and siblings. "You will never see them again," she was told in a whisper by "Blackie" (Herbert Schwarzberg), a young man her age whom she had previously known in Vienna, but who had shipped off to Riga on a previous transport, and who was now assigned by the SS as one of the baggage unloaders of the train on which she had travelled. She then provides a lengthy account of her incarceration in the Riga ghetto, where she initially lived with five others in a tiny, dark, rat infested room, and from which, every morning after the *Appell* (roll call), she was assigned to a labor squad. "We worked like horses," she remarks,

---

5 Gertrude Schneider, *Journey into Terror: Story of the Riga Ghetto* (Westport, CT: Praeger Publishers, 2001 [2nd expanded ed.), 61–66; United States Holocaust Memorial Museum (USHMM), "Interview with David Klebanow, 7 December 1989," RG-50.030*0104. Klebanow and his wife performed abortions on inmates in the Riga Ghetto.

"sometimes a week or two at the same place, sometimes for shorter." She separated military clothing that arrived from wounded and killed soldiers at the front;
she worked with construction gangs at Spilve airport; she harvested bogs for peat
stacks that were used as bricks for heating. "I was young, I worked fast, and
sometimes I helped some of the older people," Frieda notes. Occasionally, while
laboring outdoors, her work contingent would encounter Wehrmacht soldiers.
"Ninety-five percent were Nazis . . . but some put a little food and bread out for
us. This was very dangerous for those soldiers – and for us if we had been seen."

In the course of her 1992 video testimony, Frieda also mentions performing
two furtive physical actions. One took place while walking with a labor detail
from a work site. She describes herself as looking horrendous: dirty, pale, terribly
undernourished and dried out from lack of water. On the way back to the Riga
Ghetto she spotted what looked like a tiny red bead, possibly a drying but still
juicy plant seed. She managed to pick it up and take back to her ghetto room
where (and she chuckles when she tells this) "I squeezed out its red juice and put
it on my cheeks. That way I could look alive."

The second also recalls an act of self-assertion and self-care. It relates to the
time in late 1943 when the inmates of the Riga Ghetto were transferred to the
Kaiserwald concentration camp. Only about one-fourth of their total would in
fact be quartered there. The majority, including Frieda, were billeted at their
places of forced labor: various installations, factories, and farms producing materials and supplies for the German war machine. But it was upon their initial
admission to the Kaiserwald authority – after being stripped of their own clothing and given striped camp uniforms, checked for lice, and disinfected – that
the head and body hair of all newcomers was cut. Such en-masse haircutting
had previously not been carried out in the ghetto.[6] Frieda, who had possessed
(and, after the war, would again possess) luscious, thick, curly brown hair, was
shorn of hers. Instead of weeping, however, she managed to save one of the cut
curls from her head. Throughout all the subsequent days, she smilingly recalls
in her account, when she and others were made to labor for the Nazis, she wore
it proudly on her head, barely visible, stuffed on the edge of the headkerchief
women workers were compelled to wear.

•

Watching my video recordings of Frieda nowadays, I listen to her narrative,
view her facial and body motions, and I am profoundly stirred by her poignant

---

6 Schneider, *Journey into Terror*, 89–91.

testimonial act. I cannot be seen on the video. Only my voice is heard. Yet it is apparent to me that we have a relaxed, open, "just your cousin Fritzi" relationship, and I ask her many questions – many more directive ones, certainly, than I would ask in the scores of video-testimony interviews that I carried out with other Holocaust survivors and with Central European refugees that fled to South America.[7] But, much to my regret now, in these interviews I could not bring myself to ask Frieda about the "something dreadful" that had allegedly happened to her in the Riga Ghetto. In fact, I could *never* bring myself to question her about this purported profound traumatic event.

I was with Frieda in New York for a few days in February 1996 before she died. She had lung cancer. She was in bed and I held her, and we talked and talked softly in German about many things until we finally both fell asleep as dawn was breaking. But the Riga Ghetto was never mentioned. She did not talk about it, and I did not bring it up.

•

Looking back on my failure to ask Frieda about sexual violence that she may have been subjected to, I realize that I considered such a question transgressive – a crossing of the ethical boundaries of privacy. I had internalized a wariness to talk openly with others about such matters, especially across generations. Apparently, members of my family shared my reluctance. They (like me) seemed to want to focus on the "Fritzi" whom they knew and had come to love and relish: the slightly quirky relative who was a fastidious cleaner, loved Chinese food, especially sweet-and-sour shrimp, enjoyed shopping at Alexanders on Queens Blvd., played penny-pot cards with friends, laughed heartily at risqué jokes, and baked the most incredibly delicious *Nusskipferl* and *Vanillenkipferl* for every family get together. Gertrude Schneider, however – "Trudl," as Frieda called her lifelong friend – did reveal many details about life in the Riga Ghetto that Frieda always omitted. Schneider, who became a Professor of History at New York's City College, published an academic book, *Journey into Terror: The Story of the Riga Ghetto*, first in 1979 and then in expanded form in 2001. But she was also a central witness of multi-hour interviews about the Riga Ghetto and Kaiserwald that Claude Lanzmann videotaped for *Shoah* (with Schneiders' mother and sister in attendance) – fascinating, informative, sessions of which Lanzmann only used a few minutes in the final cut of his film, but whose outtakes eventually became

---

7 My interview approach with Holocaust witnesses – for the most part "minimal-interventionist" – was largely influenced by the late Geoffrey Hartman and Dori Laub, and by Joanne Rudolf of the Fortunoff Video Archive for Holocaust Testimonies at Yale University.

available for viewing on-line in the video archives of the United States Holocaust Memorial Museum and Yad Vashem.[8]

From her accounts, as well as from documents and available written, audio, and video testimonies from other witness/survivors, it has become clear that the Riga Ghetto was not only a site congregating Jewish slave labor for the Nazi Reich's war effort but also a place of sexual and gender violence perpetrated against inmates.[9] Women prisoners were forcibly aborted and, in some cases, sterilized if they became pregnant. The medical personnel generally did not use anesthesia during these procedures. And yet, since men and women were not separated in the ghetto, Schneider and other survivors also inform us that, despite written and oral prohibitions by Nazi authorities, inmates also found occasions and places to have consensual sex.

Thinking about Frieda and the "something dreadful" with these explanations in mind, I again watched all the outtake segments of Gertrude Schneider's lengthy interview with Claude Lanzmann. I came to a part where Lanzmann asks her to tell him about songs that Riga Ghetto inmates sang, and the circumstances that might have prompted them to do so – questions that immediately elicited a smile from Schneider and a visibly brightened expression. "At first, you know, when young people come together . . . you sing [and] you dance sometimes," she responded,

> There are always talented kids around and they make up songs. And one of the young boys, at the time he was seventeen, he was THE Casanova of the ghetto. Yes, he was very good-looking, no? He had grey eyes and black curls, and he had a mandolin, and he could play, you know, all the young girls . . . I think he went with three at the same time.[10]

Schneider then sings a few songs that she especially associates with occasions in ghetto life when inmates, principally younger ones, came together in small groups after a hard day of labor and entertained one another musically. Her animated performance and response to Lanzmann illuminates an aspect of the Riga Ghetto experience that our customary concentration on its horrors tends to push out of consciousness. Among these songs, however, I heard one about a ghetto seduction and its sad outcome that I found to be quite jolting. Performed in response to Lanzmann's query "You told me that there was another song

---

**8** USHMM, Gertrude Schneider video outtakes from Claude Lanzmann's film *Shoah*, | Accession Number: 1996.166 | RG Number: RG-60.5015 | Film ID: 3221, 3222, 3223, 3224, 3225, 4717.
**9** See for example listings under Riga Ghetto at the Fortunoff Archive at Yale University, the U.S. Holocaust Memorial Museum, Yad Vashem, and the Spielberg Shoah Foundation.
**10** USHMM, Gertrude Schneider video outtakes from *Shoah*, FV 3224.

about a love affair between a Latvian and Jewess . . . which ended with an abortion in a hospital," Schneider sings it with a smile and is soon joined on camera by her mother and sister who both also chuckle at its conventional representation of seduction and betrayal:

> *Ein Mädchen ging spazieren am Zaun* [A girl was walking by the fence]
> *Und war so ganz allein* [And was so all alone]
> *Doch nach einer halben Stunde* [But after half an hour]
> *Waren sie zu zweien* [There were two of them]
> *Ja, ja, ja* [Yes, yes, yes]
> *So ist traurig aber wahr* [So sad but true]
> *Nein, nein, nein, ach nein* [No, no, no, oh no]
> *Von einem Mal kann es nicht sein* [It can't be from a single time]
> *Warum kommst du heut so spät?* [Why are you coming so late today?]
> *Schatz, du weißt dass es früher nicht geht* [Sweetheart, you know that earlier was impossible]
> *Ich hab dir was mitgebracht* [I brought you something]
> *Darum bleib ich heute Nacht* [That's why I'm staying tonight]
> *Bei Dir* [With you]
> *Er hat sie dann geküßt* [Then he kissed her]
> *Wie das so üblich ist* [As is customary]
> *Der Schnaps verhalf zu mehr* [The schnapps helped greatly]
> *Sie wehrte sich nicht mehr* [She no longer resisted]
> *Ein Mädchen liegt im Spitale* [A girl is in the hospital]
> *Das Eine nie mehr vergisst* [One that cannot be forgotten]
> *Doch er hat längst sie vergessen* [But he has long since forgotten her]
> *Wie das im Getto oftmals so ist* [As is so often the case in the ghetto][11]

Listening to this song and watching the sparkle in Gertrude Schneider's eyes, I suddenly perceived Frieda's silence in a new light. I saw an opening to a different story – one more affirming and closely aligned with her account of the red-bead, the hair curl, and the concern for her appearance that they reveal. Certainly, a forced abortion, if she had one, possibly without anesthesia or a pain killer in the circumstances of her ghetto incarceration, would have been a traumatic event. But, instead of focusing on an abortion and the childlessness that it may later have brought about, the small details Frieda revealed along the way during our interviews, together with Schneider's account, suggest the possibility of pleasure, desire, and even freedom on her part. Could she have had a lover in the Riga Ghetto? Might her partner have been Blackie, whom she had already known in Vienna and mentions on several occasions?

---

11 USHMM, Gertrude Schneider video outtakes from *Shoah*, FV 3224.

It is of course understandable that she would not have wanted to talk openly about such a relationship, if it had indeed existed, or to explain her feelings at the time to her relatives – especially to us, a younger generation who had always known her as uncle Julius's loving partner.

For me, imagining this alternative story is an effort to offset, however minimally, the "something dreadful" that we in Frieda's family believed to be a shadow on her life. It opens a space for the possibility of agency, self-assertion, and even pleasure on her part. Within the vortex of horror into which she was brutally cast, it admits and recognizes in Frieda the brief presence of hope for existing otherwise – a reverie for what might yet be. It is a conjecture on my part, certainly. Yet I present it without misgivings, and offer it to her memory lovingly, as a small act of repair.

# Coda

Two weeks after I had finished my Frieda essay for this book, I was cleaning out an underused storage place in our house in Vermont when I encountered a small hard cover attaché briefcase that I did not recognize. After I managed to unlock its latch, I realized that it had belonged to Julius and to Frieda, and that it had probably been given to me more than twenty years ago, not long after Frieda's death. At that time, I may have glanced through it quickly and considered its contents largely unimportant. I had then stowed the briefcase away, like other potential family history materials of which I had become the unofficial "archivist," for future culling and possible discarding.

My first browse through the briefcase this time did not change my mind about its significance. The bulk consisted of apartment rental and insurance payment receipts, expired United States passports, and miscellaneous banking and insurance papers. But when I sifted through the papers more carefully, I also came upon an unmarked envelope in which, folded, I found an astonishing set of documents. They were copies of detailed medical reports from examinations carried out on Frieda in 1967 by doctors at St. Luke's Hospital in New York, at the behest of German authorities, responding to her application for "Compensation for Damages to Body or Health" suffered by surviving victims of the Nazi regime.

When I quickly looked through the reports I shuddered, hardly able to take in what had just come before my eyes. Frieda, who had never spoken to me about the mystery that engulfed her, and whom I had never been willing to ask about it directly, had left me documents that addressed much that we had been unable to talk about.

The set begins with a brief introductory statement of claims written by Frieda herself based on her experiences as a forced-labor inmate in the Riga Ghetto and in concentration camps in Latvia and Germany. It lists profound physical and psychological consequences of these experiences that impacted and burdened her entire adult life. A sentence in her statement that immediately drew my attention reads: "In the ghetto in Latvia I was forcibly operated on so that I cannot bear children" ["*Im Ghetto in Lettland wurde an mir zwangsweise eine Operation vorgenommen, sodass ich keine Kinder gebären kann.*"]. Pages of results from medical tests, and a physical examination pertinent to the assessment of compensation for her suffered "damages," then follow her prefatory declaration. A summary evaluation of these findings observes that, while Frieda's blood pressure, pulse, heart, and lungs were within normal limits, she suffered from chronic nervous anxiety and stress. The physical examination, however, also reveals that Frieda's lower body was marked by "two parallel scars" – one from a Caesarean operation and one from a hysterectomy.

This physical data disclosure is supplemented by a narrative explanatory account – the third part of my found document – written by Dr. Heinz Wennert, a psychiatrist affiliated with St. Luke's Hospital. Based on Frieda's life story account during her session with him, Dr. Wennert provides startling additional information:

- She was married in the Riga Ghetto in a religious ceremony; her husband (not identified) did not survive.
- She underwent a forced Caesarean abortion in the ghetto at six-months of pregnancy. [In German, the psychiatrist wrote: "*Sie war schon im Ghetto Riga einer Operation unterzogen worden, ein Kind wurde ihr nach 6 Monaten mit Kaiserschnitt genommen.*"]
- She would have liked to have children. ["*Sie haette gern Kinder haben wollen.*"]
- After marrying Julius, she had a miscarriage and was unable to conceive again. She underwent a hysterectomy in 1964.

Medical reports and testimonies claiming reparation payments have been known to exaggerate or to minimize the long-term effects of incarceration by the Nazis. But even when read with some amount of wariness about the accuracy of their every detail, the documents Frieda left – each partial, and revealing different facets of her lived experience – do, as a set, expose "something dreadful" that happened to her in Latvia.

I was stunned that the story told by the reports was not in any way reflected in the life narrative Frieda was able to share with our family – that this secret remained encased in the unpierceable skin of memory that Charlotte Delbo writes about. Delbo, a survivor of Auschwitz, states: "Auschwitz is so deeply etched in my memory that I cannot forget one moment of it . . . [It] is there, unalterable, precise, but enveloped in the skin of memory, an impermeable skin that isolates it from my present self . . . Thinking about it makes me tremble with apprehension."[12]

I had never been told by anyone, or even had it intimated, that Frieda had been married before her marriage to Julius, and in a religious ceremony in the Riga Ghetto. Oral and written testimonies that I have researched tell of only one "religious marriage" in the Riga Ghetto, performed not by a rabbi or Jewish spiritual leader but "religious" because it took place under a makeshift *chuppah* – a canopy under which a couple stand during their wedding ceremony and which, among other things, symbolizes their future home.[13] Frieda was nineteen years old when she arrived in Riga and was almost immediately told that she would never again see her parents and youngest sister alive. It is quite possible that she found some solace for her immense grief and misery from a fellow inmate – perhaps, as I speculated earlier, Blackie. She may, as indicated in the medical reports, have married during her stay in the ghetto, and perhaps done so by employing some sort of symbolic Jewish religious vow. Gertrude Schneider and other Riga Ghetto survivors identify many such heterosexual arrangements involving cohabitation by inmates, sexual relations, and sometimes pregnancy that was forcibly terminated by abortion. They called these coupling arrangements: "ghetto marriages."[14]

I can only guess what it must have meant to Frieda to find herself pregnant in Riga at such a young age and to be forced to terminate her pregnancy at a stage when the fetus may even have been viable. The terror, guilt, and remorse she may have felt then might well have haunted her throughout the rest of her life. How, in the postwar world, could she have translated the experience of the concentrationary universe that she had inhabited, and one of its consequences – her inability to have a child – except into the dry medical and legal terminology of an appeal for reparations?

Did Frieda's emotional worlds have to remain separate, with Riga encased in the skin of memory, for her to be able to survive? And, having uncovered her secret by means of the documents she left with me, what is my role now? I don't know. But I do know that a fuller portrait of Frieda will do more than illuminate her life for her family. It will also shed light on the experiences of other women survivors

---

**12** Charlotte Delbo, *Days and Memory*, trans. Rosette Lamont (Marlboro, VT: The Marlboro Press, 1990), 1–4.

**13** USC Shoah Foundation testimony by Rachel Drabkin. VHA Interview Code: 18294 02 Vol. Segment #8.

**14** Schneider, *Journey into Terror*, 61–66.

of Riga and other ghettos and camps that they themselves have been unable to relate.

My act of co-witnessing that takes Frieda's story out of the closed container in which she sequestered is inflected by my lifelong love for her, and my compassion for what she went through. I would think that by holding her story gently and conveying it to other sympathetic listeners, I can help her find the acknowledgment, recognition, and solace that she never received during her lifetime.

## Works Cited

Boder, David. "Voices of the Holocaust, Oral Histories," Illinois Institute of Technology [https://lit.aviaryplatform.com/collections/23/].

Delbo, Charlotte. *Days and Memory*. Translated by Rosette Lamont. Marlboro, VT: The Marlboro Press, 1990.

LaCapra, Dominick. *Writing History, Writing Trauma*. Baltimore: Johns Hopkins University Press, 2001.

Schneider, Gertrude. *Journey into Terror: Story of the Riga Ghetto*. Second, expanded edition. Westport, CT: Praeger Publishers, 2001.

United States Holocaust Museum Film and Video Archive. Schneider, Gertrude. *Video outtakes from* Shoah *of Gertrude Schneider's interviews by Claude Lanzmann*. Accession Number: 1996.166 | RG Number: RG-60.5015 | Film ID: 3221, 3222, 3223, 3224, 3225, 4717.

Wieviorka, Annette. *The Era of the Witness*. Translated by Jared Stark. Ithaca, NY: Cornell University Press, 2006.

Eric Kligerman

# "Cares of a Family Man": A Father's Reflections on Odradek and the Holocaust

When I was first asked to contribute to this volume, I was hesitant. In writing this essay I knew I would have to confront a double task of memory. How my path to Paul Celan's poetry, German-Jewish intellectual thought, and exploring the significance of the Holocaust is ineluctably connected to my dead brother, Bruce. It was through Celan's lyric that I first examined twenty-five years ago the negative symbiosis of German-Jewish culture.[1] But in addition to symbolizing a shibboleth that helped me enter the Shoah's traumatic ruptures, Celan's lyric also functioned as a transitional object that facilitated my engagement with familial loss. Over the years I had tried on numerous occasions to write about my brother, but only managed several fragmented pages comprised of evocative memories of him. These memories have become more diffuse and my writings were misplaced. If I were to do justice here, I would have to reflect on moments of great pain and – using lines from Celan's "Voices" – go down the *Nesselweg*:

> Voices from the nettle path:
> Come on your hands to us.
> Whoever is alone with the lamp,
> has only his hand to read from.[2]

Celan's word choice is quite revealing; the voices calling to us upon this path of pain require not a listening but an act of *lesen*: reading and (re)collecting. In preliminary steps, I prepared notes and mapped out how I came to dedicate so many years to probing the aesthetic, epistemic, and ethical implications of the Shoah. I uncovered a constellation of textual stars, including concepts from astrophysics to stars mentioned by Kant, Hölderlin, and Celan, to the asterisks my brother left in his books and papers, and finally to the German-Jewish nexus upon his grave: a Star of David beside an inscription of German poetry. Thinking about these stars, I recall a moment at Bruce's Bar Mitzvah, when during the celebration he took his three younger siblings outside to look at the night sky. Pointing out various constellations, he told us how the starlight we were seeing, while beautiful, was

---

1 Dan Diner, "Negative Symbiose: Deutsche und Juden nach Auschwitz," *Babylon* 1 (1986), 9–20, here 9.
2 Paul Celan, "Voices" in *Selected Poems and Prose of Paul Celan,* trans. John Felstiner (New York: W.W. Norton, 2001), 89.

https://doi.org/10.1515/9783110753295-016

perhaps already dead. The light was both there and absent. The real and textual stars on my map open up for me an axis of the imagination that sheds light on both my process of reflecting on the Holocaust and recollecting my brother's life and death. My work of memory entails what Celan calls *Toposforschung* – an investigation of place – that I apply to a sequence of stars connecting me to my brother and my engagement with the Holocaust.[3]

But while Celan's lyric was my entry into Holocaust studies, over the past several years my research has shifted to Franz Kafka's influence on German-Jewish intellectuals. My reflections are filtered through another star: Odradek from "Cares of the Family Man," who Kafka describes as "emanations of a star."[4] My attempts to think about this constellation of Celan, the Holocaust and my family history is reminiscent of Walter Benjamin's description of how one encounters Odradek, who "prefers the same places as the court of law which investigates *Schuld* [guilt and debt]. Attics are the places of discarded, forgotten objects. Perhaps the necessity to appear before a court of justice gives rise to a feeling similar to that which one approaches trunks in the attic which have been locked up for years that we would like to put off until the end of time."[5] In Benjamin's comparison of Odradek to a locked trunk, the past is no longer conceived as a fixed point in time. Instead, Benjamin perceives history as incomplete; the dead are not actually dead but are suspended in an in-between state whose completion is contingent on one's vigilant tarrying with the unfinished past. Flashing up at different historical moments, Odradek compels us to interrogate the past's relation to the present. My return to Celan's poetry and to memories of my brother entails *Sorge*'s three-fold meaning: it is marked by care, anxiety, and sorrow. How do I begin to read anew, let alone think about the meaning of Celan's poetry and its relation to traumatic events from my past after such a long hiatus?

The next part of Benjamin's essay provides direction for what *Sorge* entails. At the close of his Odradek analysis Benjamin concludes, "Even if Kafka did not pray – and this we do not know – he still possessed in the highest degree what Malebranche called 'the natural prayer of the soul: attentiveness.'"[6] Benjamin's use of *Aufmerksamkeit* conveys multiple meanings, including thoughtfulness, attentiveness and giving a small gift ("*Aufmerksamkeit schenken*"). The resonances I

---

**3** Paul Celan, *Der Meridian. Tübinger Ausgabe. Endfassung, Vorstufen, Materialien* (Frankfurt am Main: Suhrkamp, 1999), 10.
**4** Franz Kafka, "Die Sorge des Hausvaters" in *Erzählungen*, ed. Michael Müller (Stuttgart: Philipp Reclam, 1995), 188–190.
**5** Walter Benjamin, "Franz Kafka: On the Tenth Anniversary of his Death," in *Illuminations*, ed. Hannah Arendt, trans. Harry Zohn (New York: Schocken Books, 1968), 111–145, here 133.
**6** Benjamin, *Illuminations*, 134.

hear between Celan and Odradek are not coincidental. In his *"Meridian,"* Celan inserted Benjamin's line on Kafka's attentiveness to describe poetry's task. Characterizing how every poem is inscribed with its own January 20th, Celan depicts how poetry's attentiveness is "a concentration that stays mindful of all our dates."[7] Right after using the ominous word *"Konzentration,"* Celan inserts Benjamin's line, "'Attentiveness,' if you allow me a quote from Malebranche via Walter Benjamin's essay on Kafka, 'Attentiveness is the natural prayer of the soul.'"[8] In addition to being the fictional date when Büchner's Lenz gets lost in the mountains, Celan also has in mind the Wannsee conference from 20 January 1942, when the Nazis implemented the Final Solution. The poem, originating from a date marked by historical violence, neither transcends time nor faces the divine; instead, it remains in our terrestrial realm.

But for Celan, not only must the poet maintain this degree of attentiveness; he insists that the recipient must also be attentive to the poem and its date. A few months prior to his speech, Celan wrote, "Only truthful hands write true poems. I cannot see a difference between a handshake and poem [. . .]. Poems are also gifts – gifts to the attentive."[9] Just as writing requires attentive hands, so too must reading be marked by *Aufmerksamkeit*. Extending one's hand toward the poem, the reader pulls this "message in a bottle" from its indefinite journey and tarries with its meaning.[10] If I have learned anything from Celan, it is that our encounter with a text – whether poem, historical document or visual representation – requires from us an *Aufmerksamkeit*.

In addition to receiving a gift, the reader assumes a debt that cannot be paid off in our attempts to decipher some totalizing meaning to Celan's poetry. When probing his defamiliarized utterances or poetic ruptures, no such *Ausgleichen* – balancing of a debt – is possible for the *Schuld* we incur in our encounter with his poems. Like Odradek's spool of thread entangled across three generations, Celan's lyric embodies a transgenerational *Schuld* in debates concerning representations of Holocaust memory. This irredeemable past transfers its debt into the future. However, the intersection between poetry and historical violence is not simply about the limits of mourning. My engagement with the Shoah has evolved from questions of mourning to a poetics of justice that lies outside the term's conventional juridical sense. Similar to how Benjamin envisions the historian's task

---

7 Celan, *Meridian*, 8.

8 Celan, *Meridian*, 9.

9 Paul Celan, "Letter to Hans Bender," in *Collected Prose*, trans. Rosemarie Waldrop (New York: Routledge, 1986), 26.

10 See Celan's Bremen speech in *Collected Prose*, 35. Celan describes poetry as a *"Flaschenpost"*: message in a bottle.

as neither involving the retelling of past events nor an empathic relation with history's victims, the poet tries to enact justice for the dead. Celan was aware that a proper measurement was missing from his own lyric. He writes, "The poem does not fit the measure. Its scale is broken."[11] Whether approaching poetry as a work of mourning or search for justice, both paths involve the breakdown of poetic measure.

The structure of "Cares of the Family Man" not only sheds light on my relation to both personal and historical traumas, but it also allows me to reflect on my responsibility of transferring the past to others, whether in the classroom or at home. At this juncture between past and future, what are my *Sorge* when thinking about the meridian that passes from my brother, to me and my children? As a middle-aged father in an era of global upheavals – the spread of rightwing politics, an increase in anti-Semitic violence, the perpetuation of systemic racism and a pandemic resulting in mass death – what is my responsibility as both a teacher and parent of passing on a legacy of historical and familial trauma to my two young children? How might Benjamin's reflections on Odradek help me approach both my relation to the wreckage of history left in the Shoah's wake and to my brother's death?

For the past twenty-five years I have dedicated my teaching and research to the fields of German-Jewish studies, examining how works of the imagination take on the task of remembering and representing the destruction of Europe's Jewish communities. With no direct familial connection to the Shoah, how did I arrive here? I grew up in an assimilated Jewish family in suburban Philadelphia, where the usual markers of Jewish life were followed. I attended Hebrew school and went to Jewish summer camps. My family observed the high holidays, lit *Shabbos* and *Yahrzeit* candles, and I celebrated my bar mitzvah. My introduction to the Holocaust in the 1970s consisted of seeing a tattoo on the arm of a friend's father, listening to survivor testimonies in our synagogue and watching the television serial *The Holocaust* with my family.

But my earliest recollection of encountering scenes of genocide occurred by chance when I was around six. Before knowing how to read, I came across a book of photographs hidden on the top shelf in our family room. Years later I would find out that it was Gerhard Schoenberner's *The Yellow Star*. The cover showed a man being arrested; although the photograph was in black and white, I noticed the yellow star on his jacket. There was one image in the book that has remained with me from that day: a black and white photo of a mother walking with her children, along train tracks and barbed wire. Wrapped in scarves and

---

11 Celan, *Meridian*, 165.

heavy coats, they appeared cold. The four children reminded me of me and my siblings. The mother walked with three children, holding the youngest one's hand. The oldest child walked behind.[12] With downcast heads and bent bodies, they appeared exhausted. Unable to read the inscription ("On the way to the gas chamber"), I was at a loss for what was about to unfold in this ominous scene. On the following pages were images of mass graves.[13]

Despite not knowing their context, it was evident that the two scenes were connected; their arrangement depicted a before and after relation. Unlike Susan Sontag's description of seeing such images for the first time as a child, I do not recall being cut or wounded by them.[14] I was perplexed by what happened between the pages. Later that day, on the corner of Ashbourne Road and Tookany Parkway, as my mother drove me to little league practice, I asked her about the photographs. She told me how during the Second World War Germany placed Jews into concentration camps. I was unfamiliar with this term and asked if this was a camp where instead of playing sports, Jews played one of my favorite games: concentration. I do not remember how she responded, but on a personal level I was not too far off. The history behind these photos would eventually be the focus of my memory work and attempts to comprehend the laconic ruptures in representations of the Shoah. When I now show this image in seminars, it unsettles me more than those depicting mass graves. While I most likely identified with the youngest child back then, I now try to imagine their walk from the mother's perspective and the impending sense of doom she must have felt as she approached with her children the looming smokestacks in the distance.

Although in the years that followed I knew what happened to Europe's Jews, I never pondered its significance. My path to Holocaust studies actually began with my love of German literature and philosophy. While growing up, there was a sustained engagement with German culture in our home. Along with the Beatles my mother played Beethoven on the stereo. At bedtime she read to us stories from the Brothers Grimm. However, my strongest link to German culture came from my oldest sibling, Bruce. It is impossible for me to tell the story of how my passion for German-Jewish studies began without putting him at the center of this story.

My brother, who was the eldest of four children and seven years older than me, graduated *summa cum laude* from Harvard with a degree in physics at

---

**12** In looking at this photograph now, I realize that there are only three children in the picture. The mother is holding a bundle in her other arm.

**13** Gerhard Schoenberner, *The Yellow Star: The Persecution of the Jews in Europe, 1933–1945* (New York: Bantam, 1973), 206.

**14** Susan Sontag, *On Photography* (1977) (London: Penguin, 2008), 20.

eighteen. After college, his interests shifted away from the natural sciences and he started graduate school at Princeton in philosophy at twenty. When I asked him years later why he turned to philosophy, he said that the mysteries of bodies in motion no longer challenged him like the ontological questions pertaining to being in the world. However, his studies at Princeton did not last long. Before he turned twenty-two, he had his first mental breakdown – an acute psychotic episode – and thus began his stays in and out of psychiatric hospitals for the next ten years. During those years and up until his death, he immersed himself in philosophy and poetry. In short, his *Lebenslauf* moved in the following trajectory: from theoretical physics to Heidegger's onto-linguistic formulation of being-in-the-world, to his three dozen hospitalizations and reliance on antipsychotic medications that rendered him into what he described as a "self-strangulated monad."[15] During this period, which he referred to as one of "transcendental homelessness," he retreated into poetry.

From his first psychiatric evaluation to his last one, doctors repeatedly tried to define his illness. My father told me the story of Bruce's first evaluation, when he was six. When he asked the doctor what he thought was wrong with Bruce, the doctor moved his finger up and down the page of *The Diagnostic and Statistical Manual of Mental Disorders* (DSM) and said, "Who knows? It could be any of these. Your son could either be a Mozart or a monster. Only time will tell." When my father informed Bruce's last doctor of his death, his response was similar to that of the first psychiatrist. He said, "Your son was like a Mozart. Someone who comes around once in a thousand years. I learned more from him in six weeks, than all my years practicing psychiatry." In between these bookends, doctors applied a manifold of diagnoses as if to measure him through his illness, including paranoid schizophrenia, schizoaffective disorder and manic depression. Bruce told me his affliction came closer to what it means to "live in a destitute time." My brother's self-diagnosis, one that I have yet to read in the DSM, was actually an adaptation of a line from Hölderlin's poetry, which was the focus of an essay by Heidegger: how was it possible for one to dwell poetically in destitute times?[16] He had particular affinity for the term used to describe Hölderlin's years of madness: "*geistige Umnachtung*." While the term translates as "spiritual derangement," it also suggests that one's spirit or intellect (*Geist*) was wrapped in darkness. When he was well, one could speak with him throughout the night. But when the *Umnachtung* returned, everything around him turned dark.

---

**15** I will be using my brother's phrases from his papers.
**16** Martin Heidegger, "What are Poets For," in *Poetry, Language, Thought*, trans. Albert Hofstadter (New York: Harper Collins Publishers, 2001), 87–140. Abbreviated here as *PLT*.

I was fifteen when my brother came home from the hospital for the first time. In the evening he would sit on my bed, and we would play chess and listen to the Grateful Dead. He also had me read excerpts from Heidegger, Wittgenstein, or Rilke, asking me to reflect on their meaning. But often the discussion would turn to his psychic afflictions. At the root of his illness was Bruce's fear of brain damage, which he referred to as an aphasia anxiety: the inability to use or comprehend words. In this state of speechlessness, one retains the ability to think, but lacks the words to speak. Freud describes aphasia as "The incorrect use and the distortion of words which the healthy person can observe in himself in states of fatigue or divided attention or under the influence of disturbing affects."[17] My brother was fearful that his language abilities were being strangled by sensory stimuli, including loud noises that shocked his brain and the toxic fumes of the modern world. Technology in the form of the ubiquitous carbon monoxide fumes from cars, environmental toxins in food and exposure to heavy metals emanating from factory chimneys in Philadelphia signified for him a threat to poetic dwelling. Even such common occurrences as door slams and driving over potholes were sources of aphasia anxiety. The pollutants of modernity were both symptoms of our living in a destitute time and the cause of his all-consuming aphasia anxiety. Even though he would lament, "I can't get the words out," he was never at a loss for words, nor do I recollect his lack of eloquence.

Finally, it was the very medications he took to curb his brain damage delusions that precipitated another mode of impairment. Haldol, the anti-psychotic drug that was prescribed to him, blunted his thinking abilities and eradicated his linguistic faculties. Oddly enough, his aphasia anxiety was also a source of poetry and philosophical reflection. While searching through his belongings, I found a poem he had written about Haldol and read it with the same attentiveness as any other work of literature. His Hölderlin-inspired poem depicted how Haldol initiated the "flight of the gods," crushed his being-in-the-world and led to the "pervasive dullness of the spirit." While Proust's madeleine induced the onset of nostalgic memory, opening the cap to the amber vial "foredoomed" my brother to "Haldol's muting miasma" and "Etymicidal ennui." I marveled at his neologism that merged "etymology" with the suffix "cidal": his word demonstrated how the drug precipitated the death of language's origins. He interrupted the poem's conclusion by replacing the medication's brand name with its complex chemical structure. The medicine was another manifestation of how technology threatened the primacy of the *logos*, which was supplanted by a complex bio-chemical formula.

---

17 Sigmund Freud, *On Aphasia: A Critical Study*, trans. E. Stengel (New York: International Universities Press, 1953), 13.

I often wonder if my hyper-awareness to the linguistic and visual ruptures in post-Holocaust representations stems from our endless discussions of aphasia. His aphasia anxiety introduced to me the difficulties of unpacking Holocaust representations, where such terms as "beyond language," "unspeakable," and "unrepresentable" are used by those describing traumatic experiences or the critics who struggle to decipher their textual meaning.

Despite my recollections of checking him in and out of psychiatric clinics, witnessing psychotic episodes, and trying to persuade him on countless occasions not to kill himself, my memories are also punctuated by moments of great joy. There is one I have that condenses some of the most pleasurable and meaningful times we spent together and captures the profundity of our conversations. During one of my college breaks, we went to a Viennese restaurant near our mother's home. I do not recall how the topic arose, but as we ate lunch Bruce sketched on a napkin an image of the solar system and illustrated the physics behind the redshift of starlight. When celestial objects move away from us, the light that emanates towards us shifts to the red end of the spectrum, as their wavelengths get longer. In order to "clarify" the physics behind this process, he used Einstein's general relativity theory to explain the phenomenon. Since I could not comprehend the physics, the equations only took him so far in describing the temporal-spatial significance behind the curvature of starlight.

Without fail, he turned from physics to philosophy and poetry, which allowed him to communicate to me beyond the confines of equations. After expounding on the meaning of Kant's passage on "The starry skies above and the moral law within," he shifted to Georg Trakl's "Kaspar Hauser Song." Pulling out of a plastic bag *Song of The West*, a book I had given to him for his birthday, he focused on the following lines:

> Spring and summer and beautiful the autumn
> Of the righteous, his light footstep
> Along the dark chambers of dreamers.
> At night he remained alone with his star.[18]

While he discussed the poetic implications behind temporality's relation to space, his primary aim was to expand on the ontological significance of Kaspar Hauser's relation to language. My brother often gravitated to this poem and I assume he identified with both the potential poetic genius associated with

---

18 Georg Trakl, *Song of the West: Selected Poems*, trans. Robert Firmage (San Francisco, CA: North Point Press, 1988), 59.

Hauser – who remained alone with his star – and Trakl's own psychological anguish.

Our talk ended with an analysis of his favorite Grateful Dead song *Dark Star,* with its Prufrock-inspired lines: "Dark star crashes, pouring its light into ashes [. . .]. Shall we go, you and I while we can/ Through the transitive nightfall of diamonds?" Discussing the complexity of its instrumental improvisation and obscure lyrics, Bruce traced the intersection between cosmology (physics of black holes), philosophy (breakdown of reason) and poetry (allusions to T.S. Eliot) in these lines. But before he could continue with his analysis – a return to Einstein's relativity theory via Prufrock's question, "Do I dare disturb the universe?" from the same year (1915) – the external forces that he so dreaded impinged on us, when a busboy dropped a tray. The loud crash jarred him out of his train of thought and precipitated one of his rage-infused episodes. We walked back home, where he perseverated on the noise for the remainder of the day. Sadly, I no longer have the napkin from our conversation. I wish I could reconstruct all the connections we had discussed, but I am left with ruptures I am unable to fill in.

As the above movement through disparate texts demonstrates, my brother wanted to overcome the emphasis on formulaic thinking that reduced our relation to the world to equations. Following Friedrich Schlegel, "If you want to penetrate into the heart of physics, then let yourself be initiated into the mysteries of poetry," he showed me how the wonders behind the cosmos could be uncovered in a line of poetry.[19] During such moments, he instilled within me the value of poetic thinking, a term he borrowed from Heidegger's exegesis of Hölderlin's poem "In Lovely Blueness": "Full of merit, yet poetically, humans/ Dwell on this earth." Juxtaposing the scientific concept of measuring to poetry, Heidegger describes poetry as a measuring.[20] According to Heidegger, we measure ourselves in the space between earth and sky, mortals and gods. He stresses, "Poetry does not fly above and surmount the earth in order to escape it and hover over it. Poetry is what first brings man onto the earth, making him belong to it, and thus brings him into dwelling."[21] Going beyond the limits of any calculable measuring, poetry allows us to look to the heavens, scan the stars and posit our being-in-the-world. Our discussion that day was paradigmatic of how this upward scanning of the stars through scientific equations was ultimately replaced with poetry.

---

**19** Friedrich Schlegel, *Lucinde and the Fragments*, trans. Peter Firchow (Minneapolis: University of Minnesota, 1971), 250.
**20** Heidegger, ". . . Poetically Man Dwells . . ." in *PLT*, 221–222.
**21** *PLT*, 216.

But poetry was a double-edged sword for Bruce. In the margins of Hegel's *Philosophy of History*, he wrote, "I cannot believe that I had once grasped so much. I persist in a vacuum. The poetry makes me agonizingly aware of the human condition, my own condition. The philosophy is innocuous compared to the poetry. The poetry is a pernicious influence – an influenza." Before reading these lines, I was unaware of the etymological link between the words: originally, "influence" meant an emanation from the stars. This "influence of stars" eventually turned into something pathogenic; illnesses were linked to unfavorable astrological effects governing human affairs. "*Influenza*" derives from Italian meaning "influence" and refers to the cause of a disease.

In thinking about this etymological link, I remember watching a sunset with him around 1985. I remarked how striking the colors were as the sun sank behind the tree line. Bruce was not impressed and railed against the sunset. Our sun, he raged, was not the same as the one extolled by Plato; rather, in suburban Philadelphia the sun was filtered through the deathly haze of factory fumes. He pointed out how *pneuma* for the Greeks originally meant breath and spirit; now it was the root of pneumatic illnesses. Instead of ascending to the Platonic forms, there was only a drowning of spirit and breathlessness. "This," he said, "is how our modern world destroys itself." Whenever I teach Don DeLillo's *White Noise*, I recall this moment, when suburban residents watch "Another postmodern sunset rich in Romantic imagery."[22] But Jack Gladney, professor of Hitler studies, fails to connect its purported sublimity to the poisonous cloud hovering over his town.

My brother, however, was attuned to this toxicity, whether in reading poetry or looking at a suburban sunset. In such an impoverished era, Heidegger's questions about the poet's vocation accompanied my brother's reflections on poetic dwelling. In "What are Poets For," Heidegger writes, "To be a poet in a destitute time means: to attend [. . .] to the trace of the fugitive gods. This is why the poet in the time of the world's night utters the holy."[23] The poetic thinker does not try to flee this destitution, but rather situates him/herself within it. During destitute times, the poet remains *Aufmerksam* and listens for traces of the gods that fled. But my brother, who wished to flee this world, directed his attention to the pervasive signs of catastrophe that threatened his being-in-the-world. Every day signified for him an airborne toxic event.

Despite all our time together, I have no recollection of conversing with Bruce about poetry's relation to the Holocaust. Nor did we discuss resonances

---

22 Don DeLillo, *White Noise* (New York: Viking, 1985), 227.
23 *PLT*, 92.

between Heidegger's philosophy and National Socialism. Even when I went to college, I kept Germany's complex history and my Jewish identity separate from the literature I studied. While I read the Romantics and Modernists, from *Faust* to *The Magic Mountain*, studied the language and took seminars on Nietzsche, I never stopped to think about the Shoah. For me, Germany remained the land of "Poets and Thinkers." The turning point unfolded after graduation, when I used my bar mitzvah savings to go on a post-college grand tour of Europe. Mapping out the essential points of travel, Bruce provided me with an itinerary comprised not of the usual tourist markers such as the Louvre or Sistine Chapel. Instead, my map consisted of a poetic topography that guided me to where the writers and philosophers we so admired once lived. I travelled from the western point of Ronda, Spain, where Rilke sought inspiration in its ravine, all the way to the Greek island of Patmos. This endpoint signified for me not the place where John wrote the Book of Revelations, but instead Hölderlin's "Patmos," which I read on the monastery walls overlooking the Aegean. Bruce also gave me three texts for my travels: Heidegger's ". . . Poetically Man Dwells . . .," "What are Poets For," and a book I had not heard of, Franz Rosenzweig's *Star of Redemption*. He told me to read Heidegger before starting Rosenzweig.

No matter where I traveled, however, I inevitably encountered another topography: memorials commemorating Europe's murdered Jews. On my way to Rilke's home in Paris, I passed a plaque commemorating the round-up of French Jews. In Berlin, while searching for Hegel's gravesite, I stopped before Berlin's synagogue with its golden dome. On a wall was a memorial to its partial destruction on so-called *Kristallnacht*. During my visit to Goethe's home in Weimar, I was unsettled by its proximity to Buchenwald. In Bamberg, while walking to Hegel's home on *Judenstrasse,* where he completed *Phenomenology of Spirit*, I saw a memorial commemorating the town's deported Jews. On my pilgrimage to Kafka's home in Prague, I visited an art exhibition from Jewish children imprisoned in Terezin before they were sent to Auschwitz. At Kafka's grave was a plaque to his sisters murdered by the Nazis. In Freiburg, near the building where Heidegger taught, I saw another memorial to a synagogue's destruction on *Kristallnacht*. Wondering where Heidegger had been that evening, I was forced to think about the intersection between philosophy and these historical ruins. I left Germany shortly after, never making it to his hut in Todtnauberg. Just as the concept of poetic dwelling was upended for me, the phrase "poetry in a destitute time" took on an entirely different meaning, once I was confronted with Europe's Jewish remnants. I often wonder why Bruce gave me Rosenzweig's *Star of Redemption* and asked me to read it only *after* Heidegger. Maybe it was a subtle way of suggesting that I turn to Jewish thought. Sadly, we never had the opportunity to discuss the relation between the texts.

The proximity between the literature I loved and traces of extermination was unavoidable, as I was constantly running into memorials. Throughout Germany, there are memorials called *Stolpersteine*. These stumbling stones are raised golden blocks embedded in the sidewalk and engraved with the words, "Here lived," the victim's name, birthdate, place of deportation and date of death. We might pass over the stone or stub our toe, look down and be confronted with the task of thinking about what it signifies. Even prior to the *Stolperstein* installations, traces of the annihilated Jewish past were scattered at eye level across the landscape. I did not need to trip over them, but with a degree of attentiveness I perceived how these signs were ubiquitous throughout the cities where German poets and philosophers had lived. As I encountered traces of the dead, I was becoming aware of how enmeshed they were with the culture I venerated. This unfathomable intersection between poetry's sublime heights and the horrors of extermination compelled me to focus on how history's wounds have been transformed into works of the imagination.

After returning from Europe, I applied to comparative literature graduate programs in order to engage with a sustained study of the Holocaust in relation to poetry and philosophy. It was around this time that I first came across Celan's poetry in a New York City bookstore. His poems introduced me to the intersection between German culture and genocide, showing me how one tarried with Germany's literary tradition in the Shoah's aftermath. I was drawn to Celan's use of *Atemwende* – the neologisms, foreign and broken words – in his poems. In these breathturns, poetry swerved toward a momentary pause situated between speech and silence. The first poem I recall reading was "Aspen Tree." Although unfamiliar with Celan's background, I became obsessed with the intermingling of beauty and horror in this elegy to his mother. Without being explicit, Celan relies on the reader to uncover what transpired. The only clues he leaves are that his mother's heart was wounded by lead in the Ukraine. I read "Aspen Tree" over and over, trying to navigate between its insurmountable grief and laconic beauty. The poem interweaves nature's sublimity with human suffering: "Round star, you wind the golden loop. / My mother's heart was ripped by lead."[24] As the poem moves from dandelions, into clouds and stars, I discerned a connection between Celan's poem and Kant's passage on the starry skies and moral law. In two lines, the sublimity of Kant's stars collapsed before me. How could I read one set of stars alongside the other? The poem concludes with one final act of violence: "Oaken door, who lifted you off your hinges? / My gentle mother cannot return." With the destruction of both home and mother, Celan dismantles Heidegger's

---

24 Paul Celan, "Aspen Tree" in *Poems of Paul Celan*, 39.

concept of poetic dwelling. The upward turn to the heavens to measure one's being-in-world crashes down to earth.

I was excited to have the chance to introduce these poems to Bruce, but the opportunity never arrived. Around the time I started reading Celan, my relation to poetry was profoundly transformed by devastating events in my family. After a brief illness, my mother was diagnosed with stage four metastatic lung cancer. She died two weeks later after being admitted to the hospital. There is no space to expand here on her death and what she meant to us. Bruce captured her significance in a Mother's Day card I found after her death. In a bedside drawer, she kept a bundle of cards from her children and on top was one from Bruce. On the front was a picture of a home. Beneath its inscription, "Home is where the hearth is," Bruce included a brief Heideggerian exegesis of that prosaic phrase. Playing with its etymology, he added, "Mother/*Meter* is the *metron* of all things." Like Celan's oak door ripped from its hinges, our home too was destroyed.

Bruce killed himself several weeks after our mother died. With her death, there was no longer a home to return to, for the measure was gone. His despair, triggered by a mixture of guilt and melancholia, was too much to bear. Realizing before anyone that she was ill, he told us how our mother came to him one evening to ask for help with the *New York Times* crossword. She had been solving these puzzles with ease for years. But this puzzle was permeated with numerous fragments. From the couch he looked up at her and said, "You're dying." Although the tumor had originated in her lungs, it had already spread to the brain's language centers. Harboring a profound guilt that she had internalized his fears of brain damage, he was struck with terror that his aphasia anxiety had manifested itself as the actual affliction in her. She had lost her words.

After their deaths, I too couldn't get the words out and struggled to articulate my experience of loss. With his death, poetry and philosophy became tainted. Their associations with Bruce were too painful for me. It was only through Celan's poems where I could access my grief. They functioned like transitional objects, helping me mourn and reflect on their deaths. I specifically turned to the poems that invoked Celan's mother.[25] John Felstiner describes how Celan's lyric occupies a space between poetry and prayer. With its repeated invocations during commemorative events in Germany, he compares "Todesfuge" to a secular *Kaddish*.[26] In the thirty years since their deaths, Celan's poems have functioned for me as a form of *Kaddish*.

---

25 See "Black Flakes," "The Snow is Falling, Mother," "Wolfsbohne," and "Before a Candle."
26 John Felstiner, "Mother Tongue, Holy Tongue: On Translating and not Translating Paul Celan," *Comparative Literature* 38, no. 4 (Spring, 1986): 113–136, here 122.

As mentioned earlier, I never spoke with my brother about Celan, nor did I know if he had ever encountered his poetry. In writing this essay, I examined the last books Bruce was reading. When he killed himself, I know he was listening to Beethoven's Third Piano Concerto. His worn copy of Hegel's *Logic* and a newly purchased copy of Emil Fackenheim's *To Mend the World* lay by his bedside. The presence of Hegel made sense to me; Bruce often turned to him for what he called "spiritual illumination." On the inside cover of Hegel's *Logic*, he wrote, "Think of my brother Eric. Send him a letter." He circled my name, placing three stars next to it. On the opposite side, he continued, "What have I learned over the past years? Or, phrased differently: given that I have lost everything, why not allow myself to die now?" Beneath this, he scratched out a list with ink. I can still make out its order – A), B), C) and D) – but not the list itself.

Until now, I had never considered why he was reading Fackenheim before killing himself. Finding his copy on my bookshelf, I flipped through its pages in search of marginalia from thirty years ago. The inside covers were strikingly bare. There was no "to do list," no notes on where to eat or whom to call. In Fackenheim's preface, Bruce placed asterisks next to lines on how the Shoah arrested philosophical thought. He also marked the preface's end. Expressing his unease with how the history of Being circumvented the death camps, Fackenheim describes how Hegel's stature rises, but Heidegger's appeal had diminished for him. The last page Bruce marked was a section on post-Holocaust hermeneutics and its relation to the possible mending of German-Jewish culture. Fackenheim pondered how one could gain access to Goethe's "Wanderer's Nightsong" via Celan's *Todesfuge*. Bruce marked Celan's name with an asterisk. He had indeed heard of Celan, finding it worthwhile to consider him alongside the question of poetry after Auschwitz. Fackenheim continues, "We recall Adorno to the effect that poetry after Auschwitz is barbaric – not only the writing but also the reading of it. Is Goethe's poem, then, destroyed?"[27] This is where my brother's annotations cease. I wondered if he applied this question to Hölderlin, Rilke, or Trakl.

Two days after our mother was buried, I accompanied Bruce to the McLean Hospital. As our train approached Boston, he looked out the window and reminisced about how he had made this trip with our mother, when he started his studies at Harvard. Comparing the autumnal colors of Harvard Square from that day to the leaden grayness outside our train, he invoked his favorite literary character, Quentin Compson, who walked around Cambridge recollecting

---

**27** Emil Fackenheim, *To Mend the World: Foundations of Post-Holocaust Jewish Thought* (New York: Schocken, 1982), 261.

his complicated family history, before drowning himself in the river. He quoted Quentin's description of how his father gave him the heirloom pocket watch, "I give you the mausoleum of all hope and desire." I do not recall if Bruce made any remarks, but assume he let the line speak for itself.

In opening *Sound and the Fury*, I see what follows, "I give it to you not that you may remember time, but that you might forget it now and then for a moment and not spend all your breath trying to conquer it."[28] Unable to forget the weight of his family history, it was only by drowning that Quentin could attain a pause of breath from his obsessive recollections. My brother had a similar relation to time. It was marked by a historical dislocation in which he replaced the emptiness of the present with reflections on his past, such as summer camp, childhood trips and his college years. As he cited Faulkner, I imagine that he was also thinking about his profound grief over all that he had lost from his first hospitalization until our mother's death.

Bruce did not give me a watch, but he did leave behind hundreds of books and a cardboard box packed with his papers. While his books fill my shelves, for thirty years I have not opened this box, carrying it with me from Philadelphia, Ann Arbor, and Gainesville. Sometimes Faulkner's image of the watch comes to mind when catching a glimpse of it. But now I think Benjamin's comparison of Odradek to a locked trunk is more apt. In writing this essay I had to open the box to enter Bruce's history, as many of my memories have become distorted. My years of reading poetry and philosophy in relation to traumatic history prepared me for delving into his box and entering his history of suffering with new attentiveness. It would be hard to consider this box an archive in a traditional sense with a clear historical development. Parts of his notes were scratched out, pieces of paper torn and nothing was dated. He wrote on legal pads, napkins, and diner place mats. Although it is impossible to place them into sequential order, his papers brought me no closer to some hidden meaning, nor was I in search of one.

On one level, his handwritten fragments revealed the expanse of my brother's relation to the natural sciences, philosophy, and poetry. His Harvard thesis on laser physics sat atop his analysis of quantum mechanics' relation to Kant's *Critique of Pure Reason* from Princeton. With the exception of these two papers, the box's contents stem from the aftermath of his first psychotic break until his death. There were notepads filled with reflections on Heidegger's ontology, Hegel's philosophy of history and Nietzsche's eternal return. On a tattered legal pad, he wrote a self-analysis titled "Schizophrenia as a Form of Aphasia,"

---

28 William Faulkner, *The Sound and the Fury* (New York: Penguin, 1982), 73.

where he applied Freud and Husserlian phenomenology to uncover his illness's etiology. After three decades of not hearing his voice, I came across the terms he repeatedly used: historical dislocation, de-temporalization, and *logos*. These ideas were interspersed alongside his suicidal ideations and plans to share his "final set of memories" with his siblings. Often, he wrote clearly and with great resolve about why suicide was imperative "now"; it looked as if he was carving the page with his Bic. And yet, the notes expressing his desire for suicide were followed by another list of how he needed to "quit smoking, exercise more and find a girlfriend." I found the following note in a book on bird watching, "Please dwell within the realm of linguistic concreteness. I must temper this obsession with language, ontology and temporality! It is time to cease this fruitless retrospection and begin living." His preparations for suicide were balanced by his desire to move with ease through the world and embrace a *Gelassenheit* (letting go).

Similar to my engagement with the Holocaust, I face epistemic, aesthetic, and ethical challenges in writing this essay, not to mention psychic ones. My encounter with his belongings did not precipitate the effects of re-traumatization; rather, I felt as if I were refamiliarizing myself with a dissolving past inextricably bound to my present. What exactly could I learn from entering this archive that I didn't already know? I also face an ethical conflict and feelings of betrayal; on one page he wrote, "Destroy all my papers." The echoes are not lost. Was he addressing this to himself or his family? In telling his story, I am revealing an intimacy that he never wanted to share with a world from which he so badly wished to extricate himself.

Keeping in mind Benjamin's comparison of Odradek to a locked trunk, I approach with caution the traces Bruce left behind: books with marginalia, a box crammed with papers and multiple memories. What will become of his belongings and my memories of him when I am gone? Benjamin illustrates how Odradek embodies the past, present and future possibilities that we "would like to put off [. . .] until the end of time."[29] Fearing that we will have to engage with the *Schuld* of an abandoned history, we avoid opening this box. Odradek reminds us of a *Schuld* that has yet to be integrated or paid off. Like Odradek's spool of thread, the past transfers its debt to an indefinite horizon. When reflecting on the juncture between Bruce and my children, my thoughts turn to my responsibility in conveying to them his story. Like Kafka's "family man," I have my own *Sorge* when faced with the task of revealing to them our family history. What is my debt to my brother and how do I share his history with them?

---

29 Benjamin, *Illuminations*, 133.

Following Jewish custom, we gave our twins the Hebrew names of my mother and brother. They are familiar with the Jewish custom of *Yahrzeit*. When I light candles and recite *Kaddish*, they stand beside me and murmur along. I share with them stories about Bruce. They ask many questions; "*Who was his favorite Beatle?*" Every so often, they ask how he died. I tell them he was sick. The doctors could not help him. My hesitation in talking about my brother's mental illness and suicide with my children is analogous to how I do not share with them my work in Holocaust studies. For now, I keep details about his death and my re-search concealed, leaving unspoken the words "suicide" and "genocide." How do I fulfill my obligation to this double work of memory and pass on this two-fold *Schuld*: my brother's past and the catastrophic history I teach? As a Holocaust scholar I ask myself, when will I tell my children about the fate of Europe's Jews? When my children were around three and I told them that I was going to work, they would tap their heads and say, "Daddy thinks, thinks, thinks . . ." I won-dered, when they are old enough to ask me what it is I am thinking about, how would I respond? Little by little they are coming to know more about what I teach. I have heard my son say, "Daddy is talking about Eichmann again." They have seen Indiana Jones and already associate Nazis with something monstrous. I have not told them about the Final Solution nor mentioned the words "Ausch-witz" or "Holocaust," which remain taboos at home.

Their first encounter with Nazism occurred when they were three. While pre-paring for class one evening, I was watching Riefenstahl's *Triumph of the Will*. Although it was in German and contained no graphic images, just Nazi function-aries and Hitler giving speeches, soldiers marching and chanting oaths, I asked my wife to keep our children out of the room. How would looking at these images threaten them? There was something mechanical in my desire to censor the im-ages of swastikas, soldiers and Hitler's face. But my son escaped from his room. With such excitement he pointed at the marching soldiers and exclaimed, "Ro-bots, robots!" Despite being only three, he had reached a natural insight into the vast number of perpetrators who surrendered their independence of thought to a genocidal technocracy. There was indeed something automated in the actions of those who gave themselves over to authoritarianism.

Three years later, they now tell people, "My daddy teaches Kafka." I find this less disturbing than if they were to say, "Genocide or Hitler." One of their favorite books we read together is *My First Kafka*, which contains "The Meta-morphosis."[30] On its beautifully illustrated pages, expressionistic details adorn

---

**30** Matthue Roth and Rohan Daniel Eason, *My First Kafka: Runaways, Rodents and Giant Bugs* (Long Island, NY: One Peace Books, 2013).

the apartment. They point out when Gregor awakens that first morning, there is a clock next to his bed. Noticing the clock's absence when he dies, they deliberate what happened to time. Their favorite illustration is the one of Gregor on his ceiling. In this inverted image, the room appears upside down. Gravitating to this page, they try to imagine what the world looks like from the perspective of a human who is now a bug, cast aside by the world.

Every time I see this picture, I am reminded of Adorno's description of how Kafka's gaze was like a tortured Jew: "Kafka photographs the earth's surface as it appeared to tortured Jewish victims hung upside down."[31] Even though they are just learning to read, I am heartened by their inclination toward *Aufmerksamkeit,* poetic imagination and empathy. The epistemic, ethical, and aesthetic ruptures inherent in Kafka lead me back to questions pertaining to the Shoah. Currently, Kafka is helping me introduce my children to such topics as looking at the world from the perspective of another, systemic injustice and the reduction of humans into bare life. According to Hannah Arendt, Kafka's powers of imagination complete what reality had somehow neglected to bring to focus: "His so-called prophecies were but a sober analysis of underlying structures which today have come out into the open."[32] Her description is less about his works being prophetic than his ability to uncover the traces of impending catastrophe during the period in which he wrote.

Two years ago, while travelling to Philadelphia, I visited my mother and brother's graves with my family. It had snowed the day before. Fresh snow blanketed the ground and their granite plaques. I did not tell my kids where we were, and they waited in the car as I searched for the graves. Although I had not been there in ten years, I knew the vicinity by looking for a tree nearby. After several minutes of clearing the snow with my numb hands, I found their graves. Years earlier, when we drove to the cemetery to bury our mother, Bruce described how the prayer we would soon utter did not mention the dead, but instead exalted God's name. Even though he wondered what the purpose of saying the prayer in a world that had long ago been emptied of the divine would be, he still said *Kaddish* as he held our distraught grandmother by her daughter's grave. Standing in the same spot twenty-five years later, I recited the *Kaddish* out of reflex, but it felt empty to me.

Elie Wiesel captures this futility, when he tried to say *Kaddish* after his father died in Buchenwald and each *Yahrzeit* commemorating his death. Struggling to

**31** Theodor Adorno, "Notes on Kafka," in *Prisms,* trans. Samuel and Shierry Weber (Cambridge, MA: MIT Press, 1981), 245–271, here 269.
**32** Hannah Arendt, "Franz Kafka: A Reevaluation," in *Essays in Understanding* (New York: Schocken, 2005), 69–80, here 74.

interpret his father's last breath, Wiesel confesses, "I did not say Kaddish [. . .]. I felt a useless object, a thing without imagination."[33] On the eve of his father's *Yahrzeit* Wiesel's thoughts shift from mourning to justice in his search for new modes of commemoration. He writes, "It would be inadequate, *indeed unjust*, to imitate my father. I shall have to invent other prayers."[34] I am drawn to Wiesel's use of the term unjust to describe rituals of traditional mourning. If there is to be justice for his father, Wiesel must go beyond *Kaddish* and invent "other prayers." Contrary to being "a thing without imagination," his essay becomes his new prayer. Wiesel stresses that his pursuit of justice is future oriented: he will face the same task next year of having to invent another prayer for his father.

Celan also writes about the limitations of *Kaddish* in "The Sluice."[35] Structured around a catabasis, the poet goes beneath the poem's perforated line and crosses into a landscape of mourning searching for lost words. The poem's *Deus absconditus* is marked by the loss of *Kaddish*: the Aramaic word for "holy." However, at its conclusion the poet rescues *"Jiskor"*: another prayer associated with mourning. *Jizkor*, Hebrew for "may God remember," is the opening word to the memorial prayer during the commemorative service for the dead occurring four times a year. Read silently, it is comprised of two parts. Although critics stress its relation to remembrance, *Jizkor* also requires from the living an act of *tzedakah:* a multivalent word conveying charity and a justice that goes beyond a juridical connotation. Thus, a call to God to remember the dead is attached to the mourner's pledge to perform righteous acts on behalf of the dead. Although one could argue that by rescuing *Jizkor*, Celan finds a name for Jewish memory, I would suggest that *Jizkor's* recuperation shows how the work of memory is bound to justice. The poem does not end in silence; rather, a fragile link of communication arises between poet and reader. What might seem to be a form of *Ausgleichen* – rescuing *Jizkor* – is subverted by the lack of translation. Our encounter with this breathturn gives rise to an *Aufmerksamkeit* that makes us reflect on what remains untranslated. Instead of providing restitution for the dead, the poem concludes with a debt. A word is transmitted, but its meaning is marked with uncertainty.

Standing before my brother's grave, I read the inscription from Trakl's poem "Ein Winterabend" beneath a sheen of melted snow: "Wanderer tritt still herein; Schmerz versteinerte die Schwelle."[36] Tucked among tombs engraved

---

33 Elie Wiesel, "The Death of my Father," in *Legends of Our Time* (New York: Random House, 1968), 1–7, here 5.
34 Wiesel, "The Death of my Father," 7.
35 Celan, *Poems of Paul Celan*, 151.
36 "Wanderer, come in quietly, / Pain has turned the threshold to stone," Trakl, "A Winter Evening," in *Song of The West*, 7.

with Hebrew and English was a solitary one etched in German and addressed to me like a private message. Bruce had told me years earlier as we walked through a cemetery, he wanted Trakl's lines on his grave. Finding solace in Heidegger's ontological reading of the poem, he agreed that "language is the house of Being" and that we cannot exist without language.[37] Drawn to the power of poetry, he wanted to break from the prosaic meaning of words and listen to what lay beneath them. And yet it was his ruminations on the loss of language that became the focus of Bruce's anguish. Beckoning him across the threshold, I believe Trakl's lines signified for him the end of suffering.

I looked on the ground for rocks to place on the graves, but all lay beneath snow. For a moment I paused and then walked back to the car. My kids had never seen snow before, so I invited them to play in it. With great joy they made snowmen by the graves. There was something poetic in this act. Snow brought us back to the poem's lines about snow upon the windowsill. Unlike a rock, the material of snow would be impermanent. In front of this threshold between the living and dead, the inscription was the point of mediation between us and my brother.

Before heading to the car, I turned again to Bruce's grave. Pointing to the bronze plaque that shimmered golden in the sunlight, my daughter said, "Look, daddy, a star." Indeed, a Star of David sits between the dates of Bruce's birth and death. I never paid attention to this star, let alone remembered that it was even present. This star alongside a poetic fragment about a threshold directs me to Odradek, whose sudden appearance like a "star's emanation" upon thresholds exemplifies for Benjamin a moment of both temporal disruption and the possibility of some concealed insight. What was the meaning of this star beneath Bruce's middle name, "David"? While I had always recollected the grave's inscription, I now think about the star's significance alongside the German poem. As I begin passing on the work of memory to my children on both a personal and historical scale, a new constellation was formed that day between star and epitaph.

For years I had thought that the inscription was directed to Bruce, as if the grave had summoned him. Now I realize that those who visit could be considered the wanderers approaching in silence. My brother's epitaph is an *Atemwende*; a turn of breath marked by a linguistic disruption signifying a moment of remembrance and *Aufmerksamkeit*. In leaving behind a line of poetry, my brother consigns the visitor with a debt to interpret its meaning. Each new interpretation serves as an act of resuscitation or what Levinas calls an "extreme donation: attention – a mode of consciousness without distraction, i.e., without

---

37 Heidegger, "On Language," in *PLT*, 185–208.

the power of escape."[38] These forms of *tzedakah,* or small tokens [*kleine Auf-merksamkeit*], do not ease the dead into the other world, but prevent them from perishing again. In this transition from poetry as a space of mourning to justice, the burden of the past transfers its *Schuld* into the future.

Speaking to us from beyond the grave through an epitaph my children could not read, he left for us a poetic fragment. In turn, the fragility of memory and our impossible debt to the dead is captured in the small tokens my children left in the winter landscape that probably melted like a *Yahrzeit* candle by the day's end. The experience that day leads me to one last poetic fragment, from one of my brother's favorite poems:

> For the listener, who listens in the snow,
> And, nothing himself, beholds
> Nothing that is not there and the nothing that is.[39]

# Works Cited

Adorno, Theodor. "Notes on Kafka." In *Prisms*. Translated by Samuel and Shierry Weber. Cambridge, MA: MIT Press, 1981. 245–271. Edited by Samuel and Shierry Weber.

Arendt, Hannah. "Franz Kafka: A Reevaluation." In *Essays in Understanding*. New York: Schocken, 2005. 69–80. Edited by Jerome Kohn.

Benjamin, Walter. "Franz Kafka: On the Tenth Anniversary of his Death." In *Illumination*. Edited by Hannah Arendt. Translated by Harry Zohn. New York: Schocken Books, 1968. 111–140.

Celan, Paul. *Collected Prose*. Translated by Rosemarie Waldrop. New York: Routledge, 1986.

Celan, Paul. *Der Meridian. Tübinger Ausgabe. Endfassung, Vorstufen, Materialien*. Frankfurt am Main: Suhrkamp, 1999.

Celan, Paul. *Selected Poems and Prose of Paul Celan*. Translated by John Felstiner. New York: W.W. Norton, 2001.

Delillo, Don. *White Noise*. New York: Viking, 1985.

Diner, Dan. "Negative Symbiose: Deutsche und Juden nach Auschwitz." *Babylon* 1 (1986): 9–20.

Fackenheim, Emil. *To Mend the World: Foundations of Post-Holocaust Jewish Thought*. New York: Schocken, 1982.

Faulkner, William. *The Sound and the Fury*. New York: Penguin, 1982.

Felstiner, John. "Mother Tongue, Holy Tongue: On Translating and not Translating Paul Celan." *Comparative Literature* 38, no. 4 (Spring, 1986): 113–136.

---

**38** Emmanuel Levinas, *Proper Names,* trans. by Michael B. Smith (Stanford, CA: Stanford University Press, 1996), 43.

**39** Wallace Stevens, "The Snow Man," in *The Palm at the End of the Mind: Selected Poems*, ed. Holly Stevens (New York: Vintage, 1972), 54.

Freud, Sigmund. *On Aphasia: A Critical Study*. Translated by E. Stengel. New York: International Universities Press, 1953.

Heidegger, Martin. *Poetry, Language, Thought*. Translated by Albert Hofstadter. New York: Harper Collins Publishers, 2001.

Kafka, Franz. "Die Sorge des Hausvaters." In *Erzählungen*. Edited by Michael Müller. Stuttgart: Philipp Reclam, 1995. 188–190.

Levinas, Emmanuel. *Proper Names*. Translated by Michael B. Smith. Stanford, CA: Stanford University Press, 1996.

Roth, Matthue, and Rohan Daniel Eason. *My First Kafka: Runaways, Rodents and Giant Bugs*. Long Island: One Peace Books, 2013.

Schlegel, Friedrich. *Lucinde and the Fragments*. Translated by Peter Firchow. Minneapolis: University of Minnesota Press, 1971.

Schoenberner, Gerhard. *The Yellow Star: The Persecution of the Jews in Europe, 1933–1945*. New York: Bantam, 1973.

Sontag, Susan. *On Photography*. London: Penguin, 2008.

Stevens, Wallace. *The Palm at the End of the Mind: Selected Poems*. Edited by Holly Stevens. New York: Vintage, 1972.

Trakl, Georg. *Song of the West: Selected Poems*. Translated by Robert Firmage. San Francisco, CA: North Point Press, 1988.

Wiesel, Elie. "The Death of my Father." In *Legends of our Time*. New York: Random House, 1968. 1–7.

Karen Remmler
# Residual Remembrance: Family Genealogies and the Return of the Dead

Like one, that on a lonesome road
Doth walk in fear and dread,
And having once turned round walks on,
And turns no more his head;
Because he knows, a frightful fiend
Doth close behind him tread.

Samuel Taylor Coleridge
"The Ancient Mariner"

## Prelude

Traveling through the Catskill Mountains one snowy day, I find myself feeling strangely queasy. I feel as though I have been transported from one era to another and into a landscape where moments of time, like departure and arrival, end or beginning, are indistinguishable.

As I drive through the silent forest of pines and maples, up the snow-covered dirt road lined with craggy stone walls, I have no idea that my destination, the Daibosatsu monastery on Beecher Lake atop a mountain, will look so Japanese – like the Japanese I know from images of Buddhist Temples in Hokusai paintings. I arrive at the entrance: two heavy wooden doors skirted by a smooth white exterior and the telltale sloped roof of Japanese temple architecture. I have no idea what to do, so I simply push the heavy doors open and walk in. A woman in a blue robe, her head shaved bare, walks toward me and greets me warmly. She looks exactly like my grandmother, with her high cheek bones, sparkling blue eyes, and thin lips raised in a perpetually blissful smile. When she speaks, I hear the accented lilt of my grandmother, an immigrant whose voice comforted me and many another grandchild after altercations with harried parents and raging siblings. My grandmother was a free soul. She joined nature movements in her native Alsace and remained throughout her life an advocate for health food and bathing in the rain. When we visited grandma, we were treated to carob bars and muesli and told to go barefoot in the dew-ridden grass in an overgrown backyard; there, we played hide and seek and caught fireflies in a jar, marveling at their luminescent blinking.

https://doi.org/10.1515/9783110753295-017

I soon learned that the Buddhist nun who greets me, Fujin, grew up in Alsace a few villages away from Niederbetschdorf, where my grandmother was born. Only then does it dawn on me that I had come to within a few miles of the small village in the middle of the Catskill Mountains on the Shin Creek, where my grandparents lived soon after they came to America. The village, Lew Beach, was named after a nineteenth-century Congressman, Lewis Beach. Only now, writing this memory down, do I realize that I had always wondered as a child, about the strange pretense of a beach in the mountains, when I heard the village name in the stories my grandmother told.

In the early 1920s, shortly after they immigrated from Europe, my grandparents worked on a gentleman's farm in Lew Beach. The young Louisa had met my grandfather Walter in his home village of Röcknitz in Saxony, Germany in the devastated aftermath of the Great War. They were both free spirits, in the sense that they were both free from great expectations of family position or rank. She was the daughter of a village blacksmith; he was the youngest child in a family where twelve of sixteen children had survived to adulthood, and he had no prospects of inheritance. My grandmother had spent the war as a governess in the village pastor's household. My grandfather, drafted into the German army, had been taken prisoner by the French after the carnage at Ypres in 1915 – a capture that turned out to be what saved him from certain death in trench warfare. He had already heard about my grandmother in the letters he received from home. She would add short notes at the bottom of the letters written by the village schoolchildren, letting him know that his family was fine.

After the war ended, my grandfather returned to his home village after walking through the bombed-out cities along the way from France to Germany. Against the wishes of his mother, who had insisted he marry a wealthy widow, my grandfather fell for my less wealthy grandmother. The second he saw her he fell in love. "Even though she was not wealthy, I married a woman rich in spirit," he often said. They arrived separately in the United States in 1920, poor immigrants seeking a continent less prone to war. After a brief stint with my grandmother's relatives in Brooklyn, they headed for upstate New York, hoping to make a living from the land. They ended up in Lew Beach, working for room and board on the farm in Lew Beach that would later become Irving Berlin's summer retreat. They married in 1922 when the owner threatened to evict them, if they continued to live in sin.

Perhaps I continue to sit meditation at the monastery in the middle of nowhere around the corner from the place where my grandparents first settled because my grandmother continues to walk the mountain paths in the guise of a Buddhist nun, her once waist-long hair long ago shorn.

# The Archive

A sense of being adjacent to lifetimes that are long past, but always accessible through curated recollection, marks my own foray into the relationship between personal and collective histories of war, the Holocaust and other atrocities. In contemplating how to write for this volume, I remembered the "meaningful adjacencies" by which the names of the victims of the 9/11 terrorist attack are inscribed in the bronze edifices representing the former north and south towers of the collapsed World Trade Center. The memorial's creator, Michael Arad, hoped to arrange the dead adjacent to other victims with whom they had shared a relationship while alive. A complicated calculation, only possible with the aid of computer algorithms, grouped the names according to the location of their death on 9/11 in clusters of adjacencies signifying work, family, friendship and chance alignments based on the testimony of survivors and family. The dead are engraved in a perpetual maze of adjacencies, which they may or may not have desired while living. And yet, the notion that they are not alone in their untimely deaths allows for a semblance of repair to the interrupted lineages and genealogies that existed in lives cut short by the terrorist attack. The perpetrators of the act of terror are not inscribed on the memorial for obvious reasons. They do not belong, even as their adjacencies to the dead are profoundly and horribly recorded in the collective memory of this millennium.

What does this story have to do with my own set of collective family memories, fraught with charged lines of crisscrossing victimization and perpetration, not to mention direct and secondary witnessing? How do we write about the tributaries of family genealogies whose garbled streams are guided by the voices of ancestors all but inaudible in the vociferous din of present time? The seepage of trauma from one generation to the next demands we look back – despite the threat of turning into a pillar of salt like Lot's wife, or of sending a loved one into the Underworld, forever, as Orpheus unwillingly did. I believe that the dead live in our unconscious. They dig labyrinths of return along the broken passageways of family genealogies, myths, and memories. These dead are shrouded in loss and guilt, indicative of the porous membrane between the living and the dead. At the same time, they may radiate love and gratitude, a residue of the magic of memory and recollection.

When I was seventeen years old, I told my cousins Karin Marie and Marta that I would be writing the Remmler family history. Since that claim, I have been writing that history in one way or another, mostly through the lens of larger histories, histories more severe and deadly than those that befell my family. I write about the narrative, visual, and material memorialization of the dead in the wake of extreme violence. At the heart of my inquiry lies the affective distress that occurs when survivors are denied access to the human remains of their loved ones. This denial takes on many forms in the wake of extreme violence, ranging from total obliteration (extermination camps and crematoria), intentional disappearance and mutilation (mass graves), or political instrumentalization (virtual dissemination of the corpse without permission of the next of kin). I return again and again to the human right to a proper burial, even among those who are part and parcel of perpetrator histories. How does mourning not only of one's own, but also of the other take place? Do we even have the language to claim adjacent spaces of mourning without running the risk of blurring the distinctions of the particularities and accountabilities of the act of killing that leads to violent death? I have often struggled with these questions and wondered how to describe the need for those who experienced loss (even as members of perpetrator nations, communities, families) to first grieve others as an act of responsibility and primary (and secondary) accountability, not as quid pro quo or as self-victimization or self-righteousness, but as an act of compassion for the suffering of others.

For as long as I can remember, I have preferred the dead to the living. Not just any dead, but the dead whose lives were cut short by acts of violence. I distinctly remember lying on my back one summer day when I was nine years old in our disheveled backyard at 21 Higgins Street, North Babylon, Long Island, looking up at the lumbering clouds and wondering how many souls dallied in the wispy puffs of white and gray. I swore to myself that I would remember that moment forever and wondered if the souls knew I was gazing up at them as they passed by. "I won't forget you," I murmured.

As I began to read *Newsday* and to watch the nightly news on our first television, those imaginary, amorphous souls began to take shape, to have names, and to enter my consciousness. As I began to imbibe the war stories shared by my grandfather and father, to gaze at the somber portraits of US soldiers killed in Vietnam and see the funeral processions bearing the coffined corpses of JFK, MLK, RFK, Malcom X and countless others, known and lesser known heroes, I confused the real dead with the dead of my fantasies sparked by the television shows *Dark Shadows*, *The Twilight Zone*, and the ghost stories I read each week in my cache of mail order books for young readers. But that does not fully explain my obsession with the dead. I remember wrapping my first beloved cat, Princess, in sheets and placing her in a homemade wooden coffin and lowering her into the hole my father dug.

After engaging in my first and to this day only family ritual of properly burying the dead, I developed a morbid fascination with the photos of the dead soldiers memorialized in a 1969 issue of *LIFE Magazine* during the Vietnam War. As long as I can remember, I was fascinated by images of the dead that emerged out of this war and its wedded atrocities, flickering across the television screen, or from Holocaust documentaries that displayed the jumbled remains of victims in the immediate aftermath of the carnage in places like Auschwitz, Bergen Belsen or Dachau. But my first impressions of war emerged when I ran across two photographs. The first, taken by my grandfather in Ypres in 1915, depicts twelve German soldiers in a cramped trench. In the second photograph, my twelve-year-old father crouches between three American GIs toward the end World War II in Alsace.

Residual memories are memories that have almost faded into oblivion. Yet, a trace remains, barely perceptible, but indelible. In the realm of scientific verification, residual effects mark the afterlives of drugs, pesticides, and other synthetic human-made substances that reside in bodies, often with harmful, if initially imperceptible, consequences. The effects of residual traumatic memories often manifest as the uncanny feeling of being haunted (*heimgesucht*) by inherited experiences embedded in family genealogies. These genealogies are often severed over time and yet they return periodically, like the sharp pain emanating from a phantom limb. This trace is akin to the residual presence in a photograph, so aptly described by Walter Benjamin. In "A Short History of Photography," Benjamin writes:

> No matter how artful the photographer . . . . the beholder feels an irresistible urge to search such a picture for the tiny spark of contingency, of the Here and Now, with which reality has so to speak seared the subject, to find the *inconspicuous* spot where in the immediacy of the long-forgotten moment the future subsists so eloquently that we, looking back, may rediscover it.[1]　　　　　　　　　　　　　　(emphasis by K. Remmler)

The World War I photograph depicts twelve multi-ranked German soldiers staring out at the camera, posing within the confined space of the trench (Fig. 1). I assume that my grandfather took this photograph, since it was in his possession. The trench is clearly a precarious, cramped space, and the majority of these men will have died within a fortnight. From the looks of their shaven, mustached faces, they kept themselves intact despite the harsh conditions. The figure to the left holds his cap in his hand, his slicked-back hair parted in the middle. Further investigation would reveal the inconspicuous squint in his eye. Perhaps he and the others face the sun. The soldiers look directly at the camera lens, grimly, as though they already anticipate probable death ahead. On the verso of the photograph my grandfather's handwriting reveals the place and year: "Ypres, 1915." I imagine that my grandfather took the photograph

---

1 Walter Benjamin, "A Small History of Photography," in *One Way Street and Other Writings*, trans. Edmund Jephcott and Kingsley Shorter (London: Verso, 1992), 240–257, here 243.

**Figure 1:** Trench, from private collection of the author.

shortly before the second battle of Ypres, which took place between 22 April and 25 May 1915. My grandfather was captured by the French army after the chlorine gas launched by the German army against the French and Algerian troops had subsided. The capture saved him from certain death in trench warfare. One morning, he emerged from a deep sleep despite the barrage of mortar shells through the night and the screams and cries of the other soldiers, wounded and dying. He awoke alone. The other comrades had disappeared. He began the usual morning ritual. He threw a grenade over the side of the trench and rose to take a look. On this morning he found himself face to face with the barrel of a pistol held by a French soldier.

My grandfather did not talk about the gory details, but rather about how well the French treated him and how much he grew to distrust his fellow Germans. As a privileged prisoner of war, due to a promotion in rank conferred after his capture, living in a French castle requisitioned to hold German officers, my grandfather spent the remainder of the war listening to other German officers' plan for the next war. Consequently, he decided to leave Europe once he was released.

When my grandparents fell in love after the war, they imagined that immigrating to America would keep them from war and they would raise their family in peace. After leaving the farm in Lew Beach, following the birth of their first child, my grandparents moved to Ghent, New York and spent the next fifteen years maintaining farmhouses for room and board on defunct farms that were up for sale. Once the farmhouse sold, they moved to another until that house, too, was sold. Unable to buy one themselves, my grandmother raised the by now six children, and my grandfather taught German and Italian immigrant children in a one-school house, a position he lost with the advent of World War II. Ironically, on 30 July 1939, my grandmother, leaving four of her six children behind, returned to her hometown, Niederbetschdorf in Alsace, now a part of France, with her youngest daughter and son (my father) to attend the funeral of her beloved sister, Karin. As fate would have it, their return passage was planned for 3 September 1939, the day the British declared war against Nazi Germany for invading Poland. My father, six years old at the time, recalls sitting in an ox-drawn cart, ready to head for the train station in Strasbourg on their way to the port of Le Havre to sail back to the States, when a telegram arrived: *Ship not sailing. Stop. Return home. Stop.* My father spent the rest of the war, aged six to thirteen, in Niederbetschdorf. After the Wehrmacht invaded France in 1940, Alsace was annexed into Nazi Germany.

I grew up hearing about this tragic family separation. (It would lead to subsequent family trauma – but that is another story.) I remember my father telling stories about his adventures during the war, but it wasn't until I interviewed him ten years ago, that it became clear to me that he had witnessed the scavenging of dead American soldiers after the battle of Hatten-Rittershoffen in January 1945. I suspect that I "knew" this part of the story long before it became explicit, since my father would mention periodically his relative freedom during the war and his roaming around the fields during the day. My father witnessed a number of skirmishes, but most notably the aftermath of the notoriously savage Hatten-Rittershoffen Battle, which took place as the Germans made one last-ditch attempt to hold the Maginot line and retake parts of North Alsace from the American troops during the ferociously cold winter of 1944–1945. Depicted in great detail in Stephen E. Ambrose's book, *Citizen Soldiers*, the Americans were forced to retreat, thus leaving north Alsace once again in the hands of the German Army. My father remembers hearing American English one day and then German the next and often cites Ambrose's account as his own memory.

Shortly after the end of the battle in January 1945, my father and four other village boys aged 11 to 14 years old walked the 2-3 kilometers to Rittershoffen from Niederbetschdorf. In a matter-of-fact tone, he describes the corpses he encountered:

> The approach to Rittershoffen coming from Betschdorf was a curving, climbing road through a deep cut. There was an American tank, its turret . . . blown off and sitting on

top of the embankment. The body of the tank rested on the road in the cut. We climbed on the Sherman to see what we could see. Looking down into where the driver and codriver sit, we saw the burnt bodies, now the size of just-born babies [. . .].

As our little group approached Hatten, in the street by the first house a frozen American soldier's body lay with a hole where his knee should have been. His lower leg and foot were still attached to his upper leg by a few strands of skin. (At the time I thought that he had had bad luck to be wounded so. But as I grew up, and often thought of the soldier with the hole in his leg, I learned that such a thing, blowing a hole in a person's knee, was a form of punishment or just plain cruelty. I believe from the scene that's what had happened here.) [. . .]

As we progressed through the village, what I remember most vividly were the burned flattened houses with just the chimneys standing. Realize that this happened in cold January. Actually, one of the coldest in Europe. All the dead were frozen stiff. There was no smell of rotting corpses.

(H. Remmler, Interview, Private Notes, Compilation from
June 2009 through 11 October 2020)

The German dead had been properly removed by the triumphant German troops, and the Americans would lie on the ground until the spring thaw when local farmers buried them. My father remembers one dead soldier in particular. "Tobacco juice dribbled from his mouth and a pouch of tobacco protruded from his pants' pocket." When I asked him if he felt traumatized by what he saw, he staunchly denies any lasting damage. Instead, he insists that he had a normal childhood, just like millions of other kids who grew up during wartime, sheltered as he was, by a loving mother and a roof over his head.

The faded photograph depicts a row of four males crouching against the backdrop of a murky white sky, a smokestack rising up in the center behind them (Fig. 2). The third figure from the right facing out is my father, aged twelve years. He stares at the camera, his expression serious, unsmiling yet hinting at a sense of pride. The other three men, all American GIs, are adorned with different headgear, a cap, a helmet, and a fur hat. I am drawn to the hands of the two soldiers flanking my father. The soldier wearing the helmet dangles a cigarette between his index and middle finger of his right hand. He is wearing a watch on his left wrist. I am struck by the dirt caked on his fingers, dark enough to be blood. His mouth is slightly open. His eyes reveal a suspicious gaze, and I wonder what he has seen. My father drapes his left hand over his knee almost as though he is emulating the soldier to his left. Like the other soldier's hand, these hands are caked in dirt. The pose reveals victory, hard-won. The expressions verge on smiles, but the expressions reveal a faraway look in the eyes. I assume that my father asked the soldiers to pose with him or perhaps they were surprised to hear that he was an American from upstate New York and wanted to record the event. It is not clear who took the image.

**Figure 2:** Niederbetschdorf, private collection of the author.

Despite my father's account and insistence that he did not partake in the scavenging of the dead soldiers, for years, in my mind's eye, I could vividly see him foraging in the fields after the battles were over. He is ten or eleven and his playmates are a bit older, a bit more daring. This is not play. I imagine the stench of rotting corpses, American or German soldiers hit by enemy fire. He and his buddies look for souvenirs among the dead: a watch, a pair of boots, a dog tag or a belt. He stands shivering, not wanting to touch the dead bodies, haunted by the blackened burnt corpse he spies in a tank. Of course, this is false. My memory of his memory stands corrected by his more recent account quoted above. No stench of rotting corpses, and no souvenirs from the dead. My father was only a bystander and to this day does not remember the names of the other boys.

## Residual Memory

Looking back, my father probably inherited the residue of World War I that his father passed on to him, and I inherited the displaced psychic remnants of his experience during World War II, residual remnants that I confused with other dead, those of the Vietnam War and the Holocaust. I am not sure when I first learned about the Holocaust. I do know that World War II loomed large in my father's side of the family as the decisive historical marker of trauma, and that it created a sense of exceptionality, of specialness. Being born in France, during my father's return as a married American soldier in 1957, also helped to create my own sense of being more European than American. I became the family archivist of the war stories long before I even knew what an archive was. And I became fixated by the displaced remains captured in photographs, diaries, and letters, even before I consciously knew what death was. Growing up in the 1960s and 1970s, I had an uncanny fascination with the dead, particularly those who refused to stay dead, as though their restless ghosts had work to do before resting in peace.

Ever confusing reality with imagination, I am not sure how much I disentangled the undead in the gothic soap opera *Dark Shadows*, for example, from my own conjuring of the dead in times of war, based on the flickering images that appeared on the television or in the movies. I associated the vampires and ghosts on the screen with a feeling of comfort. I watched *Dark Shadows* with my mother every day after school. We would sit and gasp together when Barnabas, the eternal vampire, would bite one victim after another. At the same time, I learned about the dead of the Holocaust from the survivor parents of two elementary school friends. Even though no one directly referenced the Holocaust, I saw those rare photographs of murdered relatives on the mantelpieces and must have heard about the horrors of the extermination camps indirectly. I would later think I had already seen the emaciated bodies, when, years later, I became mesmerized by Alain Renais's poetic 1956 film *Night and Fog*. The documentary footage of Jews being deported from France and the pan of an empty barrack in the aftermath, once the corpses had been cleared away, seemed familiar to me, as though I had already incorporated them into my consciousness.

Even though I was born an American, in Chinon, France in 1957, I also felt shame at "being" German. My German last name exposed my German heritage. Growing up with friends who were the children of Holocaust survivors, I always hoped their parents would not ask my name or ask me about my family. I also grew up constructing multiple alibis just in case. My German grandfather "only" fought in World War I and spent most of it as a privileged POW, a commissioned officer in a converted French castle. At the end of the war, he swore off all nationalism and war and became a pacifist and left Europe because he knew the next war would come. So, he escaped even the slightest temptation to become a Nazi. (He even claimed to have encountered Hitler and his cronies in the Munich Pinacotheca, walking away from them when they tried to recruit him.) And my Alsatian grandmother could certainly be exonerated, even though she had a newly minted German passport when she left the States in July 1939 with her two youngest children, my father and his sister, to attend to the grief over the untimely death of her dear sister, after whom I am named. Family lore includes an episode of my grandmother's bravery. She helped to hide and then secretly to accompany an Alsatian deserter from the Wehrmacht near the end of the War, risking not only her life, but also my aunt's, who posed as the deserter's fiancée when they passed through numerous villages on their way to to Hagenau, where his family lived. Lore has it that the Germans tore apart the village adjacent to my grandmother's in search of the deserter, threatening to shoot anyone who harbored him. And, of course, I can always pull out the Jewish relative card and relate how I grew up learning about Jewish tradition from my Jewish uncle's father when I spent holidays in Berkeley and lit Chanukah candles and made potato latkes. At one of many family reunions as we square danced, one of my Jewish cousins jokingly called out for all Germans to create a circle and then called all Jews to the center of the circle. We laughed as we often did when we rehearsed the German Jewish "symbiosis" enacted within the extended family.

When I first visited the home of a Jewish friend, L., in elementary school, I met her mother, a Holocaust survivor. A small, slightly plumb woman, L.'s mother taught Yiddish at the local Jewish community center and always welcomed me into her home with an affectionate pinch on the cheek. "You have such pretty blue eyes and your hair, such a blonde!" This left me with a feeling of shame, even though I had not yet seen my first images of concentration camp corpses. I sensed that my German heritage mattered, even if L.'s mother didn't indicate that my clearly German last name bothered her. This initial sense of shame would grow over the years, compounded by my father's insistence that no one in Niederbetschdorf knew about the Holocaust, though his older friend Philip, who joined the SS, did seem shattered after he returned from the Eastern front. And my father was a member of the local Hitler Jugend, though he claims he and the local Alsatian boys would make fun of the Nazi salutes and of the virulent Nazi teacher, Herr Junker, behind his back.

Perhaps I decided to do scholarly work on the Holocaust while in Lüneburg at a Goethe Institute when the television mini-series *Holocaust* was broadcast on German television at the beginning of 1979. Prior to my sojourn at the Institute, as a double Sociology/German major at then SUNY Binghamton, I had wanted to write a research paper on the Holocaust for a sociology course on population control. The professor, however, insisted that the Holocaust was already "overdone" (in 1978!) and I should delve into the Armenian genocide instead, which I did. I didn't see *Holocaust* when it was broadcast to an estimated fifty percent of the German population because I was living with an elderly German couple who felt it would not be appropriate for me (at such a young age, 19!) and refused to turn to that station. It was winter and snowing, so I was too timid to try and find a place to watch it. I didn't really grasp the widespread impact the film had, until the elderly German teacher at the Institute broke down into tears the next day and told us that Germans lacked civil courage and that the film demonstrated the core of cowardice among so-called German bystanders. She insisted that people could have done more and that she herself had refused to exclude Jewish students from her class in the 1930s without repercussions.

Angelika Bammer's reckoning with her German heritage in her book *Born After* and through the analytical lens of her work on *Vergangenheitsaufarbeitung* (the working through of history) raises this dilemma about tracing convergences (not equivalencies) between traumatic collective histories, such as the Holocaust, and family genealogies of war.[2] Like others born at the cusp of World War II and the Holocaust who have written or created films about their experience of growing up under their shadows, Bammer faces her "implicated" subject position.[3] Bammer writes so beautifully. Her turns of phrase, the decisive metaphors, and the detailed descriptions of place work through multiple layers of memory and misremembrance to plumb the depths of her unconscious and the transmitted silences from her family on both sides. I admire how Bammer moves from condemnation of all things German to an empathetic stance towards her family and others whom she interviewed, encountered, or met along the way. She begins to tell one story, then lets it be for a while, and then returns to it again to dig deeper, to set the record straight as it were. The framing of her story through her relationship with her children brought me to tears, too, but also to the realization that any reckoning is an act of responsibility, not just to telling the truth, but also to leaving a legacy that names the taint so that the next generation doesn't have to struggle with it or, at least, has insight into how it came to be in the first place.

---

2 Bammer's narrative, what I would call a refracted memoir, resembles other texts that have been crucial to my own understanding of the adjacencies between the Holocaust, family and collective histories, such as: Darcy Buerkle's *Nothing Happened*, Irene Kacandes's *Daddy's War*, Alice Kaplan's *French Lessons*, Deborah Lipstadt's *History on Trial*, Leslie Morris's *The Translated Jew*, Ann Parkinson's *Emotional State*, and Paula Schwartz's *Today Sardines are not for Sale*.
3 Compare Michael Rothberg, *The Implicated Subject: Beyond Victims and Perpetrators* (Stanford, CA: Stanford University Press, 2019).

# Return of the Dead

So, how does one face family legacies head on and uncover the subtleties that often fall by the wayside when we plow the fields of memory through our scholarly writing? I learned how to recite poetry growing up under the tutelage of two grandfathers; the German immigrant, who arrived in the United States in 1920, a World War I veteran turned pacifist, and the other, a New York State prison reformer, born in Nebraska in 1900, a Methodist turned secular humanist. Both grandfathers seemed to think I would be the one to follow in their footsteps and become engaged in the pursuit of knowledge as a form of respite from the legacies of war and poverty that seeped up occasionally through the folds of family memory or, even more troubling, into the transgressions of subsequent generations. They provided me with the antidote of poetry. The poems they chose were far from comforting, and the three I remember most from my recitation lessons include Dylan Thomas's "Was There a Time," Coleridge's "The Ancient Mariner," and Goethe's "Wanderer's Nightsong." The poems are nothing short of ominous and forbidding, each with a tale of abject transgression in the midst of seeming innocence and calm.

My Quaker grandfather, the immigrant from Europe, had me recite short poems in German that in retrospect were shrouded in loss and guilt and indicative of the porous membrane between the living and the dead. I learned to recite Goethe's "Wanderer's Nightsong," its final line reminiscent of those somewhat malicious inscriptions on fungus-flaked gravestones in New England cemeteries that "you, too, will soon die, so prepare for death and follow me." The last line of Goethe's poem reads "Warte nur, balde ruhest Du auch" (Just wait, soon you too will be at rest) and elicited in me a feeling of dread, not calm.

Coincidentally, my maternal grandfather, also had me reciting poetry around the same time. The superintendent of a minimum-security prison in Coxsackie, New York, he often talked with me about the importance of not judging the person, only the crime. I remember sitting in his study, earnestly typing an essay for a third-grade paper on Pompeii, in which I described in rather morbid detail the petrified corpses left in the wake of the volcanic eruption. Upon reading the essay, my grandfather noted that I was perhaps overly obsessed with the dead. He insisted that I put these images aside and recite poetry in order to learn about literary figures of death. We started with Dylan Thomas. My grandfather played a LP recording of Thomas reading his poetry and had me choose my favorite. Too young to grasp the meaning of "Was There a Time," it remains with me to this day: "Under the skylines, they who have no arms/Have cleanest hands . . ." But it is the last line that stays with me; "[. . .] as the heartless ghost/Alone's unhurt, so the blind man sees best." I am sure I did not understand then the full weight of this poem, which I now see as being about the loss of innocence and the swell of maggots that inevitably feed on fleeting time, leaving nothing behind but regret and guilt. I do remember imagining how it could be possible to have clean hands, but no arms and I am sure now that I confused this image with the scene in the Grimm fairy tale "The Girl without Hands," the tale of a father's inept betrayal of his daughter to the devil – unbeknownst to him, until it is too late to rescind his end of the deal. The daughter eventually sacrifices her hands to save her father from the devil. I think I pictured this scene when I first heard Thomas's poem read by the poet himself, on a record I still possess but cannot play, having long ago discarded my grandfather's turntable, a de-cluttering gesture that I regret to this day.

After I mastered Thomas, we moved on to Coleridge's "The Rime of the Ancient Mariner." The doomed shipmate's fateful killing of the albatross and the subsequent dire consequences for the ship's crew, racked by thirst, stood in my child's mind for an unspeakable act of violence. The image of a wedding guest held fast by an uncanny mariner haunted me throughout my childhood and still chills me to this day. I cannot stop listening to the voice of the insistent storyteller, his voice dry and raspy in thirst. And yet, I suspect that my grandfather also taught me the art of listening to stories. The Ancient Mariner, with his glittering eye, is compelled to speak of his mistake and of the ensuing deaths upon the ship. My grandfather, a gentle man despite his daily immersion in often violent encounters between the inmates whom he hoped to rehabilitate, and the guards, whom he sought to transform into rehabilitators, also held me with his glittering eye, a glass eye that replaced the eye he lost at age five, when another schoolboy accidently poked it out with a stick on a playground. The "Ancient Mariner" is cursed to speak of his mistake and that is what I remember as a young girl transfixed by images of the dead ever returning to haunt the living. So, the killing of the albatross stood in my mind for other abject transgressions leading to violent death; the Vietnam War dead, the Holocaust dead, and foremost, the dead soldiers my father saw that frozen January 1945 as World War II was coming to an end. My father stood among the frozen dead, oblivious to the news that the Soviets liberated Auschwitz on 27 January 1945. Thus, the Holocaust dead turned to ashes appear in historical adjacency to the dead soldiers among whom he wandered. The dead are ever returning to haunt the living.

# Afterword

I write excerpts of this essay to the tune of incessant drilling. I am renting a coop apartment in Foggy Bottom in Washington, D.C. during my sabbatical as the long-term reach of Covid takes hold in the spring of 2020. Construction workers are considered essential workers, so they are drilling away at the ceiling of the parking garage. No matter that I am on the sixth floor. The noise carries, perhaps more intense as the Covid lockdown holds other noises at bay. Ear plugs diminish the intermittent traffic noise, but not the drilling that cuts through all barriers and grates on every cell of my brain, a small annoyance in this time of Covid. At least I have a roof over my head. I look out a large window and see the outline of unknown neighbors across K Street silhouetted against the glow and flash of LED and HD screens, sitting at computers, vacuuming, cooking dinner. Since the beginning of the shelter-in-place rules, I have obsessed over one apartment in particular that is directly across from mine. The two windows are wide open, no screens. I watch the wind blow the white curtains to and fro, and I imagine the rain seeping into the apartment and staining the wooden floors. Despite calling the property management company that owns the building (ah-the wonders of google address search!), and taping a note to the front entrance of the building alerting anyone going in or out about the windows, the dark room behind the open windows gapes at me unabated eight weeks into the lockdown, until I leave DC at the end of spring. I struggle to let go of my obsession with the abyss conjured by the open windows and to stop hoping that someone will close them.

The open windows reminds me of the Bardo, a Tibetan Buddhist notion of a space between life and death, a passageway for recently deceased beings who struggle to accept their passing by battling the demons who tempt them into a state of perpetual limbo, a form of non-acceptance of their death. The living can come to the aid of the dead by chanting and practicing rituals that encourage the recently deceased to continue on their way to their next incarnation or dissolution into nothingness. The dead, as they move through the Bardo, carry with them residual memories of their lives, like filmy residues that encase them and their desire to adhere to life, and like plays of light that form images of the past—as one might see in a flickering set of moving pictures—flashes of memory that recede and then burst forth, yanking the dead into a state of disarray. We, the living, remain adjacent to the dead and therefore responsible for them beyond any shared space or time.

I walk out of my apartment building and wander through the Georgetown residential streets, stop and take in the myriad reds, blues, yellows, whites, purples and pinks of tulips, irises, pansies, roses, and peonies in full bloom in the front gardens of the brick townhouses. I meander through the dirt paths in Dumbarton Oaks Park, along Rock Creek, moving aside to let the joggers and occasional gaggle of family units pass by. I reach the Potomac River, flowing muddy brown, in five minutes and walk along its embankment, listening to the seagulls squawking and the mallards quacking. Never mind the mask that keeps stifling my breath and fogging up my glasses, or that others seeking fresh air don't always keep the requisite distance. I rarely pause when I walk, though I once find an empty bench in an unkempt meadow in Dumbarton Oaks and sit there for an hour, communing with a single wren who joins me, beseeching for what, I do not know. The Oak Hill Cemetery adjacent to Dumbarton Oaks becomes my second home. I stop in front of a mausoleum in a disheveled corner of the cemetery, marveling at the coincidence of finding the location described in George Saunders's novel *Lincoln in the Bardo*. In the book, Lincoln's dead son, Willie, awakens in the tomb, not yet aware of his passing. Other dead souls convene to gently assist him in the transition to the afterlife. In real life, Willie's corpse was eventually buried in the Lincoln family plot in Springfield, Illinois, but for a temporary period, Willie, was laid to rest in this mausoleum. It is tucked away at the periphery of the cemetery, and I find it by chance after walking down a moss-covered path and then down a crooked set of slate stone steps. I marvel at the contrast between Lincoln's access to his beloved dead son's corpse and the current inaccessibility of the dead during Covid. Lincoln could go to the tomb, open the casket and hold his dead son in his arms. The plaque attached to the back wall of the tomb reads "William Wallace 'Willie' Lincoln (1850–1862). The President, sad to leave him cold and alone, visited several times and had the crypt and casket opened."

In Saunders's book, Lincoln holds Willie's corpse while unbeknownst to him, the dead gather round uneasy because Lincoln holding poor Willie prevents him from moving through the Bardo. At the same time, we learn in Saunders's book that Lincoln may have felt a tinge of guilt knowing that under his watch the Civil War left countless corpses out of the hands of their families, an experience that has become all too familiar during Covid.

Suffering the loss of Willie, and the grinding battlefield death tolls, Lincoln's uneasy displaced mourning did not extend to the enslaved people for whom the Civil War was ostensibly fought. The ongoing legacy of racism and genocide, of brutality, inequity, and neglect on the part of countless white citizens in the US today must reckon with this same failure of empathy. However, the lack of mourning, the lack of proper burial, and the lack of justice for George Floyd, for Breonna Taylor, for Michael Brown, for Trayvon Martin and many other young African Americans stands in jarring contrast to the possibility of a mourning that extends beyond one's own kin, a possibility that has yet to emerge.

When John Lewis, the eminent African American congressman, civil rights leader, and beloved friend to many communities, passed, his body lay in a coffin at the US Capitol, draped in an American flag. Afforded a proper burial of the highest degree, this body lay in sharp contrast to other Black bodies, viciously denied a proper burial since 1619 on these shores and across the land. Lewis's body represented the possibility of a more "perfect union" as President Obama put it, a tribute to a vision of a future free of racist violence or at the very least, more just and equitable. Before he died, Lewis directed the *New York Times* to publish a message from the grave, so to speak, imploring all of us, but especially young people to make "good trouble": "You must do something. Democracy is not a state. It is an act, and each generation must do its part to help build what we called the Beloved Community, a nation and world society at peace with itself."[4]

My family buried our German grandfather, Walter Remmler, next to his wife, Louise née Sturm, in the Bethpage Friends Meeting House Cemetery, not too far from North Massapequa; where, after World War II, he and his sons built the first and only family home. They shoveled dirt onto the coffin and wept together and hugged each other, years ago. Nowadays, no one from the family lives nearby, so no one visits the grave. But their stone stands, the bones beneath had a proper burial, and, although I could not be there in person, I can take comfort in the mourning. There are so many unmourned in this dreadful year. Will separation from loved ones who suffered and died alone during Covid lead to a birth of empathy towards others, for whom proper burial was and continues to be denied?

It has been over a year since I wandered around DC during the early stages of the pandemic. The open windows that remained open for the duration of my stay are most likely closed by now. And yet the space of the Bardo continues to haunt me as I read the names of the dead enmeshed with microscopic Covid virus particles and ponder those lives, amidst other lives cut short by murder, negligence, racism, poverty, and injustice. What brings the pain of not being able to properly bury one's beloved victims of injustice into a alignment with another's pain wrought by the pandemic? Or what brings one story of war and separation into an adjacent space with another's story of extermination, displacement, and survival?

---

4 John Lewis. "Together you can Redeem the Soul of our Nation," *New York Times* (20 July 2020), https://www.nytimes.com/2020/07/30/opinion/john-lewis-civil-rights-america.html 20 July 2020.

# Works Cited

Ambrose, Stephen E. *Citizen Soldiers: The U.S. Army from the Normandy Beaches to the Bulge to the Surrender of Germany*. New York: Simon and Schuster, 1997.

Bammer, Angelika. *Born After. Reckoning with the German Past*. London: Bloomsbury, 2019.

Benjamin, Walter. "A Small History of Photography." In *One Way Street and Other Writings*. Translated by Edmund Jephcott and Kingsley Shorter. London: Verso, 1992. 240–257.

Buerkle, Darcy C. *Nothing Happened: Charlotte Salomon and an Archive of Suicide*. Ann Arbor: University of Michigan Press, 2013.

Kacandes, Irene. *Daddy's War: Greek American Stories. A Paramemoir*. Lincoln: University of Nebraska Press, 2009.

Kaplan, Alice Yaeger. *French Lessons: A Memoir*. Chicago: University of Chicago Press, 1993.

Lewis, John. "Together you can Redeem the Soul of our Nation." *New York Times* (20 July 2020). https://www.nytimes.com/2020/07/30/opinion/john-lewis-civil-rights-america.html 30 July 2020.

Lipstadt, Deborah E. *History on Trial: My Day in Court with a Holocaust Denier*. New York: Harper Perennial, 2008.

Morris, Leslie. *The Translated Jew: German Jewish Culture Outside the Margins*. Evanston, IL: Northwestern University Press, 2018.

Parkinson, Anna M. *An Emotional State: The Politics of Emotion in Postwar West German Culture*. Ann Arbor: University of Michigan Press, 2015.

Rothberg, Michael. *The Implicated Subject: Beyond Victims and Perpetrators. Cultural Memory in the Present*. Stanford, CA: Stanford University Press, 2019.

Saunders, George. *Lincoln in the Bardo: A Novel*. New York: Random House, 2017.

Schwartz, Paula L. *Today Sardines Are Not for Sale: A Street Protest in Occupied Paris*. New York: Oxford University Press, 2020.

Thomas, Dylan. *The Collected Poems*. New York: New Directions, 1957.

Leslie Morris

# The Unconcealed: Family Secrets as Family History

> It is as if one saw a screen with scattered color-patches and said, the way they are here, they are unintelligible; they only make sense when one completes them into a shape. Whereas I want to say – here is the whole (if you complete it, you falsify it.) –
>
> Ludwig Wittgenstein, *Remarks on the Philosophy of Psychology* (# 257)

> The image is not a duplicate of a thing. It is a complex set of relations between the visible and the invisible, the visible and speech, the said and the unsaid.        Jacques Rancière

## My Kosher Books

I keep a strict separation of books: German books and books in English. My house is filled with books in English, but I do not allow any German books into my house; instead, they populate the shelves of my faculty campus office, staring down at me like forlorn refugees. I have hundreds of books from Germany, bought over the years while I was a graduate student in Germany. I packed them in the regulation yellow German postal boxes, tying each box carefully with string, and then placing six to eight boxes in a large bag to be sent by ship. My books came to the US the way I imagine my paternal grandparents came: on ship, battered and ragged after the long trip. Many of the hardcover books lost their spines on the journey. I have moved them from one campus office to the next. But as much as I cherish these German refugee books, they are not allowed into my home. Sometimes, a book in German is brought home to prepare for class, but as soon as I am done, back to campus it goes.

I'm not quite sure where this strict adherence to the separation of German from English books originates, but it echoes my mother's need to demarcate American from non-American, Jew from non-Jew, black from white. My mother, Jewish enough to be in danger during the war, was always on the look-out for difference (racial, ethnic, and sexual). Her favorite joke, that she would tell again and again, in a mixture of English and French, was about a cannibal on an Air France flight who asked to see the passenger list. I now understand that my mother's ambiguous status not only in this country but also in France, as her parents were Hungarian, meant that she needed to be more French than

https://doi.org/10.1515/9783110753295-018

the French. She needed to bury her Jewishness by a compulsive retelling of a French racist and colonialist joke. Black and white.

This aversion to inviting my German books into my house is not, in fact, a fear of contamination (German books as *treyf*) nor a deep-seated revulsion for all things German. I never even realized that I was doing it until quite recently. German books in the office, English books at home. Black and white. America and Europe. Despite this strict regulation and separation of books, the family narratives – my story, my mother's story – cannot be separated. They merge, in their twisted and burrowed memory-spaces, hard to grasp, slipping just out of reach. The process of re-collecting them, the attempt at anamnesis, begins with the recognition that, like Wittgenstein's scattered color patches, these memory-shards are fragmented, broken, they do not and cannot cohere into a knowable whole. The task, then, is to create, in writing, the color patches that must remain patched, and to find a form that can capture the fragmentary shards of memory and the secrets that, even when unconcealed, remain hidden. The blank page holds the secrets, with the promise, ever hovering, of revealing them.

*secret*, adj. and n.
A. *adj.*
   1. Kept from knowledge or observation; hidden, concealed.
      a. Predicatively (esp. in *to keep secret*): Kept from public knowledge, or from the knowledge of persons specified; not allowed to be known, or only by selected persons.
      **f.** Of feelings, passions, thoughts: Not openly avowed or expressed; concealed, disguised; also, in stronger sense, known only to the subject, inward, inmost. Hence said of the heart, soul, etc.
      i. Hidden from sight; not discernible or visible; unseen (chiefly *poetic*)
      p. secret life *n.* a private life of a nature concealed from the common observer; *spec.* one consisting of covert sexual dealings. (from the *Oxford English Dictionary*)

Derrida, the philosopher-poet who has fallen out of fashion but for whom I retain a secret love, reminds us that in order to possess the secret one must present the secret to oneself, thus negating the secret and breaking the original frame of secrecy: *"I must tell the secret to myself as if I were somebody else."* Is my ban on German books entering my house somehow part of the secret, or a symptom of the secret? *Das Geheimnis.* But is a secret the same as being in hiding, of being *caché* or even more, as Derrida says, *crypté?* My house is cleansed of the German but not *Judenrein*; in the crypt of my bookshelves filled with English books there is the flicker of the banished German books and the hidden traces of the French and even more *caché*, the Hungarian. Yet it was the sense of the secret, the knowing without knowing, that led me, upon my graduation from college, to leave, suddenly, for Germany, seized with a strange and sudden need to learn German and live in Germany. And as I felt that I knew something without knowing what it was, as I embarked for Germany, it was the mysterious yet compelling – the *crypté* – sounds of German that I chose as medium for the secret.

Germany was where I went because of the secret I did not yet know, and German was my secret language. I fell into the German language feverishly – mathematical beauty of the grammar, the way words could be built from more words, the pure sound of a language that was so fully removed from my mother and the French and Hungarian of my childhood. It was mine. At the time, I had a vague idea that in moving into German, in becoming a nomad stunned into a new language world, I would finally be able to write, in English, the secret I did not even know of. I framed, however shakily, this voluntary exile to German as a way of pledging allegiance to English. I had a persistent sense that I needed to write something and that it had to do with my mother and her past, but I did not know what it was. I thought of immersing in German as a way of going underground, of using German as a front in order to do the clandestine and more important work of writing in English. German was the illicit affair, it was the architectural hiding place, and I was head over heels in love with it. I gorged on the sounds of German and reveled in the magic of a grammatical structure that was not mine. I was finally separated from my mother and no longer being force-fed French. My plan, at age twenty-one, was to inhabit German for my life above ground, and to use English for the subterranean task of writing the secret that I still did not know but that I knew. *Secret* is related etymologically to excrement and seduction, both derived from the Latin verb *cernere* (crevi, cretum): to separate, set apart, to sift, to discern or distinguish an object from a distance. It also contains the idea of preserving. I would live and breathe in German, a language that would finally separate me from my French-speaking mother, and, in my youthful fantasy, what would come out would be a kind of coded English that would be touched, mysteriously, by the presence of German. What was the code to be? Again, I knew and yet did not know. But I knew that I had a pull to Germany that I could not explain.

In *Francis Bacon: The Logic of Sensation*, Deleuze describes the painter who does not need to cover a blank surface, but rather works to empty, "disencumber," and wipe clean the surface that is already full.[1] The hybrid memoir project I have begun starts with this tension between the blank surface and the stories I

---

1 Gilles Deleuze. *Francis Bacon. The Logic of Sensation*, trans. Daniel W. Smith (Minneapolis: University of Minnesota Press, 2003), 71.

cannot quite get to, but that are nonetheless there. Eleven years ago, I learned that a large number of my mother's immediate family had died in the Holocaust. I had no idea that they had ever existed. The following year, I fell into a still un-explained coma that lasted more than five weeks; the illness came immediately following a trip to Budapest in which I met my cousins who had survived. The book is experimental in form, moving from prose poems to Conceptual poetry, taking (and then erasing) the text of my fifteen hundred-page medical file to cre-ate a visual space that suggests the absence/presence that defines both the coma and the knowing/not knowing of my family history. More than an attempt to grapple with a buried family Holocaust history and the mystery of my subsequent illness, it tries to think more broadly about knowledge, memory, and the nature of consciousness. How can representation exist if the experience of the coma, which caused profound effects, is not present in conscious memory? My body, with its scars from the medical procedures, remembers my coma. And yet I cannot know my own coma. My desire is not to retrieve lost knowledge and fill in the lacunae; rather, it is the lacunae that interest me. And it is here that I begin, with a philo-sophical reflection about knowledge: I cannot know my own coma – the state of profound unconsciousness into which I fell after my return from Budapest. I also cannot fully know my mother's story of the war. And yet, what I do know is how the unconscious, as Victor Burgin has noted, plays a part in the formation of a work of art; yet as he notes, it is the form that makes the work a work of art and distinguishes it from psychoanalysis.[2] How to find a form that can contain the fam-ily history that I am in the process of unconcealing? And how to contain the coma, that place where I was but was not? The very word "unconcealed," not quite "re-vealed," suggest the process of uncovering histories, memories, consciousness.

For a number of years, my scholarly writing and my teaching have been shaped by the following set of questions about the concealed and the uncon-cealed: how does the eruption from silence into language, from the blank page into image, also mirror the process of psychoanalysis, in which the play of the concealed and the unconcealed, consciousness and the unconscious, collide in creative and new ways? How do tropes of concealment shape various forms of writing? How are experimental forms of writing a response to a new media land-scape of technological innovation, and how does technology continue to shape the forms of this writing? In my quest to look for the clues to the puzzle that I did not even know I was trying to solve, I have tried, with students and in my own writing, to explore text both on and outside the margins; text that slips off the page; paratext; writing found outside the margins, within the parentheses,

---

**2** Victor Burgin, *The Remembered Film* (London: Reaktion Books, 2004), 66.

on the body, on the wall. I have been fascinated in particular with the literary form of the lipogram – constrained writing that removes one element, suggesting with the absence (or with the rigorous discipline of removal, of concealing and then, through the act of reading, unconcealing) of a letter (Georges Perec and the "voided" letter "e") the very absence at the center of the secret that I still could not know. And less oblique than the lipogram, with its disappeared single letter, it is the literary form of erasure poetics – a poetry of creating absence from presence, with the spectral presence of that which has been erased – that pulls me, *crypté*, into the flicker of the unconscious and my coma. In my scholarly and teaching practices, I have explored new modes of writing that might capture what I could not know or see – my mother's story, my coma – and with that, create a manifesto of incompletion, text that lingers in the air between speech and silence, moving from darkwords to darkwords.

## Through a photograph, darkly

And yet, I keep coming back to the photographs, the black and white ones that cast a light on my "kosher books" and yet that still conceal the secret, the hidden knowledge, that emerges into consciousness only to recede again. Unlike the historian who stumbles, often quite literally, onto boxes of family documents hidden for decades in an attic, after a career spent in archives of other peoples' family stories, I have not stumbled onto anything. There are no buried stories of heroic survival. My family story is one of half-starts, a full life in the subjunctive, of the almost was and maybe could have beens. In so many ways, it isn't even a story, just words smothered by my mother (Sarah Kofman's *paroles suffoquées*) as other words – floods of them, nonstop chatter in French, Hungarian, and English – poured out of her and now sit deep within me, competing for space on the shelves of my mind. The story of my mother and the war was never transmitted, and yet, at the same time, a pervasive sense of darkness and deep anxiety filled the air of our house. I knew without knowing. It was always there, in plain sight, although obscured, concealed. To unconceal is impossible. To unconceal by showing the possibility of the whole that cannot ever be completed (Wittgenstein) is the task. Is the photograph the embodiment of Christopher Bollas's unthought known, the thing that seems to tell the story, and yet which for me remains blurred and part of all the stories that have yet to be told? Perhaps. The secret deep within the recesses of my mother's memory, *crypté*, was unveiled by a series of photographs that I first saw eleven years ago. My descent into the world of the unconcealed began when I went to meet my mother's half-brother in France for the first time and learned that my grandfather (divorced from my grandmother in the 1920s and estranged from my mother) was one of seven siblings. I did not know he had any siblings. My mother never told us.

*But what she did tell us about, instead of telling us about our grandfather, was the man she thought of as her father, and whom she referred to as our grandfather. Like her real father, my grandfather, he too was named Louis (and like my grandfather, originally named Lajos) and was from Budapest. But unlike my grandfather, he was not Jewish but instead from a French Huguenot family in Budapest. He and my grandmother had had a love affair years earlier, in Paris, and had remained close. He was a film chemist and had started a company in Paris, Filmolaque, that specialized in preserving – or is it restoring? – old film and in creating subtitles of foreign films. The early chemical process for creating subtitles involved pressing heated printing plates against the emulsion side of a wax-coated finished film copy. After the coating melted, the emulsion below was exposed, and the film was then washed with bleach to dissolve the exposed emulsion. Legible white letters became visible when projected onto the screen. How much of the film itself had to be erased in order for the letters to be visible? Black and white. Am I preserving or restoring my mother's memories?*

During that visit, my uncle brought out a photo, faded, of nine people; the first shock was learning that my grandfather had siblings, and immediately after, learning that only two of them had survived the camps. The others had been deported, with spouses and children, late in the fall and winter of 1944, from Budapest, to Dachau, Auschwitz, Mauthausen. *I have been to all of these places, yet without knowing.* My grandfather is standing in the back row, looking expectant, his mouth slightly open, an expression in his eyes of knowing something beyond the sepia world that bathes the family. In all of the photos I have of him, this is the expression that I most recognize, that allows me to pick him out from among the crowd of faces I do not know nor recognize. *Mon grandpère.* The two words in French, elongated, *mon grandpère*, the possessive pronoun paradoxically indicating that which I never had (but which I had with double vision), the word *grandpère* projected white onto black onto the memory screen of my childhood, and just out of reach.

If the blank canvas of the painting holds the past, the memories, that need to be wiped clean in order for figuration to come into being, then the photograph, itself birthed and bathed into the light, shocks with the suddenness and seeming certainty of knowledge, of something that did exist for that moment in the past. My grandfather and his family float here in pictorial space, looking out at the camera while holding on to each other.

A year later, after first meeting my uncle and seeing this photograph, I made a trip to Budapest to meet the cousins I never even knew had existed. During that visit, they held a gathering to meet me, their long-lost American cousin that they

**Figure 1:** Piszker family photograph, Budapest, 1917.

did know of, but had no idea how to find. My mother's first cousin, György Weil, was standing in the living room, holding a large box of what I assumed were old photographs of my grandfather and his family, photos like the one I had seen at my uncle's house in France the year before and that first shocked me into knowing my mother's story was not as it seemed. As I walked towards him, György took one of the pictures from the box and held it up to my face. Suddenly the room fell silent. Everyone was watching. But the picture he held up was not of my grandfather or his siblings or parents, but instead a baby picture of my younger sister. Silently, György held up picture after picture. They were all of my sisters and me: Nicole, age three, at her birthday party, Michele and Leslie doing a jigsaw puzzle in our Brownie uniforms in the den of our house in Glencoe, Illinois, Leslie on the swing set on the playground. On the back of each photograph was my mother's unmistakable handwriting, in Hungarian, to her aunt Lili, the one sister of my grandfather who had survived.

### Through a photograph, darkly

The world becomes silent. I hardly breathed. At the moment of seeing the box with the photographs – a box that had sat for fifty years with all the baby and childhood pictures of my sisters and me, passed from the aunt I never even knew had existed on to her son, with my mother's handwriting on the back of each photo – I was stunned into a new kind of knowledge: that two truths had always co-existed. They knew about us. For us, none of them had ever existed.

Two days after my return from Budapest, I fell into a still unexplained coma that lasted more than five weeks. Did my unconscious take me into the space of the coma so that I could reassemble the images in the dark?

Stunned into the darkness.

# The Secret *is* the History (as the Medium is the Message?)

I only met my mother's father once, when I was very young. Born Lajos Piszker – perhaps, according to one document, in Zimony, now Zemun, a district of Belgrade? Or was he born in Budapest, in the Dob Utca, just across the street from the Carl Lutz memorial, right in the middle of what is now a bustling tourist zone, marketed as the Jewish district. I have documents that suggest his birthplace was in Zemun – he was a distant relative, by marriage, to Theodor Herzl, whose family also lived in Zemun before Budapest, and whose sister had married a Piszker (or so the confusing documents collected by a cousin seem to suggest). But a quick trip to the Jewish archive in Budapest lists my grandfather's birthplace as the Dob Utca in Budapest. More documents create more confusion: my grandfather's demobilization papers from the French army in 1940 list his birthplace as Budapest, while another Hungarian document lists Zimony, Serbia as his birthplace and attests to his decision, based on the Trianon Agreement of 1921, to take Hungarian citizenship. The building in the Dob Utca – dark, dank, in disrepair – is just across the street from the Lutz statue commemorating the Swiss diplomat in Budapest who saved 62,000 Jews. That much at least is known.

Lajos became Louis after he immigrated to Paris and became a French citizen, long before the war. And that story, too, is lost, hard to unconceal. There is a picture of him in his French army uniform in 1939, looking a bit stunned; another document suggests that he had been a prisoner of war, although it is hard to decipher. What I do know: the rest of the family stayed in Budapest. Some survived. Some did not. And all of them – dead and the living – were erased, or should I say hidden (*caché*), by my mother when she came to America. But at the same time, they weren't erased, at least not for a few decades – she continued, for a number of years, to send letters and pictures of us to her aunt in Budapest. An entire box of pictures with my entire childhood in it, had sat in an apartment in Budapest for over fifty years – not hidden. In plain view. But not to us. *Crypté.*

My grandfather survived in hiding, I now know, in the south of France, working in the fields harvesting grapes. In this photo, he is shown with the workers. He is the second from the left on the bottom row, sitting upright on his knees, his face looking straight ahead, the arm of the man next to him on his shoulder. When I am first shown the photograph, in Budapest, I think how he is instantly identifiable, that he stands out from the crowd, that he is marked as different, and that he looks tentative, and also afraid. My grandfather, I learn from my uncle, was always in hiding, even long after the war. *Caché.*

**Figure 2:** Lajos/Louis Piszker in the South of France, 1942 or 1943.

I think those are grapes in the bucket on the ground, just to the right of my grandfather. I think I've seen another picture of him with his feet pounding the grapes, but that is likely an overlay from other photos of grape harvesting in the south of France that I've seen, over the years, that have nothing to do with my grandfather. But which grandfather, which Louis? The two converge in my mind. I fill in his story from my own canvas of memory-shards, and then realize that I do not know what the truth is. What I do know: I love looking at his face as he looks at the camera. These pictures are all that I have of him, *mon grandpère.*

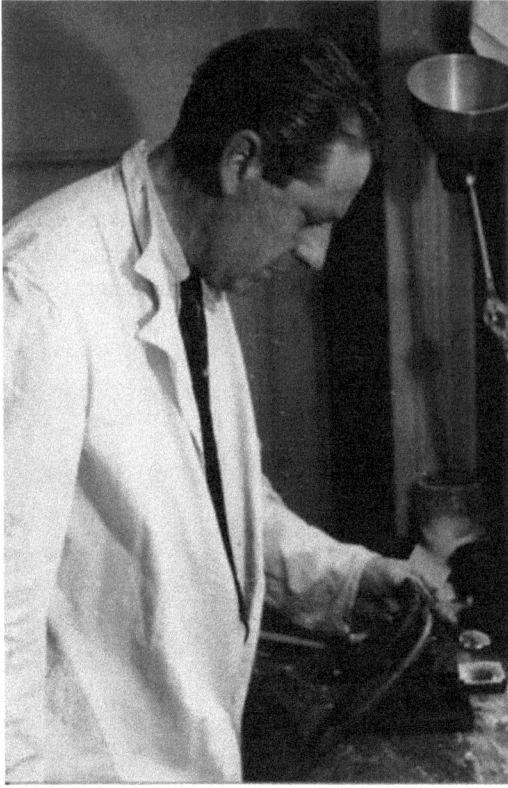

**Figure 3:** Louis Piszker in his jewelry atelier, Paris. Year unknown.

After the war, he became a jeweler and a goldsmith and had a small jewelry atelier in the Rue du Faubourg Montmartre in the 9th *arrondissement*. And he became devoted to Esperanto, part of a group of survivors in Paris who formed a sort of secret society speaking the coded *lingo* of Esperanto. I also learn, from my uncle, that he was an exceptionally quiet man and rarely spoke.

I learn, too, from my uncle, of the priest who hid my grandfather and saved his life. The handwriting on his baptismal certificate, 23 October 1940, at the age of 37, is illegible, and I cannot make out the precise location.

Again, my grandfather – tall, lanky, with the curly hair and high forehead of my older sister – looks out at the camera, looks into what I imagine is a future he cannot know. When I tell the story after my return from France about meeting my uncle and learning about my grandfather's story in the war – and I tell it again and again, like the Ancient Mariner I cannot stop telling the story to people I encounter even by chance – I turn it into a children's story, a fairy-tale of hiding, in

**Figure 4:** Louis Piszker, south of France during the war. Priest who helped him. Third man unknown.

which my grandfather has crawled out of a hole in the field, where he had been hidden by the priest. I do not know who the third man is – another Jew who is being hidden? A friend? A relative? In my re-telling of the story and in my memory of the photograph, there is only the priest and my grandfather. I have erased the third man. But here he is again. Black and white.

In the archive room of the United States Holocaust Memorial Museum, I find records of my grandfather and his family. They come up almost immediately. It is from the archive that I learn that the family lived in the war in what was then the ghetto, in the Kiraly Utca, where my great-uncle Pali Piszker had his fur shop. The Kiraly Utca is now the very heart of the commercial center of the newly restored Jewish district in Pest. I find the building, and, miraculously, it is one of the very few that had not yet been torn down to make way for the Starbucks and fancy boutiques and cafés on the street.

Sometimes I flip through the pictures so quickly that they almost become a film. That pictures of me were in the box of photographs in Budapest continues to haunt me, as if I too had a double, or secret life: in Chicago during the Cold War, and in Budapest behind the "Iron Curtain." The images haunt me. But I, too, haunt the images. We live in a symbiotic, reciprocal relationship, those pictures and I. *Europe and America. Black and White. Hidden and unconcealed.* When I first met

**Figure 5:** Kiraly utça 36, Budapest, with remnant of sign of the fur shop of Pali Piszker. Photo taken in 2012.

**Figure 6:** Left to right: Lajos Piszker, their nephew György Weil, Pali Piszker, in fur hats made by Pali Piszker in Budapest.

my cousins and saw the box of photographs, we spent hours scanning the pictures. I couldn't stop. The grief at the knowledge that my mother had kept in touch with them translated into a fervor of scanning and copying, of desperately wanting to preserve. Somehow, a number of photos ended up all together on one page – pictures of my mother in her first years in America – it turns out that she had come to the United States in 1946, not 1956 as we had been told, and had married an American GI. For citizenship? For love? I do not know. But there she is, in black and white, an American girl, as she writes, in a mix of French, Hungarian, and English, to her Aunt Lili in Budapest.

The opening shot of the collage of photos, or the film that I have not yet created and not watched in its entirety is this one, of my grandfather.

**Figure 7:** Louis Piszker in France after the war.

I believe he is walking in the streets of Paris, after the war; I see part of a sign that says Boulangerie Patisserie, but I don't know if he is in Paris or another city. Everyone seems to be walking, rapidly, and with purpose. They seem on edge, and my grandfather as well. He is in the middle of the frame, the main character surrounded by what looks like movie extras. He looks apprehensive, suspicious even. Where is he going? Why was the photograph taken? Who took it?

On the move, on the run, in hiding, carrying secrets. I rummage through the pictures on my computer, scanned and not organized, and through the package of documents I have collected about my mother and her family. We are all jumbled in there together, black and white. And how have the two grandfathers – one real, the other fashioned by my mother as our grandfather – whose names were the same, blurred and merged in my own mind? I now realize that an earlier piece I had written was a false memory that substituted one grandfather for another.

*I did meet my grandfather once when I was five years old. It was in Paris. Jardin du Luxembourg, and I am holding the stick as my magical horse and I go round the carousel, trying to capture the iron rings in the jeu de bagues. I remember my mother watching me as I caught the ring, triumphant – was she triumphant watching me play at being a little French girl? I also remember waking up one morning in Paris in an enormously high bed with my eyes glued shut from conjunctivitis. I could not see. Everything was dark. It was on that trip that we were taken to meet our grandfather. I remember being told I was going to meet him, but I do not remember the meeting. Were my eyes still shut? Did he speak to me in Esperanto?*

## Coda: Black and White

To end: an intransitive verb. But "end" is also a transitive verb. To end the story: transitive. But to end at midnight: intransitive.

I begin the memoir about my coma and my mother's elusive Holocaust family history by posing the question of how to begin a story that can never be told. To find a narrative with which to tell the story – the story about my mother, the story about my coma – can only begin as a first line among a multitude of possible first lines forever hovering in the possibility of becoming a story. These lines go nowhere but hover in their own possibility of forever becoming. "It was evening all afternoon /It was snowing/and it was going to snow," as Wallace Stevens writes. But I now realize that the reluctance to begin – or rather, to keep on beginning, over and over again, is really a fear of ending. If I cannot begin, then I can never end. And if the story about my mother's family can only begin as a series of stuttered attempts to tell the story, then it will be in a state of perpetual becoming. To wrench Wallace Stevens's line, it will always be snowing and it will always be about to snow.

I grapple with how to date the start of my coma. Where does it begin? How can I demarcate the lost weeks in the coma from the other weeks, months, and years of my life? Did it begin when I woke in the middle of the night, a week after my return from Budapest, with the oddest sensation of numbness and tingling in my arms and legs? Or does it begin in the hospital, as I was gradually slipping from consciousness? Or doesn't it also begin when I made the first trip to Budapest and Vienna with my parents, in 1978, when the first intimations of my mother's story must have fallen into my unconscious mind, sparking me to learn German and become a scholar of Jewish memory and the Holocaust?

Or did it begin with the shock of discovering the box of my childhood photographs, with my mother's handwriting on the back of each photo, sitting in a box in Budapest for 50 years? Like Freud's sense of astonishment when he went to the Acropolis and realized that it was in fact as he had read in books – "that all this really *does* exist!" I, too, found myself experiencing a revelation not unlike what Freud identifies: finding in the "real" of the photograph the reality that my mother had in fact continued her contact with her relatives. Perhaps the sense of incredulity that Freud expresses is also the beginning of the coma.

Or did it perhaps begin in Basel, where I went the day after first meeting my uncle in France and learning of my grandfather's family who had perished? It was in Basel that I met the British-born experimental poet Anne Blonstein; she was the first person I spoke with since learning the first layer of the secret from my uncle the day before – the first of many people, in those months to whom, like the Ancient Mariner, I could not stop telling the story, until the coma when I did not speak. I was in Basel to give a paper on the whispered traces of Europe in the work of Alfred Kazin and had recently read the work of Blonstein, a geneticist by training, who had created a *notarikon* translation of Paul Celan's translations of Shakespeare's sonnets, expanding the frame of each sonnet, through *notarikon*, so that the fourteen lines of Shakespeare/Celan are expanded into pages and pages of *langue cryptée*. A rabbinic hermeneutical method used in late antiquity as a mode of interpretation of the Hebrew bible, in which each letter of each word becomes the first letter of a new word, *notarikon* is itself a play with the concealment and encryption that marks the codedness of *all* poetry, and especially poetry that dwells between languages and multiple translations. And, as

Blonstein herself notes about the process of *notarikon* translation: it never ends, but rather is a potentially endless sequence of unfolding acronyms.[3] Crypté.

Later, while I was in the coma, Blonstein wrote two poems in *notarikon* for me. The first is a *notarikon* poem of the line "For Leslie Morris in your coma":

> flowers of ruffled
> letters
>     extravagant-scented
>     little idiomatic efflorescences
>
> mutely offered
>    rose-resonant
> if signs
>
> imagined now
>
> yellow or unconsciously
>    red
>
>    could outdream mental abstraction

This poem, written for and addressed to me, but not able at the time to be read by me, as I was in the coma *outdreaming mental abstraction* – outdreaming yet so deeply within, crypté. Or was I outdoing myself in dwelling in the dreamstate of the unconscious, a state of such acute mental abstraction that I had abstracted myself and abstracted consciousness and subtracted myself from the world? I was in a coma – where, in Blonstein's exquisitely wrought *notarikon* of that four-letter word "coma," I *could outdream mental abstraction*.

What has preoccupied me as much as the question of when the coma began is the impossibility of saying when it ended. Did the coma end when I first woke up? When I moved from the ICU to the neuro floor? From there to the rehab clinic? When I left the hospital for home? When I stopped feeling like Primo Levi's figure of the Muselmann, watching the faces of others watching me, as if I had some knowledge of what it is to descend to the depths of the unconscious. In those first months after I woke up, I could see the look on their faces as they watched for signs that would mark me as having emerged from an unknowable place, from "that undiscovered country, from whose bourn no traveler returns." I felt trapped in the role of the oracular patient who was there to help others navigate their own deepest fears of their own darkness, of their death.

---

**3** Anne Blonstein, "On Notarikon: A Note on the Process by Anne Blonstein." https://jacket2.org/commentary/anne-blonstein.

I live in the ever-present ongoing state of after-the-coma. There is my life "before the coma," that time of edenic innocence and grace, and the forever expanding time of "after the coma." I cannot know my own coma. To attempt to know it, to think about how I cannot know it or the full story of my mother's family, is now my task.

# Works Cited

Blonstein, Anne. *On Notarikon: A Note on the Process by Anne Blonstein*. https://jacket2.org/commentary/anne-blonstein.

Victor Burgin. *The Remembered Film*. London: Reaktion Books, 2004.

Deleuze, Gilles. *Francis Bacon. The Logic of Sensation*. Translated by Daniel W. Smith. Minneapolis: University of Minnesota Press, 2003.

Kofman, Sarah. *Paroles suffoquées*. Paris: Editions Galilée, 1987.

Stevens, Wallace. "13 Ways of Looking at a Blackbird." *The Collected Poems of Wallace Stevens*. New York: Vintage, 1990.

Tao Leigh Goffe
# The Flesh of the Family Album: Black Pacific Visual Kinship

In 2016, my mother, my sister, and I took a trip to Hong Kong in search of our ancestral roots and routes.[1] We had never been to Asia before; the vacation was a graduation gift for my sister who had just become a lawyer. We went looking for the home of my mother's deceased father. Born in Jamaica in the late 1920s to a Chinese father and Afro-Jamaican mother, Edwin was taken to live with his stepmother in Hong Kong at the age of seven.[2] In search of traces of his life, a Black boyhood in Hong Kong's New Territories, all we knew was the name of his village. Our clues were old family photographs and his last effects. With the help of a new friend from Hong Kong named Geoffrey who was able to interpret and introduce us to the village chief, we were led to the house where my grandfather grew up.

Tang-kwong, a man we later discovered was my mother's first cousin, greeted us cheerfully at the door.[3] We were surprised that he was unsurprised to see his Black relatives. Whether he knew it fully or not, he also had a Black Jamaican

---

1 Paul Gilroy, *The Black Atlantic: Modernity and Double Consciousness* (Cambridge, MA: Harvard University Press, 1993).
2 See Thomas Bird, "Back to His Roots: Chinese-Jamaican Leo Lee," *Post Magazine*, 24 May 2014; and Bernice Chan, "From Harlem to China: How an African-American Tracked Down Her Chinese Grandfather," *South China Morning Post*, 16 May 2017. Paula Williams Madison, the subject of Chan's article, documented her exploration of Afro-Chineseness in the Caribbean diaspora in *Finding Samuel Lowe: China, Jamaica, Harlem* (New York: Harper Collins, 2015), and a film of the same name (dir. Jeanette Kong, Virgil Films, 2016).
3 My sister published an article in the *South China Morning Post* on the journey. See Gaia Goffe, "How a Chinese-Jamaican's Family History Quest Led Her to Hong Kong," *South China Morning Post*, 28 July 2016, www.scmp.com/magazines/post-magazine/long-reads/article/1996068/how-chinese-jamaicans-family-history-quest-led.

---

**Acknowledgments:** I would like to thank Albert Chong with whom I was able to be in conversation at the Yale Center for British Art and who granted permission to use images of his work. Thanks, too, to Duke University Press for granting permission to reuse some portions of an essay I published in *Small Axe* in 2018. I'm grateful to Irene Kacandes for inviting me to revisit this research on photography and Afro-Asian circuits. I would like to extend thanks to the memory of the late Meena Alexander for leading a seminar on an early version of this chapter with valuable input by Renee Blake, Alexandra Chang, Ana Paulina Lee, Heather Lee, Amita Manghnani, Tami Navarro, Sukhdev Sandhu, and Marianne Hirsch. I am forever grateful to my dear sister Gaia Goffe for laying the framework for this analysis with her personal narrative of our journey to China with our mother published in the *South China Morning Post*.

https://doi.org/10.1515/9783110753295-019

grandmother. Though he had never been to Jamaica, it was as if Tang-Kwong and his children had been expecting our diasporic return. We had fancifully dreamed of but had not known if there were Afro-Chinese people like us who lived in China. Tang-kwong did not speak a word of English; we did not speak a word of Cantonese. And so, we spoke in the only language we could, the back and forth of family photographs on iPad screens, of recognition, of kin.

Weathered and torn photographs sat framed on the mantelpiece of the traditional Chinese house, which we discovered had been designated a landmark by the government, and there we saw a portrait of my maternal grandfather, who had lived a life unknown to us in Cantonese and patwa, between Jamaica, Hong Kong, and New York. Here were the muted narratives of my grandfather's boyhood in a sequence of photographs. It was a life story he would not tell us, though we had asked him many times. Now I realize it may have been because he could not tell us. How to narrate a life separated from one's mother and father, growing up during the Second Sino-Japanese War? All we had were the images that communicated an unspeakable articulation of abject and dislocated kinship. The Black Chinese homegoing was captured on camera by Bruce, a photojournalist, who had traveled with us by subway on the off chance something noteworthy would happen. None of us knew what to expect, and it was truly serendipitous that we found our family. The three of us – my mother, my sister, and I – were featured on the front cover of the *South China Morning Post*. My sister wrote an article detailing the surreal experience. The headline read "How a Chinese-Jamaican Family's Quest Led Her to Hong Kong." As incredible as our story is, it is representative of many Black subjects of British Empire who are of African and Chinese descent. A spate of news articles like ours have been published over the decades of returns, successful and unsuccessful. More important perhaps than what is discovered is the question of what each person hoped to find.

It occurred to me then that the photographic emulsion of an image is the flesh of family. Albums are the proof that visually connects families in some ways more than DNA. There are muted narratives inscribed in the flesh of old photographic portraits. In my diasporic album, the muted and mutilated stories I have pieced together are the entangled histories of Asia, Africa, Europe, and the point at which they converge in the Caribbean. My point of departure, my Black Pacific is my grandfather's itinerary from the 1930s to the 1950s, from Jamaica to Hong Kong back and forth. My research attends to a global transoceanic conception of Afro-Asia that locates the rebellious potential of Afro-Asian subjects born in the wake of the West Indian plantation. While Afro-Asia typically refers to anticolonial work of the Bandung conference in 1955, I look to other oceans for how other Afro-Asian subjects were decentering Europe in forms of quotidian life that were extra-colonial arrangements. Afro-Chinese life in the Caribbean was as much determined by colonialism as it was irrelevant to it. The photographs I examine in this chapter concern a series of collages by Afro-Jamaican Chinese artist Albert Chong. My analysis bridges what Paul Gilroy has called the Black Atlantic by considering Pacific currents towards an expanded formation of Blackness that includes subjects also of Chinese descent.

## The Hieroglyphics of Photographic Flesh

Caribbean genealogies are a "human sequence written in blood," to echo the way Hortense Spillers describes the transatlantic world order. Photographs document this sequence, but more often the absence of family images in a place like the Caribbean tells us more. The birth of the daguerreotype in 1839 happens to coincide with the abolition of enslavement in the British West Indies. The scant photographic record of abolition presents a historical challenge. Historian of photography Krista Thompson remarks, this absence should not be viewed as "a lack for which compensation is necessary, but as an intrinsic part of, and even representation of, the history of slavery and post-emancipation in the region."[4] To this convergence of events, I consider the fungibility of Asian labor that was introduced as compensation to fill the lack or labor void of abolition. Adding to this colonial timeline, the British occupation of Hong Kong following the First Opium War in 1842 primes the first Chinese subjects to arrive in Jamaica as indentured laborers in 1854. Recruited to fill the lack of enslaved African labor, Chinese and Indian men arrived in the hundreds of thousands, and yet there is a lacuna of the catastrophe of racial indenture in the European colonial archive. Gaps or vacancies offer opportunities, spaces for interrogation and reframing the notion of history, the sedimented violences of the plantation, and the hemisphere entirely.

Scarred flesh is marked in what Spillers gestures to as hieroglyphics, material and psychic. It is a traumatic grammar of the inherited racial injury of unknown horrors. Spillers describes the crisis of kinship in the shadow of transatlantic slavery as being "diasporic plight marked by a theft of the body" for African and Indigenous peoples.[5] Through the forgotten catastrophe of racial indenture, Chinese and Indian stolen bodies joined this sequence violence. They were brutally beaten

---

**4** Krista Thompson, "The Evidence of Things Not Photographed: Slavery and Historical Memory in the British West Indies," *Representations* 113, no. 1 (2010): 63. Thompson writes of the coincidence of abolition and the daguerreotype.
**5** Hortense Spillers, "Mama's Baby, Papa's Maybe: An American Grammar Book," *Diacritics* 17, no. 2 (1987): 67 (italics in original).

and often had been abducted from Asia as they were inducted into the plantation order as replacements for enslaved Africans across the British, French, Dutch, Danish, and Spanish Caribbean. Nearly three thousand depositions of this sequence written in blood are documented in the Cuba Commission Report, published in 1876 – indentured Chinese describe the horrors of being flagellated, starved, and tortured for insubordination at sea and on the plantation.[6] The brutal racialized system of Indian and Chinese debt peonage did not begin to be abolished in the British Empire until 1917. Following this period in the late nineteenth and early twentieth century waves of voluntary migrants arrived in the Caribbean as merchants and shopkeepers.[7] While these migrants importantly did not suffer the same theft of the body as indentured Asians or enslaved Africans, their racialized presence accents how the same channels of migration continue to flow from China to the Caribbean. The discontinuity of these two waves of migration is important and the attendant process of racialization as Chinese is significant in its heterogeneity, that distinct though interconnected, becomes flattened.

In the way Spillers describes the captive body severed from will, indentured laborers were dehumanized by the contracts they signed or were forced to sign, metaphorically in blood. From my archival research, more often than not I have seen a simple X mark indicating the volition of these subjects who became conscripted onto the colonial plantation. While common nineteenth-century practice, the ambiguity of the X is a hieroglyphic that points to the crimes against the flesh that were perpetrated through the violence of the European plantation owners. If what Spillers coined as *pornotroping* is the transformation of the captive body into flesh through mutilation, dismemberment, and exile – the attempt to transform the subject into chattel – then the indentured body is leased. If the enslaved body is for sale, the indentured body is for rent. Indentured people were essentially leased, as temporary property in contracts often of five years that in many cases were extended through re-indenture, thus suspending them in a space where neither consent nor consent was possible.

Sexuality is a critical rubric of the flesh in that as Spillers writes of the genealogy of the *figure* of the transatlantic Black woman of the Black woman, the

---

6 Lisa Yun, *The Coolie Speaks: Chinese Indentured Laborers and African Slaves in Cuba* (Philadelphia, PA: Temple University Press, 2008); Denise Helly, *Cuba Commission Report: A Hidden History of the Chinese in Cuba* (Baltimore: Johns Hopkins University Press, 1993).
7 While some people may have been voluntary workers, records show that crimps in South China were known to hoodwink, drug, and kidnap people, locking captives in barracoons at the docks of Hong Kong before setting sail for the West Indies. See Edlie Wong, *Racial Reconstruction: Black Inclusion, Chinese Exclusion, and the Fictions of Citizenship* (New York: New York University Press, 2015).

gendered migration of racial indenture was primarily of men. Indentured Chinese and Indian laborers were not imagined by European statecraft to be subjects with interior lives, desire, or privacy. Yet they are rendered ambivalently in the colonial discourse depending on the nation. The Chinese are at times granted a capacity for bourgeois domesticity in contrast to enslaved and newly emancipated Africans. From my observations, though, the colonial record indicates an oscillation between stereotypes of the "heathen Chinese" and the "industrious Celestial." As such the Caribbean Chinese family is desired to anchor and discipline men and yet it was impossible because women did not migrate in large numbers. Cultural theorist Lisa Lowe details the way a secret memorandum of the "Trinidad experiment" imagined the Chinese woman as a potential linchpin to propagate a racially pure family, "distinct" from the Black family, the majority population in the West Indies.[8] And yet as she finds Chinese women rarely emigrated, so this vacancy was often filled by Black women, Indian women, and sometimes by Indigenous women.

In contrast to the "bachelor societies" that formed in the United States as a result of the Page Act of 1875 and the 1882 Chinese Exclusion Act, in many parts of the Caribbean, Chinese women were encouraged to migrate too. According to Trinidadian historian Walton Look Lai, one hundred sixty thousand indentured Chinese migrated to the Caribbean in the nineteenth century, of whom the vast majority were male.[9] Many fled sugarcane plantations, breaching their contracts, and returned home if they could afford to; some chose to take their own lives.[10] But some of these men remained, settled, and started families.

Chinese futurity in the United States was determined by American eugenics and legislation. The Chinese Exclusion Act of 1882 attempted to limit the possibility of future Chinese citizens, deeming them "ineligible to citizenship." Across the hemisphere the patterns of said exclusions have not been adequately studied, but each nation-state from Canada to Panama to Brazil had its timeline of policies

---

**8** Lisa Lowe, "The Intimacies of Four Continents," in *Haunted by Empire: Geographies of Intimacy in North American History*, ed. Ann Laura Stoler (Durham, NC: Duke University Press, 2006), 200.

**9** See Walton Look Lai, *Indentured Labor, Caribbean Sugar: Chinese and Indian Migrations to the British West Indies, 1838–1918* (Baltimore: Johns Hopkins University Press, 1993). This helps explain why the pervasive stereotype of the Chinese in the Caribbean is male, "Mr. Chin." For an extensive study of how that sexualized caricature of Asian masculinity recurs through West Indian colonial literature, see Anne-Marie Lee-Loy, *Searching for Mr. Chin: Constructions of Nation and the Chinese in West Indian Literature* (Philadelphia, PA: Temple University Press, 2010).

**10** See Isabelle Lausent-Herrera, "Tusans (tusheng) and the Changing Chinese Community in Peru," in *The Chinese in Latin America and the Caribbean*, eds. Walton Look Lai and Tan Chee-Beng (Leiden: Brill, 2010).

often animated by a desire for cheap labor and xenophobia. The question of family and nation is a question of futurity. In the Caribbean, with the layers of the Indigenous, European, African, Asian (East Indian, Chinese, Syrian) presences, the subject of who has settled in the region at whose expense is a thorny asymmetrical history of violence.

In spite of being defined by a series of race-based exclusions, Chinese families in the Americas often excluded within themselves. Those who did not fit the picture of a "pure" Chinese future, because of mixed heritage or for not conforming to heteronormative gender roles, were exiled to the margins of the family frame. Therefore, the family portrait is a fraught site of representation of ambivalence, of desires and absences for the Chinese diaspora in the Americas. How to read the layers of silent narratives inscribed in family photographs? Who remains missing from the family portrait? Inasmuch as family portraits are intimate and sentimental, they also police. Albums include by virtue of exclusion. The only way then to read against the trap of the nostalgia of family photographs is to trouble the multiple exclusions enforced by the family album.

As a diasporic artist of African and Chinese heritage, Albert Chong has centered these muted narratives of Afro-Chinese entanglement in the Caribbean. In the 1980s and 1990s, he curated sequences of vernacular photography – family snapshots, passport photos, studio portraiture – to show a global audience the affect of Afro-Asian subjectivity. His artwork interrogates the very notion of blood family and the scripts of bourgeois intimacy in the context of Chinese migration to the Caribbean and Latin America. Chong's autobiographical work emerges from a moment of global transformation for the art market in the 1990s when artists of color asserted their personal geographies as a challenge to the way art is formed, how art is made, and what counts as art. Curatorial attention was brought to the Caribbean/Latin American Chinese diaspora with the joint 2017–18 exhibition "Circles and Circuits II: Contemporary Chinese Caribbean Art," by the Chinese American Museum in Los Angeles and the California African American Museum which featured Chong's work and that of many others.

Through a poetics of manipulation of the context and surface of the family photograph and as a diasporic artist, Chong remixes the album as a space of Afro-Asian inclusion. Engaging with rituals of mourning and marriage, he honors the wounded intimacies and materiality of the tears, blurs, and creases in family photos to illuminate the tensions of diasporic kinship. These marks are the scarred hieroglyphics of the photograph as wounded flesh. Chong distorts to highlight the ways colonial historiography obscures the Caribbean Chinese experience and indenture. He does so by employing techniques such as photo transfer silk-screening, video keying, and hand tinting, to suture and restitch the fragmented diasporic family. Against a simple retrieval or recovery of cultural memory, Chong is a visual storyteller and guardian. He works in the gaps of the colonial archive to trouble and not to compensate or fill the absence of what is missing about Afro-Chinese life.

Chong's mixed-media artwork negotiates the competing affiliations of race, ethnicity, nationality, and citizenship. In doing so, the work articulates the politics of what photo historians Thy Phu and Laura Wexler term "visual kinship" to examine how family photography not only illustrates the family, but also shapes the very idea of family, as a racialized and gendered structure.[11] Chong remixes the family album to challenge the racial and gender hierarchies of the colonial-era bourgeois family. He troubles notions of inheritance and the visuality of the photograph as proof of relation, by putting photos in conversation with other objects through processes of collage. He grapples with an always already mutilated history muted by the violence and convergence of racial slavery and racial indenture.

I attend to Chong's aesthetic through a kindred and enmeshed analytic, rather than through a comparative mode of analysis of the intersecting circuits of African and Asian histories. His photographic collages and writing about his process opened up for me what my grandfather perhaps could not tell me about his Black Pacific experience and Black boyhood in Hong Kong when he was alive. My methodology and the knowledge gained thereby are two of the ways in which I locate myself adjacent to historical violence. As a descendant of people who underwent these historical violences, I attempt to negotiate my adjacency. Furthermore, Asian racialization has been a process of adjacency to Black racialization in the Americas, which is at the heart of my intellectual preoccupation. In many Caribbean and Latin American countries, the numbers of people who identify or are identified as Chinese have dwindled such that it is no longer a census category. With the departures or deaths of these individuals, the history disappears. However, the visual poetics of someone like Chong counters the missing written history with a poetic archive of visuality.

Sociopolitical movements of revolution and decolonization from the late 1950s to the 1970s across the Caribbean and Latin America led to the dispersal of many families of Chinese descent to the United States and Canada. Being Chinese was often associated with merchant-class privilege and a perceived proximity to, though exclusion from, the white colonial elite. Diasporic nationalisms rooted in myths of purity often push and blur the darker skinned and non-heteronormative subjects out of focus in Afro-Asian families across the hemisphere. Against these scripts, the visualities explored here interrogate multiple

---

11 Thy Phu and Laura Wexler convened the conference "The State of the Album" on 14 April 2017 at Yale University, based on a course they cotaught on visual kinship.

exclusions by the frames of the family and history. Frames contain and delineate, and in doing so they necessarily exclude.

## Distant Cousins: Caribbean Chinese Visual Kinship

The fictive kinship of paper sons, picture brides, half-sisters, and stepmothers in the African and Chinese diasporas produces double narratives of strategic evasion from state surveillance and a continuous reassembling of intimacy. The restitching of Black and Asian kin in the wake of the violent legacies of African enslavement and racial indenture puts the family photo album, the visual proof of family, in a fraught position of performing kinship in the Caribbean and Latin America. Collecting and compiling any family album is an inherently petit bourgeois practice. "Every portrait," Allan Sekula writes, "implicitly took its place within a social and moral hierarchy. The private moment of sentimental individuation, the look at the frozen gaze-of-the-loved-one, was shadowed by two other more public looks: a look up, at one's 'betters,' and a look down, at one's 'inferiors.'" The traditional album enforces what Pierre Bourdieu calls a "hierarchy of legitimacies."[12]

Yet in the context of picture-taking for themselves, for Afro or Asian diasporic subjects, this disciplining is further complicated by the attempt to reconstruct family units splintered by migration and colonial practice. To celebrate and use the camera to humanize and record family history is to counter the way colonial-era photography was deployed as an ethnographic tool of documenting and objectifying the other. W.E.B. Du Bois, among other Black intellectuals, used the family and community album in the American Negro Exhibit at the 1900 Paris Exposition to perform this type of restorative work.[13] Speaking of Caribbean diasporic portraiture in particular, Tina Campt writes that the images were sites of articulation and aspiration.[14] She echoes Stuart Hall: "[Portraits] documented where people were at a certain stage of life, and how

---

**12** Allan Sekula, "The Body and the Archive," *October* 39 (1986): 10. Pierre Bourdieu notes as well the way family photography captures the "high points" of a middlebrow vision of domestic life. He identifies the snobbery and "hierarchy of legitimacies." Pierre Bourdieu, *Photography: A Middle-Brow Art,* trans. Shaun Whiteside (1965 repr., Cambridge, MA: Polity, 1990), 19, 97.
**13** See Shawn Michelle Smith, *Photography on the Color Line: W. E. B. Du Bois, Race, and Visual Culture* (Durham, NC: Duke University Press, 2004).
**14** Tina Campt, *Image Matters: Archive, Photography, and the African Diaspora* (Durham, NC: Duke University Press, 2012), 7.

they imagined themselves, how they became 'persons' to themselves and to others through the ways in which they were represented. The photos were what you sent home as 'evidence' that you had arrived safely, landed on your feet, were getting somewhere, surviving, doing all right."[15] Studio portraiture, in particular, performs a bourgeois intimacy of propriety that disciplines in its formal poses and clothing, encoding a politics of respectability. Artist Albert Chong seeks to unravel those codes of policing family.

Born in 1958, Chong is a wash-belly baby, a Jamaican phrase describing the last born. The youngest of nine, his relationships with his parents were mediated by a decade from when his oldest siblings were born. Chong grew up dominated by narratives of their immediate family that preceded their birth or consciousness corroborated by the family album. For as long as he can remember, Chong was imagining and searching for a window into an ancestral past. He examines the affirmation of subjecthood for his mother instilled in child studio portraiture. "I recall feeling transported back in time when looking at old family pictures," he says, "as if our most precious commodity was the experiences we had shared as a group of related people, and the proof of the connectedness and continuity was this collection of memories we called photographs."[16] The materiality of how photographs index and influence memory is central to Chong's aesthetic.

It is not enough to simply excavate or reproduce these family images. Chong positions the pictures in a way other than that intended by the original photographer. As a custodian of history, he carefully narrated context of global circulation as works of art. Aware that this is an obscured history, he is careful to include the specifics of individual Caribbean Chinese and Chino Latinx families so as to counter public fantasies about who the subjects might be. Put another way, without the context of a caption, most would probably not guess that the photograph's subjects are part of the Caribbean Chinese experience because most people who are not Caribbean have never heard of the intersection of these two cultures.

**15** Stuart Hall, "Reconstruction Work: Images of Postwar Black Settlement," in *Family Snaps: The Meanings of Domestic Photography*, eds. Jo Spence and Patricia Holland (London: Virago, 1991), 156.
**16** Albert Chong, "The Photograph as a Receptacle of Memory," *Small Axe*, no. 29 (July 2009): 128–134, here 128.

# Afro-Asian Intimacies and Ancestral Dialogues

Chong began his career with an aesthetic that was more clearly interpreted by critics as gesturing to his African Jamaican heritage. A professor of photography at the University of Colorado, Boulder, Chong is a celebrated international visual artist who has been the recipient of a Guggenheim Fellowship in Photography, among other honors. Just like the Cuban modernist painter Wifredo Lam, Chong has navigated the complex territory of the burden of representation in the art world as an Afro-Chinese Caribbean artist with a Chinese surname. His heritage as an Afro-Chinese person is representative of the undoing of the colonial imagination of a "barrier" race in the proposed Trinidad Experiment. Afro-Asian familial intimacies were formed after abolition, but not without requisite tensions and sets of exclusions based on color and class, which Chong tackles in his artistic commentary. Like Lam, Chong turns to the symbolism of Afro-diasporic spirituality, namely, Rastafarian beliefs and Santería, in his aesthetic. It was not until the death of his father in 1989 that he began to engage directly with his Chinese heritage in his art as a type of eulogy. Chong's mother and father were both born to Afro-Jamaican mothers and Hakka Chinese fathers who migrated from Southern China to Jamaica. Chong's parents worked as merchant shopkeepers, an occupation held by many people of Chinese and especially Hakka descent in Jamaica. Chong's grandfathers were part of the second wave of voluntary migration from Guangdong Province that followed Chinese indenture in the British Caribbean.

His artwork – which includes photography, sculpture, installations, video, and mixed media – draws on spiritual iconography, ritual practices, and mysticism. "Family photographs are like windows to my past and a fraction-of-a-second glimpse into the lives of the people who lived in that past," he explains. "I have continued to honor their memory through the act of veneration that is the construction of a still-life photograph focused around a picture or portrait of an ancestor. In the veneration of the past and of those who have come before us, I am African and I am Chinese."[17] Though Chong was raised Catholic, African Indigenous rituals were also an important part of his upbringing and are an important part of his art today. Chong's father, before moving his family into their new home in Kingston, employed both a practitioner of African-inspired obeah and a Catholic priest to bless the house.

His father is a frequent presence in Chong's artwork from the late 1980s through the 1990s. *Addressing the Chinese-Jamaican Business Community* (1992) is a

---

17 Albert Chong, "Receptacle," 131. On these topics, see also Lisa Yun, "Signifying 'Asian' and Afro-Cultural Poetics: A Conversation with William Luis, Albert Chong, Karen Tei Yamashita, and Alejandro Campos García," *Afro-Hispanic Review* 27, no. 1 (2008): 183–217.

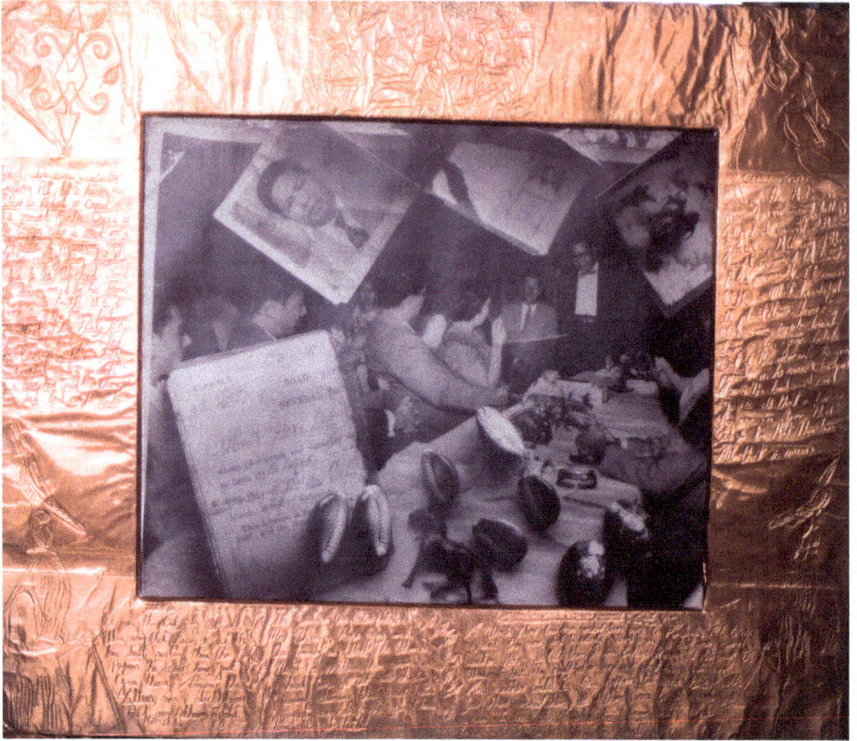

**Figure 1:** Albert Chong, *Addressing the Chinese-Jamaican Business Community*, 1992. Gelatin silver print, 30 x 40 in.

photo collage comprised of layered objects (Figure 1). A photograph of Chong's father standing at the head of a table of Jamaican Chinese businessmen, perhaps delivering a speech, forms the background. African cowry shells are scattered near a small headshot of Chong's father. To the left, an old speeding ticket of his father's lies folded in half. The government-issued fine gestures to the everyday encounters with law enforcement through which colonial subjects are made legible to the state. A miniature copy of his father's passport, which also represents his relation to the crown as a British subject, is placed askew. The irony here is that Chong's father was known by the nickname "the Justice" because of his prominence in society, which led to him becoming a local judge.[18]

---

[18] "Albert V. Chong," in *Encyclopedia of Asian American Artists: Artists of the American Mosaic*, ed. Kara Kelley Hallmark (Westport, CT: Greenwood, 2007), 36.

Despite his father's stature in the Jamaican Chinese community as a successful businessman, Chong's father was ridiculed for the dark hue of his skin, read as a marker of his African ancestry. It was common for Chinese people to use derogatory Cantonese and Hakka terms such as *Ban Lao Shee*, meaning "half brain," denoting how people of mixed ancestry were "incomplete." In the Jamaican context too, Chong remembers being called "Chiney Royal," which is arguably more ambiguous and indeterminate in origin but is likely rooted in colonial eugenicist vocabularies of racial types and breeding.[19] The derision came not only from outsiders; Chong describes the negative way his father would treat him at times because of the darkness of his skin. This was combined with the encoding of Jamaica as a pigmentocracy, where social hierarchy and skin tone are still closely aligned: "In my youth I recall being conflicted about the mixtures of my ethnicities . . . I identified with and wanted to be with my Chinese heritage. I was an early victim of Jamaican color prejudice, which stated, in no uncertain terms, that light was right. I wanted to be on the winning side, I wanted straighter hair, more Chinese features, not the kinky hair and light brown skin that defines me." Chong also recounts that growing up in Jamaica, black workers employed in the Chong family business would comment on the texture of his hair. A woman named Ella would always run her fingers through his hair and ask the same question, "'Bwoy why yu hair so bad eh?'"[20]

There is an undeniable visuality of the perceived fixity of race, which is what makes the photographic medium apt for teasing out the visual cues of reading race in physical traits, phenotypes, and physiognomy. Is it possible to read a photograph without reading race? Anthropologist Lok Siu asks this question in her reading of Chong's rephotographed collage named Aunt Winnie. "I found myself looking intently at the picture," she admits, "studying the features of her face – eyes, cheekbones, lips – framed by shoulder-length black curly, wavy hair. Over and over again, I searched for traces of her Chinese and African features; I found myself looking for race and fascinated by her racial mixed-ness."[21] This complicity of looking for markers of race is something Campt grapples with as well, in searching for photographs of Black British domestic life.[22]

---

**19** See Yoshiko Shibata, "Searching for a Niche: Creolizing Religious Tradition; Negotiation and Reconstruction of Ethnicity among Chinese in Jamaica," in *Religious Pluralism in the Diaspora*, ed. Kumar P. Pratap (Leiden: Brill Academic, 2008), 359–374.

**20** Albert Chong, quoted in Yun, "Signifying 'Asian,'" 198.

**21** Lok Siu, "Diasporic Affect: Circulating Art, Producing Relationality," in *Circles and Circuits: Chinese Caribbean Art*, ed. Alexandra Chang (Durham, NC: Duke University Press, 2017), 214–220, here 215.

**22** Campt, *Image Matters*, 127.

It is at once the problem of the colonial ethnographic gaze that seeks to separate and sort, and it is also the gaze I must use to find those subjects who have been obscured. Yet race is not a biological fact, and is rather a technology of the plantation, a system or sorting that upheld its violence. Chong contends with the question of African heritage by proudly wearing what was described as his "bad" hair in dreadlocks in a series of self-portraits he named I-traits, embracing his Blackness.

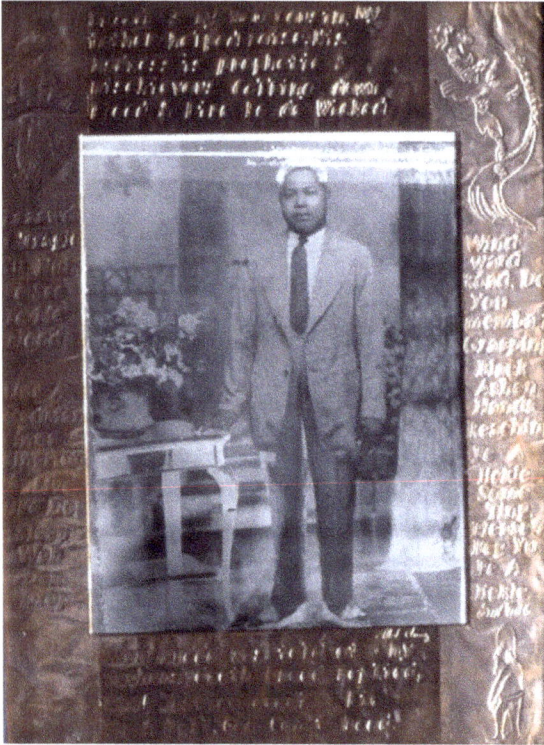

**Figure 2:** Albert Chong, *Portrait of My Father Addressing the Spirits*, 1990. Gelatin silver print, with inscribed copper mat and black wooden frame; 23 x 18 in.

This embrace of Afro-diasporic spirituality is pictured in the repetition, with a critical difference, of images of Chong's father in a number of his mixed-media photographic collages. For instance, in *Portrait of My Father Addressing the Spirits* (1990), his father is posthumously in commune with the duppy realm (Figure 2). The backdrop of the studio portrait is altered to look like a negative exposure, giving it an electrified effect. His father's black hair becomes the inverse color, a white halo. Playing with color and shade is an important part of Chong's aesthetic. How do we read color in black and white photography? As if in an X-ray, the desk and part of the curtain are transformed to a translucent white glow. The top of the photograph is damaged, with a horizontal crease. A frame within a frame, the copper mat, Chong's signature technique of mounting his collages with a border around the image, is inscribed with Jamaican vernacular and frames the rephotographed image. Chong's father is literally framed in fragments of patois – writings from a journal Chong found of his father's as well as his own musings from 1989 when he was hoping to find his father's will. There are also etchings of an African mask, a deer, and people running that signify the blending of Afro-Caribbean cosmologies, much like we see in the work of Lam.

Chong joins an Afro-diasporic materiality of mourning with Hakka Chinese ancestor worship, or Gah San. Though most Jamaican Chinese converted to Catholicism from Buddhism, Confucianism, or Taoism, the Chinese ritual of honoring ancestors endures to this day at the Chinese Cemetery in Kingston. Offerings of food, such as roast pig, are presented as a gesture to the hungry ghosts of ancestors. Historian of photography Marianne Hirsch describes Chong's artistic Gah San as making pictures into "shrine-like ritual spaces."[23] "Many [I]ndigenous people claim that picture-taking steals one's soul and see the photograph itself as suspect," Chong says. "In many of the still lifes I attempt to do the reverse by creating on the surface of the photo a shrine that is an act of reverence to restore the souls of these ancestors of mine."[24] While altering the surface of a photo would typically be seen as sacrilegious, the manipulation actually works to venerate the sacred old portrait of his father. If we see the photographic collage as troubling the transatlantic theft of the body, then what does it mean when Chong applies color to the flesh that is the photograph? Chong's reworking of old family photographs is a practice of Qingming, or tomb sweeping, a national holiday in China of tending to the dead, that brings the shadow world of obeah to the foreground.

Not only does his work communicate on many levels, but Chong also writes extensively about his strategic practice. In the essay "The Sisters and Aunt Winnie," he explains about *The Sisters* (1986): "Several years ago, my mother sent me the only remaining picture of herself as a child – an old torn and yellowing photograph of three girls. She is the smallest child in the picture; the other two girls are her cousins. She asked me to repair the picture. I could not heal it, but

---

**23** Marianne Hirsch, "Introduction," xi–xxv, in *The Familial Gaze*, ed. Marianne Hirsch (Hanover, NH: University Press of New England, 1999), xx.
**24** Chong, "Photograph as a Receptacle," 128–130.

I could rephotograph it incorporating the torn area of the image."[25] The yellow-ing of time on the surface of the photograph is layered with the blue tint. It is important that he says he could not "heal" the photograph, which references the image as an extension of the body. He becomes a type of doctor tasked with performing an aesthetic of repair that is impossible not only because of the physical deterioration of the picture but also because of his mother's psychic wounds. Elsewhere Chong says, "Without the evidence of pictures many of us would have very little proof that we ever existed."[26] At the time the original pho-tograph was taken, Chong's mother had just become an orphan. The photo-graph provided Chong's mother with a sense of affirmation. By inscribing his mother's oral history into the copper mat as frame surrounding the photograph of his mother and her cousins, Chong introduces the Afro-Chinese subject to a global public (Figure 3). "While rephotographing the picture," he recalls, "I be-came overwhelmed by the simple beauty of the image in its recording of three sisters of African Chinese ancestry as they poised themselves for history. Ja-maica was a mere sixty years out of slavery, and she was an orphan at the time."[27] Outside history and outside the family, by repositioning the girls in the tradition of found-object art, Chong transforms them into found subjects. As with the portrait of his father addressing the Jamaican Chinese community,

Chong places cowry beads ceremoniously (Figure 3). He dignifies what is everyday. This time side by side with decaying flowers, an old knife, feathers, and a fork, he mixes the decorative with the quotidian. Importantly, he makes a place setting with the utensils for the Afro-Chinese subject. The bottom part of the mat features an etching of three grown women, the sisters. Next to it reads the Jamaican proverb "Table napkin want to turn tablecloth," signifying some-one wanting an unwarranted promotion in stature. Chong also employs the practice of hand tinting on the surface of the photograph. He colors his mother's dress a pale blue, adding an accent color not for the sake of realism but to imbue a vibrant sense of the intimate beauty of the fictive kinship of cousins becoming sisters in the wake of death. Out of bereavement, the tonality of the monochrome, the past, is accented with color.

Chong views his work as conservation: "I thought of the many old histori-cal photographs I had seen, and of the fact that few contained people of color. I realized in that instant how inconsequential this photograph, and the lives that it illuminates, were to white civilization. I knew that I could not merely

---

25 Albert Chong, "The Sisters and Aunt Winnie," in Hirsch, *Familial Gaze*, 102–106, here 103.
26 Chong, "Photograph as a Receptacle," 128.
27 Chong, "Sisters and Aunt Winnie," 103.

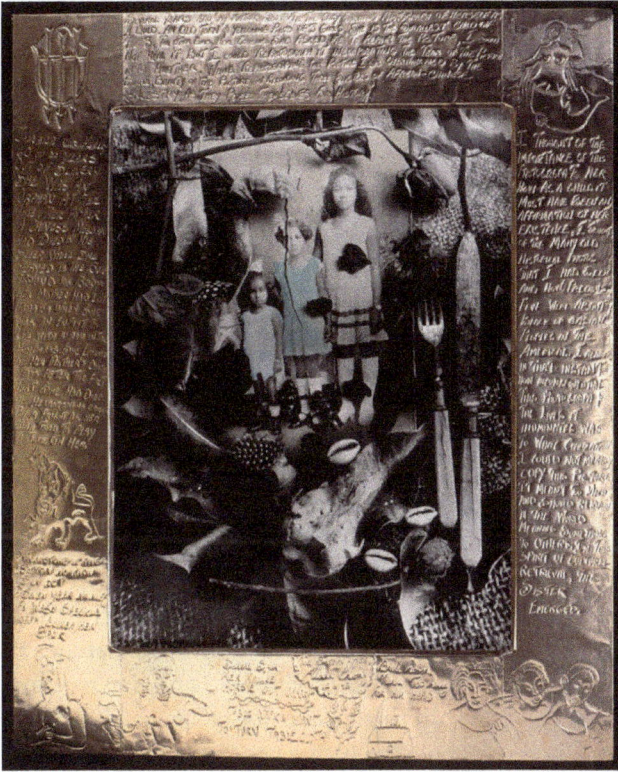

**Figure 3:** Albert Chong, *The Sisters*, 1986. Gelatin silver print.

copy this picture, that it meant too much and should remain in the world meaning something to others."[28] The sisters materialize from a fictive kinship as modern transnational subjects, gesturing to the many other Afro-Chinese children, like my grandfather and his siblings, who lived between Hong Kong and Jamaica, part of the flow between the tropical island port cities of the British Empire. The found subject that emerges, framed by found objects, is the product of what Lowe would refer to as global intimacies and the circuits opened up by a failed experiment of the British Empire.[29]

---

**28** Chong, "Sisters and Aunt Winnie," 103.

**29** In her 2015 book *The Intimacies of Four Continents* (Durham, NC: Duke University Press), Lisa Lowe maps this connection between the opulence of bourgeois intimacy and materiality of the European family and how it depended on the underside of racialized plantation labor in the West Indies.

Without Chong's essay and the framing mat, the photograph could easily become like an orphan image divorced from the Jamaican Chinese context. The artist recognizes that while photographs may appear to be unmediated representations of truth, they are, in fact, always mediated. Chong could do it justice only by representing the damage symbolic of the texturality of a torn family fabric. The deteriorating image represents the ephemerality of the photographic portrait, even though it seems to freeze a moment in time forever. His subjects are transformed from two-dimensional flat images to multimedia, multidimensional works of art, and then they become two-dimensional again when he rephotographs them. Chong frames the photographs that would likely have been forgotten and left disintegrating in a pile of papers in a basement like the photographs of my grandfather that my mother found in her childhood home in Queens.

## Impossible Photographic Detours

Against nationalism and other myths of purity or fixity, Chong meditates on a future pulled to different corners of the globe. The featured works of art attest to how diasporic conceptions of family and the multiplicity of heritage narrates alternative genealogies of Asia and Africa in the Caribbean, Latin America, and their diasporas that decenter Europe. An extension of the stolen body, the family photograph becomes a surface of both producing and challenging racialized and gendered family hierarchies. The visual poetics of Chong's remixed family album is as much about cohering a family narrative and forming a reunion as it is about the undoing of family defined strictly by blood and "purity." Beyond biology or a vertical hierarchy, an extended conception of family across continents is critical to reckoning with the existence and erasure of the Caribbean Chinese family.

Ordinary family photographs for people descended from the mutilation of the plantation carry the potential to defy the elision and segregation of the colonial order. As much as this preservation and restoration is critical to troubling the lacuna in the colonial archive – silence – Chong's poetics offers a critical mystic and material detour through a past presence. Each arrangement is a ritual that creates a diasporic world. My own journey to China as a Black woman born in England crystalizes the impossibility of true return to the point of origin or essential truth. The entangled British circuit of Hong Kong and Jamaica continues and charts new homegoings.

Albert Chong honors the families that the plantation could not imagine existing of African and Chinese heritage, forming rebellious extra-colonial potential. By drawing on Santería and forming visual shrines, Chong's aesthetic challenges the supremacy of Christianity and specifically Anglicanism in Jamaica. He exchanges the inherited lenses of colonial ordering for his own ritual deeply engaged with ecology. The materiality of what can be fashioned out of what has been torn apart exceeds the album. In this fraught moment of so-called "cancel culture," I can only understand how the nation-state excludes subjects like my grandfather rendering them cancelled. In a passport photo that my mother found of her father as a boy leaving Jamaica for China that I now have, there is a purple faded stamp that reads "CANCELLED." While my grandfather was a man of few words, this photo affirms for me his sense of self, because he kept it. The passport stamp forms a hieroglyphics of the album as flesh; it indelibly marks the colonial subject and frames him. Yet my grandfather was one of a class of exception because he was of Jamaican and Chinese parentage and could move between the countries during what was a period of Chinese exclusion from Jamaica. Chong's framing of his own family allows me to imagine that there were many more people like my grandfather who lived in this Black Pacific modern space. Another future of new, albeit fraught, familial intimacies beyond the visuality of race in the diaspora are celebrated in the diasporic aesthetic of honoring subjects who exist "cancelled" like my grandfather, (Figure 4) beyond the frame of what colonial history can accommodate.

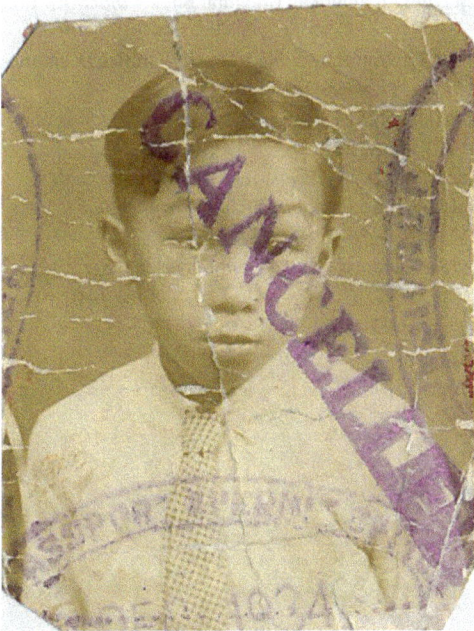

**Figure 4:** Passport photograph of the author's grandfather, 1934. In private collection of the author.

# Works Cited

Chong, Albert. "The Photograph as a Receptacle of Memory." *Small Axe*, no. 29 (July 2009): 128–134.

Chong, Albert. "The Sisters and Aunt Winnie." In Marianne Hirsch, ed. *The Familial Gaze*. Hanover, NH: University Press of New England, 1999. 102–106.

Gilroy, Paul. *The Black Atlantic: Modernity and Double Consciousness*. Cambridge, MA: Harvard University Press, 1993.

Goffe, Gaia. "How a Chinese-Jamaican's Family History Quest Led Her to Hong Kong." *South China Morning Post*, 28 July 2016, www.scmp.com/magazines/post-magazine/long-reads /article/1996068/how-chinese-jamaicans-family-history-quest-led.

Helly, Denise. *Cuba Commission Report: A Hidden History of the Chinese in Cuba*. Baltimore: Johns Hopkins University Press, 1993.

Hirsch, Marianne, ed. *The Familial Gaze*. Hanover, NH: University Press of New England, 1999.

Lee-Loy, Anne-Marie. *Searching for Mr. Chin: Constructions of Nation and the Chinese in West Indian Literature*. Philadelphia, PA: Temple University Press, 2010.

Look Lai, Walton. *Indentured Labor, Caribbean Sugar: Chinese and Indian Migrations to the British West Indies, 1838–1918*, Baltimore: Johns Hopkins University Press, 1993.

Lowe, Lisa. "The Intimacies of Four Continents." In *Haunted by Empire: Geographies of Intimacy in North American History*. Edited by Ann Laura Stoler. Durham, NC: Duke University Press, 2006. 191–212.

Siu, Lok. "Diasporic Affect: Circulating Art, Producing Relationality." In *Circles and Circuits: Chinese Caribbean Art*. Edited by Alexandra Chang. 214–220. Durham, NC: Duke University Press, 2017.

Spillers, Hortense. "Mama's Baby, Papa's Maybe: An American Grammar Book." *Diacritics* 17, no. 2 (1987): 64–81.

Thompson, Krista. "The Evidence of Things Not Photographed: Slavery and Historical Memory in the British West Indies." *Representations* 113, no. 1 (2010): 39–71.

Wong, Edlie L. *Racial Reconstruction: Black Inclusion, Chinese Exclusion, and the Fictions of Citizenship*. New York: New York University Press, 2015.

Yun, Lisa. "Signifying 'Asian' and Afro-Cultural Poetics: A Conversation with William Luis, Albert Chong, Karen Tei Yamashita, and Alejandro Campos García." *Afro-Hispanic Review* 27, no. 1 (2008): 183–217.

Yun, Lisa. *The Coolie Speaks: Chinese Indentured Laborers and African Slaves in Cuba*. Philadelphia, PA: Temple University Press, 2008.

Irene Kacandes
# "And what about your mother?"

In 2009 I published a book I decided to call a "paramemoir," because it investigated the experiences of my paternal family during the Fascist Occupation of Greece and how those experiences were stored in the memories of many individuals among two generations of my large family. I titled it "Daddy's War" to signal its opening perspective grounded in my own earliest memories – indeed I would and did label them postmemories, using Marianne Hirsch's generative term.[1]

The book produced a certain amount of interest and discussion. The question that baffled me the most, at least at first, was: "And what about your mother?" I felt confused when people asked me that, because for me, my mother was "on" every single page of the book. I might still have written something, but *Daddy's War* would have been quite a different book without my mother's involvement. This essay is my first published attempt at explaining that, though I've been crafting aspects of an answer ever since the question was first put to me.

Why did it take me so long to address this issue? Why am I writing about this now? To be sure, the Covid-19 pandemic has made me, like most of us, excruciatingly aware of how quickly we can lose a loved one. My elderly mother is at high risk. I tried to stay in close touch with her through phone calls, letters, and family zooms during this strange epoch of the pandemic when we were physically far apart from early March through late November 2020. As much as she was on my mind and in my heart causing me want to write about her, it is also the case that in 2020, precarity and violence were brought into the consciousness of much of the world's population in new ways. Specifically, the murder of George Floyd (and that murder leading to more widespread knowledge among non-BIPOC communities of a long string of lethal violence perpetrated against BIPOC individuals) and the uneven, racially inflected death rate due to Covid-19 have caused many white Americans to think about social justice in ways and to extents they simply didn't have to prior. Though as a teacher of the intersectionalities of race, gender, class, ability, age, ethnicity, geography, and more, I have had to think about my own privilege many times before, in the

---

1 Irene Kacandes, *Daddy's War. Greek American Stories. A Paramemoir* (Lincoln: University of Nebraska Press, 2009). For definition of "paramemoir," see 51. On postmemory see Marianne Hirsch, *Family Frames: Photography, Narrative, and Postmemory* (Cambridge, MA: Harvard University Press, 1997), 22.

https://doi.org/10.1515/9783110753295-020

context of 2020, I feel strongly that whatever work I undertake must engage in some way the twin issues of white privilege and social injustice. To bring these two motivations together then, I am writing this essay not only with my relationships to my own mother and father in mind, but also through the lens of choiceless choice for many Black, Brown and Native American families.

In reviewing below a couple of chapters in the incredible journey my mother and I went on together more than a decade ago during the period in which I was researching and drafting *Daddy's War*, I am expressing my thanks to her and concomitantly trying to make sure others are aware of the collaboration that I should have announced more explicitly then. My sincere thanks to my mother in the book's acknowledgments currently don't strike me as adequate, and failing to do so more fully on the verge of publishing it appears to me now as an unjust privilege or habit of academics to claim full credit for the work they do, when we all actually know that nobody writes a book without profound help of many kinds and from many types of people, including essential workers like the people who clean our offices, the technical experts who rescue us and our computers, the librarians who help us discover sources and procure access to them, people who help with formatting and editing our work. It takes a village, actually.

To be sure, the facts, stories, and memories central to *Daddy's War* are anchored in what happened to my paternal family. Mysteriously returning from the United States to Greece in 1937 with four young children but without her husband, my paternal immigrant grandmother Chrysoúla spent several years of the indigenous, relatively benign Metaxas dictatorship and then the years of the malignant Fascist Occupation by the Germans, Italians and Bulgarians, mainly in the area in which she had grown up near mythic Delphi. Though a child himself, as the oldest sibling and a male, my father John (born in December 1929) helped his family survive once the war started by taking any kind of odd job in exchange for food and sometimes money. In October 1945 the family finally managed to collect all the documents and funds needed to get back to the USA and to my paternal grandfather from whom they'd been separated for more than eight years.

My mother Lucie, born and raised in Greece, was on her native island of Andros for the duration of the war. Her nuclear family was intact, and her extended family worked together raising crops and livestock. They didn't have much, but they had enough to eat. Barely enough, but enough so that my maternal grandmother could send my mother on secret missions with food to poor neighbors. My grandmother instructed my mother to be stealthy not so much to avoid the occupiers, as to avoid her mother-in-law who did not have the same generous spirit. Andros was occupied at first by the friendly Italians and then by the threatening and greedy Germans from whom they had to hide their food supplies. There was relatively little physical violence on my mother's island, even as, to be sure, there was much hardship and fear.

My mother did not meet my father until nine years after his return to the USA. They were introduced first through correspondence and then in person in Ithaca, New York, where my father was a student at Cornell, by my mother's brother Alex, my father's best friend and co-worker in a restaurant run by a maternal uncle of theirs. Alex had brought my mother to Ithaca from a governess job in London so that she could go to college. And my father had offered her assistance already before her arrival with regard to how to apply to college.

My mother was adjacent to my father's story and to the research I did about it in two main ways. First, when she met my father, she quickly became, as her brother had been before her, a "co-witness" to the traumatic events my father had experienced during the War.[2] Knowing his language and his culture, my uncle and mother served as the sympathetic listeners my father needed to unpack the hardships and indeed near lethal events he had experienced as a child and adolescent. For similar reasons, my mother became a co-witness to my father's mother, also in the 1950s when her mother-in-law's war experiences were still quite in the front of her mind. My mother Lucie has an incredibly good memory, and she knew my father's family tree and his network of friends better than anybody who was alive when I undertook my research. (My grandparents and my father's two brothers were alas already dead when I finally went in pursuit of this chapter of family history, and my own father had some blanks due to untreated PTSD, I surmise, and, as we later determined, dementia, into which he was already slipping when I began interviewing him in 2005.) Sometimes my mother knew just what to say to jog my father's memory. And many times, her own memory was unleashed by something I asked or something my father said; she'd remember things he'd told her back in the 1950s that he himself no longer remembered.

Second, though my mother did not finish college due to starting a family with my father shortly after marrying, my mother had had a superior high school education in Greece, especially for a girl of her modest social class. She was naturally bright, and her own mother, who had not been able to attend school past the third grade, recognized her daughter's intelligence and did everything she could to acquire the resources to allow my mother to finish lykeio

---

**2** On co-witnessing, see the introduction to this volume and Irene Kacandes, *Talk Fiction: Literature and the Talk Explosion* (Lincoln: University of Nebraska Press, 2001), 89–140; Kacandes, *Daddy's War*, 262–335; and Irene Kacandes, "Die Ungnade der späten Geburt: Challenges in the Twenty-First Century for Central Europeans," *German Studies Review* 40, no. 2 (2017): 389–405.

(highest level pre-university). Lucie's literacy in Greek, meaning not just her language skills in several types of Greek, but also her general acumen and cultural knowledge, allowed me to puzzle through with her some documents and mysteries I feel pretty certain I would not have been able to manage with any other interlocutor or helper due to the trust we felt with one another.

I need to share a few personal things about myself, the significance of which will become obvious for two moments I'll reconstruct below. One: I was born with a full head of black hair. I mention this because my five siblings were born blond. Though all six of us have brown eyes like our mother – our father had very bright blue eyes – I was the only one who had dark hair and olive skin like hers. So, when we children were little, anyway, it stood out that I looked the most like her. People remarked that. I heard the remarks. If mothers and daughters easily become tangled, my own identification with my mother was powerful long before I ever learned any psychological theories.

Two: Greek is my mother tongue. It is literally the language that my mother first spoke to me and that I, for a while, spoke to her. But when my eldest sister reached school age – I am the third child – our father forced our mother to speak only English to us, and there is no doubt that English quickly established itself as my native tongue. We were sent to Saturday school at the church where we learned the Greek alphabet, but the language program was not well organized or particularly engaging, and so none of us kids became literate in Greek, though the older of us could read and write and follow a simple conversation. I'm pretty sure we all understood some quotidian phrases like "come eat" and "I'm going to spank you." Unlike many immigrant families where the next generation learned the language "back home" during summer vacation, our family simply didn't have the cash to go to Greece. Eventually my mother took four of the children there in 1967 and my younger sister and me in 1974. During that visit, I was so humiliated at feeling I was Greek and yet not being able to express myself the way I could in English, that I resolved and did eventually manage to take enough Greek classes to become fluent. After my college graduation, I spent a year on a Fulbright fellowship at Aristotle University in Thessaloniki. For a while, I had fairly sophisticated reading and speaking skills in Greek, though never native ones.

## Mystery Number One

Why did my paternal grandmother leave the USA and head to Greece in Summer 1937 with four young children and without her husband? This was the fateful event that led to their being **trapped** in Greece and to their endangerment and near extinction several times during the War years. Trapped is a word that resides in my earliest memories and that I consider the nugget of one postmemory.[3] Theories about the reason for this fatefully-timed trip abounded in the family later: my grandmother missed her own family; she and her husband wanted the kids to learn Greek properly; they were having kids too fast, and this separation would function as sure birth control; my father proved to be a troublemaker already at a very young age, and he should be transferred to another environment; they could live comfortably in Greece on the dollars my grandfather was earning, whereas his earnings didn't go very far in the USA. Without a doubt, the theory that was whispered the most frequently, however, was that her husband was fooling around, and she wanted out.

●

I've been looking for a fat bundle of letters for hours.

Ever since my father had discovered them in the wake of his father's sudden death from a heart attack at age sixty-five, my father talked about "the fat bundle of letters." The letters that proved that my grandfather had tried the whole time of the family separation to find them, to get them out of Greece, that he'd appealed to the State Department, to the Red Cross. The discovery of these letters rocked my father's world and caused him huge regret, because he'd thought of himself during the war and continued to believe after the war, that they'd simply been **abandoned** – another word that anchors a postmemory of mine. My father had spent much of his adult life relatively estranged from his father as a result of that belief. And now it was too late to rectify anything.

I'm searching every file drawer of every file cabinet my father owns. And there sure are a lot of them. In the basement. In the part of the family room that he uses as a study. In the attic. It's in the attic. The light is inadequate. Yet another file of useless mimeographed worksheets for teaching reading methods from the early 1960s. Then I feel it in the back of the drawer after the mental barrier that holds up the files. A bunch of different things: envelopes, notebooks, larger envelopes. Not much but something. Dad said it was a huge stack. This is not huge, but I take what I've found. Put it all in my briefcase. And later head back to my own home hours away. I can't even remember how or what I told my parents about what I'd found.

---

**3** Kacandes, *Daddy's War*, 1–3 and elsewhere. I begin the book with my earliest memories, what I think of as my postmemories, because they actually concern things and emotions that happened to my father, not to me, but I believe I have carried them around with me since my earliest consciousness. In that book and here, I use bold to identify them.

When I get home, I realize some of these notebooks are my father's from the immediate postwar. Some are my grandfather's. They are mostly address books or correspondence logs, but there's some writing within them I'll try to decipher later. I register for the first time a large, manila envelope, aged brown with some writing on the outside identifying the contents with my grandmother. There's an identity card and a Greek passport. Also, a wad of papers so tightly folded and so seemingly permanently stuck together that I cannot look them over. I don't know much about working with damaged documents. But if I'm going to read these, they'll have to be flatter. I open and straighten the pile as best I can. I take several of my largest dictionaries and place them on top. And then I wait. Scared to approach them again. I'm afraid I'm turning them into dust.

Weeks later, I finally lift the dictionaries. The papers are still intact, and they stay flat. I'm lying on the floor, though I have no idea why. It's not very comfortable. Perhaps my desk is in its usual messy state. I start to go through the papers. Most appear to be official documents and are partly or fully typed. Some are handwritten. Many are in καθαρεύουσα (katharévousa), a language that was artificially created from Classical Greek and the demotic to serve the Greek state when it finally became recognized as an independent nation in the 1830s. I have never studied katharévousa formally, and while I can pick out individual words, I usually have trouble getting the exact sense of it.

I call my mother.

Easily done then. Easily typed now: I call my mother.

The reality of Summer 2020 when I am first drafting this essay hits me hard in the face and knocks the breath out of me, ripping me from a fifteen-year-old memory and launching me into a new train of thought. How many people can *not* just call their mothers? How many parents cannot respond to their children's needs, profound or trivial? 2020 has been filled with the separation of people from one another. "Social distancing" is the euphemism. The reality concerns the elderly, sick, dying who cannot be visited by those who love them. When I first typed "I call my mother," George Floyd's agonized calling out to his dead mother as he was being suffocated by a police officer started ringing in my ears. What do I do with that memory? And what about the way mass incarceration, especially of African Americans, is not only separating but also devastating families? Children, even babies, being separated from their parent or parents at the US/Mexico border? These, too, have entered public consciousness in 2020 to a greater degree than perhaps ever before. What will we do with this as a nation? What will I do with it as an individual?

I can start by reflecting on the above and grafting it into what I'm doing. I can embrace it interrupting what I think I'm doing to state clearly that the ability to call mothers or fathers or brothers or sisters or children today is simply not automatic for millions of people around me today. I can acknowledge that the great blessing and indeed privilege I enjoy to simply feel a need for my mother and then find her on the other end of the telephone connection is a benefit of a demographic, not something I earned. As someone who's thought quite a lot about mortality, too, I can ponder that this ability won't be there forever, in likelihood probably not for much longer.[4]

And in the particular context of reflecting on my academic work, I can state clearly that that piece of it for sure (and many others too) would not have been the same without my mother's participation and expertise.

I call my mother.

I tell her what I'm trying to do. And I start describing and then reading each document to her. I read the Greek in front of me into the phone.

As I hear my own voice, I remember how decades before my mother had laughed, trilled really, as I read her a poem by Constantine P. Cavafy out loud during my first year in graduate school. What's so funny, I asked. You have an American accent, she replied. It struck her as odd that *her* daughter had an accent. I didn't see the humor at the time, but now I see the gap. How incongruent! This person who looked so much like her, who was so much like her, sounded like a foreigner. As I read the documents, I hear my own accented Greek. I stumble on some words. Then we start stumbling over the meaning together.

The literal meaning my mother mostly understood right away. But what it all had to do with our family history was confusing to us both. It was thrilling to puzzle over each clue together. It was clear that these papers were very connected to each other somehow, as it was clear that my grandmother must have made huge efforts to keep them safe during the extremely precarious circumstances of their wartime sojourn in Greece, preserving them even after she had returned to the United States. The papers all predated the war, though. Why did my grandmother have papers that gave her power of attorney for my grandfather? Did he trust her that much? Weren't they estranged from each other? How did she manage to get to the Consulate so shortly after she'd given birth to her son Nick and so shortly before she actually got on the boat with the kids? Older documents: Letters to banks. Bank statements. Receipts for money sent to Greece from New York, shortly after my paternal grandparents arrived there as newlyweds. But weren't they struggling financially? How could they send so much money? Was that a lot of money? Even

---

4 Steve Gordon and Irene Kacandes, *Let's Talk About Death: Asking the Questions that Profoundly Change the Way We Live and Die* (New York: Prometheus, 2017).

older documents: a loan. A loan with the house in my grandmother's village as collateral. A loan from someone in Greece to my great-grandparents. And then we reach the last one in the pile, though the most recent. It's dated after my paternal family's return to Greece but before the war broke out there. A petition to the local court to verify that my grandmother's mother was not a farmer. Not a farmer? We all knew our great-grandfather was a shoemaker. Why was it important to prove that our great-grandmother wasn't a farmer? And that was it.

My mother and I had gotten through the pile, but still didn't understand what it could all mean. It didn't fit the scenario of an unhappy couple staging a separation from one another. And it didn't fit the idea that my grandmother missed her family so much that she wanted to visit them. How could it be that my grandmother was taking her beloved mother to court? Whatever the reason, why was there no document with the final ruling on the matter about which she'd petitioned the court?

There was basically no one still alive who would have the keys to unlock this mystery. With another research trip to Greece and armed with what my mother and I had pieced together so far, I was able to consult a lawyer in Athens who knew Roúmeli, the part of the mainland my paternal family was from, and who loved local history. I've published the details of his conclusions about the documents in *Daddy's War*. What I'd like to note here is how imperative it was to me to call my mother as soon as I'd finished talking to him, to tell her his theories and how they helped me imagine a new narrative. To see if she thought my new narrative fit what she knew about my father's family. Plus, it was just so exhilarating to have these documents suddenly make sense, and I wanted her to share that thrill with me.

I called my mother.

From Greece. In the days when that wasn't so easy to do.

I told her my current theory. It *did* make sense to her, even as she had essentially never heard any of these details from her mother-in-law or anybody else. There were just the vague and persistent rumors about being **betrayed** – yet another word that belongs to my postmemories. And now I felt pretty certain that "betrayed" belonged with "house." My great-grandparents' house in the village of Chryssó, just south of Delphi, became for me what Marianne Hirsch and Leo Spitzer have so usefully called a "testimonial object" or "point of memory."[5] This

---

5 Marianne Hirsch and Leo Spitzer, "Testimonial Objects: Memory, Gender and Transmission," in *Diaspora and Memory*, eds. Marie-Aude Baronian, Stephan Besser, Yoland Janssen (Amsterdam: Rodopi, 2005), 137–164.

is how I wrote it up in *Daddy's War*, in a section in which I co-witness to my father by addressing him in a letter:

> The short version is that your mother's parents, Harálambos and Asymoúla Tsínga borrowed a large amount of money from some loan sharks in fall 1928, the very year your parents had married. They put up the house in Chryssó, which was in Asymoúla's name (it was probably her dowry, but I couldn't find proof of that) as collateral. In 1931, your father, George Kacandes, sent money from the United States to Greece to buy the loan from the loan company, and in turn, the house as collateral came to him. Something happened in 1935 that makes your father wonder if his investment is safe, for he asks a bank in Greece to determine whether the value of the house is enough to cover the loan and the interest owed on it. No existing documents indicate what could have triggered his concern at that particular time. My two best guesses are either the floods of 1935 that do much damage in the area of Chryssó or news that your mother's parents were going to give the house as dowry for their youngest daughter Loukía. In any case, the reply does not allay your father's fears, and in 1937 he gives his wife, your mother Chrysoúla, power of attorney. As soon as she has recovered from childbirth (remember, your brother Nick is born in April), she goes to Greece, arriving with you and her other three children in tow by the end of August 1937. Just a few months later, in January of 1938 your mother files a request to the local court to determine whether or not her mother Asymoúla is to be considered a farmer (αγρότισσα) or a housewife (νοικοκυρά). The reason this designation was important is because of a law that had been passed in 1937 declaring that creditors could not collect by confiscating houses that belonged to farmers; the government feared a collapse of the indebted agricultural sector if a large number of farmers became homeless. And that's where the paper trail ends; I could find no document indicating that the court had ruled on your mother's request or if they had, what their determination had been.[6]

My own mother agreed with me that the importance of this story was that it cleared my father of being guilty, guilty of his naughty behavior causing their fateful trip to Greece – something, alas, he was told by his mother in what I assume was her extreme frustration and fear when trapped there. The complicated outlines of the house's history, and its morally ambiguous penultimate chapter with a daughter planning to sue her mother for it, also made it clear, at least to my mother and me, why no one knew this story. Of course, some mysteries remain: how did my grandfather have that kind of money to send to Greece in the first place? Did my grandmother try to act further on recuperating the house or just give up because the onset of war and the concomitant cessation of the money my father's had been sending from the States brought with them more pressing issues of survival? If the latter, why did she keep the papers and yet not tell anyone about them? And speaking of these documents, how did they come into my father's possession, and why did he never mention them? Had he ever opened that envelope?

---

6 Kacandes, *Daddy's War*, Part Five, 313–314.

What's upsetting to me now, though, is that when I reread my write-up of this incident in the book I published, I realize that I mentioned the role of the lawyer, but not of my mother. In retracing my steps here, I want to flag that as unacceptable. Even if my mother was a happy and willing partner to my research, she should not remain an unacknowledged one. In my own memory, this piece of the narrative was something we puzzled over together, not something I figured out on my own. Why didn't I include that sense of what had transpired? Was it my overidentification with my mother – as if we were one and the same person? Was it my need for academic kudos, for full credit? My need to participate in the academic plague of claiming to figure things out alone despite that so rarely being the case? As mentioned above, I thanked my mother profusely in the official acknowledgements of the book. But to my mind now, our work together was an integral part of that particular episode. It was one chapter in a true collaboration that I am so happy to celebrate here.

Mercifully, I did a bit better with a different issue, clearly positioning information from my mother as the most likely answer to a question I just couldn't find answers for on my own.

## Mystery Number Two

It must have begun at some point. I must have begun on some specific day, month and year to worry about being raped. When was it and why did it start then? I was young, very young – but not so young that I wasn't allowed to walk home from school or to downtown and back by myself. I'd be walking alone along a street, and I'd be worried that some man would jump out from behind a bush. I didn't know what would happen next, but it would be bad: he would "rape" me, it would be scary, and my life would be changed forever. At some slightly later point in time, I began to connect the act of rape with the act of making love. I'm sure that at that age I still didn't know what either actually involved, but I knew that somehow the "making love" part was nice, even beautiful, and it was done with someone you really liked; the "rape" part was hurtful and frightening, and it was done by a stranger. For sure, by the time I was twelve, but I suspect at an even younger age, this "knowledge" was apparent in a prayer I recited to myself whenever I was walking alone, especially if it was getting dark: "Please, dear Lord, please let me make love before I am raped."

It took many more years before my brain registered that the "*before* I am raped" part of my mantra revealed certainty that this disaster would befall me. Even when I learned as a teenage feminist about the horrific statistics on numbers of women who *are* raped at some point in their lifetime, I had to admit that my attitude toward rape as a child in a peacetime suburban setting was peculiar. I finally began to wonder about who made the little girl I once was, so afraid of rape and so certain that it was bound to happen to her. Or maybe it's more accurate to say that I finally gave myself permission to be conscious of the fact that it was my father who had instilled this fear in me. I began to be sure that my terror had something to do with my father's War Experiences in Greece. Eventually I hypothesized that he could have been witness to someone whom he loved being raped. The idea of this was so painful to me that for years after I had made this connection, I never allowed myself to think of who it could have been. (I remark the absence of a family story about rape among my earliest memories, my postmemories.) It was only as a middle-aged adult, when I very intentionally started to piece together what had happened to my father during the war, that I let myself put names to the possible victims of the rape I now assumed my father had witnessed: Was it my grandmother? Was my grandmother raped by Germans? by resistance fighters? by a local Greek taking advantage of her status as a woman living without her husband? Was it my Aunt Pearl? Oh, my God, she was only a little girl! Had my father seen his own sister raped? Could it have been another female relative? Was my father himself raped? He was young and small, too.

In a loose affective sense, it was all these scenarios, and in a strict, factual sense it may have been none of them. While doing my research in the region in which my father grew up and asking about rape in general as an instrument of war and occupation – I didn't dare connect it directly to my family – all my interlocutors insisted that no such thing had happened there. When I reported this to my mother in a very different geographical place, it sparked a conversation that led her to remember something that I then recognized as key to the most likely source of the fear my father had planted in me. Why the singular? Because one of the other things my interviewing had turned up is that my long-standing assumption that my sisters and paternal female first cousins likewise carried this fear around with them was incorrect. They told me they did not.

Berlin. Fall 2005. I am leading a study program abroad for my university, something I loved doing and had done many times before. One of my brothers agrees to supervise my father's care while my mother comes to visit me. She's never been to Berlin before. We're both really excited about being together in this place. Almost immediately after her arrival, I am telling her about some of the conversations I was able to have during my summer research trip to Greece just prior to coming to Berlin. Our own conversation ranges widely. In rereading the transcript I created from the recording of that exchange, I am surprised to discover that my mother related to me first a whole theory about the house in

Chryssó that made sense. (Her Berlin visit was before I had discovered the wad of papers.) Then she tells me more about my paternal grandmother's family, and I confuse the names of her sister, my two great-aunts whom I had met during my first trip to Greece in 1974, one of whom I thought of as crazy; my mother is more sympathetic and explains some of the tragedies of that great-aunt's life. That segues into learning that the oldest sister had married "up," to a carpenter. And that my father would have liked to live with them and apprentice with him, but they wouldn't take him in. My mother generously attributes this to the hard times, though she also immediately adds that on Andros when she was growing up, "we could go to anybody's house, and it was home" – another difference between her war experiences and my father's . . .

At some point I announce to my mother that despite all I've learned, "one of the many mysteries remains," and I explain to her my early fear of rape and my theory that my father must have seen one of his relatives be violated. This seems to be new territory for my mother: "That I don't know. I would imagine he saw a lot of stuff going on." I try to impress on her how truly terrified he made me. And she immediately assumes that my father didn't know that what he was telling me would have that result. "He was just warning you." She had tried to convince him not to talk to us about sexual matters. That we were still too young. But he went ahead anyway in other contexts, so she didn't disagree that he could have talked to me alone at some point.

As I draft this, another connection to the present is breaking into this old story: the horrible truths about the world parents feel obliged to tell their children to keep them as safe as they can: "Don't talk to strangers"; "don't accept drinks or treats from anyone"; "don't let anyone in the house when I'm not home." We all know of an instance when the failure to do these kinds of things resulted in tragedy. But it also enters my mind while writing this, that only recently have many white Americans heard testimonies from Black Americans about telling or being told how to act around the police. It registers deeply with me at this moment, that so often those recitals must also have seemed, probably to both reciters and recipients, ill-timed. That those parents, like my mother, also wished such truths would not need to be told to young kids. Maybe even my father had compunction about drilling this fear of rape into me. I don't know.

My mind goes in another direction now, too. As awful as the number of rapes is, I wonder if the number of life-threatening encounters between African Americans and the police, the number of stops due to the crime of "driving while Black" – could we ever get such statistics! – wouldn't be even higher. I want to be clear that I am not trying to set up a competition of suffering. Rather, a lesson of 2020 at least for me has been that white Americans have to take in, indeed actively seek more information about current (and longstanding) realities of how racism plays out on a daily basis and throughout lifetimes of BIPOC individuals and communities. I remember hearing the phrase "race war" during my childhood, and that I connected it then with vague images of blacks and whites fighting each other on some kind of equal footing. My sense of what that could mean now is quite different: majority society and its well-armed institutions waging war against its citizens of color. And isn't war always the most unfair to the children?[7]

During the 2005 conversation in Berlin, my mother talks about being sent to the hills where her family has its hand-mill hidden during the war; she's supposed to grind freshly harvested wheat so that they can finally have bread after a full week without any. But the wheat's too young and humid. She cranks and cranks. "I remember it took me hours. 'We can eat bread tonight.' I was telling myself something like that. And I was going, and I was going. Until my arm – I thought it was going to fall off." But she wanted to do it; she had a role to play: "I knew I had to keep myself going without uh, being suspected of anything. A little kid."

*Daddy's War* is filled with accounts of numerous things my father did or was asked to do that were, to use a euphemism, completely age inappropriate. But that's practically a given for children in wartime. In our 2005 conversation, my mother and I remark how much my father traveled during the war, especially between his home area where the rest of his nuclear family was living, and Athens, where he had briefly gone to school before the war, and where he, like hundreds of other poor children, earned a touch of money by selling cigarettes on the street during the war. Athens. Maybe that's where he saw someone raped, I wonder aloud.

My mother's memory has been jogged. Athens. Maybe she connects it with Peter Tsardakas, an older Athenian who befriended my father and allowed him to be a kind of servant boy on a small boat he had. A boat that sailed up and down the western shore of Greece. The boat. Another point of memory. This is something my mother hasn't thought of perhaps since she was first told it in the early 1950s. I have trouble following her she's speaking so fast all of a sudden. She switches languages. She switches back. She quotes my father, and at that moment I am completely unnerved. She sounds exactly like him. Exactly. As if he were in the room with us. And he is desperate. Her jogged memory came out like this:

*Lucie.* [. . .] the other thing was when he was with Peter Tsardákas on his kaíki [small boat], and they went to Kérkyra and these islands up there, and he uh told me that uh the Germans that were occupying the vessel were uh people –. The young men would go and sell their sisters so they can get some food, some oil and bread and whatever –

---

**7** That is a premise of my book *Daddy's War* and the reason behind its dedication: "For children of all times and places trapped in Somebody's War," unnumbered.

*Irene.* The guys on the boat?

*Lucie.* To the guys on the boat. Those guys, most of them, were very young, in their teens, I mean late teens, eighteen, something like that. But they would uh – The brothers would go and offer their sisters to the Germans, so they can have, to survive!

*Irene.* Daddy told you that?

*Lucie.* Μούλεγε ότι ήτανε τόσο δύσκολα τα πράγματα εκεί επάνω στην Κέρκυρα. [*He told me things were so difficult up there in Kérkyra.*] Και δε θυμάμαι. [*And I don't remember.*] Στην Κέρκυρα νομίζω πήγανε. [*To Kérkyra I think it was they went.*] Και [*And*], you know, είχανε μεγάλη πείνα. [*They had terrible starvation*]. Και οι αδελφοί πήγαιναν τις αδελφές για να – [*And the brothers would take their sisters in order to –*] So I don't know how the transactions were going. Τους πήγαιναν σπίτι; [*Would they take them home?*] Τι κάνανε; [*What did they do?*] Δεν ξέρω. [*I don't know.*] I never asked those things.

Αυτό του είχε μείνει πολύ. Στο μυαλό του. [*That stayed with him deeply. In his head.*] You know. [Mom's voice is loud and high pitched when she quotes the next words of my father and her intonation and pronunciation are exactly like his:] Can you imagine? he used to say to me. Can you imagine?[8]

As I type this, I am reminded again of how well I actually could imagine. My father being haunted by something he'd witnessed early in his life had been haunting me since my childhood. With my mother's retrieval of something my father had told her, I could now put the pieces around my long-harbored fear together: why my sisters didn't have this same fear; why I was so very afraid and so very certain this awful fate would be mine one day; how horrified my father was by the choiceless choice that male Kerkyrans were making. How my own gender and youth could have brought up the buried image he had of those Greek girls being exchanged for food dockside. Though I still couldn't recall the exact words of his first warnings to me, I could now date them.

I had just turned nine. It must have been the summer when our mother had taken all our siblings except my younger sister Georgia and me to Greece. We stayed in the house with Dad. I became the housewife. I not only looked like my mother, now I was supposed to do all the tasks that my mother did. Shop, cook, clean, wash, iron, watch my sister. My parents had never been separated for more than a few days before that summer. My father had never gone back to Greece since he'd escaped in 1945. And in 1967 there was his wife repeating what his own mother had done, even down to the political situation (though I doubt he – or she – registered that consciously). She'd returned to Greece in the

---

8 From transcript in the possession of the author.

middle of a junta with four little kids. And there was this other kid in front of him on a daily basis who looked like her.

As I typed the words above: "he is desperate," I burst into tears. My fear. His fear. My mother becoming my terrified father for a few seconds. The unexpected solution to this decades-old mystery is liberating. And it's overwhelming. I feel like I'm drowning. I can't and don't say anything. Had I become my nine-year-old self?

In this homage and act of thanks to my mother, I want to point out that she then does what so many mothers are capable of doing. In the midst of horror, she provides a way to go on. To help her child go on. My mother adds a coda.

Δεν ήτανε όπως τους Ιταλούς που είχαμε δει εμείς. Ήτανε okay. Αλλά αυτοί ήτανε Γερμανοί. Those were Germans.

[They weren't like the Italians we had. They were okay. But those were Germans. Those were Germans.]

When she tosses me this life preserver, I grab at it and come above water again. I tell her my theories about why, despite this story, it appears that my father hated the Italian occupiers more than the German ones. And then we happily move on to her recitals of less traumatic experiences during the Occupation of her island, closing with:

*Lucie.* It didn't matter. We didn't have shoes, we didn't have clothes. But we had food and that was great.

I'm rescued. I am breathing normally again and say calmly:

*Irene.* Mom, I want to talk about this more, but we should have lunch and get on with our day.

At the end of the transcription which I hadn't reread for a dozen years before writing this essay, I see that I came up with a coda too: "[Mommy, I love you!]."

"And what about your mother?"

She is on every page.

# Essay's Coda

As nice an ending as the above makes, I won't have learned a central lesson of this exercise if I don't acknowledge in the text and not in a footnote that I didn't write this essay on my own either. I had three perspicacious readers who urged me to rethink some of its central points: fellow contributors to this volume

Professors Claudia Breger and Priscilla Layne, and my Dartmouth colleague
and friend Professor Darlene Drummond.

# Works Cited

Gordon, Steve, and Irene Kacandes, *Let's Talk About Death: Asking the Questions that Profoundly Change the Way We Live and Die*. New York: Prometheus, 2017.

Hirsch, Marianne. *Family Frames: Photography, Narrative, and Postmemory*. Cambridge, MA: Harvard University Press, 1997.

Hirsch, Marianne, and Leo Spitzer. "Testimonial Objects: Memory, Gender and Transmission." In *Diaspora and Memory*. Edited by Marie-Aude Baronian, Stephan Besser, Yoland Janssen. 137–164. Amsterdam: Rodopi, 2005.

Kacandes, Irene. *Daddy's War. Greek American Stories. A Paramemoir*. Lincoln: University of Nebraska Press, 2009.

Kacandes, Irene. *Talk Fiction: Literature and the Talk Explosion*. Lincoln: University of Nebraska Press, 2001.

Kacandes, Irene. "Die Ungnade der späten Geburt: Challenges in the Twenty-First Century for Central Europeans." *German Studies Review* 40, no. 2 (2017): 389–440.

Bettina Brandt

# Nelly and Trudie: Deciphering a Transatlantic Family Holocaust Correspondence

## Vienna, 1939, one year after the *Anschluss*

When Lilly Frey, her husband Otto and their twin daughters, seven-and-a-half-year-old Gertrude (later Trudie) and Dorothea (later Dorothy,) fled from Nazi Vienna in the late spring of 1939, they had been forced to leave behind Nelly Lehnert, Lilly's mother. A widow in her seventies from a well-established old Jewish-Viennese family, Nelly had hoped to join the Freys in the foreseeable future. Instead, Nelly was deported to Theresienstadt/Terezin, a former garrison city turned ghetto about an hour outside of Prague in what was then known in Nazi parlance as the Protectorate of Bohemia and Moravia. There, in "the Old Age Home on the Elbe" as the Nazis had deceivingly marketed Theresienstadt to German and Austrian elderly, Nelly died from starvation and disease shortly after her arrival.

Once they left Vienna, the four Freys, as a family, never spoke German again. In the immediate post-Vienna years, Lilly and Otto continued to occasionally speak German with each other, and they also wrote, of course, in German when corresponding with elderly relatives and friends left behind in Vienna after the Anschluss, or with fellow Austrian refugees and relatives in the United States.

For the twins it was a different story: they quickly unlearned their mother tongue. This process likely started during the family's initial move to England, where the Freys spent several months awaiting their American visas, and was later completed in Pasadena, California, where the Freys arrived in December of 1939 and the twins grew up.

In 2012, while preparing to move to a local retirement home and getting ready to put the family home in State College, Pennsylvania, on the market, one of these twins, Trudie Engel née Frey, by then eighty-years-old, and about to retire from Penn State, rediscovered several boxes of family papers in her attic, including the weekly German letters that Nelly had sent from Vienna to California in the three years before her deportation. Trudie was eager to find out more about her Viennese Jewish family history. When Trudie and Dorothy last said goodbye to their grandmother Nelly, they had just finished first grade. Thereafter they only recalled glimpses of Nelly: Trudie remembered that the old widow

https://doi.org/10.1515/9783110753295-021

dressed in long black skirts, was hard of hearing with a prominent limp, and Dorothy did not forget that Nelly spent a lot of time in bed snoring loudly.

To find out more about the last years of Nelly's life from the narrative presented in her German letters, Trudie started looking around for a translator in the small university town where both she and I live. When I took on "the big job," as Trudie referred to this translation task in one of her first emails to me, neither one of us knew what Nelly's letters would reveal, nor whether the content of the letters would be of historical interest to those outside of the immediate family.

Translators often conduct quite a bit of research about the author whose words they are about to carry over into a foreign tongue and a foreign culture. They read up, or, if possible, go visit the places where their author lived to familiarize themselves with the local geography as much as possible. If they are translating a writer who is no longer alive, they will study that particular historical moment to gain insight into the sociopolitical conditions under which the author's life unfolded. If they are translating ego documents (letters, diaries, travel journal, autobiography) translators might contact remaining relatives to get an additional perspective on the author. At times, translators become so drawn to their author that after translating the author's work they end up writing a book about the writer's life as well. As the translator of Nelly's letters and as professor of German and Jewish Studies, I did all of the above.

Between the fall of 2013 and the late summer of 2017, when Trudie sadly passed away, we worked together as a team. I will always be grateful for and fondly remember the friendship that developed between Trudie and myself during my regular visits to the Engel's family home on Ridge Avenue, and then later, during my weekly meetings at the local Quaker-run retirement community where Trudie and her husband Al Engel, himself a Jewish refugee from Munich and retired Penn State professor, had moved in early 2016. There, we would catch up about the work Trudie and I each had done for this project since the last time we had met over Zabar's rugelach that Trudie had ordered, or the plum cake that she would bake on these occasions. Early on, Trudie had expressed the wish to be "my assistant" and had asked for "a small task." I had suggested that she might want to reestablish contact with relatives of her own age from the extended family with whom she had been out of touch to find out what they remembered about Jewish life in Austria. Trudie followed up on this suggestion and during our weekly meetings would share the bits of information that she had gleaned from talking and eventually meeting various relatives who, like the Freys, has escaped Vienna before the outbreak of World War II.

Of particular importance was Trudie's renewed connection with Marianne Wertheim Makman, a second cousin five years her junior, now living in New

Rochelle, New York, who brought out her own family archival materials and shared some of them with us. In the summer of 2016, Trudie and Dorothy also went on a road trip to meet up with Marianne, who had been but two years old when she left Vienna with her parents, Hans and Herta Wertheim née Roth, in 1938 but who knew much more about the fate of members of the extended family than the twins did. During this family get together, Trudie and Dorothy learned a lot about the Löwy side of the family in Bohemia (the Roth in Iglau) and in Switzerland (the Josephy family in Zurich), information that was invaluable to fill in some gaps and get a fuller perspective on Nelly in Vienna and the elderly in Prague.

I was first introduced to Marianne at Trudie's memorial in late October 2017. A few months later I travelled to New Rochelle where the Makmans warmly welcomed me to their home. I interviewed Marianne who also gave me a copy of the unpublished memoir that she had written about her parents' life in Vienna, their emigration to the United States, and their new beginnings here. Generously, she also made available copies of the letters that her mother Herta had written to Irene Josephy née Löwy, Nelly's sister- in-law who had married a German Jew whose family had been residing in the Hanseatic city of Rostock for centuries when in the nineteenth century they had moved to Zurich. Both the memoir and the letters helped round out the otherwise sparse information about the last three months of Nelly's life.

During my weekly meetings with Trudie, I, on the other hand, would contextualize the most recent batch of translations that I had sent to Trudie since my last visit, identifying family and refugee friends no longer remembered and placing what was alluded to in these letters into the historical context of post-Anschluss Vienna.

Nelly's letters had presented themselves to me as an intriguing puzzle. First, many of the handwritten letters from Vienna weren't easy to decipher. Also, many of them had no dates or envelopes which made it a challenge to determine the chronology of events mentioned in the letters. Second, to protect the identity of the people she was referring to or commenting about, Nelly never used anyone's full name. Typically, she would allude to a person by their first name only ("Uncle Alex") or, and this was much more problematic in terms of trying to establish identities behind names, by the first letter of their first name only ("Miss A. has been very helpful in getting my affairs in order"). The only case in which Nelly would mention a person's full name was when she was conveying a new address of a Viennese refugee already safely outside Nazi Europe.

Third, those writing letters from Nazi Germany to their relatives both inside and outside of Europe often made use of a family code in order not to endanger others.[1] The Lehnert/Frey correspondence is no exception. By the time Trudie had rediscovered Nelly's letters Trudie's parents, Lilly and Otto, had been dead for decades. No one, in other words, remembered Nelly's code. Deciphering her code was therefore not only an additional difficulty but also a prerequisite for first understanding and then translating her letters.

Nelly masked people's identities in a number of creative ways. Most frequently – and in this regard Nelly's behavior as a letter writer during the Holocaust is not unusual – she used a (lengthy) descriptor that explained the family relationship to the addressee but omitted the actual name of the person altogether. Examples of this kind of disguising are: "Mary Ann's mother stops by frequently," or: "Otto's old flame is still in Prague." Often Nelly would only refer to a person by their profession ("my former seamstress," "my orthopedic doctor," or "your former concierge") making it essentially impossible to identify those masked in this manner.

Finally, Nelly made use of synecdoches to create imaginary new family names in the process. An example of this strategy is, for instance, when she would turn the name of the street on which a family was living into their made-up imaginary family name. This was true for instance of "the Pestalozzis," who after some research in the extended family archive turned out to be the Josephys, living near the university hospital in Zurich, Switzerland, on the Pestalozzistraße.

While a good number of names had been relatively easy to identify, thanks to Trudie's renewed connection to other family members of the generation exodus, as Walter Laqueur called the youngest generation of Jewish refugees from Nazi Germany who emigrated to the United States, figuring out some of the more obscure identities took quite some doing, including a good bit of detective work.[2]

Over the years Trudie and I together found out a lot about the life of Nelly's parents (the Pislings on her father's side, the Stiassnys on her mother side), the relatives of Nelly's husband, Berthold Lehnert born Löwy (the Löwys in Vienna, the Roths in Czechoslovakia and the Josephys in Zurich), and the Neumann family (originally from Biala) into which Hedwig, Nelly's younger sister, had married. I also uncovered the names of many of Nelly's elderly friends who, like Nelly herself, were stuck in Vienna and later perished in the Holocaust but also had been hoping to join their much beloved children and grandchildren in the

1 Walter Zwi Bacharach, ed., *Last Letters from the Shoah* (Jerusalem: Yad Vashem and Devora Publishing, 2004), 7–81, here 70–73.
2 Walter Laqueur, *Generation Exodus: The Fate of Young Jewish Refugees from Nazi Germany* (Waltham, MA: Brandeis University Press, 2001).

United States: Rosa Schück née Springer who had hoped to join her son Adalbert in Atlanta, Dora Schreck née Springer and her older sister Clementine Springer who both had hoped to join Rudolf Schreck in Buffalo, Sidonie Wolf and her father Adolf Tausski who had hoped to be brought together with their daughter and granddaughter Dr. Marianne Wolf on the Upper West side in New York City, and Heinrich and Ernestine Posamentier who had hoped to be reunited with their son Ernst and his wife Alice in Inwood.

Like Anne Frank in her famous diary, Nelly frequently wrote about the other people with whom she was forced to live together in Vienna, or what H. G. Adler, in the context of the ghetto and transit camp Theresienstadt, coined "coerced community."[3] While Anne disguised the names of those with whom she was hiding in the Annex, we nevertheless know their identities today since hers was a stable community of eight people that lived together in hiding at the same address for a period of two years. Nelly, by contrast, was forced to move four times over the course of the three-year-period before her eventual deportation. Each time having to give up more of her personal belongings and moving into ever more crowded living arrangements with always different old people, Nelly went to even greater length to protect the identity of the other eight people with whom she was living in what now are referred to as *Sammelwohnungen* or collective overcrowded apartments created to house post-Anschluss (post-annexation) Viennese Jews in different districts of the city. Nelly also made sure not to give any of the names of the dozens of elderly men and women who came to visit her regularly and who were part of her mishpocha.

I travelled to Vienna several times to familiarize myself with the urban geography into which Nelly had been born in 1868. I visited Döbling, the upscale nineteenth district where Nelly and her younger sister Hedwig had grown up with their parents, Theophile Pisling and Regina Pisling née Stiassny. I looked at the buildings (synagogues, schools, hospitals, old folk homes, commercial and residential buildings) that Nelly's uncle, Wilhelm Stiassny, and Nelly's brother-in-law, Alexander Neumann (mostly banks, some commercial and residential buildings, a few theaters) had built in Vienna and other parts of the Hapsburg Empire. All of Stiassny's synagogues in Vienna had been destroyed after Kristallnacht and many of the other buildings had since been torn down. However, enough are still standing in Vienna to give me a sense of the works of these two architects related to Nelly. To get a better idea of the street in Vienna's first district in which Nelly and her husband Berthold, a lawyer, had raised

---

**3** H. G. Adler, *Theresienstadt 1941–1945: The Face of a Coerced Community*, trans. Belinda Cooper (New York: Cambridge University Press, 2017).

their three children, and where Berthold had had his law firm, I rang people's doorbells and was shown inside the entrance hall of the residential building where Nelly and her family had been living from 1906 until 1939. Finally, I retraced Nelly's steps at the last four addresses in different Viennese neighborhoods in which she been forced to live in the three years before her deportation. Reconstructing the personal and the historical fabric that made up Nelly's life of seven decades in Vienna brought into view how, time and again, this city dramatically shifted its tone and actions vis-à-vis its Jewish minority, and how this minority had responded to it.

When I began conducting archival research about Nelly and the members of the extended network of family or friends that I had been able to identify in Nelly's letters, I was able to further contextualize Nelly's daily life and that of the other elderly Jews with whom she was interacting during those post-Anschluss years. I did research in numerous archives including those of the IKG (or Jewish Religious Community of Vienna,) which after the annexation to National Socialist Germany had come under the authority of Adolf Eichmann's central emigration bureau. There, I studied, for instance, the deportation lists, and the emigration questionnaires that all Jews asking for financial assistance for their emigration from the Jewish Community had filled out. At the Austrian State archives, I studied the "declaration of assets" that Jews who were part of Cornelia's microcosm were forced to fill out as one of many preconditions to be cleared for emigration in 1938. I similarly examined the postwar restitution papers held in that same archive that their adult children in the United States filed on behalf of their murdered Viennese elderly. These and other archival materials helped me contextualize Nelly's particular individual experience within the larger Austrian-Jewish experience post-Anschluss. They also helped me to situate Nelly and those members of her network of friends and remote relatives trapped in Vienna like herself, within the specific topographical setting of the Austrian capital. It made it possible, too, to get a sense of what was typical and what was exceptional in this particular extended family's path of emigration and destruction. Finally, this research revealed significant geopolitical historical information that allowed me to read Nelly's letters *against the grain* and helped me understand what was and what was *not* being said by juxtaposing her letters to other documents that had been produced in Nazi Vienna by victims, bystanders, and perpetrators. As both public and private archives yielded more and more documents about Nelly, her extended family, and the dozens of elderly friends that had made up her Viennese network, the seeds were planted for what since then has become a book-in-progress. At the heart of this book, provisionally titled *With Love from Vienna*, are the multigenerational responses to the Anschluss within one extended Viennese Jewish family and thick descriptions of the daily lives of elderly Jews left behind in Vienna as they were waiting in vain for their turn to come up to emigrate to the United States.

Trudie and Dorothy's parents, Lilly and Otto Frey, had the foresight not only to save Nelly's letters but also all other correspondence that was exchanged over the years among the members of an originally Vienna-based but now globally dispersed network of family and friends. The family archive shows that Otto, a former Viennese lawyer who eventually found a position in which he was helping new refugees in the United States through the American Friends Service Committee in Pasadena, spent all of his free time working on correspondence that he hoped would expedite the immigration of his mother-in-law. The time he was forced to spend as an inmate in Dachau, to which camp he had been deported from Vienna in the days following the so-called Kristallnacht, and the ultimate failure of his attempts to get his mother-in-law out of Vienna, presumably contributed, if only indirectly, to his untimely death in 1964, when he was only sixty years old.

Among Nelly's papers I also discovered a letter that Lilly, Trudie and Dorothy's mother, had written in September of 1941 to Elsi Josephy in Zurich. This German letter includes the following lines which I am quoting below in my English translation:

> The children were sick a lot last winter (. . .) but they are overall rather happy here. They have become 100% American and have completely lost their mother tongue. Though they had to change schools a lot (1 year in Vienna, 2 different schools in England, 1 school year in Los Angeles, 1 year in Pasadena), they are in the same grade as their peers. (. . .) The problem not only of our children but of all the refugee children – and through Otto's work we know quite a few of them – is that the young generation is very quickly and without any effort completely assimilated, whereas their parents can only slowly, and with great difficulty, adjust to the new life. Interesting and typical is also that all these children accept and consider everything "American," as good, whereas they want to forget all previous things and partly indeed have forgotten everything from the past and don't want to be reminded of it. Apparently, a common reaction.

It turns out that Lilly shouldn't have worried. Trudie never did forget the past and transmitting the family's Viennese-Jewish past to her children and grandchildren became one of her priorities during the last few years of her life. She frequently drew broad parallels between the plight of families fleeing from civil war and poverty today and what she, Dorothy and their cousin Hansi (now called John and living in Canada) had experienced in the late 1930s, when they were forcefully displaced from Vienna. As a trained historian (Trudie wrote a master's thesis in 1959 at the University of Wisconsin about the Harlem Renaissance), a writer of uplifting children's books, an educator, and a bridge between the generations, Trudie was interested in reconstructing and transmitting the past, both the past of others and, towards, the end of her life, that of her own extended family as well.

Finally, let me say that I am devoted to finishing the Nelly project so that we can remember not only Trudie, whose moral character can continue to inspire us all, but also those of her family members who perished in the Holocaust as well as all those who reached safety beyond Europe where they built new successful lives be they in the United States, in New Zealand, in Israel, in Australia, or elsewhere. I once again thank Trudie's family for their ongoing support that keeps up my spirit and means the world to me.

I spoke the words in the previous paragraph at Trudie's memorial service in fall 2017. Since then, and thanks to a residential fellowship from the United States Holocaust Memorial Museum in Washington D.C., *With Love from Vienna* has progressed a lot. While at the museum, I was able to finish a part of this project that had been particularly close to Trudie's heart: researching how and with the help of whom those of her extended family who managed to flee from Nazi Vienna had been saved. The twenty-two thousand refugee case files from the American Friends Service Committee that have been part of the museum's holding since 2002 held some of the answers that Trudie was looking for and also explained the many references to the Quakers in the collective post-Vienna family biographies. My research will uncover further forgotten rescue attempts, successful and otherwise, of Nelly's relatives, while tracing the global passages of this family's diaspora.

## Works Cited

Adler, H. G. *Theresienstadt 1941–1945: The Face of a Coerced Community*. Translated by Belinda Cooper. New York: Cambridge University Press, 2017.

Bacharach, Walter Zwi, ed. *Last Letters from the Shoah*. Jerusalem: Yad Vashem and Devora Publishing, 2004.

Laqueur, Walter. *Generation Exodus: The Fate of Young Jewish Refugees from Nazi Germany*. Waltham, MA: Brandeis University Press, 2001.

Atina Grossmann

# I Thought She Was Old, But She Was Really My Age: Tracing Desperation and Resilience in My Grandmothers' Letters from Berlin – Fragments

I know more about their persecution and murder than about their lives before 1938: two quite different bourgeois Jewish women who, like their adult children, would never have met if the Nazis had not brought my parents to Tehran. Letters and aerogrammes posted from Tehran to Berlin introduced the prospective mothers-in-law to each other as they endured in the city. They hoped for ever more unlikely reunions, exchanged the news conveyed in letters from the "Orient," and, the family story went, played four-handed piano together for as long as that was possible. As with so much else that went unsaid, there is no mention of a piano in the letters or at least those that I have managed – forced – myself to read for the first time in this pandemic fall 2020. That image, engraved in my childhood memory of haunting parental anecdotes, of two elderly women finding solace with each other and the music that had been part of their respectable German Jewish socialization, may yet appear in a correspondence that began its unlikely journey back and forth from Berlin to Tehran in the mid 1930s. So far, I have only delved into letters dating from 1939 to late 1941, texts constrained by censorship, exhaustion, fear, and the tension between projecting resilience and carefully expressing desperation. Yet – in their words and between the lines – the letters reveal sometimes surprising specific bits of everyday life in Nazi Berlin before the abyss of deportation and extermination, and in that sense still "adjacent" to the "Final Solution." They also offer hints to the personalities of two women whom I had always imagined as "old" – as grandmothers are supposed to be – but who were in fact younger than, or approaching my age now, as I poke about in boxes and folders, propelled in part by my own differently urgent questions about aging and precarity. There are so many letters, tightly typed or carefully handwritten on the fragile thin paper used when generating carbon copies for multiple readers on different continents and some postcards, even more cramped, sometimes with greetings from others who had remained behind. They all had new return addresses, the *Judenhäuser* (Jew houses) to which they had been forced to move after 1938 and revised names – "Sara" inserted onto the envelope, the texts inside signed "your mother." These missives are written to the children who had gotten away, to

https://doi.org/10.1515/9783110753295-022

sons, one in Tehran and one in Hartford, Connecticut, and to daughters, two in London and one in Tehran.

I don't know how many there are or even where, in which folders or basement file cabinets these deliberately saved pieces will turn up. It's embarrassing. I am a historian, my spouse is an archivist. I have used material from my accidental family archive in previous publications. These documents should have been scanned, digitized, and preserved a long time ago and yet I haven't touched them, not these, not the letters of mothers to adult children. I didn't want to let them out of my hands, but I also didn't want to read them. There are, after all, already so many samples of this genre. Holocaust Studies is replete with books documenting and analyzing family correspondence, and more collections are being unearthed now as the second generation reaches retirement age or older, especially right now as quarantine and "social distancing" have perversely provided space and time in which to "housekeep history," before it disappears into the trash, or at best, is carted off to an archive before one has had the chance to confront the most personal of pasts. The story of mothers writing letters of love and yearning to the younger generation that had escaped has already been extensively told, as has the heartbreaking, infuriating odyssey of the bureaucratic paper wars: the tangle of consulate and aid agency visits, affidavit and visa requests, ship bookings, money transfers, dashed hopes, missed connections. These precarious connections generally temporarily concluded with a returned letter, "addressee has moved," and then resumed postwar with the drip of search and trace information about transports east and estimated dates of death, restitution claims, and entries into various databases of the murdered, noting the end, with more or less precise details, in the crematoria of Auschwitz Birkenau, in Ghetto Riga or Theresienstadt.[1] At some point someone – was it my mother? – had photocopied letters from my father's mother. It must have been decades ago; the copies are oversized and themselves fading. Most of them are the "originals," insofar as papers that were produced in several copies with carbon paper meticulously placed into precious typewriters count as "original." Certainly, they are part of a genre, the trope of people turning into letters before they were reduced to ashes.

I distrust melodrama in renditions of the Holocaust ("reduced to ashes"!) and yet I find myself slipping into it, unnerved by my own resistance to diving into this material – into the abyss, the sadness, and the anxiety of these mothers.

---

1 See for example: Marion Kaplan, *Dignity and Despair: Jewish Life in Nazi Germany* (New York: Oxford University Press, 1996). See also, Anna Hájková and Maria von der Heydt, *Die letzten Berliner Veit Simons: Holocaust, Geschlecht und das Ende des deutsch-jüdischen Bürgertums* (Berlin: Hentrich & Hentricht, 2019).

And yes, trying, or trying not, to imagine their lives and their deaths after the letters stopped. Best therefore to let the elderly ladies speak for themselves. They, as I discovered once I finally starting reading, eschew melodrama and self-pity, even as they are totally clear about how much they depend on the letters and, crucially, what my Grandmother Toni, my mother's mother, referred to as the "silent greetings" (*stumme Grüsse*) that arrived as packages of rice, and *"Fett"* – (cooking oil) that my mother and father regularly posted from Tehran.

For now, as placeholder for a longer piece I will maybe write one day, I share these fragments – and some context – from both grandmothers.

Gertrud Dewitz Grossmann was the only grandparent actually born in Berlin, on 22 November 1873. I recognize the date because it is also the day my parents chose to get married in New York City in 1947, six and a half years after my father, a refugee with a J in his German passport, had decided to leave Tehran in May 1941 heading towards Bombay and a ship to San Francisco. He would be stopped at the border to British India in Quetta, interned as a "suspect enemy alien" until December 1945, finally arriving in the United States in September 1946. Gertrud had married Eugen Grossmann, who was born into a large family of innkeepers and *Schnapps* producers in Myslowitz (Mysłowice), Upper Silesia and had made his way to the Prussian metropolis before the first World War, where he achieved some affluence in classic Jewish occupations, as a textile supplier to the Kaiser and as a landlord. Together they had three sons, Walter (1896), Hans (1902) and Franz (1904), the youngest a replacement for a much-adored daughter who had died of appendicitis as a toddler. Gertrud's husband died in 1931, though not before having settled her in the elegant apartment building he had purchased in 1913 as an old age and inheritance investment for her and his children, at Fasanen Strasse 2 in Charlottenburg, in the heart of Berlin's most Jewish commercial and cultural center, near the Zoo, the Zoo train station, and the Kurfürstendamm. Walter, a physician, emigrated with fully packed lifts to Hartford, Conn. with his wife and eight-year-old son in 1938. My father, Hans Sigismund (or Hasigro as he tried to Dadaize his Germanic names), disbarred since 1933 as non-"Aryan" lawyer and endangered by his previous work for the *Rote Hilfe*, the *KPD* (Communist Party) dominated legal aid association, fled much more precipitously and reluctantly to Tehran in 1936. It was a spontaneous escape, facilitated by an Armenian girlfriend (whose divorce he had handled), with family in Iran. And Franz, who it was always said, must have been conceived with the milkman because he was blond with blue eyes and looked not at all Jewish, was the one who stayed behind, with his "Aryan" wife, and eventually five baptized children. His "privileged marriage" status notwithstanding, he was removed from his Nazi imposed position as director of the Jewish Rothschild Hospital in Frankfurt and deported to Auschwitz shortly before his mother. He, however, would survive Auschwitz, its satellite labor camp, Monowitz, and the death march to Mauthausen, returning to his family before succumbing to physical and psychic wounds in 1953. But that is another story.

Gertrud wrote from the small two room apartment at the other end of the Kurfürstendamm in Halensee which she shared with her younger sister Erna after both of them had been kicked out of their "aryanized" homes in 1938, sometimes in a beautiful but difficult to decipher script, mostly on a typewriter (which she confessed to disliking because it seemed too impersonal but was better suited for duplication and censors), to her son in Hartford and to her son in Tehran. The surviving copies seem to be the ones that traveled to the United States. Her spirits flag, her hopes fade, but never completely. Neither does her bemused irony at her sons' sometimes clueless requests – Walter still asks her to send him some of his medical equipment – or the relish with which H (these letters, typically for correspondence from and to Nazi Germany, are populated with initials) with which her middle child settled in the "Orient" describes his adventures. "I can't really report anything about parties," she notes in one letter, for "with our extended but modest socializing, everyone brings their own food, which, however, bring us no less pleasure, when we enjoy those hours gratefully and contentedly, not thinking about what will come tomorrow. We don't think carelessly but with hope and faith in the future. Work helps distract me and I certainly have enough." She keeps herself busy and seems to maintain a packed social calendar. She visits relatives and friends, including her son's Tehran acquired but Berlin born and raised girlfriend (or might it be fiancée she wonders) Erika's mother. Toni Busse now lived with her husband Heinrich in the tiny new apartment of a *Judenhaus* in the Berlin suburb of Schlachtensee near the Grunewald,[2] to which they had been exiled after the forced sale of their newly built house in the Frege Strasse in Friedenau. My grandfather, who, unlike Toni, survived in hiding, liked to call it a villa, although I, who still visited the building as a young girl in the early 1960s after it had been restituted, thought that was a nostalgic exaggeration.

Gertrud brings news and cheer and collects the latest gossip, keeping track of who is ill, who has died, who has married, and most important, who has managed to get their papers in order and is packing for departure or perhaps already gone. She worries about her young half-Jewish grandchildren and is

---

2 The house still stands today, a small villa-like building – in a quiet affluent neighborhood. On *Judenhäuser* in Berlin, see, for example, Konrad Kwiet, "Without Neighbors: Daily Living in Judenhäuser," in *Jewish Life in Nazi Germany: Dilemmas and Responses*, eds. Francis R. Nicosia and David Scrase (New York, Oxford: Berghahn, 2010), 117–148.

delighted when one or the other comes and visits from Frankfurt. She frets about her sister and new roommate Erna whose own children have also left, her teenage daughter on a *Kindertransport* to London and a son to Amsterdam where he would go into hiding, eventually to be picked up, and deported to Auschwitz from Westerbork. She is endlessly curious about the American life of her eldest grandson in Hartford, Connecticut, complimenting his letters and drawings and sending him German books so that he should not forget his native tongue. Knowing that one family at least is in safety is a comfort and source of hope for an eventual reunion in the United States, but also a painful reminder of all that she is missing. As she wrote to her sons on 13 March 1940, noting that the mail no longer arrives in any proper order but completely haphazardly: "Still, it is important and joyous (*erfreulich*) that I hear from you at all and so much good and enjoyable news. That makes me happy," she assures them "and makes it easier to tolerate the separation." In an oft-repeated ritual of reassurance, she writes: "Thank God, I have no reason at all to complain, personally I am just fine" (*mir persönlich geht es gut*). Indeed, studying a photo of her lighting the Sabbath candles and revisiting my father's attachment to the Jewish holidays of my childhood, I have come to believe that, unlike her counterpart Toni in Schlachtensee, she was an observant Jew who did draw comfort from her faith.

On 7 November 1940, she allows a more intimate glimpse at herself in a letter to Walter:

> Last night I had the most wonderful dream. You stood right before me, took both my hands in yours, and asked – [AG: a reasonable question from someone who had never noticed his mother doing serious housework and was accustomed to the services of the family housekeeper] – "why are your fingers so raw?" and I said, "that's from all the housework," and you caressed my two little fingers. This was the first time that I dreamt about my children and saw them so vividly right in front of me. This made me very happy.

Wally (Wirth), the faithful domestic servant (*das Mädchen*), another fixture of bourgeois German Jewish life (and memory), added her greetings at the end of the letter. She apparently was so truly loyal and loving that she accompanied the two sisters to their new domicile in Halensee and seems to have been the one to arrange their flight into hiding, probably with relatives of hers, when they received their deportation orders in 1942.[3]

---

**3** ITS records show that a Wally Wirth was incarcerated in Ravensbrück as a political prisoner and released on 13 November 1943. Other lists identify her as having been granted VVN (Victim of National Socialism) status in Berlin after the war. I am grateful to Jan Lambertz, USHMM, for this information and attempting to pursue further leads.

4 December 1940, Gertrud's letter marks an upcoming early January birthday; her oldest is turning forty-five years old and is still her "dear boy": "When you left two years ago, I did not believe that our separation would last so long. Think of everything that has happened during this time!" With the dignified stoicism characteristic of so many such letters, she muses, "One thinks that one cannot get over these changes, but life goes on, and we have learned to hope and to believe. Nothing can break our calm (*nichts kann uns aus unser Ruhe bringen*), nothing can shake us anymore. We maintain the cherished habits of our social life, we celebrate birthdays, like my own recent one, with homemade cake and real coffee (with beans), a rare delicacy, stored away since last year." With apparent cluelessness about the dire reality of her situation, her son in Tehran quickly responds by commenting hopefully (with, I wonder, what degree of skepticism) on how cheerful her letters were, including the sprightly rendition of her birthday party. It is the asides, the details that accompany the well-worn phrases about how well things are going, that reveal something of the everyday reality, a world in which coffee beans are saved for a year in order to properly celebrate a birthday and rough hands are just the tips of aging bodies worn by worry, deprivation, and endless futile errands associated with emigration efforts.

She does not break away from her carefully mounted cheer for long, "45 years, I can hardly believe it and I can still see you today, a feisty often all too feisty boy just like Michi [her grandson Michael], always frolicking and roaming about." And she is still, the ever-proud Mama. Erika's less housewifely mother Toni would refer to her with, I suspect, a bit of condescension as the "Gross-Mama," a play on her probably envied grandmother status as well as on her surname. Mama G. sallies forth, "And still I said to you every birthday: Stay as you are, I love all of your shortcomings . . . You will certainly receive many beautiful gifts, it is only me who comes to you with empty hands, but with an overflowing heart, thinking of the old and the new times. Perhaps there will still be a chance to see each other again, even if" – the realities do intrude as much as she quickly bats them away in her invocations of planned reunions – "the prospects are at this time very meager. That opportunity seems viable only for those who made arrangements for their departure before the outbreak of war, and even then, only for a few of them. Ultimately, things are still pretty good for us here, and so long as nothing unexpected happens" – the inserted warning – "we can be content."

Gertrud's 4 December 1940 greeting is written a year after a 3 January 1939 letter in which she had expressed relief that Walter and his family have settled comfortably into the new world, ready to revive his (but not, it should be noted, as with many female refugee physicians, his pediatrician wife Hede's) medical practice and nestled among a familiar community of fellow refugees and sympathetic authorities. At the same time, she hints that her youngest son Franz's apparent lack of urgency in pushing for his own emigration – hampered no doubt by his large family – occasionally drives her to "despair." Indeed, the Catholic "Aryan" daughter-in-law, practically deified in my family as the "good German" who refused to desert her husband even when pressed to do so by Nazi officials and family in order to spare her children, seems to be much more active than he is in seeking escape. Now in late 1940, Gertrud writes to "my beloved children," having made a copy for Hans in Tehran, as well as for Walter, to confess that "I am very sad once again to have had no news from you for such a long time." She remains eager to hear as much as possible about Walter's "new start" in the United States, persisting in the belief that he is paving the way for her to follow. Even at the relatively advanced age of forty-four he will, she is convinced, be successful in rebuilding a medical practice. Moreover, referencing sentiments that she admits at other points are more ambivalent and filled with envy as well as relief, she adds that her joy is shared among all her friends who are glad to know that another person is now in safety.

And as always, she stresses that she is just fine. The winter is severe but the small apartment she shares with her sister in Halensee is well-heated, and they have enough warm clothes. As a counterpoint she marks the welcome arrival of packages from Tehran and from other friends and relatives in Holland. She has been to the cinema several times and also to a very good concert. Whether these are, as is likely at a time when Jews are denied entry to non-Jewish cultural events, Jewish *Kulturbund* productions[4] is not further noted. Family and friends continue to provide the requisite distraction to fend off despair. Franz, her converted youngest son, has become *Chefarzt* in "his" hospital, a dubious privilege since it is precisely his role in the Jewish Hospital Frankfurt that would later make him particularly vulnerable to arrest and deportation in 1943. She can't help but be proud. What, she exclaims, would his late father whose insistent ambition it was to have his three sons enter professions rather than business (and the two doctors and one lawyer obliged) have said! At the same time, she finds it hard to conceal her irritation at the son who is geographically closest to her, still in Germany, albeit in Frankfurt, but in attitude perhaps the most

---

**4** The Jewish cultural association *Jüdischer Kulturbund*, was established in 1933 when Nazi rule expelled Jews from German cultural life. Despite ever tighter Gestapo control, especially after *Kristallnacht* in November 1938, the *Kulturbund* offered respite for Jews remaining in Germany and employment for Jewish performers until its final dissolution in September 1941. See, for example, Eike Geisel and Henryk M. Broder, *Premiere und Pogrom: Der Jüdische Kulturbund 1933–1941* (Berlin: Siedler Verlag, 1992).

distant. She worries that his large (Catholic, after all) family and perhaps his responsible position hampers his efforts to "get away (*fortzukommen*)." But she professes to believe, as she states in several postcards over the next few months, that eventually he will succeed and also depart. In the meantime, he comes to visit, bringing as a special present his oldest daughter Maya for a "lovely" four-day visit and a continual stream of supplies to which he has better access than she does. Indeed, the repeated references to receipt of packages are the clearest clue to the hardships of (not yet total) wartime for Jews in Germany.

By 8 June 1941, Gertrud's mood has noticeably darkened as the possibilities for escape become more and more unlikely. She prefaces bad news with the usual assurance that "I am still doing fine (*unverändert gut*)," and then matter-of-factly reveals that at her latest visit to the Jewish community aid office (*Hilfsverein*) she discovered that the steamer with which she was to sail to the United States on 16 January 1942 had been taken out of service. By January 1942, few if any American ships would be sailing from Europe, but she cannot yet know that. She immediately set off for the American Express office – presumably the travel agency – but could not get an appointment before the end of the week. She had immediately written to her port of embarkation, "Lisbon," and requested an earlier travel date.[5] The addressee was not further identified, perhaps it was the shipping company or the JDC (Jewish Joint Distribution Committee). She remarks, with resigned realism, that she has little hope that any of her efforts will work. "These are things," she notes stoically, "which one must expect, and I don't let them crush me." She tries for a touch of resilient sarcasm with a familiar expression, "We are used to misery (*wir sind Kummer gewöhnt*)." There is no point in trying the United States Consulate again– still open until July 1941 – it issues visas three months after the affidavit is presented, but any visa is invalidated after four months, and she doesn't know when she will get a new booking. Hinting toward another fact of life for bourgeois Berlin Jews who might still have some resources, she plans to hire an "adviser" (*Berater*), so that she doesn't have to deal with this paper war all by herself. Clearly, she is most upset about the lack of news from Hans. She had become so dependent on his weekly letters from Tehran (also part of the family archive) and, pleased as she is about his apparently finally approved journey to the United States, the lack of regular news makes her terribly anxious – with, as

---

5 See Marion Kaplan, *Hitler's Jewish Refugees: Hope and Anxiety in Portugal* (New Haven, CT: Yale University Press 2020). Portuguese ships did continue to sail; the Serpa Pinto still docked in Staten Island and Philadelphia (communication from Marion Kaplan). JDC records show that the Serpa Pinto sailed from Lisbon with destination Jamaica on 24 January 1942. See https://archives.jdc.org/the-ss-serpa-pinto-lists-a-resource-for-genealogy-research.

it turns out, good reason. For now, she hopes for the best and thinks that he will pave the way for her (be her *Schrittmacher*). She continues, as instructed, to send copies of her letters to Hans's old Tehran address as well as to Walter in Hartford– she wonders why – but perhaps he sensed that his journey might not go smoothly. And again, in a rollercoaster of emotions she adds, "My departure gives me headaches (*Kopfzerbrechen*); conditions here are becoming more and more difficult. I can't really prepare much, I wait (*harre*) and hope."

On 20 June 1941, two days before the German invasion of the Soviet Union, she posts a letter to Walter which concedes the inescapable gravity of her situation: "my mood while writing these lines is not exactly the best; dark clouds are hanging over my head because all possibilities of getting out of here in the immediate or nearer future are illusory. This waiting with no end in sight is really nerve-wracking (*entnervend*). I look into the future with great anxiety and envy the many who have still made it [out] in the last months." She seems to recognize that the doors for escape are about to, or have already closed, although as her next letters show, she refuses to give up hope. She imagines that her son Hans, who had departed Tehran in May, leaving his relationship with my mother, Gertrud's assumed future daughter-in-law, in tense limbo, will soon arrive in the United States. Optimistic about his voyage to the United States, she is mostly disappointed that, in the tangle of bureaucratic misses, her planned corollary emigration had not succeeded. He had still been sending mail from his journey to the Indian border and remarkably those letters arrived relatively steadily. In her next letter, she expresses her mounting anxiety about the unexplained cessation of communication. A letter from Meshed in eastern Iran, which he had reached after a thirty-hour bus journey from Tehran brought the last detailed news. An unexpected aspect of the Berlin side of the Iran/German correspondence is the glimpses it offers into the circumstances of refugee lives in Iran. Hans had reported, with evident pleasure, that he had been hosted in Meshed by two other German Jewish refugees, both of them physicians, one the director of the local hospital and another the staff radiologist.

Finally, there was a postcard from Zahedan in Baluchistan. Hans was some two hundred kilometers from the British Indian border in Quetta (now Pakistan) where he would be detained, an internment that would last five years. In the next paragraph of Gertrud's note to Walter she changes the subject and reports on the death of a friend who "shared the horrific (*grausam*) fate of so many parents who had to die so far away from their children." Berlin was, as historian Marion Kaplan has so powerfully described, a city with many Jewish mothers left behind, mourning, and comforted by, the absence of their children who had managed to escape. Gertrud concludes with the confession that she is depressed by her entrapment: "All my efforts (*Angelegenheiten*) are no longer relevant."

The ship with which she had hoped to leave on 16 January has been taken out of service; the company offered her the possibility of a bunk on 2 January, but it is way too late to try and reorganize all her exit papers for that earlier date. She is stuck. "Today," she writes starkly, "I am very down [using the English expression]."

7 July 1941, a postcard to Hartford addressed to "My beloved children (*Meine geliebten Kinder*)," indicates that she still hopes H. has caught his ship in Bombay and made it to his brother's home in Connecticut. She repeats, in euphemistic terms, that her affairs are suspended for the moment (*Meine Angelegenheiten ruhen*, they're "resting"). She betrays her frustration only in the remark, "that at the moment there is nothing to be done." In the understated prescription that underlies every letter and probably every moment of her day, she writes, "One has to stay calm, (*seine Nerven behalten*), which is often not easy." It was the date on which Hans was scheduled to arrive in New York after his overland, trans-Pacific sea voyage and then transcontinental train trip from San Francisco to New York. In fact, matters had taken a sharply different turn although it is unclear what she will still learn about her middle son's fate. Presumably she did somehow keep track of the family even while hiding under an assumed name in the picturesque lakeside village of Lychen in Brandenburg. She does not yet know, that not only her hopes for her son's fate but also her fantasy of being able to escape to the United States once she has two sons there, will be so thoroughly dashed. She is waiting – with great yearning (*sehnsüchtig*) – for the card he had promised to send confirming that he had arrived healthy and on time from Bombay. In fact, he had never made it further than the Iranian-Indian border at Quetta.

Turning to news of her problematic youngest son in Frankfurt, Gertrud reports her shock that he had just on 29 May become the father, with his good Catholic wife, of a fifth child. She had been kept in the dark about the impending arrival of this additional half-Jewish child in spring of 1941, further cutting off any possible escape opportunities for them. She is clearly angered by this turn of events to which she had not been privy, lest she worry, Franz assures her, and had only been informed after the fact, presented with *a fait accompli (vollendeten Tatsachen)*. And then, perhaps because she is feeling particularly anxious, awaiting the longed-for news that Hans had indeed made it to New York, she adds, "I myself don't know what will happen to me here, everything is murky and uncertain." Unlike Toni, who seems to strictly avoid mention of external events and with whom her relationship had become more bewildering now that their children seemed to have separated, Gertrud takes note, while enumerating the letters she has sent and received, that they had all been posted before the war with Russia had broken out; now in July, the situation has clearly become even more threatening and uncertain. But then she quickly reverts to form: "About myself I can only report good news."

3 August 1941 the war on the Eastern Front is in full swing and true to a culture in which birthdays were almost sacred events, she writes to Hartford in hopes of finding my father Hans there on his birthday, 26 August. She has heard nothing, is increasingly apprehensive about the total absence of reassuring news, and suffering from the uncertainty but also the loss of regular communication which had actually been easier with Tehran than with Hartford. "I can't tell you how much your so precious weekly letters have meant to me over the past five years," Gertrud confesses. She has become more open about her barely contained despair, writing with great longing and pain (*Sehnsucht und Wehmut*) about the long separation from her "beloved boy" (*mein geliebter Junge*) – who is about to turn thirty-nine – and indeed under circumstances of which she is unaware and in a place much less desirable than Tehran which he had often with affectionate frustration, declared to be "no place for a Berlin lawyer" (*kein Ort für einen Berliner Rechtsanwalt*). The darkness is more visible in this, one of the last letters that has been preserved. Yet she refuses, at least in her correspondence – I have no sense of how she comported herself in her ever more circumscribed daily life but suspect that the mixture of hope and despair was not much different – to yield to hopelessness. She had hoped to be reunited with her sons by then. Somehow, she had convinced herself that if both were in the United States, they would be able to get her there as well. "Sadly," however, "fate would have it differently, we must tolerate everything (*gedulden*) and not let our courage wane. The day will come . . . ."

As if she knew that this might be one of her last communications to her missing middle son, Gertrud writes, "Even if I leave so much unsaid, you know what you mean to me, what heartfelt (*innige*) wishes for your admittedly not easy new start I carry in my heart. But in that regard, I have no worries, you will manage, you have always mastered the situations in which you were placed which thank God I can say about all my children. And that is why I see my future with hope and without anxiety (*ohne Zagen*) in contrast to so many parents here who believe that they are a burden to their children. I know that I am safe (*geborgen*) with you. And that is why I do not let my courage sink and hope for the day when we will all be together again. It will come and it must come. Just stay healthy and don't worry about me."

On 3 August 1941, her son Hans was nowhere near Hartford, Connecticut but rather imprisoned as a "suspect enemy alien" in Central Internment Camp Deolali, ca. one hundred fifty kilometers north of Bombay. The camp, as he later reported, consisted of a "large row of straw barracks surrounded by high barbed wire" and was guarded by armed Indian soldiers under British command. He was safe from the Nazis, but nowhere near freedom, and in no position to try and fulfill his mother's dream. His mother for her part was thinking of him on the 26th, praying to "dear God" – in whom I think she, unlike Toni, did believe – that he should remain "healthy and strong, just as I am trying to be, so that we can, just as we once did in the Fasanen Strasse, all be together in love and harmony." She adds "Greetings and kisses from Erna (Gertrud's sister) and Wally (the family housekeeper who has become their support and protector)." Once again, she concludes, "my dear, faraway birthday child, be embraced in *grosser Herzlichkeit*, Your mother."

Less than a year later, Gertrud and Erna eluded the deportation order that had been sent to their Katharinen Strasse 2 address on 17 June 1942 by, as the rules had it, the rump Jewish community. At age sixty-nine, one of the few "elderly" who managed to go "underground," she slipped into hiding in Lychen, probably with the help of Wally, the family factotum.[6] Some ten months later, the sisters were denounced under unclear circumstances, arrested, and transported back to Berlin to the assembly center in the Grosse Hamburger Strasse 26 in Berlin-Mittte. There, like all Berlin Jews, they were forced to fill out the *Vermögenserklärung* (financial statement) in which Gertrud dutifully noted that she no longer had any assets but did have two sons living abroad. On 28 June 1943, the 39th Ost Transport carried her through her native Berlin from the Grunewald train station where the deportation is today commemorated with a metal plaque on the platform. 317 Berlin Jews were taken to Auschwitz on that transport. Too old for slave labor in the camp. she was presumably selected for immediate death in the gas chambers. Her sister Erna followed soon after; having lived together since being evicted from their homes in 1938 and taken the rare step – for older people – of going into hiding together, they were separated for their final horrific journey. Her youngest son, the "Aryan" looking Catholic convert with the five baptized children who lived in an ostensibly "privileged" mixed marriage might well have escaped deportation via forced labor had he returned to Berlin as his worried mother implored him to do. In Frankfurt/M whose Gestapo was notorious for its particular viciousness, he was arrested, apparently due to his prominent role as director of the Jewish Hospital. He too was deported to Auschwitz where he managed to survive as

---

**6** For one younger woman's powerful hiding story, see Marie Jalowicz Simon, *Underground in Berlin: A Young Woman's Extraordinary Tale of Survival in the Heart of Nazi Germany*, ed. Hermann Simon (Boston: Little, Brown, 2014), translated from the German edition (2014) by Anthea Bell. The records I have accessed so far do not indicate when Wally was taken to Ravensbrück so it is not clear how closely her arrest followed Gertrud's and Erna's escape to Lychen which is a mere 3 kilometers from Fürstenberg where the Ravensbrück KZ was located.

an inmate physician in the satellite slave labor camp Monowitz, and then on the death march to Mauthausen. Liberated by United States troops and after a lengthy convalescence from typhus, he returned to his family, now living in Limburg after their residence in Frankfurt was destroyed by Allied bombs in July 1945. His children were told by their mother that their father, like so many fathers in postwar Germany, had suffered terribly during the war and that they were not to disturb him with any further questions.

The first inquiry about Gertrud's fate, one of so many in a world filled with people searching for lost loved ones, came from British India, submitted by her lawyer son, my father, on 26 November 1945. "Enquirer's name" is listed as Dr. Hans Gross (the "mann" got dropped) with a return address of P. O. Box 47 (the postal address of the Jewish Relief Association), Bombay, India and seems to have been transmitted via the British Red Cross. In fact, Hans had not even made it to provisional freedom in Bombay yet and was still interned in his final camp at Purandhar near Poona when he posted the inquiry. He would finally be released on parole (pending organizing his papers for departure to the United States) on 9 December 1945, six months, as he bitterly noted, after the war in Europe had ended. The next equally unanswered search request was submitted by Gertrud's middle son Franz in 1947 after his recovery and return to medical practice in the peaceful, American-controlled Frankfurt suburb of Bad Homburg. She had, he knew, been deported from Lychen where she had been living under a false name, either to Theresienstadt or "Polen" while he laconically adds, "I myself was in the KL Auschwitz." He had last heard from her in February 1943. On 15 December 1950, Dr. med. Franz Grossmann requested via an attorney in Frankfurt on behalf of himself and his two brothers a confirmation of death from the International Tracing Service [ITS], so that they could pursue their restitution claims for the house on Fasanen Strasse from which Gertrud, the "German Jewess" as she was designated in the legal documents, had been expelled in 1938.[7]

On 11 December 1950 – the order of dates is somewhat confusing – Military Government authorities informed the Berlin court (*Amtsgericht*) that the Gestapo had listed Gertrud on 10 August 1942 as having fled her deportation order, but that no further information proving her demise was available. At almost the same time, the reconstituted Jewish community in Berlin advises her sons that their mother was listed as having gone into hiding or had "chosen" –

---

7 For a description of the Fasanenstrasse Hotel Astoria restitution case see my *Jews, Germans, and Allies: Close Encounters in Occupied Germany* (Princeton, NJ: Princeton University Press, 2007).

like no small number of Berlin Jews facing deportation – to take matters into her own hands, via a very euphemistically named voluntary death (*Freitod*). On 18 December 1950 a notice from the Tracing Service in Bad Arolsen rather interestingly lists Gertrud's last known address, not as is the case with all the other postwar documents, as the Katharinenstrasse, where she and Erna had received their deportation order but as Vogelgesang (Birdsong) Strasse 470 in Lychen. On 1 January 1951, the Tracing Service reiterates that in the absence of further information, there is no basis for a death certificate.

On 19 January 1951, a singularly unhelpful ITS informs Franz's lawyers that due to the lack of details, they cannot certify a death but should rather contact the Polish or Czechoslovak Red Cross. Finally in 1957, the ITS, still using Nazi language, wrote to my father, the by now only surviving son, at his law office in lower Manhattan, 225 Broadway. Gertrud Dewitz Grossmann had been "evakuiert" to KL Auschwitz on 28 June 1943 on the 39th Ost Transport, one of 317 Jews on the train out of Grunewald Station, not far from her son's girlfriend's mother's apartment in Schlachtensee, and without her sister Erna from whom she was separated at the Assembly Center, denied the possibility of spending their last days and hours together. The circumstances of Gertrud's denunciation and arrest remain murky to this day. When in 2002 I unearthed the "Financial Disclosure Document" (*Vermögenserklärung*) she had been forced to fill out while awaiting transport at the Grosse Hamburger Strasse Assembly Center and other documents relating to her precipitous disappearance from the Halensee apartment on 17 June 1942 (such as unpaid gas bills or the auction list for items that had been left behind),[8] I was told that the relevant Gestapo records for Lychen had been lost in the war. She did not have many assets left, and those she had managed to retain were confiscated for the expenses that a Jew marked for death would incur in Nazi Germany: 1) *Judenvermögensabgabe* (transfer of Jewish wealth), 10,000RM with a further 2650RM charged in interest – evidently she had not paid quickly enough; 2) *Sicherheit für Reichsfluchtsteuer* (Security deposit against flight), 14,000 RM; and finally 3) *Sonderbeitrag für Deportierung* (special fee for deportation expenses), a comparatively modest 600RM plus 107.90RM. In the end, the Berlin Restitution Court awarded her heirs 20,000RM for the Fasanen Strasse building, to be split three ways plus legal expenses.

In 1971, the ITS delivered its final report, confirming Gertrud's fate as well as that of my maternal grandmother, Toni Bernhard Busse. Toni had been born in the venerable German Jewish center of Fürth, near Nürnberg on 6 April 1884 and had moved to Berlin when she married Heinrich Busse, in

---

**8** See OFP files, Landesarchiv Brandenburg, Potsdam.

what her Franconian family considered a highly unfortunate mismatch to a *Saupreusse* (Prussian pig). The couple had three daughters, including my mother Erika, like my father, a middle child. In the end, the two mothers in Berlin, who would never have met under ordinary circumstances just as their children would have been most unlikely to have crossed paths, were brought together by the romance between their offspring. And as the days became darker and Toni was occupied with forced labor at the Siemens Schuckert factory, they would still find time to visit each other to share letters and news.[9]

Toni was younger, less conventional, an active partner in many of her husband's business ventures, and definitely less observant. But, like Gertrud, she was committed to maintaining a brave and resilient posture in her letters. In an 8 June 1941 letter to Tehran, she expresses concern about her daughter's feelings when her supposed fiancé takes off for his hoped-for emigration to the United States, this being the "first time in years that you are without a steady companion," but quickly adds, "you don't have to worry about me." She nonetheless cannot suppress some jealousy that the "lucky" Mama G would soon be en route to joining two of her children in a new and safe life. On 1 July 1941, while Gertrud is wracked with anxiety about her son's whereabouts, Erika's mother states, "But all thinking and worrying is useless, and I will already be satisfied if you always take care of yourself well (*gut ernähren* in the double sense of being well fed and gainfully, independently employed) and can organize your life according to your taste," even if she wistfully admits that she hopes her daughter will not have to walk that path alone. Toni's sadness and loneliness slip through: "*Ach*, there is so much more I would like to know, but what's the point. Until you can respond it's all passé anyway. So, one has to work everything out with oneself and sometimes that it is quite depressing (*und da ist's Einem oft recht schwer zumute*)." Taking her cues from her World War I veteran and former amateur gymnast husband who loved to quote the German adage that "there is no such thing as bad weather, only the wrong clothing," she briskly adds, "Today we still want to take a serious walk after several rainy days, that is the best medicine against one's thoughts, don't you think?" And then she qualifies that again by noting, "almost every path does remind me of you." After all, both she and Mama G agree that the relationship between mothers and daughters is closer than with sons. Her thanks on 8 December 1940 for the "silent greetings" of rice, lentils, beans, peas, and soap – she would so like

---

**9** For an account of women's forced labor for Siemens in Berlin, see Carola Sachse, ed., *"Als Zwangsarbeiterin in Berlin," Die Aufzeichnungen der Volkswirtin Elisabeth Freund* (Berlin: Akadamie Verlag, 1996).

some butter – indirectly betray the hardship and hunger that are part of daily Jewish life. Then she swings around again to the obdurately positive (with perhaps a touch of irony): how lucky that they had to give up their big house and move into a tiny *Judenhaus* flat – now she no longer has to climb stairs, and the building is in a lovely neighborhood, even if it is very crowded and the enforced neighbors rather too talkative.

29 January 1941, Toni has to force herself to write. She cannot bridge the distance, is tormented by how hard it is to communicate when she really does not know what the other is thinking. (Admittedly, my mother was particularly elusive.) Other sets of letters reveal that her husband, Heinrich – who survived as an "illegal" in Berlin – was energetically and with never flagging hope, scheming to find an exit, corresponding with relatives and friends or acquaintances of friends all over the world from Argentina and Uruguay to Shanghai and South Africa, England, Palestine, Switzerland, and the United States. His wife mentions none of this, even in coded terms. She does continually hint at exit routes for her daughter, suggesting that she might follow her errant partner to the United States. or elsewhere, not quite aware, it would seem, of Erika's guilty bliss at fulfilling her fantasy of living in Iran, an adventurer in the desert and bazaars.

Gertrud never seemed to give up clinging to some hope of reunion, practicing her English, vowing not to be a burden, while Toni had concluded by 1940 that there would be no escape, without fully knowing what that means. Toni Bernhardt Busse was deported in the *Fabrik Aktion* from the Siemens-Schuckert plant where she had been doing forced labor on either 28 February or 1 March 1943, and in the words of the ITS "perished" (*umgekommen)* in Auschwitz.

Somewhat to my surprise, my contribution to the theme of a volume exploring what it means for scholars examining genocidal violence in twentieth century history to be "adjacent to [current] violence" in a year marked by COVID, Trump, police murders, and the Black Lives Matter mobilization turns out to be a sense that it was these carefully composed letters by two Jewish women in Nazi Berlin that appear as most immediately "adjacent to violence" – and certainly not myself, as the historian/reader in 2020. There is no direct violence in their letters, no Nazis, no war, no bombs, no active persecution: just requests and thanks for food parcels, worries about their children's confusing love lives and health, a hectic calendar of letter writing and social visits to a long list of initials (some of which I recognize or can guess at, most of which will remain indecipherable). They are, in point of fact, "adjacent" to violence. It hasn't hit them yet. Not really. They have been moved to *Judenhäuser,* but they are still sleeping in beds in their hometown. The real violence is elided or comes later, off the page, off screen, only circumscribed by the grotesque legalese of search and trace queries and restitution proceedings, as well as the limited – no matter how informed – imagination of the reader.

# Works Cited

Geisel, Eike and Henryk M. Broder. *Premiere und Pogrom: Der Jüdische Kulturbund 1933–1941.* Berlin: Siedler Verlag, 1992.

Grossmann, Atina. *Jews, Germans, and Allies: Close Encounters in Occupied Germany.* Princeton, NJ: Princeton University Press, 2007.

Hájkova, Anna and Maria von der Heydt. *Die letzten Berliner Veit Simons: Holocaust, Geschlecht und das Ende des deutsch-jüdischen Bürgertums.* Berlin: Hentrich & Hentricht, 2019.

Kaplan, Marion. *Dignity and Despair: Jewish Life in Nazi Germany.* New York: Oxford University Press, 1996.

Kwiet, Konrad. "Without Neighbors: Daily Living in Judenhäuser." In *Jewish Life in Nazi Germany: Dilemmas and Responses.* Edited by Francis R. Nicosia and David Scrase. New York, Oxford: Berghahn, 2010. 117–148.

Sachse, Carola, ed. *"Als Zwangsarbeiterin in Berlin." Die Aufzeichnungen der Volkswirtin Elisabeth Freund.* Berlin: Akadamie Verlag, 1996.

Simon, Marie Jalowicz. *Underground in Berlin: A Young Woman's Extraordinary Tale of Survival in the Heart of Nazi Germany.* Edited by Hermann Simon. Translated by Anthea Bell. Boston: Little, Brown, 2014.

Thomas Kühne

# A Father, a Perpetrator, a Son
## Autobiographical Thoughts on Mystery and Curiosity

When I was growing up in Germany in the late 1960s and early 1970s, my dad used to send a Christmas package to an inmate of the Hohenasperg, a medieval fortress and then a well-known political prison.[1] Once in a while, my dad would also pay him a visit; the Hohenasperg was located north of the city of Stuttgart and about a ninety-minute drive from the Swabian town of Nagold, where we lived. As a soldier with the SS in the past war, the man had been tragically involved in bad things, or so we were told, for which he was tried in the 1960s and sentenced to life in prison. During the 1950s, while trying to establish himself professionally, my father contributed articles to the journal for interior design that this man edited. At that time, said man had lived under a pseudonym, Alfred Ruppert (instead of Albert Rapp), clandestinely managing to support his wife and three sons. He was a good man, according to my dad. Many bad things had happened in that war. Talking about them wouldn't make them any better, or so claimed common wisdom.

As a kid, I wasn't too interested in this man's history or my father's connection to him, nor did I know much about the SS other than that it embodied some sort of ultimate evil. The story about Rapp's two names disturbed me, and the image of a father eking out his life in a dungeon disquieted me. In my childish imagination, nurtured by all sorts of adventure literature, Rapp lay abandoned in a cold medieval cell, shackled to a stone wall. He died in 1975, about when my parents divorced. My dad moved away, and nobody talked about it anymore. At least not for a long time.

## The Father

Only the Hollywood TV miniseries *Holocaust*, watched in January 1979 by roughly a third of the West German population, refocused public interest in

---

1 Horst Brandstätter, *Asperg, Ein deutsches Gefängnis* (Berlin: Klaus Wagenbach, 1978). – I wish to express my gratitude to Thomas Kohut, Cornelia Rauh, and Mary Jane Rein for providing important feedback on an earlier version of this chapter.

https://doi.org/10.1515/9783110753295-023

the Nazi past and World War II on the murder of the Jews.[2] Before then, it wasn't the murder of the Jews that concerned Germans when they dealt with this past but rather, what the war had done to them, the Germans, including my father and his family.[3] My dad's home in Braunschweig and all of the family's belongings were destroyed by the British bombing raid in October 1944. His older brother was killed in action in spring 1945 in the Riga area. In the 1960s, war damage was still widely visible. Bombed cities such as Stuttgart or Cologne, where I was born, were not fully restored yet. Much of the damage was hidden. Most disturbing to me was the ubiquitous sight of disabled ex-servicemen – men with one leg, one arm (or a prothesis) or distorting face wounds and no voice; men sitting in wheelchairs; blind men with sticks and service dogs. These wounds were visible not only in public but even more so privately where I faced them twice a week, distributing a clerical newspaper for the elderly in my neighborhood.

My father was not one of the disabled. Yet the war weighed heavily on him. Drafted into the Wehrmacht in fall 1943, he was torn out of a happy youth (as he insisted) at the age of eighteen. He steadfastly refused to talk about his time as a soldier, except that it was the worst part of his life, apparently no better than the fourteen months he endured, increasingly sick with tuberculosis, in Soviet captivity after the surrender of his division to the Red Army in May 1945 in the Courland Pocket. I never knew him other than as fully despising any type of uniform, not only those of the new West German army, the Bundeswehr, but also those of policemen and even fire fighters. To him, a uniform and certainly the military, embodied the institutionalized betrayal of humankind.

And not only uniforms kept the trauma of the war alive. The Nazi past also mattered as the root of authoritarian mindsets that long prevailed in Germany. Coming from a social democratic family and having managed to avoid the Hitler Youth even at a time (1939–1943) when it was quasi-mandatory for boys of his age, my dad disdained, in fact was obsessed with, the dominance of *Altnazis*, or Old Nazis, in society through the 1980s. In 1975, he was bullied out of a company that was run by former Waffen SS men. He was not on speaking terms with an uncle of mine, the husband of a third cousin of my mother (and her only relative, apart from her parents). "Uncle Rolf" was a former Wehrmacht lieutenant and the epitome of a warhorse. Despite having lost a leg in the war, he prided himself on playing tennis and downhill skiing with his prothesis better than he would have with

2 Sandra Schulz, "Film und Fernsehen als Medien der gesellschaftlichen Vergegenwärtigung des Holocaust. Die deutsche Erstausstrahlung der US-amerikanischen Fernsehserie 'Holocaust' im Jahre 1979," *Historical Social Research*, 32, no. 1 (2007): 189–249.
3 Robert G. Moeller, *War Stories, The Search for a Usable Past in the Federal Republic of Germany* (Berkeley: University of California Press, 2001).

his natural leg. And he only needed a glass of wine or two to bemoan that the Third Reich hadn't completed the annihilation of the Jews. Later, my dad broke with the family of his second wife, an Austrian whose mother I remember as matching perfectly the cliché of a Nazi women (*Nazisse*, in German), both in terms of physical appearance and blatant antisemitism.

## The Perpetrator

And yet there was Albert Rapp. As a youngster I had no clue what precisely he had done, although it dawned on me that he could not have been just a simple rank and file soldier, coerced into the army, as my father had been. As my relationship to my father was distant, I wasn't too curious about his life stories including this mysterious connection, and he wasn't interested in talking about them either. My early work as a historian did not focus on the SS or other core groups of Nazi perpetrators. In college in the 1980s, I was interested in debates about Imperial Germany's potential for democratization; my 1992 dissertation explored the lack thereof in Prussia before 1914.[4] The German debate on the crimes of the Wehrmacht in the mid-1990s[5] motivated my next project, a gendered perspective on its rank-and-file as bystanders to the Holocaust. My father's contempt of soldiers may have influenced this choice.[6] My own political upbringing was rooted in West German pacifism and antimilitarism, at a time when my hometown of Nagold was known primarily for its military base, where in 1963 a recruit had been trained to death using methods preferred by Nazi elite troops – one of the early spectacular Bundeswehr scandals.[7]

Rapp returned to my mind only after I emigrated to the United States in 2004 to join Clark University's program in Holocaust and Genocide Studies and began pondering a new book project – more than ever before thinking, questioning,

4 Thomas Kühne, *Dreiklassenwahlrecht und Wahlkultur in Preußen 1867–1914. Landtagswahlen zwischen korporativer Tradition und politischem Massenmarkt* (Düsseldorf: Droste, 1994).
5 Thomas Kühne, "Der nationalsozialistische Vernichtungskrieg und die 'ganz normalen' Deutschen. Forschungsprobleme und Forschungstendenzen der Gesellschaftsgeschichte des Zweiten Weltkriegs. Erster Teil," *Archiv für Sozialgeschichte* 39 (1999): 580–662.
6 Thomas Kühne, *The Rise and Fall of Comradeship: Hitler's Soldiers, Male Bonding and Mass Violence in the Twentieth Century* (Cambridge: Cambridge University Press, 2017), 3–4.
7 "Tiefste Gangart," *Der Spiegel*, 13 November 1963: 52–56; cf. Thomas Kühne and Walter Binder Jr., "Zum 'Tag der offenen Tür' auf dem Eisberg," *Schwarzwälder Bote* (Nagold edition), 20 September 1974.

trying to make sense of my German identity and the personal history of it.[8] I felt reminded of that mysterious connection and asked my father for the name, which I had forgotten. He gave it, pretending that he didn't remember any details of Rapp's activities, except that Rapp had been sentenced for shooting five "gypsies" in Russia. But this terrible act hadn't been his fault; he had only followed orders. Rapp's hiding under a false name did not give my dad much cause for concern either. Rapp had done so to support his family.

The literature on the Holocaust and the SS didn't yet reveal much about Rapp's crimes and personality, so I decided to check the trial records in the Ludwigsburg Central Archive for Nazi Trials.[9] I learned a lot. From 1942 to early 1943, Albert Rapp had been the leader of SS Sonderkommando 7a, one of the mobile units of Einsatzgruppe A that was responsible for the murder of more than 1.2 million Jews and others in the northern regions of the Nazi-conquered part of the Soviet Union. In 1960, Rapp aka Ruppert came into the crosshairs of the West German justice system. He was arrested in 1961 and sentenced to life in 1965, not for killing five "gypsies," but for the "collaborative murder of one thousand one hundred and eighty individuals."[10]

Rapp was born in the Swabian town of Schorndorf in 1908. After earning an architectural degree, he studied law in Munich and Tübingen, a hotbed of Nazi academics. In 1936, he passed the bar exam but not to become a lawyer. Instead, he instantly joined the Security (i.e., intelligence) Service (SD) of the SS. Already as a teenager in the 1920s, he had been engaged in far rightist youth groups. In 1931, he had joined the NSDAP, then the SA where he befriended numerous fellow Nazis

---

**8** For corroborating experiences, see Ursula Hegi, *Tearing the Silence. On Being German in America* (New York: Simon & Schuster, 1997).

**9** Helmut Krausnick and Hans-Heinrich Wilhelm, *Die Truppe des Weltanschauungskrieges. Die Einsatzgruppen der Sicherheitspolizei und des SD 1938–1942* (Stuttgart: Deutsche Verlags-Anstalt, 1981), 211, 645; Kerstin Freudiger, *Die juristische Aufarbeitung von NS-Verbrechen* (Tübingen: J.C.B. Mohr, 2002), 74–79. Recent contributions include Klaus-Michael Mallmann, "Lebenslänglich. Wie die Beweiskette gegen Albert Rapp geschmiedet wurde," in *Die Gestapo nach 1945. Karrieren, Konflikte, Konstruktionen,* eds. Klaus-Michael Mallmann and Andrej Angrick (Darmstadt: Wissenschaftliche Buchgesellschaft, 2009), 255–269; Christian Ingrao, *Believe and Destroy. Intellectuals in the SS War Machine* (Cambridge, MA: Polity, 2013); Stefan Klemp, "Albert Rapp: 'Du sollst Deinen Feind aus aller Seelenkraft hassen . . .,'" in *Täter, Helfer, Trittbrettfahrer,* vol. 10: *NS-Belastete aus der Region Stuttgart,* ed. Wolfgang Proske (Gersteten: Kugelberg Verlag, 2019), 354–375.

**10** Judgement Landgericht Essen, 29 March 1965, Bundesarchiv Ludwigsburg (BAL), B 162/14174. See also *Justiz und NS-Verbrechen. Sammlung deutscher Strafurteile wegen nationalsozialistischer Tötungsverbrechen,* vol. 20, eds. C. F. Rüter and D. W. de Mildt (Amsterdam: University Press Amsterdam, 1979), no. 588, 715–815.

who later became leading figures in Himmler's SS administration and in the Einsatzgruppen. In occupied Poland in fall 1939 and early 1940, Rapp was in charge (under Eichmann) of the Germanization program, the forced replacement of Poles and Jews by ethnic Germans migrating from Russia and Romania. After appointments with the SD in Germany in the following two years, he took over the command of Sonderkommando 7a in Belarusian Klincy in February 1942. Only a few weeks later he proudly reported the killing of 1657 persons including 1585 Jews. And so it continued, until he was wounded in January 1943 and returned to Germany, serving with the SS administration in Braunschweig and Berlin. Rapp was the model of a career SS officer. Embodying the ideal of the fanatically antisemitic "political soldier" who fought unconditionally and without moral restraints for the Nazi cause, he used unbridled ruthlessness against the alleged racial enemies in the East. His men despised him as a relentless commander who couldn't care less about his troops, concerned only with his own advancement.

As is well known, the West German justice system observed utmost generosity if not negligence toward Nazi perpetrators. While many were investigated at some point (172,294 through 2005), few were prosecuted (16,740), less than half of them convicted (6,656). Most of them got away with light prison sentences. Only 981 were convicted for acts of killing, out of which less than 20 percent (182) received the maximum penalty allowed under West German law, life imprisonment.[11] Whereas American jurisprudence establishes capital crimes on the basis of objective circumstances and subsequently attaches little value to the motivations of the perpetrator, the West German system applied the "subjective theory" through the 1990s. Convicting a defendant as perpetrator requires a proven interest in the deed, i.e., if they killed or ordered the killing of people to satisfy "basic motives" such as hatred of Jews. Basic motives, however, were usually difficult to prove, and more often than not judges were not overtly interested in exploring Germans' blatant antisemitism anyway. Instead, West German prosecutors routinely returned mild verdicts by categorizing even convicted mass murderers as mere adjuncts, accomplices or assistants of the actual perpetrators, the top Nazis, all of them long dead; the adjuncts were found to have committed crimes only unwillingly, typically by following orders.[12]

Rapp was one of the exceptions.[13] Put on trial, a Nazi perpetrator could usually count on their troops and comrades to testify on his behalf, for instance, by

---

**11** Guenther Lewy, *Perpetrators. The World of Holocaust Killers* (New York: Oxford University Press, 2017), 88–89.

**12** Lewy, *Perpetrators*, 90–92, 97–102; Lewy, *Perpetrators*, 105–107 on the change from 2009 on.

**13** For another blatant example, Heinrich Hamann, Gestapo leader in Novy Sacz, see Thomas Kühne, *Belonging and Genocide. Hitler's Community, 1918–1945* (New Haven: Yale University Press, 2010), 68–76, 83, 88, 93.

deflecting any suspicions of basic motives. Not so Rapp. His former underlings were more than willing to send him to his fate. When he assumed his command in Klincy, Sonderkommando 7a had just been granted a multiweek long break to recover from months of continuous service (as killers). A different SS unit had already killed most of the Jews in Klincy. But Rapp didn't want to waste time and pushed his men to track down any racial enemies of the Nazi *Volksgemeinschaft* (community of the folk/people) remaining in the area. And he didn't stop chasing his men to chase down Jews until he was wounded. To the Essen court, there was no doubt that Rapp had acted on his own initiative, in accordance with the concept of basic instincts as defined in German judicial language.

After studying the Ludwigsburg files on Rapp, I reported back to my father, who at the age of 81 was still of sound mind and even accustomed to reading academic prose on historical topics, including my own work. I explained to him that the trial investigations, based on a plethora of testimonies of his former underlings and comrades, left no doubt of Rapp's character and mindset: he was a brutal antisemite and committed Nazi who had entered the SD in 1936 out of passion; the number of 1180 murders was certainly only the minimum of what could be proven in court, whereas Rapp's record in fact must have been much worse. My dad listened patiently to my lecture of twenty or so minutes, thought about it for a moment, and then said: well, as far as I know, he was sentenced for the killing of five individuals, not Jews, "only Gypsies," and he was ordered to do so.

This was the end of the discussion with my dad on Rapp. While annoyed by his reaction, I wasn't terribly surprised. After all, the essence of his political identity was the opposition to "the" Nazis and to Nazism. The Nazis had always been the others, not he himself, not his parents, siblings, friends.[14] He was the victim of Nazism. While the concept of friendship didn't apply to his relationship to Rapp, the threat of contamination – by sustaining some support for a Nazi fanatic and perpetrator – was unbearable to my father. Rapp could not have been such a monster. My research, on the other hand, inspired by Christopher Browning's inquiries into the social psychology of perpetrator units of "ordinary men," focused on peer group dynamics, not so much on ideological fanatics such as Rapp.[15] I decided to let go of Rapp.

---

**14** This has been a popular trope, see *"Opa war kein Nazi." Nationalsozialismus und Holocaust im Familiengedächtnis*, eds. Harald Welzer, Sabine Moller and Karoline Tschugall (Frankfurt am Main: Fischer Taschenbuch Verlag, 2002).

**15** Cf. Christopher Browning, "Twenty-five Years Later," *Ordinary Men. Reserve Police Battalion 101 and the Final Solution in Poland*, rev. ed. (New York: Harper Perennial, 2017), 238–241.

## The Son

Thinking about my "adjacency to historical violence," it would be tempting to present the strange relationship between Rapp and my father as the origin of my own efforts to creep into the mindsets of Nazi soldiers, Holocaust perpetrators, bystanders, and other people willingly or unwillingly entangled in war, atrocities, genocide. Yet, it would be a stretch and a textbook illustration of what the French sociologist Pierre Bourdieu once called "l'illusion biographique."[16] When we as autobiographers or biographers tell our life stories or those of others, we assume that this "'life' constitutes a whole, a coherent and oriented ensemble, to be understood as a unitary result of a subjective and objective 'intention.'" We present lives that follow a "chronological and logical order," with an origin and a goal. When we narrate our biography, we do so as "ideologues of our own life." In order to prove the coherence of the respective life, we select certain events, experiences, decisions, successes, and failures, and we arrange them in meaningful ways.

Historians are mandated to establish causes and consequences. As a history major in college, I was intrigued by Friedrich Meinecke (1862–1954), a model of such a historian. One of the most influential nationalist historians in Imperial and Weimar Germany, he is nowadays mainly known for his 1946 book *Die Deutsche Katastrophe* (The German Catastrophe). Without simply denying the continuities from the German authoritarian state, which he adored (just as he never hid his antisemitic resentments), from Prussian militarism, German nationalism, and imperialism to the "catastrophe" of the Third Reich, the book famously managed to nonetheless preserve those traditions by tying Nazism to common non-German, European traditions such as Bolshevism. In this way, Meinecke was able to explain away the "catastrophe" as a mere accident, an "aberration" from the true path of German history.[17] Sidelined by the Nazis, Meinecke had published already in 1941 the first part of his autobiography, which sophisticatedly selected personal experiences and arranged them

---

16 Pierre Bourdieu, "L'illusion biographique," *Actes de la recherche en sciences sociales*, vol. 62–63 (1986): 69–72, 69 for the following quotes. The translations are mine. For an English translation, see Pierre Bourdieu, "The Biographical Illusion," in *Identity: A Reader*, eds. Paul Du Gay, Jessica Evans, and Peter Redman (London: Sage Publications, 2004), 297–303. For a review of the discussion, see Ricardo Altieri, "Eine Antikritik auf Bourdieus Kritik am biographischen Schreiben," in *Work in Progress, Work on Progress*, eds. Marcus Hawel et al. (Hamburg: VSA, 2019), 41–53.
17 Friedrich Meinecke, *Die deutsche Katastrophe. Betrachtungen und Erinnerungen* (Wiesbaden: Brockhaus, 1946), 9.

into inevitable continuities embedded in the glory of the German nation-building process. On one of the first pages, the reader learns that the deep "commitment to the Prussian state" of Meinecke's protestant ancestors established a fertile ground for his career as a nationalist historian even before he was born. However, it was his "first own, entirely authentic historical remembrance" that determined this vocation irrevocably. The event at the origin of this remembrance – "the roots of my historical work" – was a pompous visit of the Prussian King William I and his Prime Minister Otto von Bismarck (who shortly after engineered the foundation of German nation state) in Meinecke's hometown Salzwedel in 1865, to which he was an eyewitness at the age of three.[18]

Fabricating my own academic biography and its adjacency to historical violence in a consistent fashion by employing my dad's connection to Albert Rapp, with the great Meinecke as the model, would face a couple of hurdles. First, it would have to be acknowledged that the Rapp episode occurred later in my life and was much less spectacular than Meinecke's face-to-face encounter with William I and Bismarck. After all, I have never met Rapp face to face. An even bigger problem results from the demise of German nationalism since 1945, without which my generation's and my own political and intellectual socialization certainly would have taken a different direction. Already Meinecke's problematic effort to save the idea of the German nation's splendor was doomed. Since then, attempts to praise, glorify, or heroisize the German past, especially the one that led to 1945, have been more and more frowned upon if not rejected outright. Instead, a "negative" approach to dealing with this past has become standard in Germany. "Negative memory" is the term for addressing Germany's reinvention of national identity and its reconstruction of national reputation through questioning and rejecting its evil past – Nazism, racist hatred, antisemitism, authoritarianism, military aggression, and genocide. Instead of heroisizing one's own collective past and its actors, "negative memory" allows for empathy if not sympathy with the victims of past evil. And it entails exploring and exposing the historical truth about past evil in all its complexities, including many Germans' compliance with the Nazi regime.[19]

---

**18** Friedrich Meinecke, "Erlebtes 1862–1901" (1941), in Friedrich Meinecke, *Autobiographische Schriften*, ed. Eberhard Kessel (Stuttgart: K. F. Koehler, 1969), 6, 13–15.
**19** Reinhard Koselleck, "Formen und Traditionen des negativen Gedächtnisses," in *Verbrechen erinnern. Die Auseinandersetzung mit Holocaust und Völkermord*, eds. Volkhard Knigge and Norbert Frei (Munich: C. H. Beck, 2002), 21–32; Aleida Assmann, *Das neue Unbehagen an der Erinnerungskultur. Eine Intervention* (Munich: C. H. Beck, 2013), 59–106.

# Sufferings

The breakthrough of "negative memory" occurred in the 1990s. While some Germans have worked toward it since 1945, and West German (and in different ways East German) governments and many politicians supported it from early on, society in most parts refused to discuss a break with the Nazi past publicly or privately and instead chose to keep silent about it. Germans did not want to be reminded of their traumatization through war, death, expulsion, captivity or of their multifaceted compliance with or their indifference toward the regime and its terror against Jews, other minorities, and the peoples in the occupied territories. Only beginning in the late 1960s did the second generation, the children of those who had been adults in Nazi Germany, question their parents' silence about the Nazi past. The reckoning of post-Nazi generations with the silence of their Nazi parents or grandparents followed a twisted road. The 1968 rebels bemoaned allegedly fascist continuities, in personnel, ideology, and institutions, from Nazism to the West German democracy and capitalism. Since the late 1970s, inspired not least by the *Holocaust* miniseries (which follows the trajectories of two fictional German families, one Jewish and one Christian), this sweeping judgment gave way to diverse inquiries into what individual parents, fathers, grandparents, but also mothers had done during the Nazi period.

What these second- and in the meantime also third-generation accounts have in common is the mode of accusation – accusation not only of the parents' or grandparents' deeds and misdeeds in Nazi Germany, but also of their denial and silence afterwards, their "second guilt" (Ralph Giordano), their inability or unwillingness to work through, disclose, and admit to their "primary" guilt, their responsibility for, complicity in or indifference toward the Nazi regime and the crimes committed in the name of Germany during the Third Reich.[20] While parts of this second generation literature is furthermore driven by the notion of a "tertiary" guilt – the idea that the children have inherited the parents' guilt or parts of it – its authors typically cope with the ballast of multilayered guilt by refocusing the narrative on the damage these moral burdens have done to them, the children.[21]

---

**20** Ralph Giordano, *Die zweite Schuld, oder Von der Last Deutscher zu sein* (Hamburg: Rasch und Röhring, 1987).
**21** *Generations of the Holocaust*, eds. Martin S. Bergmann and Milton E. Juvovy (New York: Basic Books, 1982); *The Collective Silence. German Identity and the Legacy of Silence*, eds. Barbara Heimannsberg and Christoph Schmidt (San Francisco: Jossey Bass Publishers, 1993) (German original 1988); Stephan and Norbert Lebert, *My Father's Keeper. Children of Nazi Leaders – An Intimate History of Damage and Denial* (Boston: Little, Brown & Co., 2001).

One of the most prominent examples is Niklas Frank's 1987 "reckoning" (or *Abrechnung*, as the subtitle of the German original says) with his father, the notorious Hans Frank who was hanged at Nuremberg in 1946 for his crimes as despotic Governor General in Nazi-occupied Poland.[22] Niklas, born in 1939, was raised by his mother in the spirit of pious remembrance of a sensitive, gifted, and intellectual father. The son's rage about his father's evilness and the (not much less) evil obfuscations of his mother and other family members and friends pervades the book on each page, often resorting to obscene language. That the father's evil, the Nazi past, has destroyed the son's life, is the essence of the book's message.

Another example is Alexandra Senfft's *Schweigen tut weh. Eine deutsche Familiengeschichte* (Silence hurts. A German Family History), which appeared 30 years after Frank's bestseller.[23] Senfft, born in 1961, is the granddaughter of Hanns Ludin, Hitler's ambassador to the Slovak Republic beginning in 1941. Ludin was responsible for the deportation and subsequent death of the Slovak Jews and was hanged in 1947. Senfft tells the story of the wife (her grandmother) of a Nazi perpetrator who stubbornly believed in her husband's innocence and instituted a culture of silence about guilt and responsibility in the Third Reich. The actual anti-hero of Senfft's family history, however, is her mother Erika, born in 1933. Erika lived quite the life in Hamburg's leftist intellectual in-crowd in the 1960s and 1970s – but only to increasingly suffer from and eventually collapse as a result of depression, alcoholism, and world weariness. The fact that such tragedies have occurred and still occur in all kinds of families and societies and may be related to individual choices such as an excessive lifestyle is not the subject of Senfft's analysis. Instead, her narrative focuses on the singular cause for everything that went wrong in her family: the Nazi past, or, more precisely, the original guilt of Hanns Ludin that is handed down to his children and grandchildren, not unlike the original sin of Adam and Eve that has been haunting humankind since its inception.

A dramatic example of the second generations' self-victimization was given by "Stefan," the son of an SS officer, in an interview in the 1980s with Peter Sichrovsky, a journalist born shortly after the war in 1947, son of Jewish Austrians.

---

**22** Niklas Frank, *Der Vater. Eine Abrechnung* (Munich: Bertelsmann, [1987]). English version (in some parts omitting the obscene language and imageries of the German original): Niklas Frank, *In the Shadow of the Reich* (New York: A. Knopf, 1991).

**23** Alexandra Senfft, *Schweigen tut weh. Eine deutsche Familiengeschichte* (Berlin: Claassen, 2007). Cf. "Niklas Frank, Alexandra Senfft und Malte Ludin im Gespraech mit Horst Ohde ueber autobiographische Literatur und autobiographische Filme," in *Nationalsozialistische Täterschaften. Nachwirkungen in Gesellschaft und Familie*, eds. Oliver von Wrochem and Christine Eckel (Berlin: Metropol, 2016), 213–231.

In this interview, Stefan laments over "all that talk about you Jews being the victims of the war." And he goes on: "But for those of you who survived, the suffering ended with Hitler's death. But for us, the children of the Nazis, it didn't end. [. . .] I am sure that in the old days my father brutalized Jews, but after the war there weren't any left. There was only me." Bottom line: "I'm in the same boat as you. I was the Jew in my family."[24]

To be sure, these three examples contrast with more recent, more explorative, and more sophisticated second- and third-generation analyses of their parents' Nazi past and their subsequent silence about it.[25] Nonetheless, the leitmotif of this literature is the transgenerational conversion of guilt into victimhood; the transmitter of this conversion is the silence of the Nazi generation about their guilt, responsibility, deeds and misdeeds. It is this silence that, according to the second- and third-generation narrative, has damaged them, the children or grandchildren.[26] This narrative of victimhood takes up and transforms a powerful discourse that culminated early, around 1960, in Theodor W. Adorno's and Alexander and Margarete Mitscherlich's critiques of Germans' inability to work self-critically through their own entanglement in the Nazi state, its Führer cult, and its crimes.[27] These critiques, inspired by Freudian philosophy, bemoaned rightly Germans' refusal to honor the victims of Nazi persecution. Even more vociferously did they warn of the dire consequences of this silence and denial for the new West German democracy. Old Nazi ideas and ideologies would smolder under the surface of a democratic state and await the opportunity to attack it again. Still, in 1997, after almost forty years of rather stable and increasingly celebrated German democracy, the German political scientist Gesine Schwan somberly accounted for the destruction of German families and

---

**24** Peter Sichrovsky, *Born Guilty. Children of Nazi Families* (New York: Basic Books, 1988), 137–138. (German original 1987).

**25** Katharina von Kellenbach, *The Mark of Cain: Guilt and Denial in the Post-War Lives of Perpetrators* (New York: Oxford University Press, 2013); Angelika Bammer, *Born After. Reckoning with the German Past* (New York: Bloomsbury Academic, 2019); Roger Frie, *Not in My Family. German Memory and Responsibility After the Holocaust* (New York: Oxford University Press, 2017).

**26** On this motif of "consequential damage," see Mathias Brandstädter, *Folgeschäden. Kontext, narrative Strukturen und Verlaufsformen der Väterliteratur 1960 bis 2008* (Würzburg: Königshausen & Neumann, 2020). More generally, see Erin McGlothlin, *Second-Generation Holocaust Literature. Legacies of Survival and Perpetration* (Rochester, NY: Camden House, 2006).

**27** Theodor W. Adorno, "Was bedeutet: Aufarbeitung der Vergangenheit [1959]," *Gesammelte Schriften*, 10.2: *Kulturkritik und Gesellschaft II* (Frankfurt am Main: Suhrkamp, 1977): 555–572; Alexander and Margarete Mitscherlich, *The Inability to Mourn: Principles of Collective Behavior* (New York: Grove Press, 1975) (German original 1967).

parent-child relations as well as the "damage to democracy," all allegedly resulting from "silenced guilt" in post-Nazi Germany.[28]

Rarely noticed, the discourse on the damaging effects of "silenced guilt"
to second- and third-generation Germans or even the entire society actually allowed younger Germans to reconnect with the elderly in the spirit of common suffering from the Nazi past. The war generation had always insisted on their status
as victims – as victims of a seductive leader, of a coercive or entrapping dictatorship, and most of all of a terrible war that had ended in the destruction of their
homelands, the loss of millions of fathers, brothers, mothers, wives, sisters, and
children, the traumatization of further millions of survivors by years of Allied air
raids, Soviet captivity, and expulsion, the territorial and political division of their
fatherland; and the stigmatization of their national identity.[29]

Ultimately, the enormously successful 2013 TV show *Unsere Mütter, Unsere Väter*, or *Our Mothers, Our Fathers* (*Generation War* in the Anglophone
version) concluded this transgenerational reconciliation. Set in Nazi Europe
during World War II, it tells the story of five German friends: two brothers who
serve with the Wehrmacht on the Eastern Front; a selfless war nurse; a tal-
ented singer; and her boyfriend, a Jewish tailor. Reaching spectacular viewing
rates of 24 percent, the three-part series was hailed as the ultimate cure to the
traumas of millions of German families by dissolving the tensions and conflicts caused by decades of silencing the past. As the producer explained, the
movie allowed his eighty-eight-year-old father to finally talk, as never before
possible, about the past. Note that the younger generation is now charged
with interpreting the past. It does so, however, by eventually confirming crucial features of the obfuscating and self-victimizing remembrance of the Nazi
past that the German "war generation" had produced and reproduced over
decades. While the major characters are shown as complex, the series keeps
the Nazification of German society at a distance and instead presents even all
of them, not only the Jewish tailor, as victims of a grand tragedy.[30]

---

**28** Gesine Schwan, *Politics and Guilt. The Destructive Power of Silence* (Lincoln: University of
Nebraska Press, 2001), 54, 135 (German original 1997).
**29** Dagmar Barnouw, *The War in the Empty Air. Victims, Perpetrators, and Postwar Germans*
(Bloomington: Indiana University Press, 2005); *Germans as Victims. Remembering the Past in
Contemporary Germany*, ed. Bill Niven (Houndmills: Palgrave, 2006); Gilad Margalit, *Guilt, Suffering, and Memory. Germany Remembers its Dead of World War II* (Bloomington: Indiana University Press, 2010).
**30** Aleida Assmann, *Das neue Unbehagen an der Erinnerungskultur*, 33–42; David Wildermuth,
"*Unsere Mütter, Unsere Väter*: War, Genocide and 'Condensed Reality,'" *German Politics & Society*,
34, no. 2 (2016): 64–83; Katherine Stone, "Sympathy, Empathy, and Postmemory: Problematic Positions in *Unsere Mütter, unsere Väter*," *Modern Language Review*, 111, no. 2 (2016): 454–477.

# Mystery

The topos of transgenerational suffering from the Third Reich, the Second World War, and the Holocaust offers a seductive template for framing my own biographical adjacency to historical violence much more consistently than the simple reference to the mystery about Rapp would allow. Not only the paternal part of my family history is loaded with experiences of victimhood, shielded from exploration by rigorous silence. The maternal side is so as well. My mother was even less willing than my father to reveal any details about her youth under Hitler and during the war. She was born in 1925 (a few months before my dad) into a family of Catholic believers in the village of Krojanke (now Polish Krajenka) in what was, after the Versailles Treaty, the Grenzmark Posen-Westpreußen. Krojanke was located close to the 1938 border with Poland and fifteen miles north of the city of Schneidemühl (now Piła), which, after the Nazi occupation of Poland, became a hub of the Germanization program in the Warthegau.

But unlike my father, my mother was drawn to the Nazi youth groups. She joined the Bund deutscher Mädel (BDM) as early as possible, at the age of ten, voluntarily and full of enthusiasm, as detected in her occasional cryptic remarks and those of her cousin (aunt Hilde, the wife of uncle Rolf,[31] who both came from the same region). My mother seemed to have had the best time of her life in the BDM, and, after graduating from gymnasium (high school) in 1943, with the Reichsarbeitsdienst (RAD, or National Labor Service). After she had passed away in 1984, I learned from the new *Alltagsgeschichte* (history of everyday life) of the Third Reich that it was not unusual for girls at that time who wanted to break out of stifling family life to feel empowered by the female comradery, physical exercises, and bonfire romanticism of such groups.[32] But as long as my mother was alive, there was no talk about it, only silence. Cautious questions I might have wagered were left unanswered. Remembering, let alone talking about, the first nineteen-plus years of her life was obviously disquieting for my mother. It all remained a great mystery to me. Occasional remarks and old photos suggested some happiness in Krojanke and Schneidemühl, where my mother attended gymnasium. But neither

---

**31** These two names are pseudonyms.

**32** One of the early studies was Nori Möding, "'Ich muß irgendwo engagiert sein – fragen Sie mich bloß nicht, warum.' Überlegungen zu Sozialisationserfahrungen von Mädchen in NS-Organisationen," in *"Wir kriegen jetzt andere Zeiten." Auf der Suche nach der Erfahrung des Volkes in nachfaschistischen Ländern,* eds. Lutz Niethammer and Alexander von Plato (Berlin: Dietz, 1985). For a summary of subsequent research see my *Belonging and Genocide,* 143–147.

my mom nor her parents ever entertained the idea of returning to their homeland in Poland after 1945 – not even for a short visit. This was unusual. Other refugees and expellees from the eastern German territories, including uncle Rolf and his family, traveled there routinely even before and certainly after the Wall came down.

Only one anecdote was repeated over and over. Students of the psychology of memory will easily recognize it as a "screen memory," or "Deckerinnerung," in Sigmund Freud's terminology.[33] It is told to cover or indirectly address a traumatic event that cannot be communicated directly. In this instance, the anecdote was about how my grandfather, a subaltern railroad worker and switchman in Krojanke, managed to rescue his wife and daughter from the approaching Red Army by hiding them under layers of coal in a westbound train. My irreverent questions about how the two of them had endured several days under those heavy coal blankets, how they had breathed and fed themselves, and how my grandpa later got himself out of Krojanke, the cruel bolshevists in sight, were left unanswered. This sacred story did not allow for a Q & A session. It ended happily, in any case, with the family reunited in spring 1945 in the Braunschweig region. After settling in, my mother started interning in a pharmacy, met and married my father, studied pharmacy and graduated in 1957, ready for a career that she had pursued since her gymnasium days, as her diploma testifies. In the patriarchal spirit of the 1950s, she renounced this career when I was born in 1958; she reentered it only in 1976 after her divorce.

What occurred before or during the flight to the West in winter 1944–1945? I didn't know as a child, and I still don't know. My mother died in 1984 after many years of battling cancer, and no personal documents except some basic ones including the gymnasium diploma and (remarkably) the BDM membership booklet were left. My only relatives from the Krojanke time had been uncle Rolf and his wife Hilde, my mom's cousin and my godmother. They had taken good care of my mother during the many years of her disease, despite living far away. When my mother died in 1984, I felt obliged to pay them a visit, together with my girlfriend, pushing aside ambivalent feelings about uncle Rolf's political affiliations. It wasn't a good idea. At an earlier visit, he had produced a genuine swastika flag. As he explained, he had wisely kept it so he could waive it when the Nazis returned to power. He was even more proud of what he had accomplished in early retirement: creating a "Blood Book," as he called it, sumptuously bound in precious leather and written on heavy paper in red ink that

---

**33** Sigmund Freud, "Über Deckerinnerungen," *Gesammelte Werke*, Vol. I (Frankfurt am Main: S. Fischer, 1953), 531–554.

documented his life during the Third Reich. It also served as a eulogy for it, including its efforts to "resolve the Jewish question." The visit after my mother's death was even worse than I had anticipated. Uncle Rolf couldn't refrain from reveling in his Nazism, despite knowing about not only my own but also my girlfriend's critical attitude and areas of intellectual expertise (she was writing a dissertation on everyday life in Nazi Germany).[34] Our visit ended in a huge dispute and in our early departure; no farewell to uncle Rolf, and no chance to ever talk to him again or even to aunt Hilde about the past.

Instead, my father came up with explanations of sorts, although only many years later, after he had divorced his second wife and retired, haunted by guilt feelings or only the urge to rationalize the tragedy of the mother of his two children, my baby sister and me. It all came back to the Nazis and to the war; that is, he imagined that my mother had died from skin cancer as a result of the sun exposure she experienced as a BDM and RAD girl with no sunscreen at hand (dismissing the plenitude of sun baths in the decades of my mom's adulthood after 1945). While this theory – an unintended caricature of the myth of German victimization by the Third Reich – never made much of an impression on me, another one did. It is much darker, more serious, more complicated, and maybe closer to the truth. But the "maybe" weighs heavily. Through the 1980s, it was impossible to talk about rape, let alone the kind of mass rape that Soviet servicemen perpetrated on German women when they invaded eastern Germany, retaliating for the crimes of German soldiers on Soviet territory from 1941 on, including the rape of Soviet women. Only Helke Sanders's 1992 movie *BeFreier and BeFreite* (*Liberators and Liberated*), based on interviews with German women raped in Berlin by Soviet soldiers in May 1945, initiated a public debate about these crimes.[35] At the same time, the mass rapes during the wars in former Yugoslavia broke the centuries-old Western taboo against discussing sexual violence. My father saw Sanders's movie and studied Sanders's related book, and he digested the news about the war in Bosnia, finally comprehending what he imagined had gone wrong in the marriage with my mother. My mother had been raped before or during the escape from the East in winter 1944–1945. There was no doubt about it – at least not to him. The traumatization of my mother, never addressed in her entire life, even less the subject of any attempt at healing, had burdened and eventually made impossible a healthy intimate relationship.

---

**34** Cornelia Rauh-Kühne, *Katholisches Milieu und Kleinstadtgesellschaft. Ettlingen 1918–1939* (Sigmaringen: Jan Thorbecke, 1991).
**35** Helke Sanders and Barbara Johr, *BeFreier und BeFreite. Krieg, Vergewaltigungen, Kinder* (Frankfurt am Main: Fischer Taschenbuch Verlag, 1992).

A pinch of irony seems in order when narrating these considerations and conclusions of my father. But then, I may be mocking or analyzing his escapism into the myth of German victimhood and the all too human urge to rationalize mishaps and failures of one's own life by charging them to the account of grand tragedies. Still, what do I really know? Not much. After all, I should admit that I too can subscribe to the idea of my mother's rape at age nineteen or twenty. My desire to accept my father's theory is based on my own amateur psychologizing about my mother's personality and my desire to understand the thorough melancholia, if not sadness, that characterized her for as long as I knew her. No documents, letters, or testimonies whatsoever exist, and there is no hope of ever finding such things. In the end, there are only blanks and question marks, maybe a dozen or so pieces of a mosaic of many thousand. Inserting my own adjacency to historical violence in the rich tradition of German victimization by the Nazi war and the Nazi dictatorship, is tempting, but it wouldn't suffice to establish a consistent biography of that adjacency. The mythic oil of the narrative of victimization is too obvious, and the self-pity that it fuels demands its analysis, not its autobiographic reproduction.

And yet . . . autobiographical thinking can't altogether renounce the search for some consistency, some ideas of causes, consequences, or *telos*, as even Bourdieu in his acid critique of the biographic illusion evinces. A broad range of writers may have managed to overcome "l'illusion biographique" in the mode of aesthetic fiction; Bourdieu points to Faulkner and Robbe-Grillet.[36] But narrators of their own lives face a serious dilemma if they don't want to end up with "a tale told by an idiot, full of sound and fury, signifying nothing," as Macbeth bantered at the advent of modernity, the era that mandated individuals to ponder the unique mundane meaning of their own lives.[37] Autobiographical thinking can't make do without establishing meaning. Mine can't either. I shall choose the interplay of curiosity and complexity to inject meaning into my life as a historian of violence.

---

**36** Bourdieu, "L'illusion biographique," 69–70, also for the following.
**37** William Shakespeare, "Macbeth," Act V, Scene V, in *The Annotated Shakespeare*, Vol. III, ed. A. L. Rowse (New York: Clarkson N. Potter, 1978), 460. Cf., more generally, Peter Sloterdijk, *Literatur und Lebenserfahrung. Autobiographien der Zwanziger Jahre* (Munich: Carl Hanser Verlag, 1978).

# Complexity

The mystery, the taboos, the silence, the confusion about the past of my own family and Germans at large certainly have caused damage. But they have also evoked a precious, good curiosity – the type of curiosity that has driven generations of Holocaust and Third Reich scholars, especially those who have been working since the 1980s. Questioning the previous preoccupation with the Great Men of evil, Hitler, Himmler, Goebbels and the like, and with anonymous political structures, organizations, and processes of mass crimes, academics since then have illuminated the many choices ordinary people had even under a genocidal dictatorship. Parallel to the surge of second- and third-generation inquiries into the private dimensions of that time, these scholars have probed the interwovenness of suffering from and complicity in this dictatorship. They have debated the ambiguity of evil and its humaneness. Christopher Browning's dictum that "empathy for the perpetrators [. . .] is inherent in trying to understand them," has laid the ground for a new approach to studying perpetrators in all their facets, including as accomplices, bystanders, and in the contexts of the societies that produced them, German and non-German.[38] The idea of empathy for the evil doers has widely replaced previous practices of either demonizing them or otherwise obfuscating the complexity of their motivations and the social settings that made them do what they did. To be sure, my father's shallow testimonial to the moral goodness of Albert Rapp is transparent. Yet it carries a piece of truth – a disturbing truth, though. But only the assertion of such ambiguous truths enables us to understand why individuals became perpetrators, accomplices, bystanders, and onlookers of violence, and why some of them even changed their roles during the violent dynamic.

It is this type of academic work, together with the parallel documentation of and inquiries into ambiguous agency, tenuous choices, and unfathomable traumatization of the victims of the Holocaust, above all the Jews,[39] that has laid the intellectual ground for a culture of "negative memory" that has slowly

---

**38** Browning, *Ordinary Men*, p. xx. Cf. Thomas Kohut, *Empathy and the Historical Understanding of the Human Past* (New York: Routledge, 2020).

**39** The state of research is masterly summarized by Peter Hayes, *Why? Explaining the Holocaust* (New York: W. W. Norton, 2017). See also David Cesarani, *Final Solution. The Fate of the Jews, 1933–1945* (New York: St. Martin's Press, 2016); *The Oxford Handbook of Holocaust Studies*, eds. Peter Hayes and John K. Roth (Oxford: Oxford University Press, 2010); *A Companion to the Holocaust*, eds. Simone Gigliotti and Hilary Earl (Hoboken, NJ: Wiley Blackwell, 2020).

established itself in post-Holocaust Germany, Europe, and beyond.[40] Instead of glorifying, obfuscating, or simplifying the past, this culture is committed to critically analyzing, debating and exposing the political causes, social responsibilities, and multifaceted consequences of terror and destruction. It is fueled by a deep sense of shame about the past wrongs but also by thorough confusion about its subjects. The confusion about the subjects of violence is constructive. It generates insights into the complexity of mass violence. Translated into curiosity, confusion about the past inspires questions and research, the gathering of intelligence as an ongoing, never ending process of critique and self-criticism.

# Works Cited

Adorno, Theodor W. "Was bedeutet: Aufarbeitung der Vergangenheit [1959]," *Gesammelte Schriften*, 10, no. 2: *Kulturkritik und Gesellschaft II*. Frankfurt am Main: Suhrkamp, 1977, 555–572.
Altieri, Ricardo. "Eine Antikritik auf Bourdieus Kritik am biographischen Schreiben." In *Work in Progress, Work on Progress*. Edited by Marcus Hawel et al. Hamburg: VSA, 2019, 41–53.
Assmann, Aleida. *Das neue Unbehagen an der Erinnerungskultur. Eine Intervention*. Munich: C. H. Beck, 2013.
Bammer, Angelika. *Born After. Reckoning with the German Past*. New York: Bloomsbury Academic, 2019.
Barnouw, Dagmar. *The War in the Empty Air. Victims, Perpetrators, and Postwar Germans*. Bloomington: Indiana University Press, 2005.
Bergmann, Martin S., and Milton E. Juvovy, eds. *Generations of the Holocaust*. New York: Basic Books, 1982.
Bourdieu, Pierre. "L'illusion biographique." *Actes de la recherche en sciences sociales*. Vol. 62–63 (1986): 69–72.
Bourdieu, Pierre. "The Biographical Illusion." In *Identity: A Reader*. Edited by Paul Du Gay, Jessica Evans, and Peter Redman. London: Sage Publications, 2004, 297–303.
Brandstätter, Horst. *Asperg. Ein deutsches Gefängnis*. Berlin: Klaus Wagenbach, 1978.
Brandstädter, Mathias. *Folgeschäden. Kontext, narrative Strukturen und Verlaufsformen der Väterliteratur 1960 bis 2008*. Würzburg: Königshausen & Neumann, 2020.
Browning, Christopher. *Ordinary Men. Reserve Police Battalion 101 and the Final Solution in Poland*. Revised Edition. New York: Harper Perennial, 2017.
Cesarani, David. *Final Solution. The Fate of the Jews, 1933–1945*. New York: St. Martin's Press, 2016.
Frank, Niklas. *Der Vater. Eine Abrechnung*. Munich: Bertelsmann, [1987].
Frank, Niklas. *In the Shadow of the Reich*. New York: A. Knopf, 1991.

---

**40** Koselleck, "Formen und Traditionen des negativen Gedächtnisses," 32; Jörn Rüsen, "Holocaust Memory and German Identity – Three forms of generational practices," *Textos De Historia*, 10.1+2 (2002): 95–104, here esp. 100–102 on the "inclusion of the otherness of the Holocaust."

Freud, Sigmund. "Über Deckerinnerungen." In Sigmund Freud, *Gesammelte Werke*. Vol. I. Frankfurt am Main: S. Fischer, 1953, 531–554.

Freudiger, Kerstin. *Die juristische Aufarbeitung von NS-Verbrechen*. Tübingen: J. C. B. Mohr, 2002.

Frie, Roger. *Not in My Family. German Memory and Responsibility After the Holocaust*. New York: Oxford University Press, 2017.

Gigliotti, Simone, and Hilary Earl, eds. *A Companion to the Holocaust*. Hoboken, NJ: Wiley Blackwell, 2020.

Giordano, Ralph. *Die zweite Schuld, oder Von der Last Deutscher zu sein*. Hamburg: Rasch und Röhring, 1987.

Hayes, Hayes. *Why? Explaining the Holocaust*. New York: W. W. Norton, 2017.

Hayes, Peter, and John K. Roth, eds. *The Oxford Handbook of Holocaust Studies*. Oxford: Oxford University Press, 2010.

Hegi, Ursula. *Tearing the Silence. On Being German in America*. New York: Simon & Schuster, 1997.

Heimannsberg, Barbara, and Christoph Schmidt, et. al., eds. *The Collective Silence. German Identity and the Legacy of Silence*. San Francisco: Jossey Bass Publishers, 1993.

Ingrao, Christian. *Believe and Destroy. Intellectuals in the SS War Machine*. Cambridge, MA: Polity, 2013.

Kellenbach, Katharina von. *The Mark of Cain: Guilt and Denial in the Post-War Lives of Perpetrators*. New York: Oxford University Press, 2013.

Klemp, Stefan. "Albert Rapp: 'Du sollst Deinen Feind aus aller Seelenkraft hassen . . . .'" In *Täter, Helfer, Trittbrettfahrer*, Vol. 10: *NS-Belastete aus der Region Stuttgart*. Edited by Wolfgang Proske, 354–375. Gersteten: Kugelberg Verlag, 2019.

Kohut, Thomas. *Empathy and the Historical Understanding of the Human Past*. New York: Routledge, 2020.

Koselleck, Reinhard. "Formen und Traditionen des negativen Gedächtnisses." In *Verbrechen erinnern. Die Auseinandersetzung mit Holocaust und Völkermord*. Edited by Volkhard Knigge and Norbert Frei, 21–32. Munich: C. H. Beck, 2002.

Krausnick, Helmut, and Hans-Heinrich Wilhelm. *Die Truppe des Weltanschauungskrieges. Die Einsatzgruppen der Sicherheitspolizei und des SD 1938–1942*. Stuttgart: Deutsche Verlags-Anstalt, 1981.

Kühne, Thomas. *Belonging and Genocide. Hitler's Community, 1918–1945*. New Haven, CT: Yale University Press, 2010.

Kühne, Thomas. *Dreiklassenwahlrecht und Wahlkultur in Preußen 1867–1914. Landtagswahlen zwischen korporativer Tradition und politischem Massenmarkt*. Düsseldorf: Droste, 1994.

Kühne, Thomas. "Der nationalsozialistische Vernichtungskrieg und die 'ganz normalen' Deutschen. Forschungsprobleme und Forschungstendenzen der Gesellschaftsgeschichte des Zweiten Weltkriegs. Erster Teil." *Archiv für Sozialgeschichte* 39 (1999): 580–662.

Kühne, Thomas. *The Rise and Fall of Comradeship: Hitler's Soldiers, Male Bonding and Mass Violence in the Twentieth Century*. Cambridge: Cambridge University Press 2017.

Kühne, Thomas, and Walter Binder, Jr. "Zum 'Tag der offenen Tür' auf dem Eisberg." *Schwarzwälder Bote* (Nagold edition), 20 September 1974.

Lebert, Stephan, and Norbert Lebert. *My Father's Keeper. Children of Nazi Leaders – An Intimate History of Damage and Denial*. Boston: Little, Brown & Company, 2001.

Lewy, Guenther. *Perpetrators. The World of Holocaust Killers*. New York: Oxford University Press, 2017.

Mallmann, Klaus-Michael. "Lebenslänglich. Wie die Beweiskette gegen Albert Rapp geschmiedet wurde." In *Die Gestapo nach 1945. Karrieren, Konflikte, Konstruktionen*. Edited by Klaus-Michael Mallmann and Andrej Angrick, 255–269. Darmstadt: Wissenschaftliche Buchgesellschaft, 2009.

Margalit, Gilad. *Guilt, Suffering, and Memory. Germany Remembers its Dead of World War II* Bloomington: Indiana University Press, 2010.

McGlothlin, Erin, *Second-Generation Holocaust Literature. Legacies of Survival and Perpetration*. Rochester, NY: Camden House, 2006.

Meinecke, Friedrich. *Die deutsche Katastrophe. Betrachtungen und Eruinnerungen*. Wiesbaden: Brockhaus, 1946.

Meinecke, Friedrich. "Erlebtes 1862–1901" (1941). In Friedrich Meinecke, *Autobiographische Schriften*. Edited by Eberhard Kessel, 3–134. Stuttgart: K. F. Koehler, 1969.

Mitscherlich, Alexander, and Margarete Mitscherlich. *The Inability to Mourn: Principles of Collective Behavior*. New York: Grove Press, 1975.

Möding, Mori. "'Ich muß irgendwo engagiert sein – fragen Sie mich bloß nicht, warum.' Überlegungen zu Sozialisationserfahrungen von Mädchen in NS-Organisationen." In *"Wir kriegen jetzt andere Zeiten." Auf der Suche nach der Erfahrung des Volkes in nachfaschistischen Ländern*. Edited by Lutz Niethammer and Alexander von Plato, 256–304. Berlin: Dietz, 1985.

Moeller, Robert G. *War Stories. The Search for a Usable Past in the Federal Republic of Germany*. Berkeley: University of California Press, 2001.

Niven, Bill, ed. *Germans as Victims. Remembering the Past in Contemporary Germany*. Houndmills: Palgrave, 2006.

Rauh-Kühne, Cornelia. *Katholisches Milieu und Kleinstadtgesellschaft. Ettlingen 1918–1939*. Sigmaringen: Jan Thorbecke, 1991.

Rüsen, Jörn. "Holocaust Memory and German Identity – Three forms of generational practices." *Textos De Historia* 10, nos. 1+2 (2002): 95–104.

Rüter, C.F., and D.W. de Mildt, eds. *Justiz und NS-Verbrechen. Sammlung deutscher Strafurteile wegen nationalsozialistischer Tötungsverbrechen*. Vol. 20. Amsterdam: University Press Amsterdam, 1979.

Sanders, Helke, and Barbara Johr. *BeFreier und BeFreite. Krieg, Vergewaltigungen, Kinder*. Frankfurt am Main: Fischer Taschenbuch Verlag, 1992.

Schulz, Sandra. "Film und Fernsehen als Medien der gesellschaftlichen Vergegenwärtigung des Holocaust. Die deutsche Erstausstrahlung der US-amerikanischen Fernsehserie 'Holocaust' im Jahre 1979." *Historical Social Research* 32, no. 1 (2007): 189–249.

Schwan, Gesine. *Politics and Guilt. The Destructive Power of Silence*. Lincoln: University of Nebraska Press, 2001.

Senfft, Alexandra. *Schweigen tut weh. Eine deutsche Familiengeschichte*. Berlin: Claassen, 2007.

Shakespeare, William. "Macbeth." In *The Annotated Shakespeare*. Vol. III. Edited by A. L. Rowse, 419–465. New York: Clarkson N. Potter, 1978.

Sichrovsky, Peter. *Born Guilty. Children of Nazi Families*. New York: Basic Books, 1988.

Sloterdijk, Peter. *Literatur und Lebenserfahrung. Autobiographien der Zwanziger Jahre*. Munich: Carl Hanser Verlag, 1978.

Stone, Katherine. "Sympathy, Empathy, and Postmemory: Problematic Positions in *Unsere Mütter, unsere Väter*." *Modern Language Review* 111, no. 2 (2016): 454–477.

"Tiefste Gangart." *Der Spiegel*. 13 November 1963: 52–56.

Welzer, Harald, Sabine Moller, and Karoline Tschugall. *"Opa war kein Nazi."*
*Nationalsozialismus und Holocaust im Familiengedächtnis*. Frankfurt am Main: Fischer
Taschenbuch Verlag, 2002.

Wildermuth, David. *"Unsere Mütter, Unsere Väter*: War, Genocide and 'Condensed Reality.'"
*German Politics & Society* 34, no. 2 (2016): 64–83.

Wrochem, Oliver von, and Christine Eckel, eds. *Nationalsozialistische Täterschaften.
Nachwirkungen in Gesellschaft und Familie*. Berlin: Metropol, 2016.

Part Three: **Journeys**

Angelika Bammer
# Looking for History, Finding a Life

What do I see when I see someone I don't know yet? An individual, singular –
indeed, unique – in their own way? Or a member of a particular group that they
represent to me: a woman, a man, someone handicapped; a Black person, a white
person, a fat person; a Muslim, a German, a Jew?

During the summer of 2020 in the America I know we were literally torn
apart by these questions. While different groups battled over the importance and
import of the different positions they had staked out and argued over which lives
mattered – Black Lives, Blue Lives, All Lives, or simply life itself – we were transfixed by the raw immediacy of our encounter with specific lives that seared themselves into our consciousness. We marched under banners that gathered us
together in the name of categories that we affirmed, and we called out the names
of particular individuals. "Say her name," we chanted, "say their names."[1] Our
chants reminded us of the fact that before they became symbols, or icons, or
memes, Breonna Taylor, George Floyd, Ahmaud Arbery, and the many other
Black American men and women killed by American police during this season of
grief and rage, had just been persons. Breonna Taylor, whose name at times
seemed almost to become synonymous with #SayHerName was a young Black
woman, twenty-six years old; she had a mother, Tamika Palmer; she had a sister,
Ju'Niyah Palmer; she had a boyfriend; she had a job as an E.M.T. and was planning a career in health care. She had a life before her murder ended it.

What we learned over the course of this summer of our collective grief and
rage was that categories are both powerful and mutable. They bear the weight of
history but are neither permanent nor fixed. They are situational and they are relational: to your blackness, I am white; to your Jewishness, I am German; to your
maleness, I am female. Moreover, we not only can be, but are, different things at
once. I can be a woman, a German, and Black, of-here and from-there, all in one
person. I can even embody what appear to be contradictions: I can be Jewish and
Catholic, Black and white, both at once.

How do we face the history these categories represent, acknowledge the
marks and scars they have left on us, and on each of us differently, without

---

1 The hashtag #SayHerName was coined by the legal scholar and critical race theorist, Kimberlé Crenshaw to draw attention to the particular circumstances of Black American women's
lives and deaths within a framework of intersectional and systemic racism and sexism. The
#SayHerName campaign became particularly vocal in response to the murder of Breonna Tay-
lor on 13 March 2020. See also essay in this volume by Choudury.

https://doi.org/10.1515/9783110753295-024

letting those marks and scars define us? How do we extricate persons from the categories that risk subsuming them? These questions have been at the center of my life for as long as I can remember. Born to non-Jewish German parents in the aftermath of World War II, I inherited a genocidal legacy. As a white person in the contemporary United States, I became the beneficiary of a racist system. Whiteness and Germanness mark different subject positions in relation to different histories. I try to acknowledge the different reckonings they entail in the specifics of individual encounters. This essay is about such an encounter.

•

To engage with a history that implicates me without the comforting abstractions of language, I often take up my camera. I choose images instead of words. A photograph is always of something – or someone – specific. It is of this person, this place, this moment, here and now. It is the record of a relationship I entered into. "Photographs," writes David Levi Strauss, "register the relation of photographer to subject – the distance from one to another."[2] The question is, what or whom I am choosing to relate to and what the nature of that relationship will be. Something or someone draws my attention and I respond. The glint of train tracks in the haze of a summer sunrise; a gathering of farm workers at the morning shape-up; a small Oaxan man dwarfed by the body of the cop arresting him. It begins with my decision to look, to see intentionally. But *how*, is the critical question: How close do I get? How far away do I stay? What do I focus on, what do I screen out, what do I move to the periphery? The criteria are aesthetic (how will it look?), political (what effect will it have?), ethical and emotional (how will I feel about what I am doing?). The factors determining my decisions are often unconscious, as I respond to the situation at hand at a given moment. Perhaps the light is bad, or the wind picks up, or my shutter release stops working, or perhaps taking a picture of what I am witnessing suddenly feels wrong. Yet, conscious or not, my decisions are evident in my actions, from the camera I use and the lens I pick, through my choice of focal length, shutter speed and aperture, to such things as whether I add filters or additional lights. I might even decide not to photograph.

What follows is the story of a relationship that I entered through my camera eye and the choices along the way that shaped it. It is about the photographs I took – and the ones I didn't take – and how they changed my experience of this relationship. I came expecting to confront a history defined by categories that I took as given. That is not what happened. Instead of encountering a history I

---

2 David Levi Strauss, "The Documentary Debate: Aesthetic or Anaesthetic," in *Between the Eyes: Essays on Photography and Politics,* ed. Diana C. Stoll (New York: Aperture, 2003), 3–11, here 10.

thought I knew, my perspective changed, and the categories I took as given were put in question.

•

When I met him in the fall of 2012, in a small rural town in west Germany close to the Dutch border, Günther Lehmann was seventy-eight.[3] Born in February 1934 as the second child of Franz and Ida Lehmann (his sister, Liesel, was older than he by six years), he had lived in the town of Fehlen his entire life.[4] He was baptized in the Catholic church of St. Andreas and attended the village school through eighth grade when, following the tradition of labor practices in these parts, he became an electrician like his father before him. He married, had a family, and ran a successful electrical service and supply business that he passed on to his son, Udo, when he retired. He had seen Fehlen change from the village it had been when he was born (at which time, it numbered under thirteen hundred people) to over seven thousand people by 2012, when it incorporated with the neighboring town and was awarded the designation "city." On the face of it, there is nothing remarkable about this story of a man's life. It maps the trajectory of a life lived well within established frameworks. What makes Günther Lehmann's story remarkable is what happened when he was young – a child in Nazi Germany with a mother whom Nazi law defined as Jewish.

•

Günther's parents met when his father, Franz, came to Fehlen as an electrician in the local textile mill and rented a room from Amalia Landau, a widow trying to support herself and her two daughters with the modest income from a small textile shop set up in her house. Her younger daughter, Ida Landau, helped run the store. A stranger to Fehlen and away from his family, the young electrician was soon integrated into his landlady's family. For although Ida's father had died when she was only ten, and she had grown up with just her mother and older sister, she didn't want for family. Next door, above the butcher shop they owned and ran on Bahnhof street, lived her uncle, her mother's younger brother, Abraham Frank, with his wife and children, and a few houses down, in the house that had been the Frank home for generations, were Abraham and Amalia's three unmarried sisters, Lina, Bertha, and Dora Frank.

---

**3** Lehmann is a pseudonym designed to protect Günther's actual identity. Throughout this essay, I use this pseudonymous last name for all references to Günther's family and paternal lineage.
**4** Like the pseudonymous Lehmann, Fehlen is a pseudonym to obscure the actual name of Günther's hometown.

Fehlen's Jewish community was small: two families in three houses on a single street. In a region where God himself and his chosen people were presumed to be Catholic, a community organized around the rhythm of the Catholic calendar where the church marked the center of social life, they were a drop in a mono-religious sea. Yet ever since the butcher, Levy Frank, and the cattle and small livestock trader, Herman Landau, came to this Westphalian village in the early nineteenth century, set up business and started families there, the Franks and Landaus had been part of the Fehlen community. "My parents, grandparents and great-grandparents were all born in Germany and have always lived here," Abraham Frank wrote the town administration in 1935, when antisemitic measures pushed his business to the brink of bankruptcy. "200 years of records," he added, "ought to provide ample proof."[5]

When Ida Landau married Franz Lehmann in May 1928, she didn't leave this extended-family community. Instead, Franz joined it and set up residence with his mother-in-law and wife, eventually clearing space to set up a small electrical business on the premises. What Ida left was her Jewish faith community, converting to Catholicism before her marriage to the Catholic Franz. When their children, Liesel and Günther, were born, they were baptized and raised Catholic.

Yet Nazi decrees, notably the infamous Reich Citizenship Laws (*Reichsbürgergesetze*) of 1935, commonly known as the Nuremburg Laws, not only stripped German Jews of their rights as *citizens* but denied their right to be treated as German *people*. As the Nuremberg Laws redefined the categories on which legal and social rights were based within a system of racialized "blood" laws, Ida Lehmann was defined as "Jewish" and her children as so-called "Mischlinge" (mixed-race).[6] That they were Catholic was immaterial.[7] These laws established the framework for a program of genocide. What we now refer to as the Holocaust proceeded in their wake. The story of Günther Lehmann and his family is a prism through which this larger story of the Holocaust can be refracted.

**5** Quoted in Mechtild Schöneberg, Thomas Ridder, and Norbert Fasse, eds., *Die jüdischen Gemeinden in Borken und Gemen: Geschichte, Selbstorganisation, Zeugnisse der Verfolgung* (Bielefeld: Verlag für Regionalgeschichte, 2010).
**6** In the racial mathematics of blood quantum measurement, Ida, both of whose parents had been Jewish, was considered a "full Jew" (*Volljude*), while her children were considered "half Jews" (*Halbjuden*), respectively. Officially, within the "mixed race" (Mischling) category, Günther and his sister would have been identified as "Mischlinge of the first degree," with a "Jewish" mother and a German (Aryan) father. To complicate the picture even further, according to Jewish religious custom Günther Lehmann, as the child of a Jewish mother, would be considered Jewish.
**7** See essay by Heschel in this volume.

As with the Holocaust itself, it's hard to pinpoint the exact moment where Günther Lehmann's German-Jewish story begins. There are many beginnings. In the wake of the 1935 Law for the Protection of German Blood (*Blutschutzgesetz*), there is the social pressure on Günther's parents to dissolve their marriage, for while not illegal the way new relationships between Jews and non-Jews had been decreed, it was a breach of Nazi social decorum. There is the growing economic pressure on the Franks and Landaus as the boycott of "non-Aryan" establishments push them toward insolvency and finally out of business. In 1936, Abraham and Helene Frank have to give up their butcher shop. Half a year later, they have to leave their home on Bahnhof street for temporary quarters several streets away. Amalia Landau and the Lehmann family – Franz, Ida and their two young children, Liesel and Günther – are affected also, as the comfort and shelter of their extended family next door is suddenly gone. The destruction of a family – and the small Jewish community of Fehlen – proceeds in increments, as one piece after another is lopped off. In 1938, when Günther is four, his uncle Siegfried, the twenty-five-year-old son of Abraham and Helene Frank, makes his way to Holland, where he will be shot, trying to escape deportation to Buchenwald. Shortly after her brother's flight to Holland, Siegfried's twenty-one-year-old sister Edith, Abraham and Helene's only remaining child, leaves for England, where a quota of young German Jewish women have been allowed entry with permits for work as domestic help.

The decisive blow in the destruction of Fehlen's Jews and Günther's family on his mother's side comes in 1941, when he is a first- or second-grader. Two weeks before Christmas, just before daybreak on December 11, Abraham and Helene Frank are loaded onto a bus already filled with other Jews from neighboring villages, including several of Abraham's cousins and their families. They are sent all the way across Europe to a Nazi labor camp in Riga, Latvia, where they will perish with hardly a trace. There is no gravesite, not even a death date. Their departure from Fehlen is handled with proverbial German efficiency: the single suitcase they have been instructed to pack is ready and the village policeman is in attendance as they board. No one else is present. It is over in minutes. The bus doesn't even cut the engine.

With Abraham Frank, Günther loses the man who had filled the place of grandfather for the grandfather he never knew, as his grandmother was widowed before Günther was born and never remarried. But within a few months of the Frank's deportation, Günther was to lose his grandmother too. In July 1942, the eighty-one-year-old, almost blind Amalia Landau is deported to Theresienstadt, where she will die in the spring of 1944. Deported along with her is the brother of Helene Frank, who had lived with the Franks until their deportation and moved in with the Lehmann's and Amalia Landau after the Franks were gone, their furniture and household goods sold off in a public auction, and their house appropriated as public property by the town. In the wake of the July 1942 deportations, the regional Gestapo headquarters is proudly informed that the Fehlen district is now, as Nazi parlance had it, "Jew-free" (*judenfrei*).

Günther's mother, Ida, is the only one left. Her status is ambiguous. By birth and according to Nazi "Blood Laws," she is a Jew. But, while as a Jew she is vulnerable, as the wife of an Aryan husband in a so-called "privileged mixed marriage" and the mother of children being raised as Catholics, she has the benefit of some provisional protections. The fact that her husband is critical to the town's economy and public safety in this time of war, as he is responsible for Fehlen's electrical grid and all public lighting, would likely have provided even more protection. Yet in February 1945, with the end of the war and the collapse of the Nazi regime merely weeks away (by early May, the war is officially over and the regime deposed), Ida Lehmann is deported too. She is sent to Theresienstadt, where her mother, Amalia Landau, had perished the previous spring. Meanwhile, Günther's father is recruited for the last, mad effort to save the fatherland from impending doom. Whether he is drafted into the so-called People's Storm (Volkssturm) of boys and old men fighting to defend the fatherland or whether he is assigned to the Organisation Todt to work on local emergency construction, around the time that his wife is deported to Theresienstadt, Franz Lehmann has to leave Fehlen too.

With his parents gone, his sister, Lise, dead: (the eighteen-year-old had died in 1944, with tonsillitis as the official diagnosis), and the other members of his Fehlen family dead or gone, by the end of the war ten-year-old Günther has lost almost his entire family.

In the end, the worst was averted and the nucleus of their family was spared. His father survived the war, his mother returned from Theresienstadt, and they rebuilt their life in Fehlen as a family. Both Franz and Ida Lehmann were buried in the Fehlen cemetery in the tradition of the Catholic faith that they had practiced. Günther followed in his father's footsteps as an electrician, and like his great-uncle, Abraham Frank, who had actively participated in local civic organizations and proudly carried the banner of his butcher's guild in the annual Pentecost procession led by the priest, he was an active and engaged member of the village community. A sought-after electrician, successful small businessman, and beloved member of the St. Andreas church choir, Günther Lehmann was valued – and, by all accounts, valued himself – as a pillar of Fehlen society.

•

Yet when I thought of Günther Lehmann, it wasn't in terms of his life achievements, but in terms of the history of traumatic loss that had marked his past. It was a history that I had engaged in my professional work. I taught courses on history and memory and courses on the Holocaust with a Jewish colleague. I had visited places marked by the legacy of traumatic histories – apartheid in South Africa; the Holocaust in Poland, Germany and Israel; slavery in the American South, where I lived – and had talked with people about how these pasts shaped their present lives. What did they remember, as individuals and in their communities, and what did they do, how did they live, with those memories? I had given talks and published articles on these issues and was in the process of writing a book on my struggles to confront the German history I had inherited.[8] My visit to Fehlen was both part of this work and different.

For one, this was not a place that anyone would associate with traumatic history. There had been no pogrom, no burning of synagogues, no Jewish cemeteries to be destroyed by vandals. As people told it, the Jews of Fehlen simply "vanished." What remained was a thriving village with a healthy economy and a Catholic congregation, joined by a small community of Lutherans who came as refugees after the war, when what had been the German East Prussia was realigned with and returned to Poland.

For another, the business that brought me here was not just professional. It was personal. And it was business that was not yet finished. For this was the place where German history and the history of my family formed a knot, in which the threads of Günther Lehmann's family and my own family had become entangled. I had come to Fehlen to try to sort them out.

Like Günther Lehmann, I was born in Fehlen and baptized in St. Andreas church. My mother, like his, was born and raised as a "Fehlen girl." My parents, like his, met and married in Fehlen, and my grandparents are buried in the same cemetery as Günther's parents.

But just as our histories converge in this place, they also diverge. For at the same time that Günther's family on his mother's side (including his mother) were excluded from the Fehlen community and persecuted as Jews, my maternal grandparents, Johannes and Maria Bushoff, held privileged positions in the same community as Nazi party members. My parents eventually left Fehlen and I grew up elsewhere, while Fehlen remained Günther's lifelong home. This was the place where our histories intersected, and on the ground of that intersection I wanted to understand who I was and who he and I were to each other. If we could meet, I thought, the ghosts of the past – my Nazi grandparents and his Jewish mother and her murdered family – would perhaps be present and we could face them. Or at least we would have the chance.

---

**8** Angelika Bammer, *Born After: Reckoning with the German Past* (New York and London: Bloomsbury Academic, 2019).

•

The timing was fortuitous. Upon my arrival (I had booked a room for a week in the local inn), I learned that a small memorial had recently been installed on Bahnhof street to honor the memory and commemorate the deportation and death of four members of its Jewish community, all of them members of Günther's family on his mother's side: Abraham, Helene and Siegfried Frank and Helene Frank's brother, Leopold Humberg. The project had been organized by a ninth-grade history teacher in the local school, who wanted his students to experience history in a way that didn't come from books so that they could understand their own relationship to it. His idea succeeded. As the students researched Fehlen's history and learned what happened to Fehlen's Jews, they became custodians of their own history and their view of Fehlen changed. They learned that the corner of Bahnhof Street and Coesfeld Street, where Fork's bakery sells pastries and coffee, is also the place where Abraham and Helene Frank were loaded onto a bus on the way to their death in Riga. In May 2012, four bronze memorial stones were placed in front of the houses where the Franks and Leopold Humberg had last lived.

Not everyone was happy about these stones. Günther Lehmann was one of them. "He didn't really give a reason," the teacher, Ullrich Wollheim, told me, "he just refused permission to have a memorial stone for his grandmother, Amalia Landau, placed. He didn't want the house where she had lived to be identified." It wasn't until later, after I met Günther Lehmann and began to see things from his perspective, that a likely reason emerged. The house in question wasn't just Amalia Landau's house. It was also his house that had been home to him for most of his life. It was where he had lived as a child with his parents, sister, and grandmother, and where he had lived with his own family – his wife and three children – until they moved to a bigger house close by.

That evening, I rang him up and he invited me to come to his house the following day.

•

We had first communicated earlier that year, when I called him from my home in Atlanta. We had never met, but I knew of his family from my research on the Holocaust and Nazi Germany. As I studied the history of this period in Westfalia and, even more locally, in the town where I was born, I came to think of Günther Lehmann as Fehlen's "last remaining Jew." I didn't realize until much later and in retrospect that I was applying an identity category from Nazi times to a man who refused to accept its premises. Moreover, I wasn't the only one who thought of him in those terms. "He is the only one left," a Fehlen woman offered in a discussion of the memorial stones remembering Fehlen's Jews. The uneasy silence following her words left no doubt as to what she meant by "the only one."

When I first called, he knew nothing about me (although I later learned that he had made inquiries). I knew some things about his family history during the Nazi years, but little about him, except that he was an electrician. That's what led me to him.

I went online and entered "electrician – Fehlen – Lehmann." A website, *Elektro Lehmann*, came up, with the owner listed as Udo Lehmann. Probably his son, I thought, and composed an email, asking if he could help me get in touch with Günther Lehmann. Weeks went by without response and I was not surprised. Why would he want to be in touch with me . . . he doesn't know me. Or if he does, he probably knows me as the granddaughter of Johannes Bushoff, who at the time was one of the more prominent members of the local Nazi party.

But if I want to face this history, I can't stop now. I decide to call and dial the number of Elektro Lehmann. Udo Lehmann answers and gives me his father's number. He is out of town on a trip with his wife, but will be back the following Sunday. The Monday after his return, I call.

I rehearse what I want to say and how I want to sound: casual, matter-of-fact, friendly. I don't want to let my nervousness show. But he immediately skips the introductory small talk and gets to the point. "Why do you want to talk to me?" he asks. A direct question demands a direct answer, but I am unprepared. I start stammering – about history and memory and difficult pasts . . . what we remember . . . how we forget, and I am stumbling through a fog of words. I finally stop. And suddenly the words I've been ashamed to say burst out of me: "I am the granddaughter of a man who was a member of Fehlen's Nazi party, and you are the son of a woman who was deported by them as a Jew." At that, to my surprise, he starts laughing.

"Well," he says, "when you're in Fehlen, we'll talk." I collect myself and ask where he wants to meet.

Now he is the one surprised. "You'll come to my place," he says, as if there were no other place we *could* meet. "I'll be happy to receive you," he adds.

•

Half a year later, I am in Fehlen, getting ready to be received by him. This time, I will be prepared. I get my tape recorder, slide in a new tape, check the batteries and grab my camera. It's nothing fancy – a small digital SLR with an 18–55 mm lens – but it's easy to handle, which is what I need. Battery power: full. Memory card: new. I put my coat on and head out into a bright November day so cold I can see my breath.

I pass by Fork's bakery on the corner where Abraham and Helene Frank started their fateful journey. I cross Bahnhof street, where the Franks had their butcher shop and Amalia Landau sold textiles, and where Günther Lehmann lived with his parents and grandmother Amalia when he was a child. I pass Fragemann's pub, where the furniture and household goods of the Franks were auctioned off after their deportation, and St. Andreas church, where Günther Lehmann and I were baptized. Now I, the granddaughter of one of Fehlen's former Nazi party members, am about to meet the man, whom I think of as "Fehlen's last surviving Jew." It feels momentous. Ghosts from the past are crowding into my consciousness and the debts of history are weighing on my mind.

The house where the Lehmann's live is a relatively new and modern duplex in the heart of Fehlen with a tidy patch of green lawn in the back. I check the number and ring the bell. A man opens the door, and I recognize Günther Lehmann instantly, even though the only picture I have ever seen of him was as a child. But I know this face: the high forehead and thin, straight hair over balding temples, the hooded lids, the level gaze and the smile that seems to hold a question. "You look like your father," I blurt out, as we shake hands.

He looks puzzled, "Do you know my father?"

"I don't," I admit, flushing at the inappropriate familiarity of my outburst. He is a stranger to me, just as I am to him. But in a way I feel that I know him. I know his family from my study of Fehlen history.

•

For the past few years, I had been exploring what I called memory sites, places suffused with the memory of historical trauma. I had roamed the stillness of the giant hole where the bustling World Trade Center had once towered over New York City. I had wandered the vacant field that had been the vibrant, multi-racial neighborhood of District Six in Cape Town until it was razed under a "whites only" policy. I had walked through weed-choked cemeteries in Berlin and Hamburg that lay abandoned after the Jewish communities they belonged to had been destroyed.

Wherever I went, I took my camera with me. It was a lens through which to see more clearly and a shield that could protect me from what I saw. It was a form of external memory that recorded my response to the traces that history had left in these places, the pain and shame and helpless rage I would sometimes feel.

But photographing places and memorials was one thing. Photographing people was another thing entirely.

I wasn't even sure why I wanted to photograph Günther Lehmann. I wasn't a professional photographer; my professional work had mainly been writing. No one had sent me. I wasn't on assignment. I had no one to answer to but myself. But I not only knew that I wanted a photograph, I even knew what I wanted it to be.

He would be sitting or standing and holding a photograph, either of his mother whom the Nazis deported or of his grandmother whom the Nazis killed. He would be facing the camera and holding the picture out for us to see. I had images in mind that offered possible models: Rena Grynblat holding a photograph of the little boy she lost in the Holocaust in Jeffrey Wolin's collection, *Written in Memory.*[9] The Mothers of the Plaza de Mayo in Buenos Aires carrying signs with photographs of their sons and daughters whom the military dictatorship had "disappeared." These were images of loss and violence in which the victims were not just victims, but became accusers. "See what you have done, or allowed others to do," these images said. "See what has been done to us, to our families and our communities." I imagined Günther Lehmann adopting this same pose, looking at the camera and, behind the camera, at me. I would be the witness to his accusation, the accusation of Fehlen's "last remaining Jew," and in bearing witness I would take responsibility for what my grandparents' and parents' generations had done. This was the photograph I wanted to take of him. It would express my understanding of our relationship in the context of our families' histories.

And so, before I ever met Günther Lehmann, I had made him into an image in my mind. But the man I met was not that image.

•

After taking my coat, Günther introduced me to his wife, Ria (I wondered if it was a Jewish name, but she explained that "it's short for Maria"), and we went into a glass-enclosed sun porch that overlooked the garden. The table was set for *Kaffeetrinken*, the afternoon ritual of coffee and pastries with which Germans traditionally welcome guests to their home. "Frau Bammer," the Lehmanns smiled at me, "welcome!"

As the porch flooded with late afternoon sunlight and coffee and pastries were passed around, we felt our way into a conversation. I got my tape recorder out and pressed "record" (we had talked earlier about recording the conversation and Günther had said, "no problem"). Yet now he signaled me to turn it off. "Later," he said, "we'll talk later." But I soon realized that there would be no "later." We were talking now. I would have to listen and rely on memory.

"Tell me about you," I asked him, after the preliminary chit-chat. "Tell me your story."

"My mother was a Jew," he began.

His words were startling. They spoke of things that I had been taught not to say directly. Yet they weren't surprising. This was the story that I had come to Fehlen to hear and he was offering me what I had come for. However, this wasn't the story he wanted to tell. "I don't like to dwell on the past," he admitted.

---

**9** Jeffrey Wolin, *Written in Memory: Portraits of the Holocaust* (New York: Chronicle Books, 1997), 11.

"What good does it do? Digging up the past doesn't change things. All it does is bring the old hurts back." He paused and was quiet for a moment. Then he added, "It was a dumb time." *Eine dumme Zeit*, as he said in German.

I was taken aback by his choice of words – "a dumb time" – as if he were referring to minor transgressions, not a history of genocide. But when I asked him about the memorial stones for Fehlen's Jews and why he didn't want one marking Amalia Landau's house, he didn't talk of "dumb." He talked of antisemites, who, he said, were everywhere, "even in Fehlen," and he told me of the bar next to the house where Amalia had lived, "where things can get rowdy when the guys get drunk." His account of antisemitism and drunken rowdies conjured memories of the Nazi years, when drunken SA men stalked and threatened Günther's closest Fehlen relatives, the Franks, and smashed the windows of their house in the "Night of Broken Glass," the violent pogrom known as *Kristallnacht*. And I saw that "even in Fehlen," where people still often kept their doors unlocked and neighbors would watch out for neighbors, fears lingered that trailed the past in their wake. I saw the trace of this fear in Günther's resistance to having the house marked where his grandmother and family had once lived, perhaps even his reluctance to have me tape our conversation. In the shadow of this fear, my plan to have him hold pictures from the past for a photograph suddenly felt uncomfortable.

Yet we had talked about me taking photographs and I had brought my camera. Even though I was now uncertain what I wanted to do, I got the camera out. While I adjusted the settings for the space and light, the Lehmanns arranged themselves for the photograph they envisioned me taking. Instead of posing them, I let them pose. They sat side by side and leaned in toward each other slightly. A conventional pose: a happy married couple. They looked at me, smiles in place, and I took their picture.

•

But later that evening, back at the inn, I was frustrated and discontented. These weren't the pictures I had come to get. I had envisioned pictures that told a story of loss and survival and dramatic histories, a story of the Holocaust and "Fehlen's last surviving Jew." What I got instead reminded me of Tolstoy's comment about happy families: "they are all alike" – conventional, predictable, boring.

I wanted to go back and try to get my pictures. "Come in the morning," Günther said when I called him from the inn. "I'll be here. Come when you are ready."

•

When I arrived, he was in the kitchen with a little girl who reminded me of childhood pictures of Günther: the same blonde hair and bright, blue eyes, round cheeks, and resolute posture. There was no school today, I learned, and

Günther's granddaughter, Amelie, was spending the morning with her Opa, as she called him. They were playing a board game that I had often played myself with my parents when I was a child. It was called "Don't Get Upset" (*Mensch, ärgere dich nicht*). The goal of the game was to move your "men" (four colored pegs) all the way around the board and return them safely to their home base. Whoever got their men home first had won the game, and it was over when all the men had returned home safely. The drama of the game was that you had no control over the outcome. It was a game of chance. You won or lost, were destroyed or saved, by the throw of the dice. "My men are yellow," Amelie told me, proudly. "Opa's men are green."

I glanced at the board and saw that green was losing.

"We have to stop now," Günther told Annika, "Frau Bammer and I have some business we need to finish." But I didn't want them to stop. Something was happening here that felt important, something that moved me in ways I couldn't put a name to. The thrill of the game and the evident affection between the grandfather and his granddaughter were palpable. The energy in the room was infectious. But it wasn't until later that I realized what moved me so: when Günther's grandmother was deported to Theresienstadt, he was the same age as his granddaughter, Amelie, was now. Like the little girl at the table playing "Don't Get Upset" with him, he had been just an eight-year-old child at the time.

I told them to just keep on playing. "It'll take me a while, anyway, to get ready," I explained, as I fiddled with the camera settings. For a while, I sat back and watched them. Then I turned the camera on and joined them. Amelie's cheeks were flushed with excitement.

"You are losing, Opa," she shouted, rattling the dice in her hand like a warning. "My men are coming to get you."

"Oh no," he cried, "how can I escape?" He lifted his hands to his face in mock horror, but his eyes were laughing.

Amelie threw the dice. "A six!" she yelled. "Here I come," she shouted, "I am coming for you!"

"Oh no," he moaned again, "I am doomed . . . what shall I do? I will never get home to safety!"

I had started taking photographs of Günther and Amelie by now – moving their men, advancing, retreating, rolling the dice. They were so absorbed in the game that they took no notice of me, even when I got so close that I brushed against them. The drama of their game absorbed me as well. Yet my mind was frozen by the uncanny parallels to events in the history of Günther's family – the men and women who had been taken away from their homes and never returned home again to safety. My attention was elsewhere, somewhere in the past, and I lost track of what I was doing with the camera. I had checked the aperture when I started – f/11 was the right setting, as the light was bright – but in the meantime, it had started raining and the light in the kitchen had grown increasingly dim. To compensate, I had kept adjusting the shutter speed, but in my distraction I had forgotten to widen the aperture. I hadn't noticed the rain or the change of light or the fact that I eventually slowed the shutter speed so much that most of the photographs were blurry. I had been absorbed by the life and energy, the love and laughter that filled this room of the home into which I had been invited.

When Amelie won and Günther lost, both were happy. I was, too.

•

Like the day before, I didn't get the photographs I had planned to get. But I got something more important. I came for an image, and I found a person. Looking for history, I stumbled into a life.

•

On one level, this is where my story ends. Günther Lehmann generously shared aspects of his life with me: an afternoon with his wife, a morning with his granddaughter, stories about life in Fehlen, then and now. But he didn't invite me into his life beyond these boundaries, and I didn't feel I had the right to ask. He received me, as he so memorably put it. It was a visit at my request. He didn't intend a relationship beyond that. And while he didn't express his intention explicitly (Günther Lehmann was a meticulously polite man), when he thanked me for my visit and said, good-bye, I understood what he meant. My interest in the past was my interest, not his. He chose to dwell in the present. I never saw him again, but he taught me invaluable lessons. He taught me that the categories that propose to explain history to us, like my notion of "the last remaining Jew," may not be true to the experience of those affected. And he taught me that not all lives need to be told – and preserved – as stories. For some, it is richness enough that they are lived.

# Works Cited

Bammer, Angelika. *Born After: Reckoning with the German Past*. New York and London: Bloomsbury Academic, 2019.

Schöneberg, Mechtild, Thomas Ridder, and Norbert Fasse, eds. *Die jüdischen Gemeinden in Borken und Gemen: Geschichte, Selbstorganisation, Zeugnisse der Verfolgung*. Bielefeld: Verlag für Regionalgeschichte, 2010.

Strauss, David Levi. "The Documentary Debate: Aesthetic or Anaesthetic." In *Between the Eyes: Essays on Photography and Politics*, edited by Diana C. Stoll, 3–11. New York: Aperture, 2003.

Wolin, Jeffrey. *Written in Memory: Portraits of the Holocaust*. New York: Chronicle Books, 1997.

Mark Roseman
# A Journey to Izbica and Sobibor

In 2000 I travelled to Poland with Tomasz "Toivi" Blatt. Three years earlier I had
sent a first tentative letter to his address in Issaquah, Washington. I was researching a book that involved German Jews deported to the small Polish town of Izbica
Lubielska, Tom's town of origin.[1] I already felt an extraordinary connection to
Izbica, because I had in my possession a remarkable eighteen-page document,
part love-letter, part factual report, written in pencil by one of the deportees and
smuggled out from Izbica in 1942 under circumstances that were scarcely believ-
able but happened to be true. In the report, Ernst Krombach, a handsome young
Essen Jew writing to the fiancée he had left behind, described their primitive living conditions in detail, the exorbitant price of bread, the fact that there was only
one penalty for even the most minor infractions – death, and the complicated relationships pertaining between the German and Czech deportees, and between
them and the local Polish Jews. Tom, I knew, had been born in Izbica in 1927 and
had lived there until he and his family were sent to Sobibor in 1943.[2] He had
come to my notice as a feisty defender of historical accuracy, criticizing the assertion by the journalist E. Thomas Wood that Jan Karski, the famous resistance
hero, had not visited the Belzec extermination camp but rather had been taken to
Izbica to see the violence meted out to Jews there.[3] After several exchanges with
Tom, I intimated my wish to do more extensive research on his hometown, and
Tom later invited me to travel with him to Izbica and to Sobibor.

What follows is part of my record from our trip – the train journey to Poland
from Germany, and the visit to Izbica and to Sobibor in 2000. The author of
those notes was my 42-year-old self, working as an academic in Britain. I was at
the time relatively new to scholarship on the Holocaust, having until then

---

[1] Mark Roseman, *The Past in Hiding* (Harmondsworth: The Penguin Press, 2000).
[2] Tom recounted his experiences in Thomas Toivi Blatt, *From the Ashes of Sobibor: A Story of
Survival*. Jewish Lives (Evanston, IL: Northwestern University Press, 1997), having self-
published an account of the Sobibor uprising, Thomas (Toivi) Blatt, *Sobibor. The Forgotten Revolt* (Issaquah, Washington: HEP, 1996).
[3] Jan Karski, *Story of a Secret State* (Boston: Houghton Mifflin Company, 1944), 339–340, 84;
E. Thomas Wood and Stanisław M. Jankowski, *Karski: How One Man Tried to Stop the Holocaust* (New York: J. Wiley, 1994), 128ff. The issue hinged for Tom on whether Karski's account
of the topography and a barbed wire encampment matched conditions in Izbica, which Tom
knew had not been enclosed by barbed wire. On reflection, I think that Karski may indeed
have been in Izbica. Enterprising and courageous though he was, he certainly couldn't have
been smuggled into Belzec.

https://doi.org/10.1515/9783110753295-025

worked largely on the post-war period. But not least on the strength of a book I had just written about survival and memory in the Holocaust that was then making its way to the press, I was about to take up my first chair at the University of Southampton. On a personal note, I had been largely single for the previous ten years, during which I had been single-parenting for half the week, living around the corner from my ex-wife. All of this is evident in my account.

•

At 10:30 or so in the evening Tom Blatt and I arrive in Frankfurt station. The station is smart and spacious, but by 10:30 in the evening it's descending into a mixture of international traveler and city going to seed. Tom unsuccessfully searches for a bank to buy Zlotys. I buy schnaps and plastic cups in a little store. I'd made sandwiches, but the bread has not travelled well and is dry. This mountain of dry sandwiches is a great source of amusement to Tom, particularly when I later keep forgetting to get rid of them when a trash can was nearby. For him, they are a sign of a male inability to get that kind of thing right. I had strayed into territory that was better left to women.

We hadn't pre-booked a sleeping car as Tom said it was better to get one off the conductor. He indicates to the conductor that we'd both like a (six bed!) compartment to ourselves. As he talks, he moves his hand into his pocket a couple of times, in what I thought was a rather exaggerated, almost theatrical way. When the conductor comes back, Tom gives him some Deutschmarks, and we get our compartments. In fact, not many people are sleeping on the train and the money might not have been necessary. But we get a very attentive guard. The Polish railway compartments, with their mixture of shabbiness and grandeur – the red cloth curtains, and tassels, for one thing – have about them the feel of how I imagine old Eastern Europe. Tom says we should watch our luggage because the Poles steal everything. We've just stepped off the smart, bare platform in Frankfurt station and already he is in Poland.

Tom is tired. He's just completed a promotional tour for his book which was frankly . . . and here we enter a problem with language. Several times I was talking to others about his tour and found myself saying "murderous" (or in German, "ein Mordsprogramm"); then of course, you're brought up short, at having committed this blasphemy. But for someone half Tom's seventy-three years, this two-week non-stop crisscrossing of Germany, speaking every night, would have been hard to sustain. That Tom has done it is testament to an iron constitution and a will made of something stronger still.

I had caught up with one his lectures in Cologne. I had driven down from Düsseldorf to Cologne, arriving in the early evening in an alternative, shabby but exciting cultural center, "Die alte Feuerwache," the Old Fire Station. It was a large complex with a central square including a café. The buildings surrounding the square were full of activity. A guy in his twenties sporting an earring and attitude sat at the information booth and I asked him where the talk was. Oh, you mean the thing about the refugee camp? Well, Sobibor was not actually what you would call a refugee camp, I started to say, but it seemed pointless to embark on the extended explanation that would have been necessary. For a young left-wing German, possibly so much time had elapsed since the war that refugee camps were nearer his consciousness than extermination camps. So I nodded, and he pointed me where I should go. The lecture theatre filled up slowly. Finally, Tom arrived, short, pugnacious, sharp-featured, with slightly orangey, dyed hair. He stood, stiff-backed and alert, taking warmth from the radiator while he waited for the organizers to complete their somewhat chaotic preparations. After a brief introduction, he was off. He began by saying that his language was a "misch-masch" – a bit of Yiddish, a bit of English and a bit of German. But, he assured us, after 2–3 minutes we'd understand him. There was something so engaging about this admission, and about the confidence that lay behind it, something that already said we would not be receiving a dry lecture, that it created intimacy from the beginning. Tom told fragments of his story, focusing particularly on his life in Izbica before he and his family were deported to Sobibor.

There seemed little danger of his being overwhelmed by the story he had to tell. But there were two moments that almost took over, and they both concerned the weeks before the last deportation from Izbica in April 1943 – the deportation which finally caught Tom and his family. He remembered as a sensitive fifteen-year-old overhearing his mother talking with a friend. They were discussing how the Germans "killed us." Everyone knew for months that Jews were being killed en masse. No one knew what the method was. Was it gas? Was it steam? Was it shooting? Tom conveyed the horrible clammy feeling that descended on hearing this conversation. The other moment was also a conversation involving his mother. Tom was greedily drinking some milk and she stopped him from having the whole amount. Tomorrow is another day, she said. Shortly afterwards, the final "action" took place and the Blatts were taken to Sobibor. As the Germans divided the men from the women on the platform inside the camp, Tom said to his mother reproachfully – you said tomorrow is another day. And she said to him in anguish – is that all you can say at this moment? It was all he had said. He had moved over to his father and she was gone.

Tom attributed his survival in part to the strength of his will to survive. The Germans made a selection among the men. In principle as a youngster, without skills, Tom Blatt would not have been on the list. But he prayed inwardly so strongly that he thinks the message reached Franzl, the commander, who sought him out to be his shoe-shine boy (a job Tom in fact never performed). Tom's father was not so lucky. I think Tom may have influenced his own fate, but not through telepathy. Photos I had seen showed that he had been a very good-looking boy and young man. One could still see in him an innocent self-ishness – a desire for life that made one think not of others being trampled, but just of a person full of life.

After Tom's talk, we were shown brief extracts from the feature film *Escape from Sobibor*. The film must have been made in the 1980s. It had a good cast – Rutger Hauer was the big name – and solid Hollywood production values. It was hard to believe that the real camp inmates had been so well dressed or so well-fed. One of the actors was playing Tom, or Toivi as he had then been known, since despite his youth he had performed an important role, helping to lure the SS officers into the various traps. For us, now, without our really realiz-ing it, something paradoxical was at work. The film – with all the trappings that showed it was a "real" film, and not some low-budget documentary – was up-grading Tom in our eyes. He was important enough to have an actor playing him! This was a real story!

Later, I mentioned to Tom that the actors had looked too well-fed. No, that was correct, he said. As an advisor in the making of the film, he'd had a lot of difficulty getting across the fact that this was not a concentration camp, but an extermination camp. The rules were quite different. In a concentration camp, the inmates were worked to death, or left to live and work in conditions that barely sustained life. In an extermination camp, however, everyone was killed, but those who were temporarily reprieved to process the deaths lived quite well – funded by the spoils that arrived on the transports. They bartered the gold they found, and they ate the food they found. And they did not wear con-centration camp striped uniforms. So, to compound the paradox, now I realized that the apparent inauthenticity of the film – an inauthenticity which had helped to establish the film as Hollywood and thus, in a curious way, to up-grade Tom as having deserved a role in it – was at least partly authentic.

The talk had been co-organized by the Aufbau press and a left-wing group committed to raising funds for the preservation of the museum in Sobibor. After the talk, while I waited for Tom to get through a queue of people with questions to ask or books to sign, I got into conversation with a sexy-looking guy with dyed black hair who turned out to be one of the organizers behind the event. He was a leading member of the group involved in the project. He told me that his grandfather had been a key Nazi – one of the right-hand men of Hans Frank, the Governor General in Occupied Poland. So now, here he was helping to orga-nize an event in aid of preserving the museum at Sobibor. There was something very engaging and totally non-macho, but nevertheless confident, about his manner. While we were talking, at least three very nice-looking women came up and embraced him, in an intimate way, but it was unclear whether it had sexual overtones. Laughing, after the third such encounter, I'd asked if I could join the political group too. It struck me what an odd alliance had been constructed here. Tom, with his stiff-backed, rather old-fashioned, but engagingly life-affirming masculinity and this troop, with their casual new style physicality. Tom and I went with the group to have a drink and Tom and I thrashed out the details of our trip. We would meet on 5$^{th}$ April in Frankfurt and take it from there. It would be my first time ever in Poland, in fact my first time in Eastern Europe.

And now it is 5th April and Tom is exhausted. We sit down on the bed in his sleeping compartment. For some reason, whether because we are two men in a sleeping car, or because he's assumed that I'm gay (having made the sandwiches!?), or because there is a hidden side to his identity, he starts talking to me about various encounters he's had with homosexuals. The guard comes to ask us if we'd like tea, coming back with two plastic cups with tea, and two little plastic containers of lemon juice. As we sip the tea, Tom says he's been depressed lately. He'd had a medical scare that had shaken him. But the tour had been good. It had kept him busy. He hadn't dreamed so much. He tells me that for years he's had dreams which have revolved around the logic of survival and escape. Why had he not managed to get away from the guards in Izbica in 1943? In his dream's eye he would explore the alternatives, testing this way and that, to see how he could have escaped. They were so logical, the dreams.

Tom warns me we would need our passports soon at the (East German) border. I look at him. Is he joking? Suddenly, he realizes. He was still thinking about time before the Berlin Wall. And *I* suddenly realize we are making a journey with *two* pre-histories – his life before 1945, and the journeys he's made back before the Wall came down. What continental Europe had been through! In Britain, we didn't have a clue. The schnaps bottle is surprisingly empty by now. I decide to go to bed. Tom says I should get up at 6 in the morning, because a beautiful girl, a girl with the most beautiful eyes I had ever seen, would be coming to the train in Dresden.

I go to my compartment and lie on the bed in the darkness, watching lights – some distant, some close – flash by the window. It's a glorious way to travel. It feels wonderful to be snug in one's little room, while in the night world beyond the compartment engine drivers, railway signalmen, a whole industry, organize this journey so that I can get where I want to go while I sleep. And then I think of Ernst Krombach, the young German Jew who had been deported to Izbica, and whose last journey I had followed so closely. His deportation train had taken almost the same route, crossing the German border beyond Dresden and then plunging into Poland. What will he have made of the whole nightmarish system – the engine drivers, railway signalmen, managers and SS-figures planning his nightmarish journey into uncertainty and horror? In postcards he'd managed to send to his fiancée from the train, he took care to sound sanguine. But what did those cards really tell us of what he felt and what he was doing? Until well on in the journey he was not sure even of the train's destination.

I sleep in fits and starts. When I wake it is still dark and I'm still tired. The train has stopped and I hear talking outside. I get up and look out. Tom is at the door talking to a woman on the otherwise deserted platform. I walk down and say hello. The guard is smiling. He obviously likes the idea of this old man having a twenty-year-old romantic admirer. The girl *is* beautiful, in a strange way, her eyes very large and romantic. She'd had trouble finding us, and now they only have a few minutes to chat. I forget the connection between them, she was the daughter of Polish friends, studying in Dresden, I think, and Tom had given her some advice or some help. And now here she is, on her own, on a deserted platform to see him. He gives her a copy of his book, he gives her a kiss, we shake hands, and we are off, gliding past the romantic movie scene of her walking down the platform. I go back to bed and sleep for a couple more hours.

In the morning, we discover there is no restaurant car on the train. We eat some of the dry sandwiches. Tom grimaces. And then I start asking him about his life after the war. In August 1944 he had been saved when the Red Army entered his part of Poland. He worked for the Polish secret service for a while and then, after a complicated story involving, I think, antisemitism, he fled to Germany. He spent a while there, leaving a pregnant girlfriend in Poland, and then returned to Poland to rejoin her. He worked as a "personnel" officer and then started at university. He was ejected after having refused to snitch on a fellow student and spent a very pleasant two years as a tour guide in a mountain resort. And then towards the end of the 1950s, with the campaigns against Jews, he left Poland, first for Israel and then the USA.

•

After spending a couple of days in Łodz and Warsaw, we rented a car and I drove us the three-hour trip to Izbica. The driving down the Lublin road was tough – I wasn't used to handling the shift with my right hand, and overtaking involved moving into the shared middle lane and pulling back in before the traffic coming in the opposite direction. I was telling Tom he should write the memoirs of his postwar life, but he said there were some shameful bits, it would have to be fiction. Politically or morally shameful, I asked, my mind poised between leaving girls and compromising oneself with the Communist Party. But it turned out he was talking about killing Germans. He couldn't kill children, and the fact that the Germans were capable of it had made him wonder about them. But when he found Nazis, they died.[4]

By the time we reach Piaski, halfway between Lublin and Izbica, it's dark. Tom points out where he had emerged from the forest on the run from Sobibor. He shows me the house that was the Gestapo building where a guard had gone inside, and the point where he had hidden in a ditch.

Soon after, we're stopped by the police for speeding. The transaction's all in Polish, but it's clear to me that the policeman suggested a bribe. Tom fobs him off. He tells him I'm visiting from abroad and should have a good impression of the country. That seems to have worked. Tom seems comfortable. The exchange has no unpleasant resonances for him.

There's been no advance warning but suddenly a road sign signals that we have reached Izbica. Here I am at last, in this imaginary place, whose blocks and vistas I know from an 18-page letter, written in German in neat pencil handwriting in August 1942 – and now it turns out that the place is real. Twice we miss the little road that leads to Tom's friend. Her house is isolated and dark. Dogs are barking somewhere. The woman who opens the door is not beautiful but sultry – and a bit sulky: we are late! Luckily not too late – Christina served up a three-course dinner. Soup, steak and salad, and pudding. We drink vodka. Tom will be staying there, and is hopeful, but he takes me to some kind of Catholic dormitory to spend the night. The concierge leads me to my small and rather shabby room. He opens the tap into the basin to see the water runs out. It's clearly blocked, but if you wait long enough, it empties. He nods good night and for the first time I'm on my own in Poland.

Around eight the following morning I go downstairs and stand outside to wait for Tom. Despite the absence of wind or much snow on the ground, it's bitingly cold, a very un-English kind of cold. Tom doesn't arrive. I peer into the breakfast room. Young people are sitting eating in silence at a long table. I reenter and sit down next to some girls. They carry on chewing looking straight ahead, but with a hint of uncertainty. Is this breakfast also meant for me? I turn to the young woman next to me. Does she speak English? No. But she gestures to a young woman on the other side who approaches. After several efforts at awkward miscommunication (I ask – where is the woman in charge? She responds, where is this charge?) it dawns on me. Are you supposed to be eating in

---

4 Tom never said more than this cryptic comment. I'm not sure if was referring to extrajudicial killings while in the secret police.

silence, I ask? She nods. I thank her and go outside to wait for Tom again. Tom doesn't arrive. At 8:20 I go back inside and take a doughy, unappetizing bread roll and try to impress hard butter on it. There's some rather odd-looking meat loaf which actually tastes quite nice. I have no tea. I turn to the young woman to my left and gesture, she whispers to the English-speaking girl on the other side of the table who goes and gets me a tea. Djiekuje.

I'm just beginning to relish the Zen of eating silently when Tom arrives. He refuses to talk quietly. I whisper back. What's that! I can't hear you! This hall of residence was built on the grounds of his grandparents' home, and he's not going to talk quietly for anyone. I still feel the silence of the mealtime deserves some respect. We're both cross with each other. He looks at the meatloaf contemptuously. This he doesn't eat. Because the girls next to me had gone before he arrived there's a gap between me and the remaining diners. Tom, as later occurs to me, suspects apartheid. Did they put you here? I explain. He wanders down the table to grab a tea. He tells me about a previous visit. He got talking to an old lady. She said, "it's much better without the Jews." Why? "It's much cleaner and smarter." And in a way, says Tom, she was right. "But I said to her: Do you have a cinema? No. When the Jews were here you had a cinema. Do you have a library? No. When the Jews were here you had three libraries. Do you have a station? When the Jews were here you had a station." She had no answer.

Neither do I. That the station and cinema had gone had, I felt, perhaps nothing to do with the Jews. But three libraries? It was hard to imagine a non-Jewish community of this size managing even one. (It was only later that I realized how absurd the woman's claim had been. In 1939 92% of the town's forty-five hundred inhabitants had been Jews.)

We go down to the car. Before our journey, Tom and I had both looked at some photographs from wartime Izbica. Tom reveals that this nondescript little side alley is the street depicted on a German soldiers' photo we had seen. It's hard to line up initially, until one adjusts one's sense of scale. There's a strange mixture of progress and decline. A scruffy, uneven dirt-track, lined with shacks, has been replaced by an asphalt road with solid houses. But what looked like a lively central thoroughfare has been replaced by this poky little characterless alley.

We drove over to the town's principal industrial site, established in the 1930s, a brick factory. This place was significant for me, because Ernst Krombach had gardened for the owner on his estate and had later worked in the factory. Tom produces a photo of Himmler visiting the factory with Governor Frank. You see? At first, I can't match the photo to what was before me. Again, we're both irritated with each other. But then I realize that one has to make a similar adjustment of scale. The grand structure beyond Himmler is in fact the little projecting bay in front of the main building. Now, I'm convinced. This was the brick factory where Ernst worked and where towards the end of 1942 he was probably blinded in an accident. In fact, comparing it with the photo, the factory seemed completely unchanged. Then as now there are piles of bricks in the yard. Two workmen wander over, suspicious. They ask me something. I mumble an approximation of nie mówię po polsku. Tom leaps out of the car and explains. They let us in through the gate, and Tom and the men exchange names. I take more photos. One of the men takes Tom's business card – he'd like to buy my book. After we return to the car, however, Tom says, emphatically, I don't trust them an inch. If I had seen them then, I'd have run a mile. I can see what he means. There was something about their interest that was almost scarier than their initial suspicion.

We drive on and park on the corner where the little road to the brick factory joins the main road. An old lady in a head scarf bends down to the window and talks to Tom. She wants to know why we have parked here. Next to the car is a grey trailer and a house, and between them, which I didn't even notice at first, is a gap. It looks like private property, but Tom marches through it and onto a little path. It takes us up a short steep muddy hill. Suddenly, we are fifty feet above Izbica. I hadn't properly noticed before: the village really is nestled in at the end of a little valley. When the Ukrainians were on the hills, Tom says, nobody could get out.

Tom leads me to a large gravestone, now lying flat. Until a few years ago, the stone was standing. You could see the bullet holes. This was where Engels, the infamous head of the local Gestapo, shot people. Then more gravestones. One is for a converted Austrian Jew, Gertrude Mitterbach, murdered there in an "action," with a dedication from "Dein Sohn Helmuth" – and on the stone there is a cross. The other gravestone still standing also bears Christian symbols. Tom tells me it has been put there by a priest given over to nuns by his Jewish mother and brought up a Christian. There is a stone in memory of his parents and a gravestone, engraved with fishes, ready for him to be buried there.[5] Next to these is the (symbolic) outline of three mass graves. These commemorate the two thousand murdered in the November 1942 action (though not directly in this spot.) After that round of killings only a tiny remnant of Izbica Jews was

---

5 I have this in my notes. However, the Izbica guidebook refers rather to a monument in honor of the families of all murdered Jews, erected in 1967 by Fr. Grzegorz Pawłowski (Jakub Hersz Griner) and his brother Haim Griner. Jakub escaped from Izbica in November 1942 before his parents and sisters were killed. He later converted to Catholicism and became a Catholic priest in 1958. So, I think I may have confused two things that Tom said.

left – including the Blatt family. The stone memorializing the event has been vandalized. From the old Jewish cemetery up here there is only one stone left. All the rest have been torn up – for what? On one side of the hill, however, in a little leafy world hidden from the town below is a new wooden building. Orthodox Jews from the USA have erected it in honor of the famous line of Chassidic Rabbis from Izbica. Not knowing the US Chassidic scene very well, and not yet having understood quite how overwhelmingly Jewish Izbica had been, I'm very struck that somewhere in the United States of America, completely divorced from the Christian Izbica of today, there remains a memory of another, very Jewish, Izbica, enshrined in this little hut above the town.

Tom has rushed on ahead of me – he's cold. It really is freezing. I thought of the Krombachs huddled together on days like this. I arrived in Izbica in April just as they had. Once back in the car, we drive past the main square where Tom's father's store had been situated and turn left into the little road where the family's last home was. The lower part of the wall has been hacked away as though someone had started putting in waterproofing and then forgotten about it. Tom tells me a story – either about this house, or one of the previous family homes. He had come back on a visit and asked to see inside. The owner had asked him where the gold was hidden. Mischievously, Tom had looked around as if unobserved, nodded, and murmured with satisfaction that everything was "just as it was." The next time he visited, the house had been torn apart.

Tom points out a little street where he'd been trapped by the SS having gone out on the street for a dare during curfew. I'd forgotten this story, and Tom snaps at me, wondering if I've actually read his autobiography. I say rather mildly that he is in a pretty irritable mood this morning. He's immediately more gentle. He'd not slept well, we'd eaten so late.

If you follow the little street of his family home away from the main road, it bends into another that leads right back to the same main road. As a result, the houses from the two small roads back on to each other. Behind the Blatt's home was the house where the new mayor lived, an ethnic German named Schulz. Jews had had their radios confiscated, but by listening against the wall the Blatts could hear Schulz's radio. They were taking a risk. According to testimony I had read from a man called Hejnoch Nobel, Schulz was a piece of work. He had trained his dog to attack people wearing the blue and white armband (the version of the Yellow Star the Nazis had introduced in Poland).

We drive up another small hill. Here are the ruins of the tannery, the Blatts' last hiding place. They might have escaped the Nazis' notice had not local Poles insisted that more Jews were hiding than had been brought out. At this point Tom and his family were discovered. After they were brought out, Tom had gone back briefly to retrieve something and saw that more people were still hiding. As he came out, an SS man threw in a grenade and finished off the rest. Even then, Tom managed initially to slip away. We drove further up the hill to a little wooden barn. Here he had hidden but had been betrayed by a non-Jewish boy, a former classmate at school.

At the top of the hill the road turns into a dirt track. This was the track that led eventually to the farm of Bojarski, a farmer who had sheltered Blatt for a while after his escape from Sobibor but had later tried to murder him. We don't try to drive there. It's too muddy. And the further we leave the main road, the more uncertain Tom becomes. I should be careful making notes and taking photographs. Under the Communists he had felt more secure. Now the people feel they could do anything. Once he'd been photographing his house and a local tough had told him he needed permission of the owner to take a photo. What are you doing, acting for the owner? Tom had wanted to know. If the owner wanted to say something, he would tell him that he, Tom, had a better right to the house than its current owner. The man had gone away. They think you're going to be intimidated, says Tom. On another occasion he had been in the town with a group of young Israelis and a gang of local toughs had followed them around making threats. In the end, they had been forced to take a taxi out of town.

Driving back, we are stopped by a passerby who asks for a ride and climbs into the back of the car. As a Jehovah's Witness the man is not part of the normal Catholic milieu and he becomes enormously deferential to Tom. Sometimes when we stop at houses, he leaps out and makes inquiries on our behalf. We drive out to the next village along the road. There's hardly any break between one village and the next, and yet the distance between them had been the difference between life and death. In Izbica after his escape from Sobibor, Tom was too well known and in constant danger. But in Ostrzyca he found help from his father's former friend, Mr Nizioł and then in a neighboring village from another farmer, Petla. Tom is hesitant to visit the Nizioł family now, since it had become clear that the neighbors talked when the Jew visited, and the old man had become uncomfortable. But we drive past the farm anyway. Nizioł's grandson, now himself grown, walks up the drive and shakes hands, friendly but wary. His grandfather is too ill to receive visits, he says, and perhaps it is true. We drive on to Mchy, where with the Niziołs' help Tom had gone once it became too dangerous in Ostrzyca. Here farmer Petla had given Tom extended protection as a herdsman. But he is out, so Tom shows me the barn where he had slept. Next door to Petla was another farmer who had been supportive, despite having had an enormously antisemitic wife. Ostrzyca and neighboring Mchy were "good" villages, Tom says, where he and other Jews had received protection. But just over the hill, five km away by road, is Wirkowice – a murderous place, where Jews on the run were killed by Poles or handed in. What was the difference between them, he had asked a friend. Nizioł as the biggest farmer in Ostrzyca had set the tone.

We try to find the farm in Mchy where the Rosoliński family lives, a family who had been connected to the underground and who had put Tom in touch with the partisans. Unsure once more whether the farmer would want to see him, Tom sends our admiring passenger out to check. Why is he doing all this, I ask Tom. For him, I'm a hero, Tom says, a hero of the resistance. The man had told him he'd put Tom's business card in a "holy place." Here too were echoes of past relationships – but of a rural deference – the kind of deference Tom's father, by Izbica standards prosperous and a hero of the Pilsudski army, will have enjoyed among non-Jewish Poles in the 1920s. Our friend comes back to the car. It was the wrong farm. Tom turns to me. He says he's realized why he was irritable. The place makes him nervous.

We drive back into town and drop off our passenger, but not before he asks if Tom can get the man's daughter a job in France. Tom has no links to France, but he is an international man. Surely, he will be able to help. We stop in the square so I can buy flowers for Christina, who is cooking us lunch. The store has artificial flowers in a range of lurid shades. Real flowers are limited to red roses, which seem hardly appropriate, and some faded something or others. I choose the latter. We are still too early for lunch, so we go back to my room. Tom lies down on the bed for a nap. "Ach God," he says, "This town has so much blood on its hands, it's unbelievable. They look at me, shake my hand, and in the back of my mind, always in the back of my mind, I think would I have to run away from them in those times." "The more time passes, the more angry I feel. What they have done to us! The Germans didn't know a Jew."

Christina has prepared an enormous meal. What did I think of Izbica, she asks, contemptuously. She means it's a one-horse, rundown town (which it is.) But when I say truthfully that I have never in my life been in a place so redolent of history, she says "yes?" and drops the subject.

After lunch, we drive off to Sobibor. It suddenly strikes Tom that we were driving down the same road, in the same month, as the truck that had driven him and his family from the tannery, fifty-seven years earlier, in April 1943. Another April anniversary, this one a year after the Krombachs' arrival and months after they had all been murdered. For the Blatts and the other remaining residents of the town, as they were loaded into the trucks, there was already definite knowledge that Jews had been exterminated. They had known about the Chelmno camp since early 1942, the Belzec camp since the spring and later they heard about Sobibor. On the day of the round up, the Blatts and their compatriots still nurtured a faint hope they were being taken to a labor camp to work.

As we drive along the road, Tom points out the two turnings they could have taken to other known labor camps in the area. After they passed the second turn off and still the truck stayed on the road to Sobibor there was no hope left. A collective sigh had gone up in the truck, and Tom sighs in the recall.

Tom is smoking. He says smoking makes him sleepy. Not me, I say, remembering that smoking and coffee used to be the way I got through essay-writing nights. But I gave up smoking in my early twenties. He looks at me. "If somebody had told me I would be returning here now sixty years later with an American passport in my pocket I wouldn't have believed it. So I think I can risk smoking!" "You know," he says, "life is stranger than fiction." And he means, not that the Holocaust is beyond the wildest imagination of any fiction writer. For Tom, the Holocaust is a given. He means the coming back here, now, prosperous, American, older.

We are driving through the beautiful forests in which Sobibor was based. It had been the most secret of all the camps. Though the Blatts had known about it, there were still many people who arrived there in 1943 who had no idea what it was. Finally we reach the station of Sobibor. Seeing the signal plate of Sobibor – once again, the sudden sense of the reality of place, takes my breath away. It's not just that it proves that a place called Sobibor existed. It's also that since Lanzmann's *Shoah*, I have known what Sobibor looks like. And here it is, just as it "really" is on film.

Sobibor is still a working station.[6] Beyond it are little ordinary houses. Two children are shooting each other in a cops and robbers game. The main line is single track. But then, just beyond the tiny station, there is a set of points (switches), and back from the points comes a little siding (loop line), which ends next to where we've parked the car. Unlike the station fifty yards away, the siding was built for the camp and ends within what was then the barbed wire perimeter. Next to the siding, within what was the camp, there is now a row of ordinary houses. A couple are gardening in their front garden and look up at us, expressionless, as we walk along the raised embankment in front of their house where once the initial selections had been made. Among these houses is the one that had been the chief commandant's house in the camp, still painted in the same green it had been then.

Now we're retracing the steps of those who had arrived from the trains. Unlike the mixed concentration and extermination camps, the prisoners were not sorted into fit and those who were to be gassed. In Sobibor there was no camp apart from the killing machine. But the men were separated from women to

---

6 It was still in 2000.

facilitate the preparation for death, and sometimes small numbers of workers were selected to replenish the ranks of the processors. This was where Tom had joined the men. And this was where some spark or connection induced the SS officer to select Tom.

Near the siding is a plaque and a primitive wooden building housing a limited museum. It is effectively Tom's museum. Before there had been a Kindergarten in it, but Tom persuaded the authorities to have a museum there instead. It was facing severe financial problems because the Polish government had cut the funding, and Tom was trying to raise funds to put it on a more secure footing. Tom had also managed to get the new chapel excluded from the official camp perimeter – though this was notional since virtually nothing of the camp remained.

Because so little has been preserved, Tom describes to me what had been whereby pointing to the bits of field and showing where the fences had been. Now, we're walking along a path through the woods. When the Dutch transportees had arrived in the summer of 1943, on normal trains, accompanied by doctors and nurses, with no notion of what they were arriving at, they would have left their luggage on the platform. They did this without anxiety because their suitcases were all labelled. But now we're in the area where Tom and other members of the reception team had to get the women to leave their handbags. This caused much more disquiet, with their private things and valuables, unmarked. We carry on walking down the track – it is relatively new, having been put in to mark the camp's fiftieth anniversary. We have to imagine the original track was more of a dirt track and a little over to the left, says Tom. I look over to the wood to the left and try to imagine the arrivals walking there. The path turns, and here there is a monument, a statue of an agonized figure. And beyond it we come to the area where the gas chambers were. Tom is upset – there had been a glass case set in the wall through which you could see the ashes, but it was no longer there. Some visitors have recently lit some memorial candles at this little wall. But today, on this cold April Saturday, we are the only people here. We turn to the right, onto a green meadow. This was the worst part of the camp, Tom says. This was Lager III where the crematoria were. The workers here were completely cut off from the rest. When you entered this part of the camp, you never came out. So that when the revolt took place, the resistance had to leave those in Lager III behind to their fate. Now, there is nothing, only the meadow. Tom kicks over the soil and scrabbles around. In his hand he picks up three bits of white and one blackened bit of stone. Only it isn't stone, it's organic matter. I recognize the filigree pattern from chicken bones.

Tom died in 2015, aged 88. I deeply regret having lost contact with him over the years. To Tom's chagrin I also never wrote the promised longer piece about

Izbica, and though I produced a radio program for the BBC about the deportations there, I feel guilty not to have returned the favor for his generosity.[7]

•

Holocaust scholars, me included, worry what will happen when the last survivor is no longer with us. We fear that the Holocaust will lose immediacy and moral weight or, worse, that the fact that it happened at all may no longer register on public consciousness. A recent Claims Conference poll shows how many younger adults have never heard of the Holocaust.[8] The many signs of reviving antisemitism only add substance to this concern. Yet thinking now about our trip, this assigned role as public oracle barely scratches the surface of what the interactions with someone like Tom offered, and what we are losing, have lost.

Our trip together was one of the more memorable events in my life. It made such an impression that it is hard to believe we spent less than twenty-four hours in Izbica and Sobibor. Looking back now, I realize that what made it so powerful was the intersection of two separate spaces: the social and geographic one outside Tom Blatt – the Polish world, home and alien, friend and enemy, a permanent provocation – and the remembered one within, the experiences of memories of terror and loss, self-assertion and survival, some of which could never be conveyed. For a few hours, I glimpsed, I felt, the landscape that he saw and felt, part topography, part history. In Sobibor, indeed, the topography existed only in his imagination. What made it so powerful was precisely the absence of public remembrance. There was no tourism, at least on that wintry Polish spring day. The absence of public signposting put a premium on what Tom remembered, and at the same created a bleak echo chamber for what he had to say. The environment of neglect was emotionally mood-setting, reinforcing the sense of an unbroken past: and nothing could be more authentic than kicking over the tough terrain and exposing mute, bleached fragments of bone to cold sunlight.

It wasn't that Tom could claim to be the custodian of *the* truth. The Poland I saw was through his eyes, and no doubt there was another Poland I could have seen, had I been in the company of someone else. I could have learned more about the incredible pressures this most brutal of occupations imposed on the non-Jewish population. We also traveled together in a particular moment. Growth had been strong in Poland in the 1990s, but the GDP per capita would

---

7 "The long road to Lublin" for BBC Radio 4, broadcast 24 January 2002. Accessible on BBC World Service as part of its "Essential Guide" at: https://www.bbc.co.uk/programmes/p035xbj6.

8 http://www.claimscon.org/millennial-study/.

double over the next decade and with it much in the social landscape would change. Jan Gross's *Neighbors* had not yet appeared.[9] The given knowledge from which our conversations proceeded would have been different if Gross's book had already been making waves in Poland (and different again amid the recent concerted efforts by the government to write Polish complicity out of history). The power of Germany's commitment to commemoration has made itself felt in the meantime in new sites of memory and a joint project with a local middle school to care for the Jewish cemetery.[10] I doubt that the weight of the past would have been any less palpable, however. It was present in every interaction, be it hostile, wary or admiring. And it was present in every word and every gesture from this generous-hearted, feisty, fragile, remarkable man.

# Works Cited

http://www.claimscon.org/millennial-study/.
Blatt, Thomas Toivi. *From the Ashes of Sobibor: A Story of Survival*. Jewish Lives. Evanston, IL: Northwestern University Press, 1997.
Blatt, Thomas (Toivi). *Sobibor. The Forgotten Revolt*. Issaquah, WA: HEP, 1996.
Gross, Jan Tomasz. *Neighbors: The Destruction of the Jewish Community in Jedwabne, Poland*. Princeton, NJ: Princeton University Press, 2001.
Izbica Guidebook. Accessed 17 October 2020, http://shtetlroutes.eu/en/izbica-przewodnik/.
Jankowski, Stanisław M. *Karski: How One Man Tried to Stop the Holocaust*. New York: J. Wiley, 1994.
Karski, Jan. *Story of a Secret State*. Boston: Houghton Mifflin Company, 1944.
"The long road to Lublin." BBC Radio 4, broadcast 24 January 2002. Accessible on BBC as part of its "Essential Guide." https://www.bbc.co.uk/programmes/p035xbj6.
Roseman, Mark. *The Past in Hiding*. Harmondsworth: The Penguin Press, 2000.

---

**9** Jan Tomasz Gross, *Neighbors: The Destruction of the Jewish Community in Jedwabne, Poland*. (Princeton, NJ: Princeton University Press, 2001). Since then, of course, our knowledge of Jewish-non-Jewish relations under Nazi occupation has exploded.
**10** Izbica Guidebook, accessed 17 October 2020, http://shtetlroutes.eu/en/izbica-przewodnik/.

Grazyna Gross
# Mrs. Kraus: A Short Story from a Central European Girlhood on the Run

Dear Mrs. Kraus,

You do not know who I am, we have never met. So, I must first introduce myself: I am a 13-year-old girl living with my mother and her male friend in your apartment without your permission, not that you would have been able to give it to us. I want to assure you that we will be moving out in a few weeks, and once the war is over you will be able to move back in, provided you are still alive.

You may be wondering who brought us to your apartment. If I knew who that man was and what he did for a living, I might be able to figure out how we ended up at your place. Since I know only a few words of his language, I understood very little of what he was saying. You were a widow, I learned, and crazy besides. At some point he looked at me and said, "She should be going to school." To say something so stupid, *he* must have been the crazy one. What school? Did he know that I could not speak Czech? How much did he know about us?

Your apartment is very nice. I really appreciate it. Five years ago, I lived in one similar to yours. The outbreak of the war put us in motion, from a beautiful town in the asshole of Europe, as that area is called by some, to this famous center of culture. Nowhere along the way did I sleep in an apartment as nice as yours and that means a lot because I have stayed in so many different places.

I take the streetcar to the center and walk around the old town. Everything here is oversize and grandiose: the castle, the churches, the river and its bridges, the main boulevard and the fancy apartment buildings that line its side streets. The last town I lived in had all of the above, but everything was smaller, as if built for children. There, every landmark came with its own story that I heard again and again till it became a part of me. Here, too, there must be stories, but no one can tell them to me.

What were you like when you were still living here? I am not asking what you are like now because once they took you away you stopped being who you had been here. I keep snooping around your apartment for clues and so far, have come up with only one photograph of a handsome mustachioed man who was probably your husband. That it was taken in Sarajevo explains the Turkish decor of your apartment. You must have had other photographs, but either they were thrown out when your apartment was being looted of its real valuables or

https://doi.org/10.1515/9783110753295-026

you took them with you when you had to leave. I took our photos and carry them in my knapsack. If you were here, I would show them to you. Someone has told me that people, when they have to flee, take with them pictures of their "once upon a time" life so that they can be certain that it was not just a dream. The trouble is that photographs can't be exchanged for a loaf of bread.

Your phonograph and records have been a great pleasure for me. We never had a phonograph, not even before the war, because Papa did not like music. Later, we probably had no money for something that could not easily be carried when one is on the run. I keep listening to your records. When the phonograph broke down the other day, my mother managed to find someone who was able to fix it. Since people here are not kindly disposed towards those who don't speak their language, finding someone willing to repair it was another of Mama's great accomplishments. My favorite records, which must have been yours too, are *Der Zigeunerbaron*, *Light Cavalry Overture* and *La Campanella*. I play them over and over again. Can you hear music where you are?

Your beautiful apartment may be small, but I think it would be perfect for one or two people. There are three of us living in it now, so I sleep on a cot in the kitchen, Mama and her friend on the couch in the main room. That is how we staked out our space. Up till now I was not supposed to know that Mama and the man slept in the same bed. That they do is no longer a secret, and we have no neighbors who would care. When not touring the city, I spend most of my time in the main room listening to music, writing letters to my friend, letters that will not be mailed, or trying to read two books in two different languages, neither of which I understand. With no one to talk to I am often just plain bored. Did someone steal most of your books? I know a lot about looted apartments because I have lived in one. When we moved into yours, only the heavy furniture and a few books were still here.

I have been away for a while, Mrs. Kraus. Did you miss me? My mother took me to where we had come from: a very long train ride, bombs on the way, blackouts for most of the evening, people afraid to go out even when they had to in search of food. My aunt did our shopping. She was the only one brave enough to leave the apartment. Still, I wish I could have stayed, but my mother wouldn't let me. She is not yet ready to give me up – though, every time she leaves me here alone for several days, I wonder what is on her mind. With three different sets of identity papers, all of them false, what mess would I end up in if she disappeared completely?

Do you know your neighbor? I do not. The janitor told us that an ancient man lives next door. He never goes out, and a housekeeper takes care of him. I see her sometimes on the balcony when she puts out the garbage. She never looks at me, not even with a quick glance. If she did, I would try to give her a little smile or even a nod, but it is obvious that she does not want to have anything to do with me. So, when she comes out, I go in. Now, after my mother, her besotted friend, and I got into a fight late one evening, the woman must be thinking that we are true barbarians. The noise we made was not pretty.

I am giving up on one of your books, the one in German, because, after trying to guess the meaning of words connecting those I recognize, I still cannot make any sense out of the first paragraph. Well, maybe I'll try again one day – unless time runs out. It has been much easier for me to decipher the book about clairvoyants and supernatural events, thanks to the similarity of our Slavic languages. I am trying to find the location of the psychic events that took place in your city. Mozart's haunted house, the traffic circle where someone fell under a streetcar as predicted, and that church, the famous church you must know about. No one can tell me what really happened there. Perhaps no one knows. People slow down when walking past it. I think that must be how they express their respect for those who died there.

Of course, I don't believe in the occult – or at least I am not supposed to. But that does not keep me from wishing that my brother would visit me one night, even though I would be terribly afraid of his ghost.

Once upon a time at a dinner party, when the good times were already slipping away, one of the guests insisted on looking at Papa's palm to see what the fates had in store for him. That done, he had the nerve to tell Papa that he would not survive the war. Did I ever cry! According to the palmist, Mama was going to end up on the other side of the ocean where people do not die in camps. I did not want to know what was going to happen to me and ran out of the room. What about you, Mrs. Kraus? Did you go to any of the seances described in the book to find out how much longer you are going to live? I wish we could meet in your apartment and have a talk. I am sure you have some interesting stories to tell.

You may be wondering what the three of us are doing here. Waiting, I suppose. Waiting for the front to approach from the East, and, when it comes too close for comfort, we will resume our trip to the West which, at the moment, is not yet ready to receive us. When Mama is working, I am left alone in your apartment, sometimes for days. Her friend is gone then, too, but I doubt that he is traveling with Mama. He is not brave enough for that. Since it is not safe for me to wander the streets when Mama is away – what would I say if stopped by police? – I do not go out, not even to the bakery. No, I am not starving, but after a few days of eating nothing but stale bread, I have to agree that whoever said "Man cannot live by bread alone . . ." was right.

When Mama and her friend are not away on business, we sometimes go to a restaurant for lunch. A famous place, you probably know it: it's on an island near the bridge. A vast dining hall waiting for the privileged to arrive for a sumptuous evening meal. At noon the place is empty except for one big table near a window where a group of men, twelve at the most, gather to eat dumplings and talk about changing the course of history and liberating the East. Powerless dreamers, they are. I always sit across the table from a man who listens to what the others are saying but says very little himself.

If you are no longer alive, you may be able to see what is happening in your apartment. Some people claim that the spirits of the recently deceased linger in their favorite rooms before settling down for eternal rest, which can be either heaven or hell. If this is the case, then you probably know that I have been snooping around in your closets and drawers, not that there is much left in them. I must have been terribly bored when I began to rummage in the small drawers of the kitchen cabinet where people store odds and ends, the kind of stuff they may need one day but, in the meantime, do not know what to do with. One drawer contained a baby bottle. What could a baby bottle be doing among your belongings? As far as I know, you never had children and at your age you were not going to have them. The man who brought us here, the one in the dark suit, assured Mama that you had to be taken to a safe place because you were old and muddled in the head. Perhaps you were, and that would explain the baby bottle. Intrigued, I took it out of the drawer and was holding it when Mama yelled "Put it back! You don't have to touch everything that is here!" I am sure that had I taken out scissors or a roll of string, she would have said nothing. It is the bottle that set her off, and I have been trying to figure out why. Perhaps you can explain. I know that you have a doll, sometimes I even play with it, but many homes without children keep dolls for decoration because even grown-ups like to look at a doll's deep blue eyes that open and close and its beautiful porcelain face. But a baby bottle? I can't figure it out.

I would like to let you know what happened here the other day in case you are not aware of it. I was sitting at the desk in your living room writing a letter to my friend Renata, while Mama and her friend were having lunch on the island in the company of plotters, dreamers, and spies. I assure you that Mama is none of the above. She joins them so that she can get a hot meal without having to cook it herself. I stayed home because, without any new ideas, the meetings are becoming stale and boring. Besides, Mama promised to bring me back a serving of dumplings. I was surprised to hear the front door open before it was time for them to return. Mama entered the room, a tall man in a black coat behind her. Her friend, the coward, stayed in the foyer where, so close to the entrance door, one could slip out easily. I wish you could see how proudly Mama can walk. The more frightened she is, the more assured her gait is. This time she was really scared, and if her gait was not enough of a warning, her eyes, wide open for a moment, were alerting me that she did not know what was going to happen and that whatever it was going to be did not look good. Half-way across the room Mama stopped and let the man pass. He was heading straight for the desk and me while Mama watched. Was I scared, you may wonder? No, I wasn't. There was not enough time to be. When in danger I tell myself that I should just watch and not become a part of what is happening. Perhaps no one can see me then, especially if I do not move.

The man in the black coat wants to open the middle drawer of the desk, but I am in the way. I do not get off the chair, but I do lean back to give him space. If I stand up, I will miss out on seeing what is in the drawer or what he will do with what he finds there. He seems to know where he should be looking. And there it is, in plain view resting on top of some papers, one of Mama's three false passports, the most fake of them all. Not a single entry in it is correct, and that it looks brand-new, because it is only used on special occasions, also makes it suspect. The man picks it up, reads the fairytale data entered into it, puts it back into the drawer – at the far end where it should have been – and walks out of the apartment shaking his head. Mama is so upset that she has to lie down and rest. She may be trying to make sense of what has just happened. I think that if I could somehow explain this strange event, I would know how we got here and why. Will I ever be able to find out what is going on? Living in a strange city without any friends to turn to is like being stranded in a desert without any landmarks to navigate by. Yes, I am lost.

Big news here: Friday, February 13, Praha was bombed. Did you hear about it? Were you able to watch the fires at night, black hills dotted with clusters of white flames? In the morning a bomb landed in the no man's land behind your house. Except for some broken glass, the building was not damaged. In the past, when air raid warnings started and stopped, Mama and I did not pay much attention to them, thanks to our "it can't happen here" attitude. This time, as soon as we heard the whistling, we ran down to the shelter packed with people who were crying and praying. Unused as they are to the niceties of war, they were scared and that did not feel or look good. To avoid witnessing their suffering, we stayed in the hallway till the end of the raid. I, too, am afraid of bombs – they make my teeth chatter – but I would never create a scene. At dusk we listened to distant explosions that were turning another city into rubble and watched the glow of a distant fire spread over the sky.

Does this mean that the front is approaching?

In case you have not noticed, I have been sick, so sick that I'm coughing up my innards. Mama says that, if it were safe, she would take me to a doctor as I am sick enough to see one. But a trip to the doctor may expose our illegal status and God knows what else. Besides, it is very cold outside, and I am running a fever. I am using anise liqueur as a cough syrup and already drank up a bottle of it. I like it, it tastes very good.

Sick or not I'll be leaving soon. Mama has rented a flatbed truck which we will share with another couple. If nothing goes wrong on the way, it will drop us off at the border where we may have to wait till it is safe to cross it. Would it be better to stay here longer? Once the war heats up again, renting a truck may become impossible, and then we may get stuck here till it's too late. Mama always operates on hunches and psychic insights. She may even have consulted a fortuneteller about this. Her guts are telling her that we should leave now, even though I am sick. Death has not shown up in the cards, at least not yet, and so we will be leaving in a few days, perhaps even tomorrow.

I won't be writing to you anymore. This is my last letter. I'll leave it in your desk in the back of the drawer where, according to the man in the black coat, private or secret papers should be kept. If you come back, you will find it there, and, intrigued by my foreign writing, you may even ask someone to translate it for you. That is how you will learn what happened in your apartment after you were taken away. I want you to know that Mama took good care of it. Nothing got broken or damaged, at least not by us. The looting was done before we moved in. I am taking with me a photograph of your great room and will keep it as a souvenir so that years later I'll know that your apartment existed, that it was not just a dream. Mama, who is very practical, may be taking some knick-knacks which, unlike the photograph, probably have some value and could be exchanged for a loaf of bread when the need arises.

Thank you very much for your hospitality and for listening to me when I was home alone at night. Even if you survive, I doubt that we will meet. Good luck to you, Mrs. Kraus.

Adam Zachary Newton

# In Hopes of Failing Better: An Academic Afterlife; Or, Enormous Changes at the Last Minute: On Altered Lives, Refuges, and Refugees

## Part I

### (M)untergang

For the cover image of my final academic book, I chose a droll self-portrait by Yosl Cutler, satirist, artist, and co-founder of the first Yiddish puppet theatre in the United States. All angular and elongated modernist lines, the illustration depicts Cutler wedge-shaped at his desk, stretched linear legs tapering to bulbous bare feet underneath; one hand wields a fountain pen, the other teases out a long strand of frizzled hair dangled over a big toe (on the foot that *doesn't* resemble a thumbs-up), while a housecat at lower left, the pointed exclamation point of his tail in synch with every other slanted line in the drawing, proffers an inkwell. Writer as burlesque.

The drawing doubles as ex-libris and frontispiece for *Muntergang*, Cutler's 1934 crazy-quilt of stories, poems, and surreal images, which in the words of one critic, stages "a collision between hard left Yiddish politics and Jewish folklore and tradition." (That the author was killed in an automobile accident just one year later sharpens the edge of the metaphor just a little.)[1] In the preface, Cutler calls the work "a kleyn bikhl mit di greste kleynikaytn in der velt" [a little book with the greatest trivialities in the world].[2]

---

1 Edward A. Portnoy, "Zuni and Yosl's Lost-and-Found Adventure," *Tablet Magazine* (6 April 2020) https://www.tabletmag.com/ sections/arts-letters/articles/yivo-painting-yosl-cutler. See also Portnoy's article on Cotler's puppetry collaboration with Zuni Maud, "Modicut Puppet Theatre: Modernism, Satire, and Yiddish Culture," *TDR: The Drama Review* 43.3 (T 163) (Fall 1999), 115–134 and his chapter on Modicut from *New York's Yiddish Theater: From the Bowery to Broadway*, ed. Edna Nahshon (New York: Columbia University Press, 2016), 222–257.
2 *Muntergang* (Farlag Signal Proletaren, 1934), 5. The book's cover shows a photo-realistic, proletarian Cutler with rolled up sleeves, dungarees, and work shoes stoically traversing a rocky landscape while his cartoonish, murine shadow grins from just ahead.

https://doi.org/10.1515/9783110753295-027

The title itself bespeaks a mash-up, fusing the words *munter* (courageous, cheerful) and *untergang* (downfall), as if to put a wryly smiling face on the Great Depression and other harbingers of civilizational collapse.[3] Its general import is gestured at rather than explained. In one of the book's very few iterations, for example, from the short story "In der kinigraykh fun milbn" [In the kingdom of mites], a mother bookworm rallies her seventy thousand children, volunteered by their father to the mite-king for military service, to wage insurrection against the monarchy instead. "Nemt, khevre, di kley-zayin," she exclaims, "un geyt mit a muntern gang" [take up your weapons, gang, and forth with a cheerful step].[4]

Beyond any satirical thrust, Cutler's title likely rings different bells for different twentieth/twenty-first century ears, depending on whether they bend towards Yiddish or German. At one pole, the mother-mite's exhortation all but *collides* tragic-ironically with the first stanza of "Yugnt Himen" [Youth Hymn], an anthem written a decade later for the Vilna Ghetto Youth Club by poet, partisan fighter, and Yiddish song compiler Shmerke Kaczerginski (1908–1954): "Undzer lid iz ful mit troyer, – Dreyst undzer munter-gang, Khotsh der soyne vakht baym toyer, – Shturemt yugnt mit gezang:" [Our song is filled with grieving/Bold is our cheerful step/Although the enemy stands guard at the gate, – /Storm it, youth, with song][5]

At the more German-inflected pole cluster some well-known fore-echoes (minus Yiddish *munter*), like Oswald Spengler's *Der Untergang des Abendlandes* or Felix Theilhaber's *Der Untergang der deutschen Juden* (referenced briefly in Kafka's diaries). Of more recent vintage, Oliver Hirschbiegel's controversial 2004 film about Hitler's last days, *Der Untergang*, probably takes pride of place, along with its second life in meme culture. Yet, taken on its own terms as a kind of "puncept" *avant la lettre*, Cutler's title puts its particular spin on the poetics of adjacency: skirting the edge of plausible deniability perhaps, or else bridging the ambiguous space between encroaching shadow and sanguine overlay.

Truth be told, it was the illustration that first caught my eye. I chose the image because it seemed to offer me a witty, modernist, Jewishly inflected visual metaphor for the comedy of academic practice; but also, and in all candor, as a very inside joke, since it let me gently satirize myself. Still, the personal salience of Cutler's title word – as portmanteau – grew on me. Confronting a foreshortened academic career when the exigencies of consolidating my complexly blended

---

**3** Portnoy, "Zuni and Yosl's."

**4** Portnoy, "Zuni and Yosl's," 147.

**5** The musical transcription and full lyrics are included in Shirli Gilbert, *Music in the Holocaust: Confronting Life in the Nazi Ghettos and Camps* (Oxford: Oxford University Pres, 2005), 76–79.

family took priority over late-stage professional longevity, I wished to sign off on an affirmative note.

Call it an autumnal *unter* in professional circumstances refashioned as principled *munter,* as I trusted to providence, perhaps as quixotically as I wished for the book itself, and the winds of change. In proportion to the degree of radical displacement and compound loss I was experiencing, a measure of radical acceptance was in order anyway.

## Bending "Remains" Toward "Remainder"

Some thirty years ago, around the same time I glimpsed the lineaments of a prospective career, a core question, a feminist question was put to the Academy's conditions of discursive mastery by Hélène Cixous and Catherine Clément: *"What remains of me at the university, within the university?"*[6] Back then, dogged by a lingering sense of accidental arrival, what I nevertheless saw for myself were horizons of possibility, even if ground for "new" identity had already been tempered by a keen sensitivity to the contingency of place and position. The university's own walls – more specifically, the discourse and speech genres I acquired within them – did not at this apprenticing moment quite suggest to me comparisons with an ideological apparatus whose "enveloping membrane" felt like "a net or closed eyelids" (145).

A late-arriving enthusiast, although I now better understand the obligation for every university citizen to "constantly try to scratch and tear" (145) at that ideological membrane, what I descried then, intellectually at least, was an expanse of open spaces. These many years later, through a chain of contingencies, a version of the same question became, for me, inescapably personal, *nonideological,* and concrete: a question not about what remains "at" or "within" but *after* the university, a question about no longer *being* an academic while continuing, curiously enough, to write like one. A strange sort of adjacency.

*"The non-master must be imagined"* (144), Cixous and Clément insist, in the second of three italicized pronouncements. Mastery's "increase in value," its collusion between knowledge/power depends in part on adopting a sovereign voice as one's own, its authoritativeness so persuasive that it must be actively,

---

6 Héléne Cixous and Cathérine Clement, *The Newly Born Woman,* trans. Betsy Wing, intro. Sandra M. Gilbert (Minneapolis: University of Minnesota Press, 1986), 145. The question is provocatively reframed in the plural in a timely and important essay by Michelle Boulous Walker, "The fragility of collegiality: 'What remains of me at the university, within the university?'" *Parrhesia: A Journal of Critical Philosophy* 30 (2019): 18–28.

*imaginatively* contested. "*Give me the password*" (146), the authors mockingly demand in the third. Imagining privileged access to mastery thereby mocks. How fitting that the trend three decades later in Information Technology is to link together one's university ID and password under the general heading, "credential."

Personally speaking, this was a discourse I always felt more *mastered by* than fully master of. It's no accident that my final book begins with a treatment of Kafka's "Report to the Academy" as a multi-purpose allegory (though it probably makes for too long a prologue). Those prologues, like the periphrases, parentheticals, and pleonasms I came to favor, signaled a hinge, or let's say an expressive merger, between the university and me. Becoming gradually habituated to such a voice did not readily allow for carefree modulation to some other "house style."

Moreover, for the special kind of literary-philosophical practice to which I was drawn, the art of populating one's argument with an array of voices – part jazz combo, part *Marionettentheater* – seemed always just outside my confident grasp. I'd typically end up with a text more populous than needed, less organic than desirable. Just as I'd customarily fall back on a favored quotation from Bakhtin: "The word in language is half someone else's," a tricky and only partially successful balancing act I'd assay, conference to conference, class to class, book to book.

I became a kind of anti-Houdini: what should, in part, have been performative escape artistry morphed into a kind of self-fettering instead. *Ausweg* [a way out], the key word from "A Report to the Academy," became the ever-elusive password. Cixous and Clément's question was thus never all that alien to me. Impelled by their feminism, the critique of the center that they mount from the margins – contesting the given, the received, the authoritative – turns out to be curiously germane to the course I have plotted in both my lived experience and this essay. (Or not especially curious at all: if feminism has universally applicable critiques to make, that actually shouldn't be a wonder to us, one of its central claims being that such interrogations aren't relevant to women only.)

"What remains" connotes a remnant or trace; or else, what lingers, abides in diminished form. "Remainder," a term more recently elaborated by Jonathan Lear and Eric Santner, gestures toward more redemptive possibilities, less a matter of immurement than openness: "an opening to a 'beyond' as the very thing that places us in the midst of life, in proximity to our neighbor";[7] and also, "a peculiar kind of possibility the disrupts the field of possibilities, and thus is experienced as coming from 'outside' and 'beyond.'"[8] As Lear develops

---

**7** Eric Santner, *On the Psychotheology of Everyday Life: Reflections on Freud and Rosenzweig* (Chicago, IL: University of Chicago Press, 2001), 66.
**8** Jonathan Lear, *Happiness, Death, and the Remainder of Life* (Cambridge, MA: Harvard University Press, 2000), 161.

his argument through an elegant double-reading of Aristotle and Freud, so Santner performs a similar operation through penetrating treatments of Freud and Franz Rosenzweig. As *readings*, they exhilarate someone whose work at and within the university placed the ethics of reading at its very core. Formulations such as these deeply resonated with, even thrilled me – they were more than mere passwords.

So, if a question about *what remains* understandably dogs me now, alternate spaces of *remainder* are what dare me now to stake any current or future claims. Despite my still feeling beholden to, even hampered, by the discourse, such spaces have animated my scholarship from the very start. So, whatever may remain of me *after* the university – a rather different formulation than Cixous and Clément's – might thus, in the best sense, not be bounded by the university at all (as a place), but rather be situated quite outside and oblique to it (as a "possibility").

Not "of me," then, but *in me* would be my personal problematic. How did the university make/leave its mark on me? Could such effects mean anything good or secure anything good for me in a new, post-academic life? How might I export a concept like adjacency into meaningful practice in the world beyond ivy walls? Such questions mark not only a bookend to the jumping-off point for a journey begun years ago but also the occasion for a very personal *durkhmuntern* [to "good cheer" one's way through] as I reorient it now.

## The University in Me

The notion of "continuing to write like an academic" feels anything but incidental to my immediate purposes, as the rhetorical habits and tics of academic discourse to which I'd (too) long become (too) accustomed, once again take shape in front of me. A long-standing double-bind as practitioner, feeling both inside and outside at the same time, guild-member and skeptic-dissenter, is most keenly evident on the plane of rhetorical style – writing, so to speak, in spite of myself.

Other contributors to this volume have expressed their awareness of intentionally or unintentionally displaced family genealogies in their academic work and their compensatory desire to alter the conditions whereby personal memory can be made adjacent to academic theorization. I have felt frankly less conflicted in this regard, even though my scholarship has often focused on the exigencies of telling and listening, on the non-narratable within the narrated. Friction, intersection, or more often, overlap, between the personal and professional has taken other forms for me – a specific matter of professional upheaval rather than of family genealogy or events of historical trauma.[9]

In my 2005 book *The Elsewhere*, for instance, it was a fairly straightforward matter to relate the uncanny discovery of a partial translation of the Strzegowo *Yizkor bukh,* which I'd heretofore known only in Yiddish, including a three-page spread, "Mayne Umgekumene" [My murdered ones], devoted to my grandfather and great-uncles, one of whom, along with his wife and son, met his end at Treblinka.[10]

I was prompted in turn to share the differently uncanny story of my father's name-change from Novogrodsky to Newton. This latter, more anodyne family history was something I wished to relate and reflect upon critically in the context of a broader argument about place, name, and belonging. Nor is that to say that family trauma of a quite intimate sort did not exert a formative influence of its own, the informing prelude to all ensuing episodes of picaresque.[11]

At the same time, though, I'm obliged to confess that a primary motivation behind my wish to contribute to this volume connects directly to the object of Cixous and Clément's interrogation of academic discourse in its presumptions of mastery: the desire to express myself otherwise, in a different voice.[12] Elsewhere in his writing Lear privileges "the call of another's words" as a paradigm for an ethics of reading and telling, through whose interpellation one descries new possibilities of tone and genre.

For me, this notion represents a very compelling *déjà entendu,* given how palpably it transports me to the very different kind of social space I occupied before acquiring my university passport. Phases of a life both before and after the university (re)-situate me in the counterspaces of work and action they've adventitiously introduced to me.[13] No longer institutionally embedded as I am

---

**9** I borrow these categories from the seminar call for papers by Leslie Morris and Karen Remmler, "Private Matters: Expanding the Margins of the Lebenslauf," for the 2018 meeting of the German Studies Association.

**10** Esther Novogrodsky, "My Lost Ones," in *The Life and Death of a Polish Shtetl*, eds. Feigl Bisberg Youkelson and Rubin Youkelson, trans. Gene Blustein (Lincoln: University of Nebraska Press, 2000), 45–47.

**11** Marc Fisher, "The Master," *The New Yorker* (25 March 2013), https://www.newyorker.com/magazine /2013/04/01/the-master-2.

**12** "The Call of Another's Words," *The Humanities and Public Life*, eds. Peter Brooks with Hilary Jewett (New York: Fordham University Press, 2014), 109–115. See also Judith Butler, *Giving an Account of Oneself* (New York: Fordham University Press, 2005).

**13** I have in mind Hannah Arendt's distinction from *The Human Condition*, 2nd ed. (Chicago, IL: University of Chicago Press, 1998), where "action" represents the arena in which one discloses oneself to others.

(at least for the foreseeable future), I'd wish to project the contribution I make to this volume from the place that I find myself now: outside the university, with diminished allegiance to its habitus, its rhetoric recalibrated to "the lower frequencies."

It is precisely there, in fact, in that outside and beyond, in that space of re- mainder, on the lower frequencies, that the personal import of adjacency as a nearness to historical violence was driven home to me in the small town of Clarkston, Georgia. Self-proclaimed "Ellis Island of the South," Clarkston is home currently to almost six thousand refugees from over fifty nations in Africa, Central America, the Middle East, and Southeast Asia. And I only discov- ered it, a stone's throw away, when my own sense of displacement became keen enough to make me change my circumstances.

## Adjacency in Theory

Before making that pivot to such different precincts (and I hope, idiom), let me just say something additional about my enduring intellectual stake in the poet- ics and rhetorical praxis of adjacency. Commonly associated with figures like Eve Kosofsky Sedgwick and Judith Butler (and as I note in my last book, a paral- lel philosophical lineage through Giorgio Agamben and Jean-Luc Nancy),[14] the term was re-introduced by C. Namwali Serpell as one of "seven modes of uncer- tainty" for a practical literary criticism of the Novel, in specific connection with Toni Morrison's *Beloved*:

> This is what I mean by adjacency, a mode that sets disparate entities beside each other in an enclosed space, allowing them to brush up against and interrogate each other . . . a tenuous, momentary contiguity that offers a countermodel . . . . Cued to a contingent uni- verse, adjacency promotes Kairos, a sense of what is enough, a discernment of that main- tains a vivid sense of possibility-what is discretely opportune – that is readjusted according to a sense of discretion, of what is appropriate.[15]

I myself came to "adjacency" intuitively and very early on in my career as a way of staging ethics literary-critically. When I refer periphrastically to "a jumping-

---

**14** Judith Butler, "Precarious Life, Vulnerability, and the Ethics of Cohabitation," *The Journal of Speculative Philosophy* 26, no. 2 (2012): 134–151; Steve McCafferey, "Parapoetics and the Ar- chitectural Leap," in *A Time for the Humanities: Futurity and the Limits of Autonomy*, eds. James J. Bono, Tim Dean and Ewa Plonowska Ziarek (New York: Fordham University Press, 2008), 161–179, here 163.
**15** C. Namwali Serpell, *Seven Modes of Uncertainty* (Cambridge, MA: Harvard University Press, 2014), 133, 144.

off point" and "very different kind of social space," what I actually have in mind is the taxicab I drove for nearly a decade before I pivoted to graduate study and teaching. Any theorizing of adjacency I have subsequently performed followed directly from that employment, that occupation, and that vehicle for conversation, through which the (sometimes ethical) interaction of driver and fare came to crystalize for me the lived experience and exigency of "tenuous, momentary contiguity."

Whether juxtaposing African American and American Jewish literary traditions, aligning literary memoir from Eastern Europe and the Levant, making proximate the "meaningful adjacencies" of the 9/11 memorial and W. G. Sebald's prose, hinging the philosophical outlooks of Levinas and Yeshayahu Leibowitz or the postwar legacies of Bruno Schulz and Emmanuel Ringelblum, all my inscription during my time within the university has been bound up with some version of *neighboring*.

According to Serpell, adjacency "permits us to conceive of a communal ethics that preserves the integrity of the one and the many" (132). Appositely enough, that underscores both the genesis and express purposes of this volume of essays, even if the sense of "adjacency" in our title gestures quite elsewhere.[16] As adjacency, *muntergang*, and remainder severally locate a vantage from which I aspire to write here, I turn directly to their correlates in the real world.

# Part II

To be a doctor of philosophy no longer satisfied us; and we learnt that in order to build a new life, one has first to improve on the old one. A nice little fairy-tale has been invented to describe our behavior; a forlorn émigré dachshund, in his grief, begins to speak: "Once, when I was a St. Bernard . . ."[17]

---

16 That would be closer to Judith Butler's claim in *Giving an Account of Oneself* that violence "delineates a physical vulnerability from which we cannot slip away, which we cannot finally resolve in the name of the subject, but which can provide a way to understand that none of us is fully bounded, utterly separate, but, rather, we are in our skins, given over, in each other's hands, at each other's mercy" (101).

17 Hannah Arendt, "We Refugees," *Menorah Journal* 31.1 (January 1943): 69–77.

## Clarkston, Georgia

The postage stamp-sized small town of Clarkston lies approximately six and a half miles west of my residence in an unincorporated subdivision of Dekalb County. Where I live is called Atlanta, but only in the sense of being part of a metropolis sprawling in all directions. The older, historically black neighborhoods of central Atlanta, like Sweet Auburn where Dr. King was born and is interred, or Pittsburgh where Rayshard Brooks was shot and killed by police in the summer of 2020, lie much farther to the South. So, in comparison with a city where de facto residential segregation remains the norm, Clarkston's diversity comes as a revelation.

Founded in the mid-nineteenth century as one of Atlanta's first *true* suburban communities because of access to the Georgia Railroad, Clarkston was named after a railroad official, and one of its more prominent features remains the CSX-Norfolk Southern tracks that traverse the length of the town. Following a trend that began in the 1970s with the influx of Vietnamese refugees, and crystalizing in the 1990s, that same proximity to the Atlanta metro region presented a new rationale for various refugee asylum and resettlement programs geared to an expanding immigrant population that now represents almost half of the city's total. Since 1980, more than sixty thousand persons have been resettled there.

On one side of the railroad tracks, in the small city center whose two strip malls consist entirely of ethnic food markets and restaurants, hair salons and beauty stores, and budget houseware/hardware stores, sits the repurposed 1960s-era service station of Refuge Coffee, Co., created for – and run by – refugees, its chalkboard sign announcing "welcome" in Amharic, Somali, Swahili, Arabic, Sudanese, Vietnamese, and Nepali. I racked up many hours there both before and after my tutoring sessions, making notes and trying to figure my way. On the other side of the tracks, under the overpass, in what used to be the old Clarkston High School is the Clarkston Community Center, which hosts adult programming and schooling for young children. The Refugee Family Literacy Program, a nonprofit school overseen by Friends of Refugees, rents space here; and through a series of fortuitous, extra-curricular connections, I volunteered to tutor refugees in ESL and civics.

Ordinarily, I would not have been able to provide those services, since the classes and individual tutoring sessions there are composed entirely of women who come from a variety of traditional cultures and whose ideal of modesty enjoins strict gender separation. While women from Syria, Somalia (many of them devoutly Muslim), and Myanmar/Burma learn to communicate in English, their small children attend the pre-school in the same building.

Such strictures were not unfamiliar to me, due to my own participation in traditional Jewish communal life, but they obviously did not permit much of a place for me to be helpful in this particular enterprise. (As for how I may have signified in my *kippa* I'll refer to the title refrain of a favorite country song, and "just let the mystery be.") Good fortune or irony intervened, however, affording me an opportunity to serve as I had hoped. As I was very quickly to confront, leaving behind an academic career qualifies as very much a first-world problem. And yet, it was displacement itself that erected a kind of bridge.

Tam Mla Wah, an ethnic Karen nearing forty and the father of five children, is a refugee from Burma/Myanmar. The almost seven million Karen (Kayin) people reside on the southeast edge of Burma and northeastern Thailand; the area known as the Karen State, one of several ethnic states in Burma/Myanmar, cuts across both countries. The Karen are culturally and religiously diverse, sufficiently so that, as a unifying term "Karen people" is probably misleading. The Karen language comprises multiple branches and dialects, of which the primary subgroups are Sgaw Karen, Pwo Karen (its own four dialects almost mutually unintelligible), and Pa'o Karen. (Properly speaking, Ta Mla Wah would probably identify himself ethnically as "Pwa Ka Nyaw" or "Kanyaw" rather than "Karen.") Most Karen people are Buddhist, with twenty-five percent of the total population identifying as Christian, and the smallest percentage either animist or Muslim.

The Karen conflict, one of the world's longest running civil wars, has its roots in the nineteenth century, as ethnic social structures were steadily altered through the intervention of American missionaries and British colonialists. To this day, it remains a complicated entwining of politics and religion. Ta Mla Wa, like the other ninety-five percent Karen refugees who come to the US, is Christian and Sgaw-speaking. Before he came to Clarkston in 2015, he lived in Mae La Oon, one of the nine refugee camps along the Thai/Burma border. The camps, which together house about two hundred thousand, are supervised by refugees themselves in coordination with NGOs in conjunction with UNHCR, the RTG (Royal Thai Government), and security personnel. With strategic operations that include summary executions, burning of villages, forced relocation and portage, and recruitment of child soldiers, the Karen continue to endure systematic human rights abuse at mortal risk.[18]

Tam Mla Wah's mother and one sister followed him to the United States; another sister and brother-in-law remain in the Mae La Oon camp, the latter

---

18 Regarding the 2021 Myanmar coup d'état, see https://teacircleoxford.com/2021/02/04/hope-and-heartbreak-karen-communities-in-the-wake-of-the-coup/.

having been recruited by the Free Burma Rangers, a "multi-ethnic humanitarian relief," solidarity organization run by Christian evangelicals (quite separate from the KNLA, the military branch of the Karen National Union).[19] Karen families tend to be large, so there are likely more siblings I'm unaware of. When we became acquainted, while his wife Eh Paw was covering third shift at a sanitation company, Ta Mla Wah was not working, which allowed his two youngest children, two and four years old, to attend a morning class at the CCC. (The couple would later reverse the arrangement.). The other three children, ages six, eight, and ten, were enrolled in public school.

Stay-at-home dads alike with a shared window of opportunity, Ta Mla Wah and I would be seated opposite one another in an adjoining public hallway jury-rigged for ESL lessons amidst the hum and buzz of children's voices and an array of languages spoken in the immediate background. The last time I felt so conversationally confined, subject to the everyday-unforeseen, was when I looked into the rearview mirror of my cab. After years of theorizing adjacency in the interim, here it was again in concrete form, in proximity to my neighbor.

## Adjacency, Once Removed

Notwithstanding the fact that I had enrolled in the first module of an online TESOL certificate program, I had only the barest idea of what I was doing. After four years in the US, Ta Mla Wah's aptitude in English was still extremely basic. He tested at a beginner's level. Reading and writing were developing skills; speaking lagged much farther behind (final consonants, absent in S'gaw, being the most troublesome). We had friendly rapport in our sessions from the start, though, and I followed the textbook and workbook closely, with occasional impromptu lessons in grammar using flashcards and a whiteboard.

I did notice a struggle with retention both between and within our week-to-week sessions. After a mistake or memory lapse, Ta Mla Wah would offer up an embarrassed chuckle that I'd come to recognize as a culturally favored propensity for *durkhmuntern*. I'd also noticed it with the other two men I tutored, Thaun who was Chin and studied civics with me, and Lokman, a Burmese Indian who was disabled, having lost both of his arms in a factory accident.

---

**19** On the FBR, see Alexander Horstmann, "An American Hero: Faith-Based Emergency Health Care in Karen State, Myanmar and Beyond," *Religions* 10, no. 9 (2019): 503; https://doi.org/10. 3390/rel10090503, and the more propagandistic "Free Burma Rangers Annual Report 2019" https://www.freeburmarangers.org/wp-content/uploads/2020/06/FBR-Annual-Report-2019. pdf, and *Free Burma Rangers: The Movie* (Deidox Films, 2020).

After Ta Mla Wah seemed to require review in excess of what I may have anticipated, I grew curious enough to ask one of the supervisors whether she was aware of any cognitive deficit that could account for the difficulty. What I learned was painful to discover. In his twenties, after his village had been burnt and Ta Mla Wah fled with his family, he sustained a blow to the head from a rifle butt wielded by Myanmar military, smashing his skull. (I'm ashamed to admit my mind instantly conjured a scene re-enacted in dozens of war movies by anonymous extras.) He'd never received medical treatment, I was told; so, whatever the exact nature or severity of the injury, he may well have incurred brain damage. This was nothing Ta Mla Wah confided in me, of course, nor was it any window into precarious life that I was expecting.

But why? Among other points of background knowledge, I'd familiarized myself with the basic facts of the still-ongoing civil war in Myanmar and state-sponsored violence against the Karen people and other ethnic Burmese minorities, Karenni, Arakanese, Chin, Kachin, Shan, and Mon, which escalated in the 1990s. Rebel militias in border states like Karen, Kayah, and Shan have fought against Burma/Myanmar's military since independence in 1948. According to one reputable account, almost four thousand villages have been burned since 1996, with more than a million people left homeless and uncounted dead.

Even a partial context for understanding Ta Mla Wah's past lay beyond my resources – fringe figure that I was, perched on the margins of the Clarkston Community Center logging but two or three hours a week in volunteer work. My puzzlement pacified, "so that explains it," is what I told myself. I was fairly stunned all the same, though I never knew with certainty whether this act of violence played a role in his present performance as a student.

Upon reflection, I concluded that perhaps more than anything else, it was the benign setting of the Center itself, a renovated early 1900s building that presented to my eyes a kind of gauze curtain, or blind, over the many individual stories of flight, persecution, and worse that lay directly behind it. Ambient clamor aside, it's hard to imagine a more congenial scene of impromptu one-on-one teaching, especially during Atlanta's temperate fall season. Indeed, the center's aesthetic environs bear a certain resemblance to what many people probably think of as "gracious Southern architecture": white windowpanes against red brick, lush landscaping, light-washed corridors.

Surely unintended by the building's renovators, accidental echoes of plantation architecture impinge as well, as when a home's genteel façade and elegant interiors stood in stark contrast to the coercive, dehumanizing scenes and structures that lay just beyond. The atmosphere inside the Community Center is so serenely welcoming that whatever conditions of brutality stand behind or over the presence of individual refugees and asylees, they remain almost entirely *in* or obscured by the background. Whether my surmise that my student had sustained a traumatic brain injury was correct or not, all of what I just depicted preceded the coronavirus pandemic, which has altered everything in its wake.

## Return of the Repressed

Much later, what was triggered for me by this kernel of knowledge about Ta Mla Wah's past was also the stuff of reported knowledge. By rights, the association should have been much more intimate and immediate. I recalled my father's military service in Burma. David Novogrod enlisted in the Second World War as a private in the Army's long-range penetration special operations jungle warfare 5307th Composite Unit more famously known as Merrill's Marauders or Unit Galahad, which fought in the China-Burma-India Theater (CBI). His official unit status was Cavalry; but upon disembarkation, the 360 Australian Waler horses reassigned from the 112th Cavalry could not be deployed for jungle warfare, and so he and his comrades became infantrymen. When the unit was disbanded in August 1944, of 2,750 to enter Burma, "only two [men] were left alive who had never been hospitalized with wounds or major illness."[20]

Beyond learning that he had indeed contracted a parasitic infection, my father never shared any information about his wartime service with me. I have his medals and his Eisenhower jacket. I know nothing about any interactions he may have had with Burmese civilians, or indeed how he felt about finding himself in such alien environs. Did he know who the Karen were, for instance? Did he follow developments in Burma immediately after the war, when the country gained independence a year after India/Pakistan and in the same year as the State of Israel? Did his brief flirtation with the CPUSA a decade earlier, or his own father's history as a refugee, dispose him to the pathos of "useless suffering"? Did he never look back? All unanswerable questions.

All of this is also to say that once I discovered more about him, Ta Mla Wah became connected to me and my own family history, in my thoughts, by some quite unexpected filaments. Either all my work of the past thirty years had prepared me for such a confluence, or I was on my own.

---

20 Charles N. Hunter, (Col.), *Galahad* (San Antonio: Naylor Press,1963), 215.

## Levinas, Interrupted

I must also attest that the shock I experienced upon hearing about this terrible incident in Ta Mla Wah's life prior to immigrating to the US was *not connected* to the call of another's words; he did not "give an account of himself" or narrate a story, nor did I ever prompt him to. (A scene of address, a rhetorical, performative dimension over and above a strictly narrative one, wasn't really possible anyway, given how almost insurmountable the language barrier was.[21])

Whether his individual experience qualifies as an instance of what Julia Kristeva calls the refugee's "scorched happiness," I could not know.[22] His cheerful mien certainly led me to believe otherwise. At any rate, Ta Mla Wah's own individual history of violence lay behind a scrim: reported to me, but not presented, made present, in any way. We did our language work together, and tact took care of the rest. I won't deny that my attitude towards him if not my practice as tutor altered after hearing about his past life in Burma. That we were thrown together in the scene of teaching makes all the difference, of course. I didn't meet Tam Mla Wah at Refuge Coffee, where details about his past life might have surfaced in conversation. Nor did we ever really parley beyond the mutual pride we expressed about our young children, his Jay and Kelly and my Manu.

Were I still professorially aligned, I could say that I seized on the intervention of imparted knowledge as an instance of "break," the category that Lear identifies in relation to unconscious activity, but which may also become the stuff of agency and choice, "the possibility for new possibilities."[23] Likewise, I could re-enlist Eric Santner who locates "the point at which we truly enter the midst of life, when we truly inhabit the proximity to our neighbor, assume responsibility for the claims his or her singular or uncanny presence makes on us."[24]

---

**21** "Is the final aim to achieve an adequate narrative account of a life?" asks Judith Butler. "And should it be? Is the task to cover over through a narrative means the breakage, the rupture, that is constitutive of the "I," which quite forcefully binds the elements together as if it were perfectly possible, as if the break could be mended and defensive mastery restored?" (*Giving an Account of Oneself*, 69).

**22** Julia Kristeva, *Strangers to Ourselves*, trans. Leon S. Roudiez (New York: Columbia University Press, 1991), 3.

**23** Jonathan Lear, *Happiness, Death, and the Remainder of Life* (Cambridge, MA: Harvard University Press, 2000), 118.

**24** Eric Santner, *On the Psychotheology of Everyday Life: Reflections on Freud and Rosenzweig* (Chicago, IL: Chicago University Press, 2001), 8.

Thus do we become unbound, open and vulnerable, answerable not to socially symbolic institutions but to the human Other. I had sat across from or next to Ta Mla Wah for several weeks. Perhaps, another space or zone had opened up, a different mode of responsibility and proximity. The truth is, such frames of reference only come to me now, artificially, as I find a way to move this essay to conclusion. And I did promise myself and my readers another voice.

*Tha Ba Pwoa*, S'gaw Karen for "sadness, heartbreak," was nowhere expressed, hinted at, or exposed in my time with Ta Mla Wah. Punctuated by noisy corridor-traffic, children's recess, a pinging cellphone and hurried conversation between Ta Mla Wah and his wife, the interhuman, to filch a word from Levinas, was just *there*, bare and prosaic. Curative help, merciful care, analgesia (signature phrases also drawn from Levinas)?[25] The tutoring itself occupied that overdetermined rhetorical space. A call for aid? Inasmuch as I was helping Ta Mla Wah negotiate the ins and outs of speaking, reading, and writing English, anyone in my position would be responding to such a call – although it's just as true that my particular identity and present circumstances made me rather singular on the refugee family literacy landscape.

Had two and a half decades of scholarship, or my life before that, prepared me in any way to do this work, to have these experiences? When I implied earlier that my time in the Clarkston Community Center was vaguely reminiscent of the taxi-driving interval when I became the involuntary interlocutor for other people's stories – *drafted* into listening, as it were – the difference between the two bookends still remains significant. I was in my thirties then; I'm now in my sixties. Then, I drove while I listened in the vehicular analogue to Sebaldian prose fiction, all vagary and peregrination. Now anchored in one place in Clarkston, Georgia, I gave Ta Mla Wah my undivided attention, as we reviewed the names of household objects or months of the year. Most crucially of all, no storytelling/hearing was ever transacted between the two of us; "narrative ethics" was just an abstraction, an academic phrase I'd once deployed.

And while the proper name "Levinasian" might be enlisted to describe my encounter with Tam Mla Wah, with its entailments of teaching, facing, hospitality, Saying over Said, and sonority, I resist the (I confess, still powerful) urge to make that citational move. For one thing, it betrays the purposes for this half of the essay that I'd already laid out. For another, what point does it really score,

---

25 "Useless Suffering," *Entre Nous: On Thinking-of-the-Other*, trans. Michael B. Smith and Barbara Harshav (New York: Columbia University Press, 1998), 80. See also Dermot Fagan, "The Century of Useless Suffering: Alain Finkielkraut Reads an Essay by Emmanuel Lévinas," *Jewish Studies Quarterly* 9, no. 2 (2002): 193–203.

what reward does it secure? For someone securely berthed, identifying things in such a way continues to authorize and validate a professional identity and voice. As aptly as it might apply to the specifics of Ta Mla Wah's situation, channeling Levinas's voice would still amount to an artifice of discourse, however, just as any adjacency to historical violence to which I could lay legitimate claim is embodied here, in my highly refracted account, at a near distance.[26] I miss the exhilaration. But no password is needed.[27]

Because it is so pat, the one cliché I particularly wished to avoid, rhetorical as much as positional, was the putative role-reversal where the tutor becomes the one taking tuition (despite Levinas's oft-quoted insistence that the Other is a teacher).[28] Moreover, on top of my own predispositions, the little I knew about Free Burma Rangers made me chary of the presumptions of "white saviors," whether in extreme forward areas in the Karen refugee state, or where I was situated, academic afterlife and all, in very deep background. I suppose this is what I intended for the title of this essay, "failing better": namely, an adventitious stumble into a common ground of resilience.

---

26 Noteworthy, however, are the spate of recent publications that connect Levinas's philosophy to various contemporary and global refugee crises, chief among which is Nathan Bell, *Refugees: Towards a Politics of Responsibility* (Lanham, MD: Rowman & Littlefield, 2021). Also pertinent to this context is the essay from which I drew the epigraph for Part II by Hannah Arendt: "We Refugees," *Menorah Journal* 31, no. 1 (January 1943): 69–77.

27 Theory into practice? That would be a little too facile. Theodor Adorno's final publication, "Resignation," *Critical Models: Interventions and Catchwords*, trans. Henry W. Pickford (New York: Columbia University Press, 1998), 289–293, in which he offers a spirited defense of critical consciousness on the sidelines without really resolving the apparent contradiction of thinking disembodied from political action, has been adduced by several contributors and respondents during this volume's preliminary "workshop" stages. But Adorno's philosophical pessimism doesn't quite address the very personal predicaments I've tried to sketch here. "Open thinking points beyond itself," his essay on "Resignation" proudly concludes while still projecting a discourse from both another time and a privileged institutional place. And while the entirely individual sessions I spent with Ta Mla Wah might have borne some thin resemblance to "pseudo-activity" of the *Mach es Selber* sort (Adorno's derisory name for do-it-yourself praxis), there was nothing particularly activist about my labor. I made some headway, but I also floundered a lot.

28 "The absolutely foreign alone can instruct us," wrote Levinas in *Totality and Infinity*, together with many other such observations across the span of his work. See, however, the sharp critique by Jordan Glass, "The Question of the Teacher: Levinas and Hypocrisy of Education," *Symposium* 20, no. 1 (Spring/Printemps, 2016): 129–149.

## Coda

My stint as ESL/civics tutor at the Clarkston Community Center ended abruptly when Ta Mla Wah's daily schedule changed after he was hired as a laborer at two local chicken processing plants in rural Georgia. As meat processors around the country began to shut down because of COVID-19 outbreaks, Fieldale Farms and the other plant in Georgia remained open for business, even while hundreds of Georgia poultry workers tested positive for the coronavirus.[29]

Once shelter-in-place became the order of the day and with the Community Center temporarily shuttered, I directed my energies elsewhere, taking up the task of teaching my four-year-old to read. When Ta Mla Wah and I were able to reconnect, masked and socially distanced (an ironic phrase in this context, to be sure), I learned that he may have developed symptoms of bio-toxicity due to the harsh chemicals he'd been working with. He was, as far as I could ascertain, Covid-free. But he did seem somewhat changed, perhaps at more of a remove.

I felt an increased obligation to pick up where we left off, even though I suspected the tutoring would be compromised through contact via screens, which was, indeed, the case. His only hope for employment in another kind of environs was greater communicative facility in English. The jobs he'd worked since I last saw him, I suspected, in addition to the stresses of life during Covid, had taken their toll. According to the National Center for Farmworker Health, the subpopulation of agricultural workers to which Ta Mla Wah now belonged, while "racially and culturally diverse [are] universally vulnerable to labor abuses and health and safety hazards."[30] He'd found refuge in the United States from persecution in Burma, and yet his life circumstances fell prey to different structures of oppression here, introducing yet one more permutation of adjacency to conditions just outside my ken.

All told, our time together was dramatically punctuated. In a time of pandemic, with whatever instructional role I might play in Ta Mla Wah's life even more curtailed than previously, it does appear that for the present, I find myself yet again in a condition of "after" rather than "at" or "within." But with my narrated transition from one teaching/discursive space to another still in force, I would still prefer to think of this afterwards also along the lines of *remainder* rather than *remains*. I began composing this essay still thinking of such radically different spaces as the professoriate and the Clarkston Community Center

---

**29** "Update: COVID-19 Among Workers in Meat and Poultry Processing Facilities – United States, April–May 2020" (10 July 2020) https://www.cdc.gov/mmwr/volumes/69/wr/mm6927e2.htm.
**30** http://www.ncfh.org/uploads/3/8/6/8/38685499/fs-poultryworkers.pdf.

as disjunct and dichotomous. During the course of writing it, I've come to perceive more of a convergence.

What I want to tease out at the end here is how my encounter with Tam Mla Wah returned me to an anterior time and place *before* the university, when I was all but completely preoccupied by the exigencies of storytelling. In all candor, I am honestly struggling to tell this story, with how to tell it *as a story*. Really, a double yet incommensurable story: my forfeiture of guild membership and professional position as what feels, in part, like failure; and the effect on me of a shocking revelation about someone else's past. A twinning of a career exploring such theoretical precincts as narrative ethics, the face and space of the other, on the one hand; and a wholly non-theoretical, concrete space of adjacency to historical violence, on the other.

Now, despite having equated losing one's professional berth with downfall (*untergang*), at the same time, I recognize (in theory) that it's not a question of success or failure at all. Rather, and quite definitively for me, considering all I've invested intellectually in the ethics of witnessing, it's a question of how I tell the story. One supreme satisfaction that stories provide is a climax, redemptive or otherwise. The sense of an ending reinforces the wondrousness of made narratives, especially the verisimilar kind. Ta Mla Wah's immigrant story, its before-and-after (including the brief interlude of our tutoring sessions), for the present at least has settled into the quotidian drudgery (and potentially worse consequences) of working in a chicken processing plant. The story of my career has, perhaps likewise, not ended where I imagined it would.

Was I a competent ESL tutor? I can't say. Adequate, at least. Those skills were brand-new for me, learned on the fly. What I can vouch for is that I carried this twenty-five-year career and all things it signified for me to Clarkston, Georgia, in the fall of 2019. None of its elements necessarily made me better at second language teaching. But they did establish me, formatively, as *me*. Or rather, as the institutional persona I'd left behind was nowhere on display, never referenced; since it didn't oversee my teaching – so completely different from any kind of teaching I'd been involved in before – if it wasn't present *in* it somehow; nevertheless, I was who I was, having logged those many years, no less than the years logged in a taxicab which preceded aspiring to and finding a life vocation.

Somehow, Ta Mla Wah and I intersected, were thrown together. How else to account for the uncanny filaments that led me back to stories about my father and grandfather? Each of us traced his own arc across radically disparate spaces. But maybe the better way to tell both his story and mine would be to focus on *phases* rather than *spaces* – or as I have had occasion to formulate it elsewhere, a matter more of picaresque than *Bildungsroman*. Living a life in phases doesn't obliterate the thing that came before – a cautionary tale, differentially encoded, for each of us. I can't and won't speak for Ta Mla Wah in this respect. Being between two worlds (as the title of a Levinas captures the life of Franz Rosenzweig) may just have to be where I am for the present.

The true story of my life involves, after some early years of vagary in the wilderness, thinking about narrative ethics at a near distance to theory; somewhat later, theorizing it while family life remained at a far distance; and, finally, at a late moment, exchanging that professional berth for family happiness, only to recommence thinking about it – about tellability and acoustics, about force and effect, about being tied and untied by story – at a near distance all over again. This peculiar cyclicality, too, may just be how a life gets lived. With Ta Mla Wah, however, I discovered a new/old mode of responsibility – part pedagogical, for I was teaching once again; but part that of witness, of being proximate.

Neither Tam Mla Wah nor I were employed when we crossed paths. How he may have thought about it, I also never inquired. When I reconnected with him some months later, he'd found the kind of work available to recent refugees like himself in this region of the country, exposing him to numerous health and safety hazards. Perhaps the truest thing I can say in this regard is that what I witnessed in my pre-Covid sessions with him, unfailingly eager and for lack of a better word, *munter,* was resilience, even if, upon re-encounter, that initially brighter light appeared somewhat dimmed. For myself, I can certainly attest to what felt like an unmanning out of season, which is perhaps what throws me back into the feminist question that begins this essay in earnest. But maybe the better way to tell the story – his and mine alike in tandem – is the formula I stumbled into above: discovering a common ground of resilience amidst the concatenation of events, episodes, phases that make a life.

I'm tempted in these final sentences to fall back on an old speech habit, the allusion or block quotation, even when slyly smuggled through the rhetorical magic of *paralipsis* (a university password if there ever was one). Instead, I'll let our volume's title force my hand. Adjacency means several things for me personally, as I hope I've made clear enough by now. The bookends, as I called them, to my found vocation, the quasi-ethical spaces of confined conversation, of emergent or impeded stories, were adjacent to it, one at the entrance portal, the other at the egress; much like the discourse of the university I adopted as my own was adjacent to quite different modalities of speech and reception; just as, in turn, my sojourn upon leaving academia led me to an adjacency in person, if not in actual fact, to the shadow of historical violence.

I've confided all such permutations of nearness within the covers of this volume to the specialized readership for whom it's intended. Yet, the interlude with Ta Mla Wah points me toward a different horizon of choice, a differently lived mode of responsibility. *"Pok t'nei l'vara,"* reads the Talmudic injunction about the bounds of received discourse (in a tractate that deals specifically with borders and boundaries, no less): "go teach this outside."[31] I'll see whether I can.

# Works Cited

Adorno, Theodor W. "Resignation." *Critical Models: Interventions and Catchwords*. Translated by Henry W. Pickford, 289–293. New York: Columbia University Press, 1998.
Arendt, Hannah. *The Human Condition*. Second edition Chicago: University of Chicago Press, 1998.
Arendt, Hannah. "We Refugees." *Menorah Journal* 31, no. 1 (January 1943): 69–77.
Bell, Nathan. *Refugees: Towards a Politics of Responsibility*. Lanham, MD: Rowman & Littlefield Publishers, 2021.
Boulous Walker, Michelle. "The Fragility of Collegiality: 'What Remains of Me at the University, Within the University?'" *Parrhesia: A Journal of Critical Philosophy* 30 (2019): 18–28.
Butler, Judith. *Giving an Account of Oneself*. New York: Fordham University Press, 2005.
Butler, Judith. "Precarious Life, Vulnerability, and the Ethics of Cohabitation." *The Journal of Speculative Philosophy* 26, no. 2 (2012): 134–151.
Cixous, Héléne, and Cathérine Clement. *The Newly Born Woman*. Translated by Betsy Wing. Minneapolis: University of Minnesota Press, 1986.
Cotler, Yosl. *Muntergang*. New York: Farlag Signal Proletaren, 1934.
Fagan, Dermot. "The Century of Useless Suffering: Alain Finkielkraut Reads an Essay by Emmanuel Lévinas." *Jewish Studies Quarterly* 9, no. 2 (2002): 193–203.
Fisher, Marc. "The Master." *The New Yorker* (25 March 2013). https://www.newyorker.com/magazine/2013/04/01/the-master-2.
Gilbert, Shirli. *Music in the Holocaust: Confronting Life in the Nazi Ghettos and Camps*. Oxford: Oxford University Press, 2005.
Glass, Jordan. "The Question of the Teacher: Levinas and Hypocrisy of Education." *Symposium* 20, no. 1 (Spring/Printemps, 2016): 129–149.
Horstmann, Alexander. "An American Hero: Faith-Based Emergency Health Care in Karen State, Myanmar and Beyond," *Religions* 10, no. 9 (2019): 503.
Hunter, Charles N. (Col.). *Galahad*. San Antonio: Naylor Press, 1963.

---

**31** Tractate *Eruvin* 9a. During the pandemic, when so much of life shifted to virtual spaces blurring at home and outside, it became possible for me to tutor a Somali woman in preparation for her the USCIS naturalization interview. On the day I submit the final draft of this essay, she passed the exam but received word that her case is subject to further scrutiny. I want to hope this micro-instance betokens an altered horizon for refugees and asylum seekers in what Emerson called "this new yet unapproachable America." Yet, truth demands that I end this meditation as she will end this day: in a state of in-between.

Kristeva, Julia. *Strangers to Ourselves*. Translated by Leon S. Roudiez. New York: Columbia University Press, 1991.

Lear, Jonathan. "The Call of Another's Words." *The Humanities and Public Life*. Edited by Peter Brooks with Hilary Jewett, 109–115. New York: Fordham University Press, 2014.

Lear, Jonathan. *Happiness, Death, and the Remainder of Life*. Cambridge, MA: Harvard University Press, 2000.

Le Breton, David. "From Disfigurement to Facial Transplant: Identity Insights." *Body & Society* 21, no. 4 (January 2014). https://citeseerx.ist.psu.edu/viewdoc/download?doi=10.1.1.930.1126&rep=rep1&type=pdf.

Levinas, Emmanuel. "Useless Suffering." *Entre Nous: On Thinking-of-the-Other*. Translated by Michael B. Smith and Barbara Harshav, 92–102. New York: Columbia University Press, 1998.

McCafferey, Steve. "Parapoetics and the Architectural Leap." In *A Time for the Humanities: Futurity and the Limits of Autonomy*. Edited by James J. Bono, Tim Dean and Ewa Plonowska Ziarek, 161–179. New York: Fordham University Press, 2008.

Nahshon, Edna, ed. *New York's Yiddish Theater: From the Bowery to Broadway*. New York: Columbia University Press, 2016.

Novogrodsky, Esther. "My Lost Ones." In *The Life and Death of a Polish Shtetl*. Edited by Feigl Bisberg Youkelson and Rubin Youkelson. Translated by Gene Blustein, 45–48. Lincoln: University of Nebraska Press, 2000.

Portnoy, Edward A. "Modicut Puppet Theatre: Modernism, Satire, and Yiddish Culture." *TDR: The Drama Review* 43, no. 3 (T 163) (Fall 1999): 115–134.

Portnoy, Edward A. "Zuni and Yosl's Lost-and-Found Adventure." *Tablet Magazine* (April 6, 2020) https://www.tabletmag.com/sections/arts-letters/articles/yivo-painting-yosl-cutler.

Santner, Eric. *On the Psychotheology of Everyday Life: Reflections on Freud and Rosenzweig*. Chicago, IL: Chicago University Press, 2001.

Serpell, Namwali. *Seven Modes of Uncertainty*. Cambridge, MA: Harvard University Press, 2014.

Brad Prager
# "Is this really necessary?": On Atrocity Images in the Classroom

In 2012 I was invited to give a talk at Ithaca College. My presentation centered on Yael Hersonski's documentary *A Film Unfinished* (2010) for which the director filmed survivors of the Warsaw Ghetto while they watched footage that had been filmed by Nazis in the Ghetto, footage that was for the most part intended to be propaganda meant to convince its viewers that the Jews had abhorrent lifestyles and that they had rendered even their own living spaces uninhabitable. The Nazis' vile project, which they fortunately never completed, was a cruel and dehumanizing attempt at propaganda, yet the footage taken by their cinematographers managed to capture scenes of Jewish life in the Ghetto. That footage, the fragments of the propaganda film that survive, contains sporadic glimpses into realities of Jewish life, and it grants a bit of insight into a painful chapter in Jewish history. These traces of the past include scenes of Jewish rituals, of Jewish culture, and of Jews' deaths, and, were it not for the film, those events would have disappeared in time. The survivors interviewed in Hersonski's documentary respond accordingly, recognizing that the images that we and they now see contain scenes of suffering distorted by the Nazi lens, while they simultaneously acknowledge that these images can be aptly described as the last remaining traces of a great many Jewish lives.

As a film scholar, my talk that day centered on reading this Israeli documentary in light of its implicit treatment of German national cinema, and I argued that attending to the documentary filmmaker's strategies helps us interpret the strategies implicit in antisemitic Nazi propaganda films.[1] My argument was predicated on close readings of *A Film Unfinished*, and it included an examination of sequences and stills from that film. In the documentary's latter half, Hersonski includes footage that many would describe as atrocity footage: we see Jews disposing of the dead for purposes of sanitation. The Ghetto's improvised cemeteries were at times no more than mass graves into which bodies were thrown. These bodies, now little more than skin and bone, pile upon one another, as did the bodies at Buchenwald after the liberation. This is no way to bury the dead, and it is difficult, even today, to look directly at these sequences. My presentation included one of those images as a still frame capture, which was for a time projected on a screen behind me,

---

1 The talk was later published as "The Warsaw Ghetto, Seen from the Screening Room: The Images That Dominate *A Film Unfinished*," *New German Critique* 41, no. 3 (123) (2014): 135–157.

https://doi.org/10.1515/9783110753295-028

where I could not see it while I spoke. In the midst of my talk a woman's voice interrupted: would I please take the image down or replace it with another one? I was looking out into the darkened auditorium, and I might have felt similarly to her had I been on the other side of the podium. Her request was, of course, reasonable. On the other hand, looking at atrocity images is part and parcel of the work I and others in my field do. My transgression, if it was one, was quickly corrected, but should any attendee have been surprised to encounter such an image at my talk? The violence that humans are capable of doing to one another is a part of this history, if not the very point of studying the Holocaust. Still, I had to ask myself the question: had I done something wrong?

I recalled that incident a few years later, when I was invited to teach some sessions at the Holocaust Educational Foundation's Summer Institute on the Holocaust and Jewish Civilization in Evanston, Illinois. Those seminars are targeted at current and prospective college and university teachers, including professors and graduate students, and the guest faculty each teach in their areas of specialization, which include history, literature, film, and gender studies, among others. I was there to teach about the cinematic history of Holocaust representations, but I also audited the literature sessions, which were led by Erin McGlothlin of Washington University in St. Louis. Because I frequently begin my own Holocaust film courses with compilation films including *Death Mills* (1945), *Nazi Concentration Camps* (1945), and *Memory of the Camps* (1946), I began my portion of the seminar with some of this material as well. These films, consisting largely of footage taken by Allied cinematographers at the end of the war, include images of piles of corpses, of naked and emaciated survivors, as well as shots of accumulations of the personal belongings of victims, such as their glasses and shoes. The films are generally constructed in a way that is intended to convey the shock that goes hand-in-hand with encountering these atrocities for the first time. That initial encounter is part of the Holocaust's history and, insofar as it was part of the Allied soldiers' experiences, it is part of American history as well.[2]

Nearly all of this footage is difficult to watch, to be sure, but it is also important. Even the permanent exhibition of the United States Holocaust Memorial Museum, in the hopes of connecting with its diverse visitors, begins with the question of what the Western armies first encountered in 1945, when the camps were initially opened. The images posed the question to the Western world:

---

2 On the experiences of the US soldiers entering the camps, see John C. McManus, *Hell Before Their Very Eyes: American Soldiers Liberate Concentration Camps in Germany, April 1945* (Baltimore: Johns Hopkins University Press, 2015).

how could this have happened? Framing the matter in this way has, at least historically, been considered pedagogically essential. Knowing what people have done to one another and what they are capable of doing to one another is an obligation for students as they begin to make their own informed political decisions. It is part of learning from the past, and, as the world becomes arguably more horrific and more violent, atrocities continue to happen. Compassionate encounters with the history of violence should unquestionably play a role in our pedagogy and in our scholarship.

But these are sensitive matters, and everyone brings to the table their own set of sensibilities, so I generally prepare students prior to any screening of atrocity images, and I did so that summer, warning participants that what they were about to see would by most standards be considered difficult to watch, and that, if anyone felt that they should not be exposed to atrocity images, then it would not be required of them. Such warnings are necessary and sensible precautions. Even there, as part of a course designed for those who wished to teach the Holocaust at college and university levels, some of the participants argued against the obligation to include this material, wondering what purpose such images serve. Some students asserted that the only appropriate response to material of this nature is silence, and that no good could come of showing images that serve only to stifle discussion. Paradoxically, we then had a lengthy and memorable discussion about whether such images inhibit dialogue.

The many problems associated with screening atrocity images – whether they reproduce the Nazis' dehumanizing gaze, whether they objectify Jewish victims, and whether they traumatize or retraumatize contemporary viewers, to list only a few – are not problems that are restricted entirely to visual representations. Although visual material seems to have a special claim on our imaginations, the medial distinction is by no means absolute. Some of these same students also struggled, in the course of Erin McGlothlin's sessions devoted to literature, with graphic passages from Ian MacMilan's *Village of a Million Sprits: A Novel of the Treblinka Uprising* (1999), which includes, among other seemingly transgressive imagery, a description of the sexual thoughts and fantasies of a Jewish prisoner at Treblinka. For some students, this too crossed a line. Whether in the form of documentary film or fictional literature, some texts and images associated with the Holocaust pose challenges to viewers and readers. The question of appropriateness is a longstanding one. Documentary images, the naked and wounded bodies that either show what humans do to one another, or imagined words meant to conjecture what humans may have done to one another in the absence of witnesses and historical records, each have their place on college syllabi. Questions such as "why?" and "is this really necessary?" have long been asked. It is the answers rather than the questions that have changed over time.

Everyone encounters limits as to what they feel is appropriate and what makes them uncomfortable to watch. I can hardly look very long at Nazisploitation films such as *Ilsa, She Wolf of the SS* (1975), and I have difficulty envisaging such a film's pedagogical merit.[3] (I do not, however, instruct others *not* to watch such things). And sometimes I am made uncomfortable by atrocity's aestheticization: what is one to make of redemptive narratives or of acceptable Holocaust photographs, among which I would include some of those taken by Margaret Bourke-White? When *Life* magazine published her well composed photographs of the concentration camps in May 1945, they were accompanied by the caption, "Dead men will have indeed died in vain if live men refuse to look at them." Bourke-White was a remarkable photojournalist who was responsible for extraordinary and groundbreaking work, but if we cannot see the atrocities in photographs of atrocious historical events, what, then, are we meant to be looking at?

At the US Holocaust Memorial Museum, I have seen footage of Polish deportations that in its cruelty and brutality embarrasses any attempts to dramatize such scenes. However painful those documents are to watch, I am grateful for having seen them because I feel that they left me a little more aware of the dynamics of deportation, and more knowledgeable about the gaps that separate these archival documents from their supposedly realistic counterparts in melodramatic feature films. Steven Spielberg's dramatizations seem insubstantial in comparison. Footage of that sort also informs us about the horizons of expectation that were experienced by those who were about to be deported: one cannot draw certain conclusions from looking at images. They are, after all, merely collections of frozen moments in time onto which, in the absence of testimony, viewers project their own thoughts and ideas, and much of what we can say about them is speculative. I believe, however, that it is still important to watch them, to see whether they speak to us, and to try to come to terms with whether those who were on the way to Auschwitz were conscious of the fate that awaited them. Most important of all is that such footage confirms the absolute uncertainty in drawing conclusions about others' states of mind.

Among the most important images in my Holocaust pedagogy has been the Theresienstadt propaganda film that was originally entitled *Theresienstadt. Ein Dokumentarfilm aus dem jüdischen Siedlungsgebiet* (*Theresienstadt: A Documentary Film from the Jewish Settlement Area*). The history of that film, which falls in general terms along lines similar to that of the Warsaw Ghetto

---

3 On the definition of Nazisploitation, see Daniel H. Magilow, Elizabeth Bridges, and Kristin T. Vander Lugt, eds. *Nazisploitation!: The Nazi Image in Low-Brow Cinema and Culture* (New York: Continuum, 2012).

film, is well known: in 1944 the Nazis briefly prettified the streets of the Theresienstadt Ghetto, in part clearing up the ghetto's overcrowding through additional deportations, in preparation for a visit by Red Cross inspectors. The Nazis then produced a propaganda film using those streets and other parts of the ghetto as false evidence of the grand opportunity Hitler had given the Jews to establish a thriving community. Only a fragment of the film, which was directed under duress by the German Jewish actor and filmmaker Kurt Gerron, remains. The term "directed" here is, of course, problematic, because Gerron, who was gassed to death at Auschwitz only a few months later, was forced to work behind the camera in the vain hope of prolonging his life. In light of the work's compulsory character, it is difficult though perhaps not impossible to think in terms of a filmmaker's distinctive handicraft.[4] The Theresienstadt fragment includes scenes of musical productions, public lectures, and even soccer games. On its surface, some of the film's images seem to simply document daily life. In keeping with the style of Pathé newsreels from the wartime era, a narrator explains what we see, and Theresienstadt, with all of its culture and industry, is made to look like a lively city on the go. Other sequences, including footage of public bathing facilities, seem strangely invasive or ethnographic. The question that confronts the contemporary student of Holocaust film is how to go about encountering and interpreting such images, made, as they were, by persons who were persecuted and imprisoned, forced to behave as though they were content and as though the circumstances in which they lived were normal.

All of the film's scenes of Theresienstadt contain images of people who have been torn from their homes and are suffering. As Irene Kacandes expressed it to me when I presented a reading of this fragmentary document at a 2004 conference of the Society for the Study of Narrative Literature, one may honestly say that these images are portraits of prisoners in the process of being tortured. The Theresienstadt Ghetto was, as Ruth Klüger points out, not so much a ghetto as it was another concentration camp. Klüger, who was there for a time as a child before being sent to Auschwitz, writes, "a ghetto doesn't normally mean a prison, but that part of town where Jews live. Theresienstadt, however, was the stable that supplied the slaughterhouse."[5] In this sense, the

---

**4** On the history of the film, see Karel Margry, "'Theresienstadt' (1944–1945): The Nazi propaganda film depicting the concentration camp as paradise," *Historical Journal of Film, Radio and Television* 12, no. 2 (1992): 145–162, and, Brad Prager, "Interpreting the visible traces of Theresienstadt," *Journal of Modern Jewish Studies* 7.2 (2008): 175–194.

**5** Ruth Kluger, *Still Alive: A Holocaust Girlhood Remembered* (New York: Feminist Press at the City University of New York, 2003), 70.

film, if we call it a document, is a document *as* violence. Its production was itself a violent act and anyone today should be conscious of that fact should they decide to show or watch it. However, the film also contains images of Jewish life. Art was created in Theresienstadt and, oddly enough, even soccer games were played. Although we might come to know what life in Theresienstadt was like on the basis of the vivid descriptions in H. G. Adler's definitive account, were it not for this film we would not be able to see the Jewish prisoners' faces, in some cases the faces of numerous identified and as-yet unidentified prisoners, who lived for many months behind the ghetto's walls.[6] The film is witness not only to the forced conditions under which it was made, but it is also evidence of the cultural life experienced and created by those who lived there in their final days and months. It includes, for example, fragmentary images of the now famous children's production of *Brundibar*, the opera composed by Hans Krása, mentioned in many survivors' testimonies and adapted sixty years later by Tony Kushner as an act of Holocaust remembrance.[7] The faces of the Theresienstadt prisoners watching the film's soccer game are meaningful as physical traces precisely because they were made under duress and because their sometimes ambivalent expressions now seem to me as though they are conveying hidden messages. Is it not meaningful that we, fully conscious of the conflicts, see these prisoners as they simultaneously conceal and reveal their distress?

While one can surely heed the many complexities involved, looking away from this and other, similar documents does not seem to me to be a preferable choice, neither as a pedagogue nor as an empathetic person. As Roland Barthes asserts throughout *Camera Lucida* (1980), photographs capture singular moments and are in this way uniquely bound to their referents.[8] They capture something that, like lived experience itself, is fleeting and cannot be recreated. As impressions made by light reflecting off the real persons who once stood before the camera, photographs are a remnant of those persons, they are tactile

---

6 For details about daily existence in Theresienstadt, see H. G. Adler, *Theresienstadt 1941–1945: The Face of a Coerced Community*, trans. Belinda Cooper (New York: Cambridge University Press, 2017 [orig. 1955]). Identifiable prisoners in the film include Leo Baeck, Karel Ancerl, and Carl Meinhard, among others.

7 See Charles Isherwood, "Tony Kushner and Maurice Sendak Adapt 'Brundibar,' a Czech Children's Opera," *New York Times*, 9 May 2006, https://www.nytimes.com/2006/05/09/theater/reviews/09brun.html.

8 See Marianne Hirsch, *Family Frames: Photography, Narrative, and Postmemory* (Cambridge, MA: Harvard University Press, 1997), who writes, "Barthes intensifies the indexical relationship when he speaks of the photograph as a physical, material emanation of a past reality; its speech act is constative: it authenticates the reality of the past and provides a material connection to it" (6).

traces. As John Berger articulated it in a now famous quotation, "a photograph is not a rendering, an imitation or an interpretation of its subject, but actually a trace of it. No painting or drawing, however naturalist, *belongs* to its subject in the way a photograph does."[9] The Theresienstadt film contains the last physical traces of those who were deported from Theresienstadt and died; these are the impressions, akin to footprints in the sand, that the prisoners left on the filmed material itself.

Claude Lanzmann, the French director of *Shoah* (1985), avoided including any historical, archival footage of prisoners and victims in his lengthy, epic Holocaust documentary. He seemed to have many reasons for this: archival footage looks historical and makes the Holocaust appear to be an event from the past. He wanted his film to be about the presence of the Holocaust, lest it be treated as something over and done with. The testimonies of the survivors in his films generate an impression of the event's presence or contemporaneousness by speaking on location, at the sites where atrocities took place, about the persistent presence of those events in memory, rather than by pointing viewers in the direction of time gone by. Moreover, according to Lanzmann, archival footage is often dehumanizing; if it was made by the perpetrators, then it reduced those in the footage to the objects of the perpetrators' gazes.[10] We should not see Jews as their tormentors' saw them. However reasonable Lanzmann's and others' concerns are, and however admirable Lanzmann's consistently asserted and maintained standpoint may be, archival Holocaust footage serves an important evidentiary function: it is the least controvertible, least negotiable, and least questionable confirmation of what happened. Were there only witness testimony, fewer people would acknowledge the truth; film and photography are, after our own eyes, the second-best witnesses.

One wonders what it would be like for someone to watch *Shoah* had they never seen images of the camps' liberations. Probably few such viewers exist. The experience of watching Lanzmann's documentary, with its prohibition against archival footage, is largely predicated on its viewers already knowing what is contained in those images, of having that series of images already churning around in one's head.[11] The provocation of *Shoah* thus has at its heart the paradoxical notion

---

9 John Berger, *About Looking* (New York: Knopf Doubleday, 2011), 54. Italics in original.

10 The concept of the perpetrator gaze is crystalized by Marianne Hirsch in "Nazi Photographs in Post-Holocaust Art: Gender as an Idiom of Memorialization," in Alex Hughes and Andrea Noble, eds. *Phototextualities: Intersections of Photography and Narrative* (Albuquerque: University of New Mexico Press, 2003), 19–40; here, 34.

11 On Lanzmann, Spielberg, and the varying uses of archival images, see Karyn Ball, "For and against the *Bilderverbot*: The Rhetoric of 'Unrepresentability' and Remediated 'Authenticity' in

that we ought not look at atrocity images because we have already seen them, and, perhaps, because we continue to see them. Curiously, late in his life and nearly 30 years after the first release of *Shoah*, Lanzmann included excerpts from the Theresienstadt propaganda film – archival footage, produced at the Nazis' behest, in which the black and white past seems squarely contained in the past – in *The Last of the Unjust* (2013), the director's own re-edited interview of his 1976 filmed encounter with Benjamin Murmelstein, one of the Jewish elders in Theresienstadt.[12] Did Lanzmann make a decision to include the footage, in violation of his own strictures, because the Theresienstadt propaganda film was directed by Gerron and therefore not truly a product of the Nazi gaze, or can we treat his decision as an acknowledgement that the Jews who can be seen in that film tell us more about their lives when we try to look at them, and that there is, in this case, less to be gained by looking away?

Just as Lanzmann reached a different conclusion in 2013 than he did in 1985, today differs from yesterday. Some students take Holocaust courses for empathetic reasons – there are a good number who have read *The Diary of Anne Frank* and would like to know more – while others take such courses for biographical reasons (for example, they or someone they love are Jewish). My students tend to come from backgrounds that differ a great deal from my own, and syllabi that were once effective, in different times and for different locations, have to be reconsidered. The first weeks of my courses often include footage of the liberation of the camps, although I have had to rethink those inclusions as I have come more and more to feel that they generate a palpable disquiet in the classroom, which I sometimes interpret as shock. For a time, I would have thought that all American students, in their high schools or elsewhere, had encountered images of Holocaust atrocity, but this is no longer the case, and one might conjecture that high school curricula have been tending away from the subject.

I was exposed to images of the Holocaust at a very young age and have always imagined that it helped me come to terms with history and historical violence; that it gave me a stronger basis for thinking critically about such matters later in life. If some contemporary students seem to be squeamish, this might be

---

the German Reception of Steven Spielberg's *Schindler's List*," in *Visualizing the Holocaust: Documents, Aesthetics, Memory*, eds. David Bathrick, Brad Prager, and Michael D. Richardson (Rochester, NY: Camden House, 2008), 162–184.

**12** On the film, see Tobias Ebbrecht-Hartmann, "Double Occupancy and Delay: *The Last of the Unjust* and the Archive," in *The Construction of Testimony: Claude Lanzmann's* Shoah *and its Outtakes*, eds. Erin McGlothlin, Brad Prager, and Markus Zisselsberger (Detroit, MI: Wayne State University Press, 2020), 207–232.

ascribed to a lack of exposure, but one might also propose that some contemporary squeamishness has to do with sexuality. Even when it comes to *Schindler's List* (1993), in which glimpses of naked bodies play a significant part in that film's alleged authenticity, one detects evidence of discomfort among students, frequently male ones, who wonder aloud whether such images are "necessary."[13] Male nudity is always a particularly fraught subject where students' sensitivities are concerned, and this particular assortment of discomforts may be what comes to the surface, at least in part, when students are confronted with atrocity footage. Many stories I have heard about students who experience discomfort in film courses revolve around films that contain sexual violence. A current graduate student described having been shown a particularly violent scene from *A Clockwork Orange* (1971) in a freshman class and reported that she lost confidence in the faculty member who screened it because he failed to properly introduce it. This is hardly an isolated anecdote, and the most common examples pertain to the lines around sexual assaults, including films that were formerly common staples of film courses such as *A Clockwork Orange* and *Blue Velvet* (1986). For better or worse, American students who might not ask to be excused from war films will be made uncomfortable by sexual violence.

When it comes to students' discomfort with the films from Buchenwald and Bergen-Belsen, one has to ask whether the sexuality in these films – the naked bodies of survivors and of the dead – might be more upsetting to them than the fact of mass murder.[14] The obvious response is that both may be upsetting, and that students' concerns should surely be attended to. One wonders how the two responses interfere with one another, and whether it would be productive to disentangle the twin conflicts. Where precisely does the "revulsion reflex" enter into it? If, for Susan Sontag, looking at photographs of camp liberations, images of the survivors and the dead, left something inside of her "still crying" afterwards, it was likely not the nudity that had abraded her soul.[15] I feel my response is closer to Sontag's than to that of my own students, and I see these images in terms of the violence done to the victims' bodies – in terms of the

---

**13** The idea of "necessity" is a strangely frequent detour taken by student responses. Students often describe aspects of a work that run counter to their tastes as "unnecessary," speaking as though works of art should be subject to standards of efficient delivery. Responding to this student perspective is an ongoing pedagogical challenge.

**14** Sven Kramer, "Nacktheit in Holocaust-Fotos und -Filmen," in *Die Shoah im Bild*, ed. Sven Kramer (Munich: edition text + kritik, 2003), 225–248, provides a thoughtful reflection on the problem of nudity in Holocaust films.

**15** See Susan Sontag, "In Plato's Cave," in *On Photography* (Picador: New York, 1977), 3–24; here, 20.

struggles undergone by those bodies, which appear before us, starved and terrorized. One might not, in other words, *see* their nudity, and whether or not the filmed survivors are covered by the tattered rags that remained of their camp uniforms has little impact on my perception of them.

One might consider, as an instructive example and as an illustration of historical difference, Nick Ut's famous Pulitzer Prize-winning 1972 photograph known as "Napalm Girl," featuring a nine-year-old nude female after having been being burned by a napalm attack. At the time and in that context the photograph's nudity was accepted, even revered, and it was published on the front page of the *New York Times*, but how would that same photograph be received today? And, moreover, it would be fair enough to note that survivors, who rarely have control over their depictions – to say nothing of the lack of control that the dead have over these matters – might feel differently about the circulation of their likenesses. How visual documents are used, both when they are created as well as long after the fact, are very contemporary concerns, ones on which many students tend to home in. But things are different for students than they once were: those who are attuned to this question may be so concerned because many of them have grown up in a world where they needed to consider the possibility of unauthorized circulation of their own personal and private images on social media and elsewhere.

We need not look all that far to find a recent example on which to draw: the journal *American Anthropologist* placed on the cover of its March 2020 issue an archival photograph of Margaret Mead among a collection of skulls with which she had returned from her fieldwork in New Guinea. The journal was criticized for the decision, and its editor in turn apologized, changing its cover and permanently erasing the error, to the extent that such erasures are possible in the digital age. Deborah Thomas, the journal's then editor-in-chief, had been well intentioned and had only meant to hold the image up to critical scrutiny. She explained in an interview, "even our critiques of white supremacy can end up reproducing that which we are fighting against. Our use of the image of Mead illustrated exactly that."[16] Those skulls, like the bodies at Buchenwald and Bergen-Belsen, were human remains, and owing to the potential of those images to participate in the very violence that scholars and others hope to critique, any use of such images should be subjected to scrutiny. Most readers would likely

**16** See "An Interview with the Editor of *American Anthropologist* about the March 2020 Cover Controversy," *americananthropologist.org*, 29 June 2020, http://www.americananthropologist.org/2020/06/29/an-interview-with-the-editor-of-american-anthropologist-about-the-march-2020-cover-controversy/. I am grateful to Dagmar Herzog for referring me to this.

agree: the publication of the cover was a misstep, and it was better that it was withdrawn.

Context, however, matters. One might inadvertently come across a journal's cover, whether on a now antiquated magazine rack or, more likely, where it is infinitely reproduced in thumbnail form in an electronic search. Cover images are used for marketing, but what takes place in a classroom might arguably be different. Were I an anthropologist who wanted to discuss that very image of Mead in a course – someone who wanted to ask meaningful questions such as "What might Mead have been thinking?" "Who died to make this image possible?" and, above all, "What, if anything, can be learned from this?" – I would provide a framework for the image's viewing. Chip Colwell, the editor of the anthropology journal *Sapiens* noted, as a response to the controversy over the cover, that one should make efforts "to ensure that individuals do not *unwittingly* view photographs of human remains," adding, on behalf of the journal, "when we do elect to use images of human remains, we include a caution at the start of the article stating: 'Please note that this article includes image(s) of human remains.'"[17] People should not be unwittingly subjected to Holocaust images, and as teachers, our job is by no means to make students uncomfortable. If there is a new resistance to looking at images that hurt, then it most likely comes along with the rise in mental health awareness over the past few years, as students have been trying to take control of mental health in way that they had not before. But where the classroom is concerned, one hopes that we would avoid talking about images without looking at them. To be sure, no student should be made to see what they do not want to see. At the same time, no historian or future historian ever benefited from turning away from their objects of study.

Each encounter, whether at Ithaca College or at Evanston, presented me with an opportunity to reflect on my pedagogy. In my course's most recent iteration, rather than starting with the liberation footage, I began with a 38-minute documentary film produced by the United States Holocaust Memorial Museum and directed by Raye Farr entitled *The Path to Nazi Genocide* (2013). That film, which provides a broad narrative about the growth of antisemitism between the wars and about Hitler's consolidation of power, contains, towards its end, some atrocity images, and for some students it still seems to have been a bit much. The film includes some of those images, and it then concludes with footage of survivors

---

17 See Chip Colwell, "Is It Ever OK to Publish Photographs of Human Remains?" *Sapiens.org*, 11 March 2020, https://www.sapiens.org/culture/photographing-human-remains/. Italics added.

warmly greeting the liberators and having their first bites of bread, none of which manages to erase the atrocious images of brutalized bodies. It is difficult to imagine teaching the Holocaust without difficult images, whether in visual or textual form. Their limited inclusion, in Farr's film and in my syllabi, seems to me to be an apt, logical, and necessary reminder of what happens in a society in which race-based hatred is encouraged, unleashed, and goes uncontained.

# Works Cited

Adler, H. G. *Theresienstadt 1941-1945: The Face of a Coerced Community*. Translated by Belinda Cooper. New York: Cambridge University Press, 2017 [1955]).

Ball, Karyn. "For and against the *Bilderverbot*: The Rhetoric of 'Unrepresentability' and Remediated 'Authenticity' in the German Reception of Steven Spielberg's *Schindler's List*." In *Visualizing the Holocaust: Documents, Aesthetics, Memory*. Edited by David Bathrick, Brad Prager, and Michael D. Richardson, 162–184. Rochester, NY: Camden House, 2008).

Berger John. *About Looking*. New York; Knopf Doubleday, 2011.

Colwell, Chip. "Is It Ever OK to Publish Photographs of Human Remains?" *Sapiens.org*. 11 March 2020. https://www.sapiens.org/culture/photographing-human-remains/.

Ebbrecht-Hartmann, Tobias. "Double Occupancy and Delay: *The Last of the Unjust* and the Archive." In *The Construction of Testimony: Claude Lanzmann's* Shoah *and its Outtakes*. Edited by Erin McGlothlin, Brad Prager, and Markus Zisselsberger, 207–232. Detroit, MI: Wayne State University Press, 2020.

Hirsch, Marianne. *Family Frames: Photography, Narrative, and Postmemory*. Cambridge, MA: Harvard University Press, 1997.

Hirsch, Marianne. "Nazi Photographs in Post-Holocaust Art: Gender as an Idiom of Memorialization." In *Phototextualities: Intersections of Photography and Narrative*. Edited by Alex Hughes and Andrea Noble, 19–40. Albuquerque: University of New Mexico Press, 2003.

"An Interview with the Editor of *American Anthropologist* about the March 2020 Cover Controversy." *americananthropologist.org*. 29 June 2020. http://www.americananthropologist.org/2020/06/29/an-interview-with-the-editor-of-american-anthropologist-about-the-march-2020-cover-controversy/

Isherwood, Charles. "Tony Kushner and Maurice Sendak Adapt 'Brundibar,' a Czech Children's Opera." *New York Times*, 9 May 2006. https://www.nytimes.com/2006/05/09/theater/reviews/09brun.html.

Kluger, Ruth. *Still Alive: A Holocaust Girlhood Remembered*. New York: Feminist Press at the City University of New York, 2003.

Kramer, Sven. "Nacktheit in Holocaust-Fotos und-Filmen." In *Die Shoah im Bild*. Edited by Sven Kramer, 225–248. Munich: edition text + kritik, 2003.

Magilow, Daniel H., Elizabeth Bridges, and Kristin T. Vander Lugt, eds. *Nazisploitation!: The Nazi Image in Low-Brow Cinema and Culture*. New York: Continuum, 2012.

Margry, Karel. "'Theresienstadt' (1944–1945): The Nazi propaganda film depicting the concentration camp as paradise." *Historical Journal of Film, Radio and Television* 12, no. 2 (1992): 145–162.

McManus, John C. *Hell Before Their Very Eyes: American Soldiers Liberate Concentration Camps in Germany, April 1945*. Baltimore: Johns Hopkins University Press, 2015.

Prager, Brad. "Interpreting the Visible Traces of Theresienstadt." *Journal of Modern Jewish Studies* 7, no. 2 (2008): 175–194.

Prager, Brad. "The Warsaw Ghetto, Seen from the Screening Room: The Images That Dominate *A Film Unfinished*." *New German Critique* 41, no. 3 (123) (2014): 135–157.

Sontag, Susan. "In Plato's Cave." In *On Photography*. New York: Picador, 1977.

Claudia Breger
# The Affects of Reading

I would like to begin with two anecdotes from my early years as a scholar. The first one revolves around a fuzzy memory of a secondhand information snippet: according to my advisor, one of my committee members emphatically insisted during the closed proceedings at my dissertation defense at Humboldt University Berlin in 1996 that "charging [Johann Wolfgang v.] Goethe with racism," that really was going too far. I obviously didn't think so: it was about time for the discipline to tackle that side of the quintessential national poet. My dissertation had traced the emergence of intersecting modern discourses of race and gender in German-language literary and ethnographic representations of so-called "gypsies" around 1800. As I may need to explain, I used this derogatory label in quotation marks to unfold a constructivist argument underlining that the name did not initially apply to the ethnic minorities of Sinti and Roma alone. Today, that discourse-centered framing feels at least borderline insensitive, which might be another story. Or is it part of the larger story I pursue in this chapter?

Anecdote two: During a postdoctoral fellowship in Munich and my first teaching position in Comparative Literature in Paderborn, Germany, I fully plunged into the theory (or Theory, as some would spell today) conversations of the 1990s. One of the resulting articles resulting was a piece in which I deployed queer and postcolonial theory to trash Slavoj Žižek, the then very hip Marxist-Lacanian philosopher, effectively charging his increasingly theological thought with antisemitism, misogyny, and totalitarian implications. I still believe that I was onto something (and others have since developed resonant critiques),[11] but I probably shouldn't have been surprised when I lost some (Theory) friends over this intervention. And I was in for a shock when the journal *Diacritics*, upon agreeing to publish the article despite concerns about its polemical tone, asked Žižek for a response, to which I was then invited to briefly respond once

---

1 See, e.g., Marcus Pound, *Žižek: A (Very) Critical Introduction* (Grand Rapids, MI: Eerdmans, 2008), and its (in turn very critical) reviews, such as Alessia Ricciardi's in *Notre Dame Philosophical Review* 7, October 2009 (https://ndpr.nd.edu/news/iek-a-very-critical-introduction/).

**Note:** I would like to thank my parents, Ulrike and Manfred Breger, and my aunt, Sabine Woysch, for working with me on this, and for letting me write it up for publication, including the painful aspects. Thank you also to Irene Kacandes and Priscilla Layne for their feedback and encouragement on an earlier draft.

https://doi.org/10.1515/9783110753295-029

more. In writing my piece, I hadn't exactly made mental room for a human-to-human dialogue.

Some of my inability to imagine myself in an actual exchange with a famous contemporary thinker I was critical of can, of course, simply be attributed to my junior status at the time. But perhaps some of it also points to the larger questions I want to reflect on here: questions about my own scholarly trajectory in relation to paradigms of co-witnessing,[2] the legacy of the Holocaust and other (more or less adjacent) contexts of racism, collective violence, and trauma. In a nutshell, most of my scholarship over the past two and a half decades is – directly or indirectly – written in an engagement with these different contexts of collective violence and trauma. At the same time, the ethos of my scholarly readings was not always one of emotional, supportive co-witnessing, in particular, not early on. More recently, I (along with many others) have been actively grappling with such questions of scholarly and personal ethos – tone, affect, stance – and become more inclined to advocate empathy, respect, nuance, and complexity. But this essay will not be a straightforward academic *Bildungsroman,* or therapy narrative. In many ways, I am still in the process of sorting out these questions, and yet newly again in 2020, after almost four years under the Trump regime, in the midst of a pandemic with its heightened feelings of anxiety, anger, and depression, and vis-à-vis ever more pressing questions of racist violence. How do I develop forceful political critiques while taming paranoia? How does the ethos of respectful reading fit with the imperative of outrage, or the potentially productive role of anger in facilitating change highlighted by Audre Lorde early on?[3] Where and how is empathy urgently needed vs. misguided?[4]

Of course, I was not the only aggressive, radical young academic out there in the 1990s whose reading habitus was heavily slanted towards suspicion and, at moments, paranoia. As Eve Sedgwick, Rita Felski, and others have forcefully argued since then, late twentieth-century critical theory had hardened the "hermeneutics of suspicion" (a term coined by Paul Ricoeur for the modern tradition building on Marx, Nietzsche, and Freud) into an "imperative framing" that

---

**2** On this concept, see the introduction to this volume.
**3** Audre Lorde, "Uses of Anger," *Women's Studies Quarterly* 9, no. 3 (Fall 1981): 7–10.
**4** Christina Sharpe emphasizes the stark asymmetries in her students' capacity to show empathy with victims of the Holocaust vs. those of slavery, while also underlining that "no mode of (white) empathy [. . .] can replicate the daily strain of knowing that as a black person you can be killed for simply being black." Christina Sharpe, *In the Wake: On Blackness and Being* (Durham, NC: Duke University Press, 2016), 11, 16.

widely succeeded in casting alternative stances as "naïve, pious, or complaisant."[5] While these dominant practices of academic critique may have prided themselves on clear-sighted, more or less sovereign "detachment," their – my – endeavors of textual interrogation, unmasking, subversion, and destabilization were arguably heavily fueled by negative affect.[6] To be fair to the feminists within ourselves, we may have never fallen into the sovereignty trap entirely: there was certainly an acknowledgment – long before today's "post-critical" turn – of the ways in which positionalities and experiences shape reading practices. And we may want to supplement the negativity diagnosis by underlining, as recent discussions have equally done, that the heightened criticality of 1990s' academic reading practices was variously counteracted or undergirded by the very different affects of non-academic reading, including downtime fan practices and feminist and queer icon crushes.[7] Closer to the topic at hand, there were my experiences of being moved by the autobiographical texts of Ruth Klüger and Audre Lorde,[8] and much earlier, when I was barely an adolescent, Judith Kerr's *When Hitler Stole Pink Rabbit* (1971); *The Diary of Anne Frank* (1947); and Janina David's *A Square of Sky* and *A Touch of Earth* (1964, 1966).

As indicated above, the debates about critical and postcritical reading practices have significantly reshaped my academic engagements. In the past decade and a half, my work has contributed to thinking through questions of empathy and affect from different angles: both critically in relation to power hierarchies and sociosymbolic conditionings and affirmatively as constitutive forces that build relations and offer political openings. My most recent book, *Making Worlds: Affect and Collectivity in Contemporary European Cinema*, foregrounds films that (in Jacques Rancière's words) reshuffle "the sensible" by developing intricate configurations of layered, often diverging affects – or in my wording, affectscapes – in complex

---

**5** Eve Kosofsky Sedgwick, *Touching Feeling: Affect, Pedagogy, Performativity* (Durham, NC: Duke University Press, 2003), 125, 126. See also Rita Felski, *The Limits of Critique* (Chicago, IL: University of Chicago Press, 2015). Drawing on Melanie Klein, Sedgwick associates the practices of paranoid reading with affects of "hatred, envy, and anxiety" and contrasts them with the "reparative" practices of "love" (128).

**6** Felski, *The Limits*, 6, see 2, 5. See also Felski, *Uses of Literature* (Hoboken, NJ: Wiley Blackwell, 2008), 3. By way of terminological explication, many affect scholars distinguish (fluid, indefinite, transsubjective, nonconscious) affect from (more definite, subjective) emotions. I share the particular interest in affect's respective conceptual affordances, but emphasize the affect-emotion continuum over oppositional delineations.

**7** See, e.g., Robyn Warhol, *Having a Good Cry: Effeminate Feelings and Pop Culture Forms* (Columbus: Ohio State University Press, 2003); Felski, *Uses*, 12 (with reference to John Guillory).

**8** Ruth Klüger, *weiter leben: Eine Jugend* (Göttingen: Wallstein, 1992); Audre Lorde, *Zami: A New Spelling of My Name* (Trumansburg, NY: The Crossing Press, 1983).

scenarios.[9] In trying, for example, to make sense of people's complicated attachments to violent collectives, in creating unexpected, shifting audience alignments, or in imagining transcultural connection and solidarity in the face of persisting legacies of hate and inequality, I argue, these films encourage us to see, hear, feel, and think again about present, past and future forms of affective belonging.

The invitation to contribute to this volume inspired me to ask biographical questions about my own scholarly trajectory. This piece, then, is a – preliminary – attempt to bring my theory to my family histories, and these family histories back to the theory. Recent (U.S.-centric) contributions to the postcritical turn have argued that the late twentieth-century ethos of academic critique "arose from a Cold War state epistemology."[10] The academic persistence of this hyper-critical habitus, Christopher Castiglia charges, indicates unresolved attachments to state-fostered suspicion and paranoia ("communists" and "homosexuals" could "be anywhere"), attachments at odds with most critics' manifest leftist commitments.[11] I am uncomfortable with how this historical contextualization seems to delegitimize critique wholesale and inclined to lend an ear also to the opposite argument for the continued importance of critique to any politically minded scholarship precisely in our neoliberal times.[12] As indicated above, Castiglia's Cold War contextualization also needs to be supplemented with the reminder that critique's intellectual genealogies in the "hermeneutics of suspicion" extend back to the nineteenth century. However, I want to probe a story partly resonant with Castiglia's metacritical narrative by shifting its geopolitical framework and linking the argument about critique's underlying negativities to the affectscapes of my own childhood and adolescence in West Berlin. Geographically at the very center of the Cold War, this coming of age was overshadowed – well, certainly not by U.S.-sponsored suspicion alone. In particular, I am wondering how the reading practices I grew into were shaped by the "psychic deformations of violent histories" that (in Gabriele Schwab's words) affect perpetrator along with victim families, "albeit in different ways and with different responsibilities."[13] How were my reading methodologies inflected

**9** Claudia Breger, *Making Worlds: Affect and Collectivity in Contemporary European Cinema* (New York: Columbia University Press, 2020).
**10** Christopher Castiglia, *The Practices of Hope: Literary Criticism in Disenchanted Times* (New York: New York University Press, 2017), 2.
**11** Castiglia, *The Practices of Hope*, 12, see 13.
**12** See, e.g., Bernard E. Harcourt and Didier Fassin, eds., *A Time for Critique* (New York: Columbia University Press, 2019).
**13** Gabriele Schwab, *Haunting Legacies: Violent Histories and Transgenerational Trauma* (New York: Columbia University Press, 2010), 72.

by the "crippling 'inability to mourn,'" the "coldness," "rigidity" or "lack of emotion" that so famously pervaded postfascist German culture?[14] Respectively, how were they shaped by the "strong," albeit "often extremely negative emotions" uncovered by Anna Parkinson's more recent attempt at "'thawing'" this affective landscape through a closer look at "postwar phenomenology"?[15]

My rhetorical emphasis on these questions *as questions* is deliberate; I intend for this piece to serve as a methodological exercise in caution that, in part, counterbalances Schwab's psychoanalytic certainty. Even as her findings continue to resonate with my own interpretative intuitions, I find Schwab's analysis in places too definite today. But perhaps, it is possible to complicate the narrative without altogether discarding its relevance to my reflective endeavor. One layer of complexity is generational. Born in 1969, I was taught to read critically by teachers and professors with passionate attachments to antifascism and social justice in the wake of the student movement. Pace Castiglia, I do think that this emergence of academic critique *against* repression, in an intentional movement *towards* mourning does make a difference. My own parents, to be sure, largely missed out on the 1968 social revolution, as I detail below. But to foreshadow, and towards this volume's exploration of links between histories of violence, they spent the academic year 1967–1968 in New York City – my father on a fellowship at New York University, my mother working as an *au pair*. Having joined a march after the assassination of Dr. Martin Luther King, Jr., they brought home a "We shall overcome" record, a vinyl single that would accompany my childhood as a prominent memory object of their time in the United States. Arguably, this biographical detail resonates with larger sociopolitical phenomena, namely the postwar German tendency to perceive and work through questions of racism primarily in relation to the United States rather than the local context.[16] Posing as many questions about political identification and solidarity as answering them, the little "We shall overcome" record gestures at some of the complexity and unevenness of anti-fascist memory work in my parents' (and my own) generation.

Another layer of complication is methodological. Parkinson's affect-theoretical attempt to "thaw out" the emotional landscape of postfascist Germany is not

**14** Schwab, *Haunting Legacies*, 75, with reference to the Mitscherlichs; the other quotes are from Anna Parkinson, *An Emotional State: The Politics of Emotion in Postwar West German Culture* (Ann Arbor: University of Michigan Press, 2015), 5–6, with reference to Hannah Arendt and Theodor W. Adorno (1959).
**15** Parkinson, *An Emotional State*, 1, 5.
**16** See, e.g., Heide Fehrenbach, *Race After Hitler: Black Occupation Children in Postwar Germany and America* (Princeton, NJ: Princeton University Press, 2005).

inconsistent with the earlier critical verdict regarding its coldness. The widespread inability, or unwillingness, of Germans to express the normative emotions demanded by official postwar culture, she simply insists, did not equal the absence of affect, but indicated "a flood of overwhelming affect to be kept at bay" in an assemblage of defensive disavowal, ambivalence, numbness brought about by destruction, shame, and anxiety.[17] This insistence on the negative affect "under" or within more or less cool demeanors does, however, perform a shift in perspective to more fully engage the experience of living in the wake of mass murder, as and among perpetrators and victims, or (with Michael Rothberg) from a range of positions of deep historical implication.[18] My own background overall positions me in adjacency to perpetrator rather than victim histories, a difference vis-à-vis many other contributors to this volume that needs to be marked, even as I pursue complexities and, in 2020, re-read my family histories in less starkly oppositional terms as a textbook case of "implication" in Rothberg's sense, with its connotations of entanglement, involvement, close connection, and being "folded into" events.[19] In reflecting on these histories in relation to my academic reading practices, I wonder how I can engage – and honor – affective experience without slipping into apology or becoming complicit in lies, embellishments, and silences. Navigating that challenge may be a crucial element of finetuning postcritical method for the context at hand.

Furthermore, I do not want to short circuit – or summarily explain – personal with collective experience in tracing impacts of affective negativity. Not all affect scholars would agree to delineating affectscapes along national lines in the first place, as they distinguish political emotion from underlying "unqualified" and "irreducibly bodily," more or less biological affect.[20] In my own work, I do underline that even comparatively indistinct "floods" of affect are historically shaped – here: incited and molded in the entanglements of fascist and post-fascist discourse, violence, and bodily memory. Emphasizing the historical load of affective negativity in postfascist Germany should not, however, translate into diagnosing a homogenous national habitus. Affective assemblages are uneven, heterogeneous, changeable, and always also emerge from transnational flows. With an eye on these latter flows, my biographical as well as scholarly interest in the German context amounts to a balancing act also

---

17 Parkinson, *An Emotional State*, 8, see 10, with reference to William Ernest Hocking.
18 Michael Rothberg, *The Implicated Subject: Beyond Victims and Perpetrators* (Stanford, CA: Stanford University Press, 2019).
19 Rothberg, *The Implicated Subject*, 1.
20 Here Brian Massumi, *Parables for the Virtual: Movement, Affect, Sensation* (Durham, NC: Duke University Press, 2002), 28.

insofar as I affectively remain bound to some of the exceptionalism I grew up with – the Shoah is not just any genocide – in tension with my scholarly insights into the Holocaust's position as "part and parcel of the pervasive history of Western colonial and imperial aspirations" and my ethical and political commitments to "linking" different past and present "violent histories."[21]

With these caveats, there certainly is plenty of affective negativity to report on – mesh ups of disavowal, ambivalence and anxiety, defensive reserve, numbness, shame, anger, contempt, and hate – in my readerly reconstruction of the personal-collective histories I grew up with and against in post-Holocaust Germany. My own phenomenological experience of the affective fallout included confrontations with my father's anxiety, affective volatility, and moments of depression, along with his attempts to calm all of these feelings through orientations at protestant work ethics, frugality, and respectability; and my mother's emotional reserve and heightened criticality. Of course, I inherited/internalized most of these feelings and coping mechanisms myself. As a kid, I was scared by the tensions and fights between my parents, deeply haunted by worries that they would break up, including the anticipated shame associated with the status of having divorced parents. At fifteen, I had turned into a ferocious feminist advocate of my mother breaking up with my father. To no avail: they actually stayed together, and I have since been credibly reassured – and experienced – that things got much better.

At holiday parties and other family gatherings, my paternal grandfather would start talking fondly about his war experience after a couple of drinks, complete with a range of racist comments about "the French" and "the Poles." As a young adult, I would summarize my perpetrator background by freely mentioning my "Nazi grandfather." My reading strategies were significantly more generous towards my grandmother, who had told him to shut up with some regularity and directness on those occasions. Probably, both evaluations were too simple. In response to the invitation to write this piece (and inspired by reading Angelica Bammer's *Born After* earlier this spring),[22] I finally followed up with my parents and an aunt about parts of my family's histories that I knew little about or remembered only vaguely. At this point, the research process is far from complete, limited to long-distance communication in a summer without transatlantic travel. But even this initial foray has uncovered enough baffling evidence to sharpen my questions about how I want to read: to do justice

---

**21** Schwab, *Haunting Legacies*, 72.
**22** Angelica Bammer, *Born After: Reckoning with the German Past* (New York: Bloomsbury, 2019).

to the complexity of my parents' and grandparents' lives but also to the lives of others they judged, exploited, and in whose premature deaths they were involved directly or indirectly through their (our) positionalities.

Quick fact-checking indicates that my grandfather actually served for just a few weeks in World War II, in a supply unit in Lothringen in spring 1940: he was born in 1904, thus relatively old and also considered essential as owner of a farm. The lance corporal badge on the uniform in the photograph of him holding little Manfred, my toddler father, certainly doesn't indicate major military honors. The farm did rely on forced Polish labor during the war. I was always suspicious vis-à-vis the family claim (I remember as unanimous) that the laborers were treated well and relations hospitable. When I now ask my father for detail, he acknowledges more complexity, including the hierarchies indicated by the fact that the workers addressed his father as "boss" (*Chef*) while being called by their first names. Also, he says, there were significant differences between the two Polish workers on the one hand, and two Ukrainians on the other, who spoke less German and were considered to be "withdrawn" and "less cooperative." But my father also sends me a picture of the male Polish worker called Johann – probably, he adds, actually Jan – in a white Sunday shirt, as he gently supports little Manfred guiding a horse by a rein. My grandfather's first name is Johannes; the Germanizing of Jan's name created a father double in sound also. As my father relates, my grandparents remained in touch with Johann/Jan, who initiated the correspondence after the war, resulting in occasional care packages to Poland. (Had Jan hoped for these? Or what were his memories?) During the war, my father also relates, the Polish workers spread out across the village would socialize in Johann's apartment, which was illegal; my grandfather would indirectly warn them of imminent Nazi patrols. My father's fondest memory of Johann, however, is that he once served as Santa Claus, with wide-eyed little Manfred afterwards pitying him all evening for having just missed that big event.

In 1986, when I was preparing for a trip to Poland involving symbolic volunteer shifts at the Majdanek memorial site, my grandfather would warn me to be "careful with the Poles." (Or was that "the Jews" when I later went to Israel? I remember similar comments on both occasions.) A couple of years before the Poland trip, I had once staged a pointed exit in adolescent rebellion in response to my grandfather's tipsy storytelling. My father asked me to let it go, citing my grandfather's age. Much later, my grandmother, then in fact getting frail, would shock me by approaching me, out of any context I could determine, with the all-too-well known defense claim: *Wir haben nichts davon gewusst*, "we didn't know about 'it.'" Despite or because of her resolute attempts to keep my grandfather in check, I had not expected her to resort to this cliché gesture. Do I believe her that they were not, or barely, aware of the Holocaust during the war? In any case, I grew up with the affective dislocations of half-buried, "tacit knowledge" in the war's aftermath.[23] When I now, in 2020, ask my father about his war memories, one of the first impressions he volunteers is that during a short-lived attempt to flee the farm in spring 1945 (when he was six years old), they saw a death march walk by: primarily women, he describes, watched over by uniformed guards. No, he responds to my insistent follow-up questions, of course, he didn't really understand what was happening; and the adults were not talking about it, or not in front of the children.

My grandfather, he confirms, was not a party member, although he later joined the SED (Socialist Unity Party) after the Brandenburg farm had become part of the GDR. Just for strategic reasons, "of course." When the collectivization of agriculture began in the early 1950s, in theory still voluntary, but under various pressures, the family fled to West Berlin (in a moment in which borders were still largely uncontrolled). For my father, this was the sudden end of a childhood in the company of animals but eventually a path to higher education; for my grandfather, it entailed a permanent loss of status: "Chef" became a school janitor.[24] My grandmother, Naëmi, who had more education, would find work giving religious instruction. My grandfather's problem, my father narrates, was that shortly before their escape, the roof of a barn had caught fire, after an apprentice, a strong smoker, had been in there to get hay. The apprentice testified that he had not smoked in the process, and my grandfather was charged with arson; it was known, he was told, that he had just been to West Berlin. Short-tempered, my grandfather responded: "You think I got the matches there because the local ones don't burn." Of course, my father comments, that was not "smart." While he tires easily of some of my other questions, my father seems to enjoy telling this particular story, and I find myself chuckling with fondness for the grandfather grump from whom I had emotionally distanced myself so forcefully during his lifetime. But even as we both appreciate the memory of rebellion, my internal suspicious reader still wonders – less about who set fire to the barn than about whether my grandfather never had to strategically join the NSDAP because his overall political loyalty, or affective support, was less in question under Hitler than under socialism.

Different strategies of reading are key also to an adjacent flashpoint of my longer-term trajectory of making sense of family histories. According to a set of

---

**23** Schwab, *Haunting Legacies*, 79.

**24** Schwab comments on the "crippling sense of inferiority" pervading postwar society (*Haunting Legacies*, 76). From that angle, I was obviously too hard on my grandfather.

documents from my father's family that surfaced early in my adult life, my father's uncle had, in 1935, assembled evidence to demonstrate seven generations of "Aryan" ancestry. No one seemed to know much about the context, and I gave up asking questions quickly myself. In my spontaneous reading, which became part of the larger perpetrator background narrative, this great uncle doubtlessly wanted to join the SS. Later, however, there was also the seduction of deploying suspicious depth hermeneutics towards an alternative, possibly more benign genealogy: inspired by my Theory readings about the hybridity and performative nature of all identities, I linked the archival evidence of these documents to the ambiguity surrounding my last name, Breger, about which I had been questioned by the mothers of Jewish friends as well as Israeli border guards. In light of the name's commonality among Eastern European Jews, might this great uncle have been pressured to prove racial credentials his contemporaries held in doubt? When I check with my father again in 2020, to be sure, it turns out that the uncle in question was a maternal uncle of his, by the name of Gnirk, not Breger. My father's best guess is that he assembled the documents when taking over a farm in the *Sudeten* territories, as part of Hitler's politics of colonizing Eastern Europe. My grandfather, my father mentions now, was later asked to take over a farm in the Ukraine, yet managed to resist the request. He did not want to live there.

In which ways do the everyday racisms that have been so deeply engrained in (German, European, U.S.-American) modern histories make Nazis? In the end, was my youthful verdict just close enough – or when and for which purposes is it important to read more carefully and insist on complexity? My father further adds that a cousin of his grandfather, on the Breger side, had done similar ancestry research, although before the Nazi period, and without securing the long-term evidence he was looking for: the older documents he found were in a Slavic language. If we take the point about the ethnic hybridity underlying ideological constructions of collective identity to be obvious enough, the more interesting question is in which ways such hybridity actually impacted the lives of my ancestors. Since I have no respective evidence at all for my father's family, this might be the point to transition to my mother's story.

Ulrike was born in Silesia in September 1944; a few months later, my grandmother fled westwards with the infant and her two toddler siblings, Sabine and Ludwig, alongside other family members. In addition to the paternal Nazi grandfather, that is, my young angry self also had a reactionary family branch politically aligned with revanchist Silesian emigre associations to show for – questionable things were indeed said at family gatherings over the years. As late as 2015, I found myself affectively quite challenged (while academically impressed) by Yuliya Komska's more differentiated look at these histories in *The*

*Icon Curtain.*[25] Looking back even farther, the family of my grandmother's mother had migrated to Silesia in the nineteenth century from the Tirolian Zillertal, as protestants fleeing respectively having been expelled by the Austrian Emperor as part of a policy of religious separation in the mid-1800s. My grandmother met her future husband in their Lutheran church (as did my mother later). While both of the Silesian families spoke and identified as German, my maternal grandfather's name was Woysch, which clearly resonates as Polish in my ears. My father explains what I would have said myself but wanted to hear from my mother: that Polish names in German families and the other way around were completely normal, after centuries of intermarriage under – in my own words – Prussian colonialism.

My only memory of this grandfather is of a formal portrait photo in my grandmother's apartment: he died of dysentery in a Soviet POW camp in 1945. When I now ask my mother about him, pointing out that I know next to nothing about her father's background, her first response is: I don't know anything, either. Then she puzzles together some. Her father's family owned a leather wholesale business. They seem to have led an upper-middle class life; Ulrike mentions skiing vacations and her own sense that without the war, she would have had a very different childhood. This is crucial to my mother's larger story: in response to my later questions about growing up as a refugee in West Germany, she says that she does not remember discrimination, but poverty as the defining puzzle piece: missing out, for example, on the daily milk treat the other kids enjoyed in school. Apparently, my grandfather's sisters also liked to emphasize that their mother (whose birth name was Martin) was from a family of manor owners with some connection to Gustav Freytag, the – in my critical summary – canonical nineteenth-century novelist of German colonialism in Poland. The Woysch family was educated as well. My grandfather, Hans, studied theology for a few semesters before being called home to take over the business because his father had passed; his no-good artsy brother took off to Berlin. (Yay you, I cheer on silently.)

Despite all the class credentials, it seems that my grandmother's relatives never warmed up to Hans Woysch and his family; tensions would persist into the postwar years. In response to my follow-up questions, my mother shares that anti-Polish racism was definitively prevalent in her mother's family, although she does not remember the connection to her father's family being made directly.

---

**25** Yuliya Komska, *The Icon Curtain: The Cold War's Quiet Border* (Chicago, IL: University of Chicago Press, 2015).

Hans's sisters, my aunt Sabine fills in later, would firmly resist the suggestion of any Polish heritage. My next question to my mother is fraught with my own hesitation: despite all my academic work on the subject, the ways in which race became taboo in post-45 Germany still exert a pull towards unspeakability in the family.[26] Am I over-reading yet again anyway? But I do ask: Could the Woysch family's relatively darker physical appearance have played a role also?[27] Clearly taken aback, my mother reluctantly considers the possibility, but then associates an adjacent set of family "taboos" (I don't think I had actually mentioned that word). There definitively were whispers, she says, not only around her father's "degenerate" brother (her original word choice, *aus der Art geschlagen*, cites the eugenic discourse that she probably overheard as a child), but also her father's sister, aunt Gretel, who was apparently depressed, or "psychologically unstable," and who was sterilized during the NS regime.

Now the shock response is on my end of the conversation. I am fairly sure I had not known about this until 2020. I try to imagine the life of this great aunt whom I never met: she died in the 1960s from cancer. What impact did the silences – and whispered knowledges – have on my mother and her siblings? (Later, I learn that my aunt does remember childhood comments that stung about her own "Jewish" appearance and the "degenerate" shape of her arms and earlobes.) And how does all of this impact my own autobiographical narrative? I do know that it doesn't provide me with a new victim genealogy. My mother also remembers a hushed family conflict on the occasion of aunt Eva, aunt Gretel's younger sister, bringing a copy of Alfred Rosenberg's *Der Mythus des 20. Jahrhunderts* into their postwar household. This seems to have been at a request for historical readings by the adolescent Sabine, who, in 2020, reflects back on the postwar culture of silence: her rare attempts at asking questions, "provocatively" and "accompanied by affect," would "unsettle" the adults,

---

26 On the disappearance of the language of race in German post-1945 discourse, see, e.g., Rita Chin and Heide Fehrenbach, "Introduction," in Rita Chin et al., *After the Nazi Racial State: Difference and Democracy in Germany and Europe* (New York: Columbia University Press, 2009), first 3.

27 I may have been over-reading. Sabine offers yet another puzzle piece: My great-grandfather had left the Lutheran church in converting to Adventism. The remaining family stayed, but his conversion left a "stigma" (*Makel*) on all of them. [Sabine suspects that her grandmother's manor owner family accepted her marrying a Woysch because she had a "Makel," too: she was supposed to have become a little "strange" (*eigenartig*) after a bad childhood fall.] Then again, racialized perceptions of physical appearance do play a role in another of Sabine's memories: one of my grandmother's siblings, who had died as a child, was idealized as a "special," "blond" kid.

trigger "embarrassment" and "defensive" counterattacks. According to these adults, my grandfather had been politically anti-Nazi. But Ulrike and Sabine also relate another story that I probably had heard before, although not clearly remembered: the last house that the Woysch family lived in in Legnica before 1945 was a house that they had bought from a Jewish family on their way into emigration. According to my grandmother, the sale had been a favor to the previous owners, facilitating their emigration, and she was very upset when the survivors sued in the 1960s on the grounds of unconscionability, complicating the financially struggling family's reparation claims. It seems, my mother says, that the experience activated and reinforced my grandmother's antisemitic stereotypes. (Antisemitism, Sabine contextualizes, was running rampant in the Lutheran church.) Ulrike reports this with analytical distance, but adds, eager to defend her parents: they probably couldn't pay more than they did when they bought the house; whatever those skiing vacations (that I mercilessly remind her of now), they were not rich.

I also try to imagine the life of my grandmother, Martha. She was 36 when she got married in 1941, definitively on the older side by contemporary standards. She had worked housekeeping jobs until then and helped out with her siblings' children. Towards the end of her life, my grandmother talked about a summer season gig on a ship on the Baltic Sea; Sabine thinks this was a fantasy of hers. My grandmother would have liked to get an education, my mother emphasizes; she "learned quickly" and wanted to be a teacher, but that was not in her father's plans (Sabine's words: "unthinkable" in a farmer's family). In my mother's retelling, my great-grandfather insisted on Martha learning needlework to take care of his shirts; he served as the village's mayor and allegedly always wore white shirts. If this sounds a bit cliché, it gets worse. Ironically (or unsurprisingly), my grandmother passed down her education trauma to my mother. While Sabine, the oldest daughter, got to attend the *Gymnasium* leading to the university entrance qualification in the three-tier German school system, my uncle was struggling in school, and my grandmother designated Ulrike, who was a year younger but had effectively skipped a grade in the one-room-village school, to help get him through the second tier, the *Realschule*. And again, there was no money; one educated girl in the family seemed plenty. My mother then trained to become a pre-school teacher. When she later wanted to go back to school at the outset of her marriage, my father apparently said: but we want to have children now. While I had known the first part of the story early on, this second part my mother shared only when I was working through family history in therapy as an adult. She initially found it hard to emotionally relate to the child who, by arriving too soon, had effectively put an end to her hopes for more education.

Education was not lacking in my life; both of my parents made sure about that. Whereas TV access was restricted, books were plentiful and deliberately chosen, facilitating the above-mentioned experiences of co-witnessing, and/ or imaginatively putting myself in the shoes of Anne Frank and others, long before the Holocaust was covered in history class. Although protective of their parents, my parents did want to break the silence for my sister and me. Unlike my father's younger siblings, Manfred and Ulrike had not become part of the 1968 student movement who took the cause of remembrance to the streets. But to my surprise, my father reports that he had learned about the Holocaust in his Lutheran church youth group in the 1950s, led by a perhaps exceptional pedagogue with a part-Jewish background, of which he only became aware after her death.[28] My mother did learn some things about the Holocaust in her vocational school in the early 1960s in the regional city, Darmstadt, before moving to Berlin; and she remembers reading Anne Frank's diary as well, although she does not when. As indicated above, my parents' urge to co-witness, and align in anti-racist solidarity was activated in part by the experience of racism and the Civil Rights Movement during their year in New York. Time and again, my father would relate his shame and upset, with unfiltered emotion in his voice even after many years, about an encounter with an elderly African American woman, who had stepped down from the sidewalk to make room for the white stranger. My parents did later begin to translate the solidarity mandate into the local context as well. In 2015, both of them, by now retired to a small town in the Harz mountains, became personally involved in supporting locally placed families from Syria.

Needless to say, none of this prevented the ongoing resurgence of various everyday racisms over the years, racisms attesting to a lack of more sustained processes of working through the complex intersections of affect, memory, and discourse in social institutions as well as personal relations. Over the years, I would re-encounter the legacy of these racisms in myself, in the midst of my own activism and memory and reconciliation work. After one of the trips to Poland in the 1980s, organized in association with *Aktion Sühnezeichen* by the

---

28 He pointed me to a brief biography: https://www.selk.de/download/Kirche-und-Judentum _4c.pdf. An echo of these silences in the next generation: I was stunningly unaware of the background of one of my classmates almost all the way through high school, where the Holocaust was a central part of the curriculum. But the default assumption of Christian homogeneity, and inability to address treacherous differences, was so dominant still that I didn't learn about Elina's immigration history and faith until I encountered her as part of her synagogue delegation meeting up with my church group at some interreligious dialogue event.

church I would leave in early adulthood only,[29] my group got in trouble for insensitive language and insufficient reflexivity about late night bar encounters in our report: a sobering reminder that the serious attempt to work through the past did not yet equip us for doing justice also to present intersections of ethnicity, gender, class, and economic inequality between East and West. Later, there were the painful conflicts around racism in Berlin's feminist-queer circles and intimate relationships, in my own experience prominently including the challenge to account for the ways in which my affective experience of someone else's anger unfolds in a racialized context.[30] Not to mention account for my own anger.

Clearly, it was easier to go the discourse-analytical route on the academic terrain. I do not mean to retrospectively discard the importance of my own or anybody else's scholarship in this vein: especially in the 2020 world of new identity-based fascisms, I fervently want to hold on to those 1990s' insights into the historicity of ethnic and racialized identities. At the same time, I also want to acknowledge that the approach facilitated remaining at a safe distance. Here is one more haunting memory snippet, unrelated on the surface, but intimately embedded in local histories of racism as complexly charged with tropes of hospitality: me, probably ten or eleven years old, near-panicking in the entrance of a Turkish family's apartment in Berlin Neukölln.[31] My sister had been invited to her classmate's birthday party, and my mother, who had been helping this classmate with her homework in a non-native language at our house, had sent me up to fetch my sister. She was parked in the second row, as I eagerly emphasized, making fervent excuses for not being able to stay myself and share some food and drink. I believe parking was in fact an issue in the neighborhood, but I

---

**29** To my parents' credit, this was no longer the Lutheran splinter church they and their parents had met in, but the mainstream German protestant church to which they had switched, a comparatively liberal and in parts of Berlin downright radical leftist church.

**30** See, again, Lorde's generative piece on "The Uses of Anger," which responds to racialized perceptions of Black women's anger by situating anger as a response to racism and emphasizing its productivity, in contrast to hate: "Anger is the grief of distortions between peers, and its object is change" (8).

**31** The hospitality tropes in play here include stereotypes of ethnicized (rarely German) hospitality, on the one hand; and on the other hand, the asymmetrical "logic of hospitality" that framed official discourses of labor immigration as "guest work" in postwar West Germany – a logic that, Damani Partridge reminds us, is not easily transformed into mutual relations of responsibility and solidarity. "Articulating a Noncitizen Politics: Nation-State Pity vs. Democratic Inclusion," in *Refugees Welcome? Difference and Diversity in a Changing Germany*, eds. Jan-Jonathan Bock and Sharon Macdonald (New York: Berghahn, 2019), here 266.

was also quite terrified by the repeated invitations, the bunch of adults talking on top of each other in a language I did not speak, and the table overloaded, it seemed, with unfamiliar foods I would come to appreciate only later in life. As soon as I had escaped, I felt shame for failing to connect, aware of the hospitality and openness shown to me while utterly overwhelmed by it. (Except for those obligatory family gatherings, there was not much adult socializing in my parents' house: the "We" pronoun on that record from New York in some respects indicated a gaping absence in our daily lives, along with their memory of a moment of connection.)

That was a long time ago. But of course, the "history of ghostly hauntings by the phantoms of a silenced past"[32] unfolds in the longue durée, in layers, twists and turns. As I came of age, the impulse to flee took turns in different directions: a major ocean provided just about enough distance from the oppressive atmosphere of the German provinces, and even in comparatively cosmopolitan – or ever more touristy – Berlin, I increasingly struggled to feel connected. Affective responses, clearly with class layers, but coded and experienced in national terms: gut reactions of intense negativity in response to the tone of certain interactions, outfits, haircuts, and gestures, for example on the train to visit my parents. Even after four years of Trump, I (thankfully) don't quite have those kinds of responses at a convenience store in the rural Midwest or the economically disadvantaged, Republican-leaning parts of upstate New York not yet flooded with Brooklyn hipsters and fellow Upper Westside weekenders-turned-into-pandemic-residents.

Academically, the flight impulse aligned with these personal feelings more recently, in a perhaps just logical twist, but creating a new conundrum for the German Studies scholar. As my reading methods have changed, I have found myself virtually unable to stand many of the literary texts that I loved to take apart critically earlier. In the introduction to *Making Worlds*, my wise internal theorist responds to the "heated debates about whether, or how, to seek dialogue with right-wing perspectives" with a call for finding "a path" in-between "naïve faith in the power of love or rational deliberation on the one hand, and the refusal of all engagement on the other."[33] As a literary reader, I would much rather just not engage with Peter Handke, Botho or his son Simon Strauß any longer. And it's not just the notorious right-wingers, either, or the straight white (West) German and Austrian males alone. I have been struggling with the

---

32 Schwab, *Haunting Legacies*, 72, with reference to Abraham and Torok.
33 Breger, *Making Worlds*, 17.

affective negativity I feel confronted with in Thomas Bernhard, Elfriede Jelinek and many of Feridun Zaimoğlu's texts alike; and a graduate student pointedly commented on my apparent dislike for Heiner Müller in a recent seminar. Can I (should I, do I want to) find new approaches to these texts, relearn to appreciate their wit, complexity, or provocative force? Or is there something to be said for listening to my affective response to this historically amassed affective negativity?[34] Towards the imagination of different futures, is it time to re-center the German(-language) literary canon around the warmth and part joyful, part mournful imagination of Else Lasker-Schüler, Irmgard Keun and Emine Sevgi Özdamar anyway?

Of course, I shouldn't simplify at my own expense. In the book and adjacent recent work on cinema, I do chart paths of engagement with challenging perspectives and affectscapes. My readings trace experiences of affective reshuffling, including the imaginative productivity of partial alignments with authoritarian perspectives or of moments of unexpected tenderness in the midst of violent worlds, for example in *The White Ribbon* by Michael Haneke, whom I thus halfway defend from his notoriety for spectatorial cruelty, or in *Western* by Valeska Grisebach.[35] I do believe that the affective labor invited by these films can make a political difference: *The White Ribbon* and *Western* invite us to open up to their historical protagonists respectively East German men with flags and stereotypes to a degree that attenuates (short of canceling) critical judgment with a probing of tangential connection to their pain and hope. However, I also reflect on the limits of such careful undertakings: their failure to articulate forceful resistance to hegemonic conditions of violence and the importance of alternative aesthetic strategies that imagine different futures more radically, strategies I trace, for example, in the defiant cinematic imagination of Aki Kaurismäki.

As I write in 2020, my attempts to evaluate, balance, and entangle divergent vectors of reading are very much ongoing. Over these past few months, there has been way too much anger, anxiety, and despair around and within me to place much trust in the power of more complex, compassionate, co-witnessing reading strategies alone – if also ever more urgent need for them? Asymmetrical

---

**34** As indicated above, I do not mean to suggest a coherent national habitus. While one can likely immigrate into collective affectscapes, too, the poetics of negativity flagged here has many sources – including aesthetic avant-garde commitments to shock and rupture.

**35** Breger, *Making Worlds*; Clauda Breger, "Cinematic Assemblies: Latour and Film Studies," in *Latour and the Humanities*, eds. Rita Felski and Stephen Muecke (Baltimore: Johns Hopkins University Press, 2020), 300–327.

life chances created by socio-economic inequality are on ever clearer display, White House ramblings at their most unabashed yet, and the police keep taking Black Lives. Sometimes, such as in the case of George Floyd, the *New York Times* provides backstories for making sense of the aggression, the failure, or the conflicts of individual officers. Back in Germany, the degree to which police and military overlap with neofascist networks is starting to receive overdue attention: the postwar "climate of a relentless authoritarianism" and its repressions keep making their cross-generational reappearances.[36] My internalized resistance to these authoritarian legacies makes me shake my head in disbelief (and perhaps feel a bit too superior) when I read in May 2020 that Amy Cooper called the police on a birdwatcher in Central Park who had asked her to leash her dog. Less spontaneously in affective tune with my leftist political collective, however, I feel some of the same disbelief when my Facebook friends want to call the police on the kids on their block (likely white, based on zip code) who were violating social distancing protocols with a ball game. Can we be angry and still demilitarize the internal SWAT teams in reading – perhaps not organized political crime but complicated fellow humans?

Or in reading textual configurations. In the German media, the debate on the antisemitism charges against Achille Mbembe, occasioned by the link he made between Israeli histories of institutionalized anti-Palestinian violence and South African Apartheid, has been spinning ever more out of control.[37] Here, the question as to how hard it can be to read a little more carefully may sound willfully naïve, if not plain sarcastic, in the face of organized political attempts at delegitimizing all criticism of Israeli policy as antisemitic – as well as the legacies of actual antisemitism across Germany's political spectrum, including in leftwing, anti-Zionist circles. The German Mbembe "affair" has been particularly haunting to me for how it is embedded in increasingly hegemonic pushbacks in German-language media against the critical theory legacies of the 1990s, especially postcolonial and gender studies.[38] Per this backlash, postcolonial and queer critiques were plain "identity politics," leading

---

**36** Schwab, *Haunting Legacies,* 75; see Katrin Bennhold, "Body Bags and Enemy Lists: How Far-Right Police Officers and Ex-Soldiers Planned for 'Day X.'" *New York Times,* 1 August 2020 (https://www.nytimes.com/2020/08/01/world/europe/germany-nazi-infiltration.html?action=click&module=Top%20Stories&pgtype=Homepage).

**37** E.g., Michael Rothberg, "The Specters of comparison. On the Mbembe Affair." *Latitude,* May 2020 (https://www.goethe.de/prj/lat/en/dis/21864662.html); Aleida Assmann, "Antisemitismus: Ein neuer deutscher Sonderweg?" *Berliner Zeitung,* 18 July 2020.

**38** See, e.g., Jens Jessen, "In der Sackgasse zwischen Weiß und Schwarz," *Die Zeit,* 27 May 2020 (with a partial defense of Mbembe, but in very questionable terms).

in a more or less direct line to Trump on the one hand, and today's "cancel culture," on the other hand.[39]

Against that backdrop, I will conclude by asking whether part of the solution to the challenge of balancing and entangling diverging reading strategies might be in fortifying the postcritical ethos with some of critical theory's analytical insights (and vice versa). Thus, we can re-read the texts from the 1990s more carefully and respectfully to reconstruct the intricate analyses of identity that underlie their (sometimes overly deconstructive) emphases on hybridity and performativity. For example, Homi Bhabha and Stuart Hall reflect on "the construction of cultural authority within conditions of political antagonism or inequality," in a field in which "race" (and religion) are "constantly crossed and recrossed by the categories of class, of gender, and ethnicity;" Eve Sedgwick attends to the ways in which "each act of identifying as involves multiple processes of identification *with*" and "*as against,*" all "fraught with intensities of incorporation, diminishment, threat, loss, reparation, and disavowal;" and Paul Gilroy encourages us to move "away from 'race' . . . toward a confrontation with the enduring power of racisms."[40]

In 2020, I aim to bring these insistences on reading historically, intersectionally, and in a process-oriented way, with an eye for tensions, layers, and divergences, to my classrooms, where students with radical social movement affiliations clash with others worried about identity politics and cancel culture. At the end of my recent graduate seminar on "Aesthetics and Politics Today," the students and I compiled an open-ended list of adverbs ("reading how"). In the absence of closure, perhaps this list can stand in for my ongoing task of exploring affectively charged modes of reading in the spirit not of simply overcoming, but of working with, through, and against the affective negativity and the realities of disconnection surrounding us in each of our (variously distant and adjacent) positions. Thus, I hope to keep reading suspiciously and generously, coolly and

---

**39** E.g., Albrecht Koschorke, "Die akademische Linke hat sich selbst dekonstruiert. Es ist Zeit, die Begriffe neu zu justieren," *NZZ*, 18 April 2018; for a measured perspective on the "cancel culture" debate, see, e.g., Franziska Schutzbach, "Wer oder was wird ‚gecancelt'?" *Republik*, 14 August 2020.
**40** Homi Bhabha, "Culture's In-Between," in *Questions of Cultural Identity*, eds. S. Hall & P. Du Gay (London: Sage Publications, 1996), 58; Stuart Hall, "New Ethnicities," in *Critical Dialogues in Cultural Studies*, eds. David Morley and Kuang-Hsing Chen (New York: Routledge, 1996), 444; Eve Kosofsky Sedgwick, *Epistemology of the Closet* (Berkeley: University of California Press, 1990), 60; Paul Gilroy, *Postcolonial Melancholia* (New York: Columbia University Press, 2005), 9.

warmly, angrily and hopefully, vigilantly and lovingly, reflectively and boldly, closely and literally, imaginatively and precisely, multivectorally and radically, carefully and defiantly, sensitively and rigorously, curiously and in solidarity.

# Works Cited

Assmann, Aleida. "Antisemitismus: Ein neuer deutscher Sonderweg?" *Berliner Zeitung*, 18 July 2020.

Bammer, Angelica. *Born After: Reckoning with the German Past*. New York: Bloomsbury, 2019.

Bennhold, Katrin. "Body Bags and Enemy Lists: How Far-Right Police Officers and Ex-Soldiers Planned for 'Day X.'" *New York Times*, 1 August 2020 https://www.nytimes.com/2020/08/01/world/europe/germany-nazi-infiltration.html?action=click&module=Top%20Stories&pgtype=Homepage, 25 September 2020.

Bhabha, Homi. "Culture's In-Between." In *Questions of Cultural Identity*. Edited by S. Hall & P. Du Gay. London: Sage, 1996, 53–60.

Breger, Claudia. "Cinematic Assemblies: Latour and Film Studies." In *Latour and the Humanities*. Edited by Rita Felski and Stephen Muecke. Baltimore: Johns Hopkins University Press, 2020, 300–327.

Breger, Claudia. *Making Worlds: Affect and Collectivity in Contemporary European Cinena*. New York: Columbia University Press, 2020.

Castiglia, Christopher. *The Practices of Hope: Literary Criticism in Disenchanted Times*. New York: New York University Press, 2017.

Chin, Rita, and Heide Fehrenbach. "Introduction." Rita Chin et.al., *After the Nazi Racial State: Difference and Democracy in Germany and Europe*. New York: Columbia University Press, 2009, 1–29.

Fehrenbach, Heide. *Race After Hitler: Black Occupation Children in Postwar Germany and America*. Princeton: Princeton University Press, 2005.

Felski, Rita. *The Limits of Critique*. Chicago: Chicago University Press, 2015.

Felski, Rita. *Uses of Literature*. Hoboken, NJ: Wiley Blackwell, 2008.

Gilroy, Paul. *Postcolonial Melancholia*. New York: Columbia University Press, 2005.

Hall, Stuart. "New Ethnicities." In *Critical Dialogues in Cultural Studies*. Edited by David Morley and Kuang-Hsing Chen. New York: Routledge, 1996, 441–449.

Harcourt, Bernard E., and Didier Fassin, eds. *A Time for Critique*. New York: Columbia University Press, 2019.

Jessen, Jens. "In der Sackgasse zwischen Weiß und Schwarz." *Die Zeit*, 27 May 2020.

Klüger, Ruth. *weiter leben: Eine Jugend*. Göttingen: Wallstein, 1992.

Komska, Yuliya. *The Icon Curtain: The Cold War's Quiet Border*. Chicago, IL: The University of Chicago Press, 2015.

Koschorke, Albrecht. "Die akademische Linke hat sich selbst dekonstruiert. Es ist Zeit, die Begriffe neu zu justieren." *NZZ*, 18 April 2018.

Lorde, Audre. "Uses of Anger." *Women's Studies Quarterly* 9, no. 3 (Fall 1981): 7–10.

Lorde, Audre. *Zami: A New Spelling of My Name*. Trumansburg, NY: The Crossing Press, 1983.

Massumi, Brian. *Parables for the Virtual: Movement, Affect, Sensation*. Durham, NC: Duke University Press, 2002).

Parkinson, Anna. *An Emotional State: The Politics of Emotion in Postwar West German Culture.* Ann Arbor: University of Michigan Press, 2015.

Partridge, Damani. "Articulating a Noncitizen Politics: Nation-State Pity vs. Democratic Inclusion." In *Refugees Welcome? Difference and Diversity in a Changing Germany.* Edited by Jan-Jonathan Bock and Sharon Macdonald. New York: Berghahn, 2019, 265–287.

Pound, Marcus. *Žižek: A (Very) Critical Introduction.* Grand Rapids, MI: Eerdmans, 2008.

Ricciardi, Alessia. "Žižek: A (Very) Critical Introduction." *Notre Dame Philosophical Review* 7 October 2009 (https://ndpr.nd.edu/news/iek-a-very-critical-introduction/).

Rothberg, Michael. *The Implicated Subject: Beyond Victims and Perpetrators.* Stanford, CA: Stanford University Press, 2019.

Rothberg, Michael. "The Specters of Comparison: On the Mbembe Affair." *Latitude*, May 2020 (https://www.goethe.de/prj/lat/en/dis/21864662.html).

Schwab, Gabriele. *Haunting Legacies: Violent Histories and Transgenerational Trauma.* New York: Columbia University Press, 2010.

Sedgwick, Eve Kosofsky. *Epistemology of the Closet.* Berkeley: University of California Press, 1990.

Sedgwick, Eve Kosofsky. *Touching Feeling: Affect, Pedagogy, Performativity.* Durham, NC: Duke University Press, 2003.

Sharpe, Christina. *In the Wake: On Blackness and Being.* Durham, NC: Duke University Press, 2016.

Schutzbach, Franziska. "Wer oder was wird ‚gecancelt'?" *Republik*, 14 August 2020.

Warhol, Robyn. *Having a Good Cry: Effeminate Feelings and Pop Culture Forms.* Columbus: Ohio State University Press, 2003.

Žižek, Slavoj. *Against the Double Blackmail. Refugees, Terror and Other Troubles with the Neighbours.* London: Allen Lane, 2016.

Darcy Buerkle
# Falling Down on the Job/On Revulsion (November 2016–September 2020)

> It almost looks, as if analysis were the third of those impossible professions in which one can be sure beforehand of achieving unsatisfying results. The other two which have been known much longer, are education and government.　　　　　Sigmund Freud[1]

> As progress tales lose traction, however, it becomes possible to look differently.
> 　　　　　Anne Loewenhaupt Tsing[2]

A fractured text follows, citing its adjacency to violence in the particular and perpetual state of moral injury perpetuated in the United States beginning in 2016. More: this text was composed in Summer 2020 during a pandemic, an eventual mass casualty event inseparable in the United States from authoritarian neglect. Form and voice take refuge here in the fragment as a sign of the times, grasping at unspokens, leaning into previous refusal and interweaving an early history of Freud and grief that has long fueled my thinking.

The sites of my witnessing have been various during this time. But the classroom has been a consistent place of encounter, and among the most shattering. The person I had been as a teacher could not hold. In this state, I write here about my own life and its adjacencies in a stutter, still mining my hitherto coded scholarship for signs of slippage – places where the illegibility of the code into which I wrote can be cracked.[3] As much as I may think that I write in the After, I write in the/a Before times; in September 2020, unfolding as it does as an after-time to November 2016 and with a daunting horizon of November 2020 in sight, at last: "Forever," Emily Dickinson wrote from her bedroom, "is composed of Nows."[4]

---

1 Sigmund Freud (1937), "Analysis Terminable and Interminable," *The Standard Edition of the Complete Psychological Works of Sigmund Freud*, Volume XXIII (1937–1939): Moses and Monotheism, An Outline of Psycho-Analysis and Other Works, 209–254, here 247.
2 Anna Lowenhaupt Tsing, *The Mushroom at the End of the World: On the Possibility of Life in Capitalist Ruins* (Princeton, NJ: Princeton University Press, 2015), 22.
3 I lay out some aspects of the psychoanalytic contributions regarding suicide that I refer to here in *Nothing Happened: Charlotte Salomon and an Archive of Suicide* (Ann Arbor: University of Michigan Press, 2013).
4 Emily Dickinson, "Forever – is composed of Nows (690)," in *The Poems of Emily Dickinson* (Cambridge, MA: Harvard University Press/Belknap Press, 1998).

https://doi.org/10.1515/9783110753295-030

The forever now will inevitably sound like a more innocent Before by the time this is read: before the 2020 election, and before more than 205,000 dead by COVID-19 and before the next televised Black death in the streets of America, but also after the 2016 election, and after 204000 COVID-19 deaths, and after Eric Garner said he could "not breathe" and then George Floyd said he could "not breathe" and we do not know what Breonna said at all. Now: after the dog-whistle at the first presidential debate to "stand by" and after dozens of White House staff fall ill. Now, again: before 209000 dead by COVID-19, and multiple members of the administration diagnosed with COVID-19.

Group gatherings hosted by the editor of this volume in preparation for this writing revealed, significantly, the limits to which individual scholars were willing to go. There are things, we all decided, we would not discuss, we would not reveal. Not beyond a footnote. The same could be said of my pedagogy, which has long relied on the conviction that my teaching should not refer to or rely on autobiography, however much that autobiography has fueled my thinking. This stance has become untenable.

You could say, my privacy has been taken from me. Or you could say, the fear that underlay that resolute devotion to the privilege of privacy has splintered out of conviction, even as it is an effect of utter revulsion. Either way, a resolute determination that carried me is broken now.

Especially in this moment of persistent provocation and violence in the United States, in this long moment in which each day arrives with the promise of reminders of how very negligible violence remains for millions and violence against women and people of color is sport – negligible enough for ample votes in the electoral college to place a perpetrator in the White House, for enough votes to confirm a Supreme Court Justice. It is laughable. By which I mean: the men involved and their vociferous accomplices cannot seem to stop laughing. "Oppressive regimes, microaggressions, and relational traumas that occur daily in Trump's America," psychologists have recently argued, "leave deposits akin to pathological parental projective identifications."[5]

"The hallmark of a traumatized patient is her inability to tell the story or her refusal to transmit the knowledge of death and loss that has lodged itself inside her."[6]

---

**5** Tracy A. Prout, Leore J. Faber, Emma Racine, Rebecca Sperling and Rebecca F. Hillman, "Clinical Encounters with Children in the Trump Era," *Journal of Child Psychotherapy*, 45, no. 2 (2019): 193.
**6** Leanh Nguyen, "Psychoanalytic Activism: Finding the Human, Staying Human," *Psychoanalytic Psychology* 29, no. 3 (2012): 314.

This text lays out my immediate and teacherly past by following an autobiographical trace in the form of a gesture of evasion in some of my earlier writing, and links it to the acuity of the present political situation as I understand it and as it manifests itself in my classroom.

So, this is not going to be a story about me, exactly. It is a story about how *a position of the refusal to tell*, to draw explicitly on my own life has failed me – and it is a story I will tell by, as one does, relaying another story about the inscription of autobiography at the heart of another and well-known and symptomatic text, Freud's "Mourning and Melancholia." It is a story about how that critical text emerged from a site that went completely missing in its final version; emerged, that is, from the scene of the classroom, and from the urgency of teaching as vocation.

•

Composed in 1915, and usually read as a text in conversation with the mass death in World War I, Freud's well-known essay "Mourning and Melancholia," appeared in 1917.[7]

But that story of origin is not quite right.

Instead, Freud meaningfully forecast his thinking about melancholia's mechanics years earlier, in a series of conversations at the Wednesday Society in April 1910 – when the group was still all men, when they still met in the Berggasse 19. These conversations were recorded in two places: first, in the Minutes of the Vienna Psychoanalytic, and second, in a book entitled "On Suicide," which was based on those conversations.[8] The vestiges of those gatherings reappear, barely marked, in the eventual and lauded essay, "Mourning and Melancholia." Based on this essay, the difference between mourning and melancholia is often cast colloquially as a matter of healthy versus unhealthy grief, sometimes in order to make an emancipatory claim on the latter, or to "refuse diagnostic categories of normative grief altogether."[9]

The men in attendance at the Wednesday Society in April 1910 had just come from the Second Psychoanalytic Congress, which had been framed by the significance of antisemitism insofar as it was actively blocking the advance of

---

**7** Sigmund Freud, "Mourning and Melancholia," *The Standard Edition of the Complete Psychological Works of Sigmund Freud*, Volume XIV (1914–1916), 237–258.

**8** Paul Friedman ed. and trans., *On Suicide: With Particular Reference to Suicide among Young Students* (New York: International Universities Press 1967).

**9** Zoë Wool, "Mourning, Affect, Sociality: On the Possibilities of Open Grief," *Cultural Anthropology* 35, no. 1 (2020): 40–47, here 44.

psychoanalysis. At this meeting, Freud had handed over the international leadership to Carl Jung, and in his appeal to the Viennese remarked that: "Most of you are Jews, and therefore you are incompetent to win friends for the new teaching. Jews must be content with the modest role of preparing the ground . . . The Swiss will save us, will save me, and all of you as well."[10] The material that underlay the psychoanalytic exploration of melancholia in 1910 was the same material that led Freud to name Jung president of the International Psychoanalytic: antisemitism. But at the Wednesday Society's meetings, the much less theoretical conversation about student suicide pivoted on the antisemitic humiliations and suffering of Jewish schoolboys.

David Ernst Oppenheim led the 1910 student suicide discussion. It had been his idea to examine the subject. As a teacher himself, he explained, the "subject concerned" him.

And as a teacher, he had come to the topic of student suicide as addressee in the proliferation of journalistic writing about student suicide. The newspapers, to his chagrin, blamed the overzealous rigors in school for student despair. As the notes from an earlier meeting a few weeks before reflect, Oppenheim's related reading of a recent article by Hermann Swoboda in the *Österreichische Rundschau* that took student suicide as its topic "seemed promising."[11] Swoboda's rendering of student suicides allowed for both "conscious and unconscious" motives – but either way, for him, those motives were always a matter of sexuality. Just a year later, in 1911, Swoboda would ruminate on his friend Weininger's 1903 death by suicide and publish a small and penetrant volume called *Otto Weiningers Tod*.[12]

Swoboda was not a benign invocation among the psychoanalysts. His exchange with Weininger about bisexuality was inscribed in *Sex and Character*, but the origins of these insights about bisexuality caused significant repercussions on Freud's relationships with Swoboda. More importantly, echoes of those tensions ended Freud's close friendship and exchanges with Wilhelm Fliess.[13] What Oppenheim may have known about any of this rancor is unclear. But for his presentation on student suicide, he would nonetheless choose to (try to) sidestep queer sexuality and share in detail another text at the next Wednesday gathering rather than Swoboda's.

---

**10** Fritz Wittels, *Sigmund Freud: His Personality, His Teaching & His School*, trans. Cedar and Eden Paul (New York: Dodd, Mead & Company, 1924), 140.

**11** Hermann Swoboda, "Schülerselbstmord," *Österreichische Rundschau* 22 (1910): 375–377.

**12** Hermann Swoboda, *Otto Weiningers Tod* (Vienna: Hugo Heller, 1911).

**13** *The Complete Letters of Sigmund Freud to Wilhelm Fliess, 1887–1904*, trans. and ed. Jeffrey Moussaieff Masson (Cambridge, MA: Belknap Press, 1985).

"Oppenheim himself," the *Minutes* read, "can suspect the existence of a connection with sexual life [in student suicide]."[14] Only much later would the publication of David Ernst Oppenheim's letters to his fiancée, Amalie Pollak, fully clarify what "Oppenheim himself" might mean in this instance, beyond pedagogy.

David Ernst Oppenheim did not survive Terezin, but his widow did. She brought his papers and letters with her when she was able at last to emigrate to Australia; they tell the story of their early life together. In 1904 he had written to her with a request that she might help him "see clearly into this dark matter." The "dark matter" was his own homosexuality. He shared details of his relationships with men before meeting her, sending her years of his love letters to a man and later detailing the unfolding of other relationships with men, too. The question of homosexuality was at the center of their bond since Amalie, too, was well familiar with "the dark matter" in her own life.[15]

To be clear, then: Oppenheim, a young, queer, Jewish teacher in the well-known *Akademisches Gymnasium* in Vienna *in which over forty percent of the students were Jewish boys and only three of the teachers were Jewish men*, was moved by the attention afforded student suicide in the press through a close friend of the well-known, Jewish, implacable and misogynist queer Otto Weininger to make a presentation to the Wednesday Society. Oppenheim reported that he felt "deep rancor" about the "entire question of suicide because it concerns my own life so painfully closely." His presentation about student suicide would provide a substantive genesis for lasting psychoanalytic ideas about melancholia.

On 20 April 1910, Oppenheim presented for discussion a slim 1901 volume by Adolf Baer, *Schülerselbstmord*, in which Baer argued, among other things, against apathy: "[F]ew experience the sadness and strangeness in these events as serious and painful; only the very few are moved to think about the disgusting tragedy involved in this fact and the difficult accusation against society that it represents."[16]

As Oppenheim suggests: "the point at which Dr Baer stops is indeed the point at which the problem of the psychologist only begins. Why was the child so excessively afraid of punishment . . . that she fled from it into death?"[17]

Deflecting and actively disguising Adolf Baer's Jewishness, and his own, Oppenheim called up examples from Catholicism, exclusively. He even suggests

---

**14** Herman Nunberg and Ernst Federn, eds., *Minutes of the Vienna Psychoanalytic Society*, vol. 2: 1908–1910 [hereafter: *Minutes*] (New York: International Universities Press, 1967), 456.
**15** Peter Singer, *Pushing Time Away* (New York: Harper Collins, 2004), 19.
**16** Adolf Baer, *Schülerselbstmord* (Leipzig, 1901).
**17** Nunberg and Federn, *Minutes*, 486.

that Baer's religiosity – by implication Christianity since he refers to Baer's "strongly . . . religious principles . . . of a blissful life after death"– has something to do with encouraging suicide. Reading these assertions about religion as statements about Jewish life changes their impact significantly; one wonders what he replaced with Christian theodicy: "[I]t is precisely religious promises of a blissful life after death, as urgently preached by sentimental mothers and nurses, that are quite capable, under certain circumstances, of furthering the flight from life on earth."[18] Also missing and displaced: Oppenheim's first reason for raising the issue, namely Swoboda's insight about sexuality, rendering these critical elements that underlay his pedagogical commitments and the centrality of these questions in his own life, adjacent. Yet, antisemitism and homophobia were the central matters that he emphasized without naming them as such.

With varying sites of emphasis, the men in 1910 agreed: heredity and love are the reasons for suicide.

Alfred Adler was the first to speak after Oppenheim finished his presentation, allying himself with the experiences of schoolboys as Oppenheim hypothesized them, noting that Oppenheim had "opened up the path to a discussion in which the personal experiences of each man with regard to this interesting topic will surely be dealt with."[19] "The issues" were the experience of being Jewish in school and a lack of recognition by the beloved teacher.

Isidor Sadger commented that when the child seeks out the teacher and that teacher turns away, they have "often provided [thereby] the final impetus to a complete despair of any prospects of love . . . and therewith to suicide."[20]

Buttressed by conversation among the attendees about their experiences of humiliation in school, student suicide, Freud noted, points to "despair about the prospects of being loved . . . In suppressing the practice of homosexuality, one has simply also suppressed the homosexual direction of human feelings that is so necessary for our society. The best teachers," Freud continues, "are the real homosexuals who actually have that attitude of benevolent superiority toward their pupils . . . Just as the homosexuals are the best teachers, so the repressed homosexuals are the worst, and the strictest."[21]

The eventual published booklet on student suicide based on the record of the 1910 gathering removed key components of the meetings described here: gone was any reference to Jewishness, gone any mention of homosexuality. Oppenheimer's name, a pseudonym.

---

**18** Nunberg and Federn, *Minutes*, 487.
**19** Nunberg and Federn, *Minutes*, 526.
**20** Nunberg and Federn, *Minutes*, 514.
**21** Nunberg and Federn, *Minutes*, 495.

Paul Federn returned to the group's 1910 thinking about student suicide almost two decades later as the editor of a special double-issue of the *Zeitschrift für Psychoanalytische Pädagogik* in 1929. There, he ceases silence: antisemitism appears as the persistent weather, so, too, the urgency of desire that was missing from the earlier published remarks.[22]

1910 casts its shadow in the 1917 "Mourning and Melancholia" as ". . . the shadow of the object [that] fell upon the ego, and the latter could henceforth be judged by a special agency, as though it were an object, the forsaken object." This formulation mirrors the much more explicit rendering Freud offered in his 1910 summation about the subjectivity of the student: "The melancholic displays something else besides which is lacking in mourning – an extraordinary diminution in his self-regard, an impoverishment of his ego on a grand scale. The patient represents his ego to us as worthless, incapable of any achievement and morally despicable; he reproaches himself, vilifies himself and expects to be cast out and punished. He abases himself before everyone and *commiserates with his own relatives for being connected with anyone so unworthy.*"[23]

What was "the object" internalized, the object that could destroy? In 1910, it was the antisemitic, thoughtless teacher. Full-stop.

"I should gladly tell you more about melancholia," Freud wrote to his confidant Karl Abraham in Berlin on 3 July 1915, "but could do it *properly* [italics in original] only if we met and talked. Have I not sent you the typed [in English in original] manuscript of 'Mourning and Melancholia'"?[24]

Later, in "The Ego and the Id," Freud returns to these matters and is as explicit as he ever would be about the beginnings of the conversation about melancholia: "The ego is first and foremost a bodily ego . . . the projection of the surface . . ."[25] Reflecting on Freud's text, Judith Butler has observed that here, Freud reverses what he means by resolution such that "there is no final breaking of the attachment." The only way to "let go," is to incorporate.[26]

In describing mourning, Freud describes the conditions of successful assimilation. In describing melancholia, he gives us the failure of assimilation which, for him, could only mean diminution.

---

**22** *Zeitschrift für Psychoanalytische Pädagogik*, 3.11–13 (1929).
**23** Sigmund Freud, "Mourning and Melancholia," *The Standard Edition of the Complete Psychological Works of Sigmund Freud*, Volume XIV (1914–1916), 248.
**24** *The Complete Correspondence of Sigmund Freud and Karl Abraham, 1907–1925*, trans. and ed. Ernst Falzeder (London: Karnac, 2002), 313.
**25** Sigmund Freud (1923), The Ego and the Id. *The Standard Edition of the Complete Psychological Works of Sigmund Freud*, Volume XIX (1923–1925): The Ego and the Id and Other Works, 25.
**26** Judith Butler, *The Psychic Life of Power* (Stanford, CA: Stanford University Press, 1997), 134.

*In other words: theories of racial melancholia in critical race theory are a return to the racialized subjugations that were at the heart of first formulated thoughts about melancholia.* The experiences on which the men in the Wednesday Society were encouraged to draw were their own – the experiences of Jewish boys in secular education. "*This* is racial melancholia for the raced subject: the internalization of discipline and rejection – and the installation of a scripted context of perception: the one lost and the one losing," Anne Cheng writes.[27]

"The child," Freud wrote, "must learn to control his instincts . . . Education must inhibit, forbid, suppress and this is abundantly seen in all periods of history. A moment's reflection tells us . . . education has fulfilled its task very badly and has done children great damage."[28]

As we discussed: We all have our secrets and *we intend to keep them.*

But such a decision suggests a controlled relation to language and to voice that has itself become something of a nostalgic relic. That is: the shallowness of breath under eminent assault evidences itself involuntarily. What I cannot speak under those conditions is less a matter of control and more of an affective demand. Perhaps it is more accurate to say that I will try not to tell you about my own failings, indignities, humiliations, and subjugation in detail. But metabolizing moral injury has become a constant that literally begs the question. As I try to decide which of the many times this has been the case since 2016 that I want to include, I cannot settle on a list. There are still children in cages, the streets are still not safe for Black people and many more will die because of it, the echoes of Charlottesville killed eleven in Pittsburgh, and violence against women means nothing when it comes to the Presidency or, as I have already mentioned, but it bears repeating, the Supreme Court.

The laughter: narrated in the testimony in 2018 by Christine Blasey Ford is audible, almost always.

(There it is again: In the White House Rose Garden backslapping for the Supreme Court in Fall 2020.)

Janine di Giovanni writes that "moral injury rather than PTSD or depression [has] emerged as the biggest psychological challenge confronted by journalists covering the migration crisis." Di Giovanni asks in a recent article what "the long-term effects of revulsion" will be in what we hope will be an after, a new inescapable but nonetheless still distant adjacency.[29] Beginning with Jonathan

---

**27** Anne Anlin Cheng, *The Melancholy of Race* (Oxford: Oxford University Press, 2001), 17.

**28** Sigmund Freud, (1933). New Introductory Lectures on Psycho-Analysis. *The Standard Edition of the Complete Psychological Works of Sigmund Freud*, Volume XXII (1932–1936): New Introductory Lectures on Psycho-Analysis and Other Works, 148.

**29** Janine di Giovanni, "On Moral Injury," *Harper's Magazine* (August 2020): 69.

Shay's 1994 book, *Achilles in Vietnam*, the concept of moral injury as distinctly different from trauma – though not to its exclusion – has gained currency in psychiatric and philosophic literature on combat veterans.[30] Shay distinguishes his work from others in the definition of violator – for most writers, the violator is the self. For him, the *violator is the powerholder*: "From my observation," he writes, "where leadership malpractice inflicts moral injury, the body codes it as physical attack, mobilizes for danger and counterattack, and lastingly imprints the physiology every bit as much as if it had been a physical attack . . . Moral injury deteriorates . . . character, ideal, ambitions and attachments begin to change and shrink . . . When social trust is destroyed, it is replaced by the settled expectancy of harm, exploitation, and humiliation from others."[31] Moral injury is a condition of the witness who as a witness participates in the betrayal of their own moral universe. Moral injury, Jane Tillman writes, "implicates the social order."[32]

In these After/Before times, I think about who I am to my students as I stand before them. I think about my resistance to self-disclosure; the adamance with which I have insisted on this restraint and the illegibility that has produced. "You have to understand: My PTSD has everything to do with moral injury. It was not from killing or seeing bodies severed or blown up. It was from betrayal, from moral betrayal."[33]

I should also say: the people in my "historically women's college" classes are women or (gender) queer or both. To face them, I face myself now. And while I am thinking about that, they cry out. I mean this literally: in the last four years, the otherwise contained and studious crowd in my classroom – when I could still go to a classroom before COVID-19 – more and more, they would cry out, incredulously and seemingly involuntarily. Standing at the lectern, I have witnessed a kind of repetitive affective seizure, a take-over in a time that disallowed otherwise for some of us: I can play back all the instances in which the palpability of visible and audible and public gasping, incontrovertible and public grief unroll as the record of that class. No one is laughing.

**30** Jonathan Shay, *Achilles in Vietnam* (New York: Simon and Schuster, 1994).
**31** Jonathan Shay, "Moral Injury," *Psychoanalytic Psychology* 31, no. 2 (2014): 182–191, here 186.
**32** Jane Tillman, "The Intergenerational Transmission of Suicide: Moral Injury and the Mysterious Object in the Work of Walker Percy," *Journal of the American Psychoanalytic Association* 64, no. 3 (2016): 541–567, here 544.
**33** Nancy Sherman, "Recovering Lost Goodness: Shame, Guilt and Self-Empathy," *Psychoanalytic Psychology* 31, no. 2 (2014): 217–235, here 219.

No one here is laughing. But they hear the laughter still. It's the kind that Christine Blasey Ford described, the kind that rang in the head of the queer Jewish teacher Oppenheim in 1910. The kind that lacerates, the kind that won't stop. We all hear it: Me too.

Be serious, I say. Be serious now.

# Works Cited

## Archives

Notes from Wednesday Society (Otto Rank). Sigmund Freud Collection, OV 3–1. *Library of Congress Manuscript Division*. Washington, DC.

## Published Sources

Baer, Adolf. *Der Selbstmord im Kindlichen Lebensalter*. Leipzig: Thieme 1901.

Buerkle, Darcy. *Nothing Happened: Charlotte Salomon and an Archive of Suicide*. Ann Arbor: University of Michigan Press, 2013.

Butler, Judith. *The Psychic Life of Power*. Stanford, CA: Stanford University Press, 1997.

Cheng, Anne Anlin. *The Melancholy of Race*. Oxford: Oxford University Press, 2001.

Cvetkovich, Ann. *Depression: A Public Feeling*. Durham, NC: Duke University Press, 2012.

di Giovanni, Janine. "On Moral Injury," *Harper's Magazine* (August 2020).

Eng, David, and Shinhee Han. "A Dialogue on Racial Melancholia." In *Loss*. Edited by David Eng and David Kazanjian, 343–371. Berkeley: University of California Press, 2003.

Freud, Sigmund. "Mourning and Melancholia," *The Standard Edition of the Complete Psychological Works of Sigmund Freud*. Volume XIV (1914–1916) (London: Hogarth Press, 1953–1974), 237–258.

Freud, Sigmund. (1923). The Ego and the Id. *The Standard Edition of the Complete Psychological Works of Sigmund Freud*, Volume XIX (1923–1925): The Ego and the Id and Other Works, (London: Hogarth Press, 1953–1974) 1–66.

Freud, Sigmund. (1933). New Introductory Lectures on Psycho-Analysis. *The Standard Edition of the Complete Psychological Works of Sigmund Freud*. Volume XXII (1932–1936): New Introductory Lectures on Psycho-Analysis and Other Works, (London:Hogarth Press, 1953–1974) 1–182.

Freud, Sigmund. *The Complete Letters of Sigmund Freud to Wilhelm Fliess, 1887–1904*. Translated and edited by Jeffrey Moussaieff Masson. Cambridge: Belknap Press, 1985.

Freud, Sigmund. *The Complete Correspondence of Sigmund Freud and Karl Abraham, 1907–1925*. Translated and edited by Ernst Falzeder. London: Karnac, 2002.

Friedman, Paul, ed. and trans. *On Suicide: With Particular Reference to Suicide among Young Students*. New York: International Universities Press, 1967.

Nguyen, Leanh. "Psychoanalytic Activism: Finding the Human, Staying Human." *Psychoanalytic Psychology* 29, no. 3 (2012): 308–317.

Nunberg, Herman and Ernst Federn, eds. *Minutes of the Vienna Psychoanalytic Society*
    Volume 2: 1908–1910. New York: International Universities Press, 1967.

[Oppenheim, David Ernst]. *Über den Selbstmord insbesondere den Schüler-Selbstmord*.
    Wiesbaden: Verlag J. F. Bergmann, 1910.

Prout, Tracy A., Leore J. Faber, Emma Racine, Rebecca Sperling, and Rebecca F. Hillman.
    "Clinical Encounters with Children in the Trump Era," *Journal of Child Psychotherapy*, 45,
    no. 2 (2019): 191–208.

Shay, Jonathan. *Achilles in Vietnam*. New York: Simon and Schuster, 1994.

Shay, Jonathan. "Moral Injury." *Psychoanalytic Psychology* 31, no. 2 (2014): 182–191.

Singer, Peter. *Pushing Time Away*. New York: Harper Collins, 2004.

Sherman, Nancy. "Recovering Lost Goodness: Shame, Guilt and Self-Empathy."
    *Psychoanalytic Psychology* 31, no 2 (2014): 217–235.

Swoboda, Hermann. "Schülerselbstmord." *Österreichische Rundschau* 22 (1910): 375–737.

Swoboda, Hermann. *Otto Weiningers Tod*. Vienna: Hugo Heller, 1911.

Tillman, Jane G. "The Intergenerational Transmission of Suicide: Moral Injury and the
    Mysterious Object in the Work of Walker Percy." *Journal of the American Psychoanalytic
    Association* 64, no. 3 (2016): 541–567.

Tsing, Anna Lowenhaupt. *The Mushroom at the End of the World: On the Possibility of Life in
    Capitalist Ruins*. Princeton, NJ: Princeton University Press, 2015.

Wittels, Fritz. *Sigmund Freud, his personality, his teaching, & his school*. New York: Dodd,
    Mead & company, 1924.

Wool, Zoë. "Mourning, Affect, Sociality: On the Possibilities of Open Grief." *Cultural
    Anthropology* 35, no. 1 (2020): 40–47.

Mita Choudhury
# Tears and Empathy: Possible Methodologies for Studying Sexual Violence

How does an early modern European historian position themselves adjacent to violence that took place two hundred, three hundred, four hundred years ago? Is it possible or even relevant non-academics and even some academics might ask. Yet, in early modern Europe, violence "was part of the discourse of early modern interpersonal relations."[1] Early modern views of sexual violence – domestic violence, rape, abduction, and seduction – were shaped by ideals of masculinity, faith, family, and property. Accusations of sexual assault were further complicated by rank, status, and function. Thus, the sins of a peasant were not the sins of a nobleman. And what about a priest, the absolver of sins in the Catholic world? For a French historian like myself, ecclesiastical and police records in the Archives Nationales and the Bibliothèque de l'Arsenal in Paris suggest how institutions – legal procedure and men in positions of authority – shaped the narrative of sexual violence, a narrative that reflected a larger culture in which victims and their accusations were often dismissed or ignored. How then do we uncover a crime that in many ways seems undocumented? These are some of the questions I am tackling in a book-length project on sexual abuse in the early modern Church (1610–1774). One of my central goals is to move away from representations of rape and to uncover the voices of victims long erased, in other words to write a history of sexual abuse privileging victims, not perpetrators, individuals, not institutions. The questions for this essay are if and how can I write a victim-centered violence when the experiences of my subjects are so chronologically distant from me.

To my mind, the possibilities for bridging this chronological gap and archival silences about sexual violence originate not with the documents, the cornerstone of historical training, but with the world I inhabit in the early twenty-first

---

1 Julius R. Ruff, *Violence in Early Modern Europe* (Cambridge: Cambridge University Press, 2001), 2. See also Gerd Schwerhoff, "Early Modern Violence and the Honour Code: From Social Integration to Social Distinction?" *Crime, Histoire & Sociétés/Crimes, History and Societies* 17 (2013): 27–46; Graeme Murdock, Penny Roberts, and Andrew Spicer, eds., *Ritual and Violence; Natalie Zemon Davis and Early Modern France* (Oxford: Oxford University Press, 2012); and Joseph P. Ward, *Violence, Politics, and Gender in Early Modern England* (New York: Palgrave Macmillan, 2008).

https://doi.org/10.1515/9783110753295-031

century. In recent years, the question of ecclesiastics and sexual assault has gained urgency with revelations of sexual abuse and cover-ups within the Catholic Church in the United States, in France and other European countries, in Chile, Argentina, the Philippines, Kenya, South Africa, and so on. Allegations and trials have included high profile individuals like archbishop George Pell of Melbourne and cardinal Philippe Barbarin of Lyon, both of whom purportedly protected pedophile priests. While such corruption has been going on for decades, what seems new is the committed and coordinated efforts of victims to speak out. For example, in 2015 François Devaux and Bertrand Virieux founded the organization *La Parole Libérée* as a support group for victims abused by the priest Bernard Preynat.[2] Their cause was amplified in François Ozon's film, *Grâce à Dieu* (2019), a French version of the Academy-Award winning *Spotlight* (2015). But it was the power of one documentary, *The Keepers* (2017), detailing sexual abuse and murder in a Baltimore high school, that led me to this topic. Released by Netflix, this series was followed by events aired and discussed in the public sphere, events like #MeToo in October 2017, the Brett Kavanaugh hearings in September 2018, and the Jeffrey Epstein scandal in July 2019, as well as revelations in August 2018 by the Pennsylvania Grand Jury on extensive clerical abuses. These disparate phenomena as well as survivors' stories in documentaries and films led me to confront my own experiences, which evoked rage, grief, and empathy. It is that state of empathy that influences my investigation and brings me closer to the sexual violence of the distant past.[3]

The following essay describes this journey of emotion and empathy, a journey which creates a closeness with my subjects and makes me a witness as much as it does an interpreter of the past. It is an open admission that I come to this topic as a feminist with a deep, personal engagement. Empathy influences my investigation into the silences that surround sexual abuse. Empathy is what brings the subaltern subject to the fore.[4] The rich studies of slavery, which have given voice to those marginalized because of race and gender, have equipped

---

2 La Parole Libérée. Pour tous les enfants victimes de violences sexuelles, accessed 7 October 2020, https://www.laparoleliberee.org/. Preynat headed the scout group of Groupe Saint Luc between 1970 and 1991.
3 Finnish historian Satu Lidman makes a similar statement in her study of gender-based violence: "It cannot be a question of simple, cold observation of research objects, a research must demonstrate empathy." *Gender, Violence and Attitudes: Lessons from Early Modern Europe*, trans. Eva Malkki (London: Routledge, 2018), 10. See also Thomas A. Kohut, *Empathy and Historical Understanding of the Human Past* (New York: Routledge, 2020).
4 I use the terms "subaltern" as it had originally been coined by Antonio Gramsci to identify those who are left out of history. Marcus E. Green, "Rethinking the subaltern and the question of censorship in Gramsci's *Prison Notebooks*," *Postcolonial Studies* 14 (2011): 387–404.

me with compelling tools for examining sexual violence in the past. This reflection on the connection between empathy and historical research also demands that we consider how existing practices within the historical profession can re-affirm and reify certain hierarchies of power. As Satu Lidman notes, "examining violence, whether it has taken place today or in the past, is an ethical task demanding special sensitivity."[5]

# A Historian's Journey

I had encountered alleged clerical abuse when working on the Girard/Cadière trial, the subject of my book *The Wanton Jesuit and the Wayward Saint* (2015). In 1731 twenty-one-year- old Catherine Cadière accused her Jesuit confessor and spiritual advisor Jean-Baptiste Girard of seduction, bewitchment, and heresy. My analysis focused on how contemporaries located the trial within the period's politico-religious conflicts and how the affair opened a new window of dissent in early eighteenth-century France. Indeed, attacks against the Jesuits as well as the convoluted legal proceedings in Aix-en-Provence overshadowed Cadière and Girard who were at the heart of the scandal. As for the actual relationship between Cadière and Girard, I dealt with the question of whether there had been sexual abuse rather cursorily.[6] I essentially mimicked Cadière's contemporaries who used her victimhood as a vehicle to articulate their own positions. Beginning in 2017, a confluence of events led me to re-evaluate my assessment of the Cadière affair. Documentaries, films, and contemporary events moved me to tears as I listened to victims recount their experiences, even on occasion identifying with them. This state of empathy transformed into a decision to bring out the voices of victims and survivors in the distant past.

The possibilities for this topic opened up when I attended a lecture by religious studies scholar Robert Orsi at Vassar in April 2017.[7] The talk focused on his extensive interviews with American survivors of abuse suffered at the hands of priests. Orsi highlighted the power and reach of the Catholic Church, which

---

**5** Lidman, *Gender, Violence and Attitudes*, 10.

**6** Mita Choudhury, *The Wanton Jesuit and the Wayward Saint: A Tale of Sex, Religion, and Politics in Eighteenth-Century France* (University Park: Pennsylvania State University Press, 2015).

**7** Robert Orsi, "Religious Dimensions of the Catholic Clergy Sexual Abuse Crisis," The Annual Frederic C. Wood Lecture, 29 March 2017, Vassar College. See also Robert Orsi, "What is Catholic about the Clergy Sex Abuse Crisis?" *The Anthropology of Catholicism*, eds. Kristin Norget, Valentina Napolitano and Maya Mayblin (Berkeley: University of California Press, 2017), 282–292.

used its wealth to buy off claimants and relocate priests to stifle any scandal. Such widespread abuse was not simply the result of institutional might and vows of celibacy. It was also the product of faith. Being a good Catholic was not simply about top-down indoctrination but involved a more complicated set of beliefs about how to lead a Christian life which centered the priest as the direct channel to God. The power of faith, Orsi and I agreed, was not restricted to institutional pressures to conform and obey but also in the interiorization of doctrine as true belief and therefore intrinsic to the believer's identity. Cognizant of these ingrained beliefs, priests deployed threats of sin and damnation to control their victims. Thus, for devout Catholics, it was almost unimaginable to question a priest, and it seemed more instinctive to blame oneself as a sinner. I agree with Orsi's argument that in order to give victims full voice, it is crucial to take their faith seriously, to understand how victims found themselves in a painful conundrum in which abuse also represented a test of their faith.[8] Such an explanation does not blame the victim, but rather exposes the scaffolding on which victimization can occur.

A few months after Orsi's lecture, I stumbled across the aforementioned Netflix documentary *The Keepers*, which became the determining catalyst for my pursuing a topic on clerical sexual abuse in seventeenth- and eighteenth-century France. This series, released in May 2017, explored the themes of faith, institutional power and arrogance, and networks of complicity, themes addressed by Orsi. Directed by Ryan White, the documentary begins with two women, Gemma Hoskins and Abbie Schaub, and their quest to determine who had murdered their beloved teacher Sister Cathy Cesnik at the all-girls Keough High School in Baltimore. Cesnik had disappeared in November 1969, and her body, which included a bludgeoned skull and strangulation marks, was discovered two months later. The scenes of these indomitable women pouring over microfilm, meticulously cataloguing and crosschecking their findings would make any historian proud; research was Abbie's province, while Gemma doggedly followed up with interviews including police officials, journalists, and eventually, survivors.[9] However, what begins as a murder investigation turns into unimaginable horror and endless grief. Over the course of seven heartrending episodes, we learn, along with Abbie and Gemma, that the school chaplain and psychologist Father Joseph Maskell had abused over 100 girls at Archbishop Keough. Wielding spiritual and psychological authority, the chaplain

---

8 I am currently developing this topic using Judith Butler's idea of performativity. Mita Choudhury, "Performativity and Faith: Rethinking Judith Butler and the *Convulsionnaires of Saint-Médard*," Western Society for French History, Reno, November 2017.
9 *The Keepers*, episode 1, "The Murder," directed by Ryan White, aired 19 May 2017 on Netflix.

not only systematically assaulted these girls, he essentially acted as their pimp, subjecting them to abuse by other clerics, members of the Baltimore police, its business leaders, and the city's municipal officials.

What struck me almost immediately was that unlike *Spotlight* (2015), *The Keepers* revolves around the courageous individuals who chose to come forth and to claim their right to speak, which among so many other things, had been stolen from them.[10] Forty women were filmed, although only five of them made the final cut. Thanks to the series and the efforts of Abbie and Gemma, these women found networks on Facebook that provide support as well as guidelines for activism and pursuing justice. The series' central narrative belongs to Jean Hargadon Wehner whose story went back to her freshman year when she revealed to Father Magnus, Maskell's lieutenant and partner-in-crime, that she had been sexually abused by an uncle.[11] Over the course of *The Keepers*, Jean reveals how Maskell used this knowledge to rape her on numerous occasions as well as to force her to have sex with various other men. Maskell extorted her silence with threats to tell her devout Catholic family and policeman father, impressing upon her that their sinful encounters were all her fault. The teenage Jean finally confided in 26-year-old Sister Cathy Cisnek, who was accessible and empathetic.[12] And then Cisnek was murdered. One of Jean's memories included Maskell taking her to see the nun's maggot-ridden body and warning her that this is what happened when people spoke up. To maintain her sanity and live her life, Jean repressed these memories, which only began resurfacing when she was in her 40s. Indeed, there are moments during *The Keepers* when it becomes painfully clear that new memories emerge while she is on camera. In the early 1990s, Jean and another Maskell victim, Teresa Lancaster, took the Archdiocese of Baltimore to court. Their efforts were defeated on the grounds that the statute of limitation had expired, a ruling which reflected the Church's pervasive influence over the Baltimore criminal system. It would not be an exaggeration to say that it was a conspiracy of silencing, designed to vilify the victims and protect the Church. More than twenty years later, *The Keepers* turned the tables, restoring voice to the victims while placing the Archdiocese in the hot seat.

---

**10** Alex Abad-Santos, "Netflix's new true crime doc, *The Keepers*, isn't Making a Murderer. It's far more Haunting," *Vox*, 22 May 2017, accessed 5 October 2020, https://www.vox.com/2017/5/20/15643380/the-keepers-netflix-sister-cathy-cesnik-murder-mystery-review.
**11** Anne Thompson, "Why Neflix Bingeing Brought Fresh Power to the 50-Year-Old Mystery of 'The Keepers,'" *IndieWire*, June 26, 2017, accessed 5 October 2020, https://www.indiewire.com/2017/06/netflix-the-keepers-documentary-sister-cathy-nun-expose-emmy-1201846839/.
**12** *The Keepers*, episode 3, "The Revelation."

*The Keepers* is, of course, not the only film or documentary to showcase abuse within the Catholic Church. In many respects, *Spotlight* paved the way for other film narratives of sexual assault in the modern Church. Indeed, François Ozon's recent film *Grâce à Dieu* (2019) makes several nods to the earlier Oscar winning film, although its focus is also on the survivors.[13] The Spanish documentary *Examination of Conscience* (2019) and two crowdfunded documentaries *Tell No One* (2019) and *Hide and Seek* (2020), directed by Polish directors Tomasz and Marek Skielski, are also examples of victim-centered narratives that showcase the trauma and courage of survivors.[14] In addition to the Skielski documentaries, YouTube hosts other documentaries about church sex abuse cases, indicating how survivors have an urgent need to tell their stories.

Each story has its own poignant details, but as survivor, activist, and child psychologist Miguel Angèl Hurtado states in *Examination of Conscience,* "the victims always tell the same story. There is always the same pattern of covering up. It's like the film you know the ending of."[15] Thus, *Hide and Seek* follows the story of two brothers, Jakub and Bartek Pankowiak and the abuse suffered in their own home, at the hands of a parish priest, Arkadiusz Hajdasz; the bishop of Kalisz Edward Janiak responded by moving the priest, a pattern seen in other documentaries. The documentary *Sex Abuse in the Church: Code of Silence* (2017), directed by Martin Boudet, tracks how the Church transferred European pedophile priests to Latin America and to Africa. Blurred faces and muffled voices show how the fear of retaliation continues to haunt survivors in *Sex Abuse in the Church* and in *The Keepers*, which is particularly heartrending because Maskell died in 2001.[16] The depth and persistence of pain are revealed when

**13** *Grâce à Dieu,* directed by François Ozon (2019; France, Belgium, Mandarin Films, Scope Films).
**14** *Tell No One,* directed and written by Tomasz Sekielski, aired 11 May 2019 on YouTube, https://www.youtube.com/watch?v=BrUvQ3W3nV4. *Hide and Seek,* directed and written by Tomasz Sekielski, aired 16 May 2020 on YouTube, accessed 7 October 2020, https://www.you tube.com/watch?v=T0ym5kPf3Vc. Marek Sekielski has served as producer. See Jan Cienski, "Poland's church abuse scandal becomes political," *Politico,* 13 May 2019, accessed 7 October 2020, https://www.politico.eu/article/poland-catholic-church-sex-abuse-scandal-politics-tell-no-one-film/.
**15** *Examination of Conscience,* episode two, directed by Albert Solé, written by Meritxell Llorens and Albert Solé, aired 25 January 2019 on Netflix. See Ruth Riddick, "Clerical Sex Abuse in Spain," *Conscience,* 1 April 2019, accessed 7 October 2020, https://consciencemag.org/2019/04/01/betrayal-global-story-local-voices-as-netflix-explores-clerical-sex-abuse-in-spain/.
**16** Ryan White discusses his own fears in an interview with *The Guardian.* Rebecca Nicholson, "The Keepers; 'I've dealt with survivors and they're sickened by the church's response," 15 July 2017, accessed 6 October 2020, https://www.theguardian.com/tv-and-radio/2017/jul/15/keep ers-church-response-netflix-baltimore-abuse. *Sex Abuse in the Church: Code of Silence,* directed

survivors confront priests in *Tell No One* and *Examination of Conscience*. But we are also shown how survivors come together and realize that they are not isolated. Interestingly, before filming *Grâce à Dieu*, Ozon had wanted to make a documentary, but another survivor of Lyon, Alexandre Dussot-Hezez insisted on a feature film to make the story more universal.[17] Offscreen, Miguel Hurtado, along with François Devaux of *La Parole Libérée* and other survivors, are the founding members of the international Ending Clerical Abuse (ECA).[18]

This activism, to my mind, is a part of our current zeitgeist in which victims are finding their voices, each other, and the public is listening, or, at the very least, being forced to listen. During the fall of 2017, the #MeToo movement exploded, and we witnessed an avalanche of stories of women (and men) who had experienced some form of sexual violence. The patterns were similar to those associated with clerical sexual assault: shame and fear that induced silence, trauma that covered up memories, and the wearying knowledge that allegations could very well invite disdain and disregard. Power, akin to that of the priesthood, protected men like Harvey Weinstein and Jeffrey Epstein despite decades of suspicion. The world of the Church collided with the political arena when the Supreme Court nominee Brett Kavanaugh testified before the Senate in September 2018. In defending himself against Dr. Christine Blasey Ford's accusations of sexual misconduct, an irate Kavanaugh insisted that he was a virtuous Catholic who diligently went to mass with his family and engaged in public charitable acts. From Kavanaugh's perspective, it was impossible that he engaged in such behavior and how dare Blasey Ford and the left even think of accusing him. Kavanaugh's tearful performance of Catholic virtue and claim to impunity, as I argued in an NPR interview, was tied to his Jesuit education, an

---

by Martin Boudot, 2017. https://www.amazon.com/Sex-Abuse-Church-Code-Silence/dp/B0718SMTRY/ref=sr_1_1?crid=1L5GJELMDUYG9&dchild=1&keywords=sex+abuse+in+the+church+code+of+silence&qid=1601989682&sprefix=church+codes+of+%2Caps%2C160&sr=8-1.
**17** François Ozon, "'People are Afraid of Cinema': François Ozon Takes on Church Sexual Abuse," interview by Rebecca Rosman, *Weekend Edition Saturday*, NPR, 26 October 2019, accessed 7 October 2020, https://www.npr.org/2019/10/26/773585708/people-are-afraid-of-cinema-fran-ois-ozon-takes-on-church-sexual-abuse. See also "Invités: Melvil Poupaud, Swann Arlaud, Slexandre Dussot-Hezez et Pierre-Emmanuel Germain," interview by Yann Barthes, *Quotidien*, TF1, 18 February 2019, accessed 7 October 2020, https://www.tf1.fr/tmc/quotidien-avec-yann-barthes/videos/invites-melvil-poupaud-swann-arlaud-alexandre-dussot-hezez-pierre-emmanuel-germain-thill-gra-a-dieu-de-francois-ozon.html.
**18** Ending Clergy Abuse: Global Justice Project, accessed 7 October 2020, https://www.ecaglobal.org/founding-members/. See also BishopAccountability.org. Documenting the Abuse Crisis in the Roman Catholic Church. http://www.bishop-accountability.org/.

insight I gained from my own work on early modern Jesuits.[19] As men of God, the Jesuits regarded themselves above such accusations, a line of defense used in the Cadière trial. Not only was it unthinkable – so the Church has argued – for servants of God to commit such sinful acts (a claim that blames the victim), but members of the laity had no right to make them accountable. By evoking his Catholic lifestyle, Kavanaugh borrowed the mantle of clerical privilege, using his faith and masculine status as weapons against Blasey Ford.

Although Blasey Ford and Kavanaugh did not attend the same school, the high school scene Blasey Ford described is reminiscent of a certain private school culture. In June 2020, Lacy Crawford published *Notes on a Silencing*, which describes the assault she experienced at the elite preparatory St. Paul's School in New Hampshire; her decision to write this memoir nearly three decades after the events had taken place was influenced by another St. Paul's student Chessy Prout who in 2014 claimed that a senior Owen Labrie had raped her when she was fifteen.[20] For both Crawford and Prout, one of the most demoralizing elements of their ordeals was how St. Paul's silenced them both. Echoing Hortado, Crawford notes: "It's not a remarkable story. In fact, it's ordinary."[21] She's right. I know this firsthand because I spent a year teaching at another elite New England boarding school, the Hotchkiss School. While there in 1988–1989, I had an overwhelming sense that something was not right when I observed certain teachers interacting with students. It was deeply uncomfortable to watch, but as a powerless intern, a woman of color in a very white Protestant community, I had no space to express concerns. And apparently, neither did the students who were seduced, molested, and victimized by individuals who, back in the 1980s, enjoyed almost unchecked positions of power in a school located in the remote northwestern corner of Connecticut. Then, in 2018,

---

**19** Interviewed by Alison Dunne of WAMC in October 2018 on the Brett Kavanaugh hearings; the interview for Midday magazine and 51% https://www.wamc.org/post/vassar-professor-speaks-aspect-kavanaugh-hearings. See also Libby Berry, "Brett Kavanaugh's religious credentials are not enough for some Catholics," *Boston Globe*, 6 October 2018, accessed 7 October 2020, https://www.bostonglobe.com/news/politics/2018/10/06/brett-kavanaugh-allega tions-roil-catholic-faithful-still-grappling-with-clergy-abuse/PoimxWFIPzvwJqGPA5ZGnN/ story.html.
**20** Sarah Lyall, "Assaulted at 15, a Writer Looks Back and Comes Forward," *New York Times*, 27 June 2020, accessed 7 October 2020, https://www.nytimes.com/2020/06/27/books/lacy-crawford-notes-on-a-silencing.html.
**21** Lacy Crawford, *Notes on a Silencing: A Memoir* (New York: Little, Brown, and Company, 2020), 10.

Hotchkiss achieved national notoriety when the school issued a detailed report of sexual abuse, most of which had taken place in the 1980s.[22]

In hindsight, I don't believe it was at all coincidental that my own memories of assault surfaced while teaching at Hotchkiss. Many allegations against teachers or priests emerge decades after the event because victims suppress their memories, a strategy of survival and self-preservation. I am one such survivor. When I was seven, my parents' marriage was in trouble; my mother and I were in India without any clear sense of what was next as we shifted from one relative's home to another. What happened to me during this time is not at all atypical. It was the story of someone who was much older, a relative who took advantage of their power and my vulnerability. My abuser knew that I was uprooted, and he also knew that shame and confusion would keep me silent, so much so that the memories of what happened were repressed for nearly two decades. As I have come to understand over the years, the impact of this episode (or episodes) was profound even though the memories resemble snapshots rather than a movie reel. One of the consequences was a sixth sense, an intuition, knowledge which comes from experience. I was never comfortable around another relative, and in my early twenties, learned that he had harassed a cousin. When a childhood friend told me that she had been systematically raped and assaulted for years by a family friend, I wasn't surprised just as I wasn't surprised by my cousin's revelations. So, I feel I "knew" what was happening at Hotchkiss, even without proof.

In my twenties, my approach to what had happened to me was detached. It was not until I was in my 50s, the time when many victims begin coping with abuse, that I felt the full impact of what I had experienced. I had just watched *Spotlight*, and Stanley Tucci's line, "If it takes a village to raise a child, it takes a village to abuse one" echoed in my head. I sobbed uncontrollably for what seemed like hours. I am quite certain that my assault was known because it didn't take place in a remote or hidden place. Up until this moment, I just carried shame and blame, which made themselves felt in various forms. Now, as a mother, I felt full anger at the adults who failed to protect me. Anger at a culture of networks that did not allow such subjects to be discussed and in which families, like schools and churches, protected the community and not the individual. In the end, my story is not one of a kind. How many readers have seen Mira Nair's *Monsoon Wedding* (2001)? It is a universal story, as Alexandre Dussot-Hezez has argued. Is it any

---

22 Elizabeth Harris, "At Hotchkiss, Sexual Misconduct and 'Missed Opportunities' to Stop It," *New York Times*, 17 August 2018, accessed 7 October 2020, https://www.nytimes.com/2018/08/17/nyregion/hotchkiss-school-sexual-misconduct.html.

great surprise that in August 2018, after reading the *New York Times* article on Hotchkiss, I woke up in the middle of the night screaming?

The waking from what must have been a nightmare points to the range of emotions I, and many others, experienced. We follow with horror and rage the kinds of abuse that were perpetrated against unsuspecting, trusting victims. No matter how many times I have watched *The Keepers*, there is a visceral, almost physical, shock in hearing Jean Wehner talk about how Father Magnus masturbated in the confessional while she described her uncle's abuse. Why subject myself to watching such scenes? To bear witness. I have come to understand how the violence a victim experiences is twofold – the violence of abuse and the violence of silence. When we ask how this could happen, how this was possible, we need to look at the complicity of not just institutions but also of cultural attitudes, which are often internalized. Accusations of abuse and rape are often greeted with disbelief and skepticism, as suggested by the thousands of tested and untested rape kits stored in police precincts throughout the United States.[23] Understandably then, victims remain silent, doubting they will be believed, expecting derision not sympathy, inaction rather than justice. In some ways, one can say that this dynamic is as true today as it was in the early modern era. Only since 2017, does that silence imposed on victims seem to be breaking apart with the waves of accusations against the Church and revelations that have been a part of #MeToo. I can't count the number of conversations or moments of painful self-reflection, of collective and individual grief of which I have been a part. Out of this emotional maelstrom, my determination to write this story of victims and survivors in early modern France has only strengthened.

## The Ethics and Methodology of Researching Historical Abuse

Can empathy be a starting point for critical inquiry, and what are the methodologies available to us, especially to historians of pre-modern Europe? My emotional responses will undoubtedly prompt some to argue that I can hardly be

---

**23** Barbara Bradley Hagerty, "An Epidemic of Disbelief. What new research reveals about sexual predators, and why police fail to catch them," *The Atlantic*, August 2019, accessed 7 October 2020, https://www.theatlantic.com/magazine/archive/2019/08/an-epidemic-of-disbe lief/592807/.

neutral in writing a history of sexual assault that is victim-centered. I may be faulted for identifying too much with my subjects and questioned as to how such an endeavor can be productive given my biases. The failure to be detached, the argument follows, would make it impossible for me to write an objective history, a view that is the scholar's death knell because behind it is a challenge to the legitimacy of my project and my standing as an historian.

Echoing Satu Lidman, I would argue that empathy can be a productive point of entry and analysis, one that is used in conjunction with and not instead of more traditional historical methods. Furthermore, an absence of neutrality or objectivity does not necessarily entail a polemical goal behind an argument. Empathy represents a point of access and a state of awareness. Recognizing the "biases" or the "baggage" we bring to our research subjects can illuminate how we operate within sites of power. It leads us to think about how the archives themselves represent structures of authority, designed to benefit dominant groups. Through empathy, we may question how we understand historical action by challenging ingrained notions such as agency. Instead, we might consider looking for presence that is often embodied and not textualized. Beginning with a state of self-reflection, empathy potentially allows us to pursue different topics and ask different questions of our sources than we might otherwise consider. At the same time, historical analysis, with its emphasis on context, prevents us from a false appropriation of other people's subjective experience by recognizing its uniqueness.

For most of my career as a French historian focusing eighteenth-century Augustinian Jansenism and politics, I subconsciously felt immune from such self-reflection. Six years in a Catholic school did not lead me to seek vengeance against nuns in my first book *Convents and Nuns in Eighteenth-Century French Thought and Culture* (Ithaca, 2004). A bit naive and not just a bit arrogant. However, my examination of the Cadière affair in *The Wanton Jesuit and the Wayward Saint* illustrates a true moment of blindness. Building on the first monograph, this microhistory focused on the political culture of eighteenth-century France with an emphasis on religion and gender. I briefly addressed the possibility of rape but quickly opted for the idea that the attraction between the Jesuit Jean-Baptiste Girard and his penitent Catherine Cadière was mutual. After all, how could we know what really happened between the two within the confines of the confession? In *Scandal in the Parish: Priests and Parishioners Behaving Badly in Eighteenth Century France*, Karen E. Carter acknowledges the necessity of considering sexual relations between clergy and laity, whether they were consensual or coerced. Yet in her conclusion, Carter states: "I suspect that many of the accusations against him [the curé Nicolas Hyacinthe Vernier] regarding sexual misconduct were false, although I find I am less certain about that than I am about other things."[24] Why this hesitation? Lacy Crawford offers a clue when describing her dilemma about how to label her own sexual assault, which involved forced fellatio that gave her herpes. It wasn't "rape," Crawford reasoned to herself, because "rape was serious, and I thought – and wanted to think – that what happened to me didn't really count. I did not understand how the boys' violation was of me, rather than only a part of me."[25] Crawford's words and her story echo how historians approach rape and sexual assault – it is both not serious enough or too serious. We have been taught to read over or around rape because it is distasteful and even considered biased.[26]

My own repelling from the uncomfortable possibility of rape in *The Wanton Jesuit and the Wayward Saint* was not just cultural but also professional. How we emphasize one form of violence and remain silent about another is predetermined within the historical profession. For example, why are there so many military histories of battles and strategy and yet so few studies of domestic violence, given the prevalence of the latter? One topic has received endless attention, forming its own cottage industry, while the other has only been a topic of discussion and research in recent decades. This privileging of certain forms of violence reflects how the historical profession engages not just in unearthing the past but also in silencing it. According to Michel Trouillot, "one 'silences' a fact or an individual as a silencer silences a gun. One engages in the practice of silencing. Mentions and silences are thus active, dialectical counterparts of which history is the synthesis."[27] Revisiting the rape account of a nineteenth-century Japanese woman, Amy Stanley discusses how she avoided this narrative in her first telling of the story. As Stanley admits: "But I, too, was letting my own social conditioning inform my reading of the sources. After all, the story of a woman who engages in rebellious sexual behavior, regrets it, and then makes up a story to explain herself is its own kind of cultural narrative. It's mine. It's ours."[28]

---

**24** Karen E. Carter, *Scandal in the Parish: Priests and Parishioners Behaving Badly in Eighteenth-Century France* (Montreal: McGIll-Queen's University Press, 2019), 237.

**25** Crawford, *Notes on a Silencing*, 11.

**26** Christine Rose, "Reading Chaucer Reading Rape," *Representing Rape in Medieval and Early Modern Literature* (New York: Palgrave, 2001), 21–60, here 25.

**27** Michel-Rolph Trouillot, *Silencing the Past: Power and the Production of History*, fwd. by Hazel Carby (Boston: Beacon Press, 1995, 2015), 48.

**28** Amy Stanley, "Writing the history of sexual assault in the age of #MeToo," *Perspectives on History*, 24 September 2018, accessed 20 December 2018, https://www.historians.org/publica tions-and-directories/perspectives-on-history/november-2018/writing-the-history-of-sexual-as sault-in-the-age-of-metoo. See also Julie Hardwick, "Sexual Violence and Domesticity," *The*

As Stanley suggests, perhaps it was my combined discomfort and training that prevented me from addressing whether there had been sexual violence in the Cadière story. As a woman hoping to succeed as an academic, how could I ask questions that others might not take seriously or challenge a methodology that involved extrapolating way beyond my sources?[29] There was no #MeToo while I was writing *The Wanton Jesuit and the Wayward Saint*, and even if there had been such a movement, would I have been willing to engage with it, especially since I was preparing to come up for promotion to full professor?[14] Moreover, as a woman of color, I work in an area that is outside of what my skin color designates as my "natural" or "destined" field. In other words, I operate within certain professional hierarchies (as well as those of race, gender and sexuality, and class).[30] These structures and our positionality within them shape the subjects and methodologies of our discipline. Trouillot succinctly sums up how scholars shape the "truth" and "facts" through the choices they make and the structures they inhabit: "The naming of the 'fact' is itself a narrative of power disguised as innocence."[31] Only now, bolstered by the security of rank, do I feel I have permission (bestowed by invisible authorities) to use empathy in pursuing the uncomfortable, to crack open the silences.

The road of uncovering begins with our sources. Karen Carter's work on scandal and parish priests is meticulous in its research, but her reluctance to discuss the possibilities of sexual violence suggests her unwillingness to imagine a reality beyond the sources. I certainly don't challenge that as a methodology, but what I do question is how we interrogate such sources in terms of their reliability. How are we reading the documents when we choose who to believe and who not to? I have found my own inspiration on how to delve into this opaque topic in recent scholarship on slavery. In *Dispossessed Lives*, Marisa Fuentes argues how the silencing of the enslaved women in colonial Barbados reflected the power structures of eighteenth-century colonial rule, structures which are affirmed in the creation and organization of the archives themselves.[32] We cannot discon-

*Routledge History of the Domestic Sphere in Europe, Sixteenth to Nineteenth Century*, eds. Joachim Eibach and Margareth Lanzinger (London: Routlege, 2020), 239–243.

**29** For one approach on how striving for legitimacy shapes scholarly production, especially in terms of gender, see Emily Callaci, "On Acknowledgments," *American Historical Review* 125 (February 2020): 126–131.

**30** Mita Choudhury, "Race, Privilege, and Identity in French History," *H-France Salon* 12, 1, 4 (2019): 1–2. https://h-france.net/h-france-salon-volume-12-2020/#1201.

**31** Trouillot, *Silencing the Past*, 114.

**32** Marisa J. Fuentes, *Dispossessed Lives: Enslaved Women, Violence, and the Archive* (Philadelphia: University of Pennsylvania, 2016), 5.

nect the authorities presiding over the procedures or the witnesses from a world in which a woman's word was automatically suspect. It is worth bearing in mind that in early modern France, the acknowledgment of violence against women operated, at least explicitly, within a narrow band. As Anna Clark has suggested, "Myths about rape are thus engendered and manipulated by ruling elites at particular historical moments."[33] The lack of evidence or conviction as determined by a court does not mean that sexual violence did not take place. As Jessica Johnson has argued, there are histories, such as those of Jamaican escaped slaves, that are hidden in plain sight within our sources.[34]

For early modern historians like myself, we remain dependent on ecclesiastical court records and police files, institutions designed to maintain power, which deliberately and/or unintentionally silence certain voices, often to safeguard others.[35] Indeed, Catherine Cadière's family claimed that the ecclesiastical courts in Old Regime France, pressured witnesses and threw out testimony that might incriminate the Jesuit Girard. In a similar vein, Jean Wehner encountered church officials and lawyers who manipulated the system in their efforts to protect the Church. Laws, like the statute of limitations for bringing accusations of sex crimes to court, can shield those in power. Looking back to the past, we can scrutinize the terminology – *rapt de séduction*, *viol*, or *inceste spirituel* – and consider how such crimes were framed, how these terms position perpetrators and victims. How do we access and analyze a form of violence that in certain circumstances did not have a name, let alone go unreported and uncontested? When we examine court documents, we must interrogate the procedure, beginning with the collection of evidence as well as the adjudication process. Empathy fits into this process from the awareness of how authority and procedure can intimidate and silence.

Indeed, historians have been trained to detect signs of action and visibility and not silence, often in quest of agency. In his famous essay on agency, Walter Johnson contests and complicates the term agency, especially as it used in the context of enslaved and dispossessed people. Johnson argues: "the term 'agency' smuggles a notion of the universality of a liberal notion of selfhood, with its emphasis on independence and choice, right into the middle of a

---

**33** Anna Clark, *Women's Silence, Men's Violence; Sexual Assault in England 1770–1845* (London: Pandora and New York, 1987), 5.
**34** Jessica Maria Johnson, "Who's HiPS? Plain Sight Histories of Slavery," *William and Mary Quarterly* 76 (January 2019): 15–18.
**35** Julie Hardwick's recent study provides an important methodological template for early modernists. *Sex in an Old Regime City: Young Workers and Intimacy in France, 1660–1789* (Oxford: Oxford University Press, 2020).

conversation about slavery against which that supposedly natural (at least for white men) condition was originally defined. By applying the jargon of self-determination and choice to the historical condition of objectification and choicelessness, historians, have, not surprisingly, ended up in a mess."[36] As Johnson's argument suggests, humanity is recognized when connected to tangible and visible activity, which also presupposes a self that is coherent and therefore, can be cohesively narrated. In her work on Black women, rape, and migration, Darlene Clark Hines also challenges these notions of agency, arguing that "only with secrecy thus achieving a self-imposed invisibility, could ordinary Black women accrue the psychic space and harness the resources needed to hold their own in the often one-sided and mismatched resistance struggle."[37] Instead of looking for agency, we may consider looking for presence, a process that can start with that intuitive empathy.

As Fuentes and others have shown, we can use the archives to access the presence and voices of victims by paying attention to fragments and not ignoring them in search of coherent narratives. For example, the lack of naming effectively erases the violence as well as the victim, as we are reminded when we emphasize "say her name."[38] From the perspective of early modern authorities, women's grievances in and of themselves did not always merit investigation and did not trouble officials who were more concerned about the potential scandal created by disclosure and publicity. In eighteenth-century Paris, this omission was pronounced in the police records that identified monks and priests who were not in an "odor of saintliness" and associated with unnamed "femmes du monde." Thus, the 55-year-old abbé de Courbe came to the attention of the police in 1752 when frequenting the brothel "La Richard" on the rue Meslée. The police spy noted that one of the "women" was an 11-year-old virgin whom de Courbe initially deflowered with a

---

**36** Walter Johnson, "On Agency," *Journal of Social History* 37 (October 2003): 115.

**37** Darlene Clark Hines, "Rape and the Inner Lives of Black Women in the Middle West," *Signs* 14 (Summer 1989): 915. See also *Signs*, 45, no. 3 (2020) for a symposium on Clark Hines' groundbreaking essay.

**38** In 2020, the phrase "say her name" is linked to the murder of Breonna Taylor by police who broke into her apartment in Louisville, Kentucky, but it was established in 2014 by the African American Policy Forum (AAPF) and Center for Intersectionality and Social Policy Studies (CISPS) led by Kimberlé Crenshaw, accessed 29 November 2020, https://aapf.org/sayher name. See also Kimberlé Crenshaw, Andrea J. Ritichie, et al., *Say Her Name; Resisting Police Brutality Against Black Women* (New York: African American Policy Forum, Center for Intersectionality and Social Policy Studies, 2015). On the issues with naming, see, for example, Sue Peabody, "Madeleine's children: slaves from Isle Bourbon (present-day Réunion), Eighteenth and Nineteenth Centuries," *Clio, Women, Gender, History* 45 (2017): 166–178. This entire issue is devoted to naming.

candle; eventually, he focused on her 16-year-old sister.[39] Since prostitutes by definition could not be raped, neither woman was named; their identities and their experiences were not central to the inquiries of inspector Meusnier. Nevertheless, the
references to bodies being handled remind us that they did exist. It is up to us as
practitioners to resist taking the lead of our sources and instead, focus on the
victims.

Such examples illustrate how we – often out of necessity – privilege textuality over subjectivity and embodiment. Thus, in her study on clerics "behaving
badly," Myriam Deniel-Ternant mimics police records and focuses on the clerics,
effectively ignoring the victims who worked in brothels. Arguably, this approach
affirms the structures that kept women silent.[40] To uncover violence, I would
suggest we get beyond textuality. Instead, we must read bodies, in the forms of
inarticulate expressions of pain and distress or visible markers of violence. Moreover, we need to be able to identify trauma, which can defy linear narration and
is often characterized by the fragmentation of memory and indeed, of the self.
Historians can explore a society's language of emotion, both verbal and nonver-
bal, to unpack how individuals may articulate their trauma. The articulation of
such trauma can be historically specific as Michel de Certeau has argued in his
examination of mystics and possessions. Thinking about my own work on
Cadière, I wonder if ecstasies and possession, which made Cadière a celebrity in
her native Toulon, were complicated responses to sexual assault. Cadière's faith
was central to her sense of self as well as to her public identity. Mystical trances
could have been a means of processing trauma with the familiar language of
faith at the same time that she protected herself through disassociation. Moreover, for Cadière, who lived far from the world of the Enlightenment and was
raised to revere priests, mysticism and possession could possibly allow her to
preserve her relationship with Girard and reaffirm his sanctity. With this speculation, I am not making the case for a definitive causal connection between sexual
assault and religious expressions. Rather, my point is to show the possibilities
for inquiry when historians position themselves to ask direct questions about
sexual assault as opposed to reading around it, ignoring fragments and bodies.

My training as an early modern historian drilled into me how vital it was to
be vigilant against reading the present into the past. This essay then goes
against the grain and instead, describes how a personal journey rooted in my

---

**39** "The abbé de Courbe, rue Licorne, 12 janvier 1752)," Bibliothèque de l'Arsenal, Archives de
la Bastille, 10246, f. 39.
**40** Myriam Deniel-Ternant, *Ecclésiastiques en débauche (1700–1790)* (Ceyzérieu: Champs Vallon, 2017).

contemporary world has shaped my project on early modern abuse in the French Church. Significantly, it highlights how the sources of engagement and empathy were non-academic and arguably freed me to conceive of a victim-centered history. It suggests how despite chronological distance, empathy can bring us adjacent to violence by presenting a different set of priorities in terms of topics, approaches, and questions. We early modernists have a fragile sources base, the result of deliberate destruction or accidents as well as of larger social conditions such as illiteracy or the availability of writing tools. There are gaps, which lay bare that historians also have recourse to imagination to fill in missing information. Writing a microhistory as if some scholarly authoritative body was hovering over me, I kept my imagination in check so that I would not be accused of being presentist or not being objective.[41] But imagination and empathy also allow us to uncover the ways in which victims of sexual violence were silenced and the ways in which archives perpetuate that suppression. My argument is not to dispense with established methods of historical research and analysis. Rather, it is to be more aware of how these methods replicate structures of power. We should become more aware, too, of how notions of objectivity can have gender and racial hierarchies embedded within them. Empathy can be a critical tool for placing victims of violence, even those who lived in centuries past, at the center of inquiry. It can shed light on spaces that have deliberately been kept in the dark and help us see our own blindness.

# Works Cited

## Primary Sources

"The abbé de Courbe, rue Licorne, 12 janvier 1752)," Bibliothèque de l'Arsenal.22 Archives de la Bastille, 10246.

## Secondary Sources

https://aapf.org/sayhername.
Abad-Santos, Alex. "Netflix's new true crime doc, *The Keepers*, isn't Making a Murderer. It's far more Haunting," *Vox*, May 2017. https://www.vox.com/2017/5/20/15643380/the-keepers-netflix-sister-cathy-cesnik-murder-mystery-review.

---

**41** For a discussion of presentism, see Francis Dolan, *Marriage and Violence: the Early Modern Legacy* (Philadelphia: University of Pennsylvania Press, 2009), 15–17.

Berry, Libby. "Brett Kavanaugh's religious credentials are not enough for some Catholics," *Boston Globe*, 6 October 2018. https://www.bostonglobe.com/news/politics/2018/10/06/brett-kavanaugh-allegations-roil-catholic-faithful-still-grappling-with-clergy-abuse/PoimxWFIPzvwJqGPA5ZGnN/story.html.

BishopAccountability.Org. Documenting the Abuse Crisis in the Roman Catholic Church. http://www.bishop-accountability.org/.

Callaci, Emily. "On Acknowledgments," *American Historical Review* 125 (February 2020): 126–131.

Carter, Karen E. *Scandal in the Parish: Priests and Parishioners Behaving Badly in Eighteenth-Century France*. Montreal: McGill-Queen's University Press, 2019.

Choudhury, Mita. "Race, Privilege, and Identity in French History." *H-France Salon* 12, 1, 4 (2019): 1–2. https://h-france.net/h-france-salon-volume-12-2020/#1201.

Choudhury, Mita. *The Wanton Jesuit and the Wayward Saint: A Tale of Sex, Religion, and Politics in Eighteenth-Century France*. University Park: Pennsylvania State University Press, 2015.

Cienski, Jan. "Poland's church abuse scandal becomes political," *Politico*, 13 May 2019. https://www.politico.eu/article/poland-catholic-church-sex-abuse-scandal-politics-tell-no-one-film/.

Clark, Anna. *Women's Silence, Men's Violence; Sexual Assault in England 1770–1845*. London: Pandora and New York, 1987.

Crawford, Lacy. *Notes on a Silencing: A Memoir*. New York: Little, Brown, and Company, 2020.

Crenshaw, Kimberlé, and Andrea J. Ritichie, et. al. *Say Her Name; Resisting Police Brutality against Black Women*. New York: African American Policy Forum, Center for Intersectionality and Social Policy Studies, 2015.

Deniel-Ternant, Myriam. *Ecclésiastiques en débauche (1700–1790)*. Ceyzérieu: Champs Vallon, 2017.

Dolan, Francis. *Marriage and Violence:Tthe Early Modern Legacy*. Philadelphia: University of Pennsylvania Press, 2009.

Ending Clergy Abuse: Global Justice Project. https://www.ecaglobal.org/founding-members/.

*Examination of Conscience*. Episode two, directed by Albert Solé, written by Meritxell Llorens and Albert Solé. Aired 25 January 2019 on Netflix.

Fuentes, Marisa J. *Dispossessed Lives: Enslaved Women, Violence, and the Archive*. Philadelphia: University of Pennsylvania Press, 2016.

*Grâce à Dieu*. Directed by François Ozon. France, Belgium, Mandarin Films, Scope Films, 2019.

Green, Marcus E. "Rethinking the subaltern and the question of censorship in Gramsci's *Prison Notebooks*." *Postcolonial Studies* 14 (2011): 387–404

Hagerty, Barbara Bradley. "An Epidemic of Disbelief. What new research reveals about sexual predators, and why police fail to catch them." *The Atlantic*, August 2019. https://www.theatlantic.com/magazine/archive/2019/08/an-epidemic-of-disbelief/592807/.

Hardwick, Julie. *Sex in an Old Regime City: Young Workers and Intimacy in France, 1660–1789*. Oxford: Oxford University Press, 2020.

Hardwick, Julie. "Sexual Violence and Domesticity." *The Routledge History of the Domestic Sphere in Europe, Sixteenth to Nineteenth Century*. Edited by Joachim Eibach and Margareth Lanzinger. 237–253. London: Routlege, 2020.

Harris, Elizabeth. "At Hotchkiss, Sexual Misconduct and 'Missed Opportunities' to Stop It." *New York Times*, 17 August 2018). https://www.nytimes.com/2018/08/17/nyregion/hotchkiss-school-sexual-misconduct.html.

*Hide and Seek*. Directed and written by Tomasz Sekielski. Aired 16 May 2020 on YouTube. https://www.youtube.com/watch?v=T0ym5kPf3Vc.

Hines, Darlene Clark. "Rape and the Inner Lives of Black Women in the Middle West." *Signs* 14 (Summer 1989): 912–920.

"Invités: Melvil Poupaud, Swann Arlaud, Slexandre Dussot-Hezez et Pierre-Emmanuel Germain." Interview by Yann Barthe. *Quotidien*, TF1, 18 February 2019. https://www.tf1.fr/tmc/quotidien-avec-yann-barthes/videos/invites-melvil-poupaud-swann-arlaud-alexandre-dussot-hezez-pierre-emmanuel-germain-thill-gra-a-dieu-de-francois-ozon.html.

Johnson, Jessica Maria. "Who's HiPS? Plain Sight Histories of Slavery." *William and Mary Quarterly* 76 (January 2019): 15–18.

Johnson, Walter. "On Agency." *Journal of Social History* 37 (October 2003): 113–124.

*The Keepers*. Episode 1, "The Murder." Directed by Ryan White, aired 19 May 2017 on Netflix.

Kohut, Thomas A. *Empathy and Historical Understanding of the Human Past*. New York: Routledge, 2020.

La Parole Libérée. Pour tous les enfants victimes de violences sexuelles. https://www.laparoleliberee.org/.

Lidman, Satu. *Gender, Violence and Attitudes: Lessons from Early Modern Europe*. Translated by Eva Malkki. London: Routledge, 2018.

Lyall, Sarah. "Assaulted at 15, a Writer Looks Back and Comes Forward." *New York Times*, 27 June 2020. https://www.nytimes.com/2020/06/27/books/lacy-crawford-notes-on-a-silencing.html.

Murdock, Graeme, Penny Roberts, and Andrew Spicer, eds. *Ritual and Violence; Natalie Zemon Davis and Early Modern France*. Oxford: Oxford University Press, 2012.

Nicholson, Rebecca. "The Keepers; 'I've dealt with survivors and they're sickened by the church's response." (15 July 2017). https://www.theguardian.com/tv-and-radio/2017/jul/15/keepers-church-response-netflix-baltimore-abuse.

Orsi, Robert. "What is Catholic about the Clergy Sex Abuse Crisis?" *The Anthropology of Catholicism*. Edited by Kristin Norget, Valentina Napolitano, and Maya Mayblin. Berkeley: University of California Press, 2017, 282–292.

Ozon, François. "'People are Afraid of Cinema': François Ozon Takes on Church Sexual Abuse." Interview by Rebecca Rosman. *Weekend Edition Saturday*, NPR, 26 October 2019. https://www.npr.org/2019/10/26/773585708/people-are-afraid-of-cinema-fran-ois-ozon-takes-on-church-sexual-abuse.

Peabody, Sue. "Madeleine's Children: Slaves from Isle Bourbon (present-day Réunion), Eighteenth and Nineteenth centuries." *Clio, Women, Gender, History* 45 (2017): 166–178.

Riddick, Ruth. "Clerical Sex Abuse in Spain." *Conscience*, 1 April 2019. https://consciencemag.org/2019/04/01/betrayal-global-story-local-voices-as-netflix-explores-clerical-sex-abuse-in-spain/.

Rose, Christine. "Reading Chaucer Reading Rape," *Representing Rape in Medieval and Early Modern Literature*. Edited by Christine Rose and Elizabeth Robertson, 21–60. New York: Palgrave, 2001.

Ruff, Julius R. *Violence in Early Modern Europe*. Cambridge: Cambridge University Press, 2001), 2.

Schwerhoff, Gerd. "Early Modern Violence and the Honour Code: From Social Integration to Social Distinction?" *Crime, Histoire & Sociétés/Crimes, History and Societies* 17 (2013): 27–46.

*Sex Abuse in the Church: Code of Silence*. Directed by Martin Boudot, 2017. https://www.ama zon.com/Sex-Abuse-Church-Code-Silence/dp/B0718SMTRY/ref=sr_1_1?crid=1L5GJELM DUYG9&dchild=1&keywords=sex+abuse+in+the+church+code+of+silence&qid= 1601989682&sprefix=church+codes+of+%2Caps%2C160&sr=8–1.

Stanley, Amy. "Writing the history of sexual assault in the age of #MeToo." *Perspectives on History*, 24 September 2018. https://www.historians.org/publications-and-directories/ perspectives-on-history/november-2018/writing-the-history-of-sexual-assault-in-the-age-of-metoo.

*Tell No One*. Directed and written by Tomasz Sekielski, aired 11 May 2019 on YouTube, https:// www.youtube.com/watch?v=BrUvQ3W3nV4.

Thompson, Anne. "Why Neftlix Bingeing Brought Fresh Power to the 50-Year-Old Mystery of 'The Keepers.'" *IndieWire*, 26 June 2017. https://www.indiewire.com/2017/06/netflix-the-keepers-documentary-sister-cathy-nun-expose-emmy-1201846839/.

Trouillot, Michel-Rolph. *Silencing the Past: Power and the Production of History*. Foreword by Hazel Carby. Boston: Beacon Press, 1995, 2015.

"Vassar Professor Speaks to an Aspect of Kavanaugh Hearings." WAMC, https://www.wamc. org/post/vassar-professor-speaks-aspect-kavanaugh-hearings.

Ward, Joseph P. *Violence, Politics, and Gender in Early Modern England*. New York: Palgrave Macmillan, 2008.

Leo Riegert, Jr.
# Blood, Boden and Belonging

## "The Cut-Across"

"My cousin told me what you said about me," snarled Bunny. On my daily walk from school to the Reservation Business Committee (RBC) building where my mother worked as the tribal accountant, I had risked the cut-across path. Bunny had not been in school that day, and now he blocked my way. The weeds around us released a pungent odor and were nearly as tall as we were. They blotted out the exit to the street and the American Indian Movement (AIM) protest banners hanging on the other side.

There was often no relationship between the nicknames of my classmates and their appearances, but Bunny's name fit. He was tiny, smaller even than I; rumor had it that he had stunted his growth sniffing glue. He stared up at me, maybe high, definitely defiant and angry. Not because I had, in fact, said anything about him, but because – as I think I sensed even then – of what the ancestors of people who looked like me had done to the ancestors of people who looked like him.

The day had begun a few hours earlier in my elementary school on the White Earth Indian Reservation in northwestern Minnesota. All of us had at least a slice of Native American ancestry. Or in my case, more like a sliver. Some of us arrived by school bus, from tidy farms or modest lake homes in the surrounding countryside. Others, like Bunny, walked to school from scrubby, government-built houses in the village of White Earth, the ramshackle hub of reservation life. In school, however, we were just a muddle of bodies and noise, under the cover of which my larger, stronger, and less brainy friends copied my homework, in exchange for future protection from the likes of Bunny.

Now though it was just the two of us, me and Bunny, the cover of friends and the school building far away. I was silent, knowing that fighting back would only make it worse next time when he brought along the aforementioned cousin, and anticipating what inevitably came next. Nonetheless, the violent intensity of Bunny's fist slamming my face stunned me, as it did every time. My nose tingled as blood began to flow, and I wondered: What had *I* done? It was what I got for trying to take the cut-across.

Getting beat up by Bunny on the cut-across path really did happen, and I had numerous similarly painful encounters and near-misses there during my elementary school years. However, re-reading this short piece, written a year or

https://doi.org/10.1515/9783110753295-032

so ago, I realize I embellished the account in at least two ways that are telling of my own preoccupations: I don't remember if there were, in fact, AIM banners hanging on the other end of the path, and I don't, in fact, have any idea what was going through Bunny's mind. Though clearly of Native American appearance in a school with mostly Native students, Bunny himself was a kind of outcast because of his size and the fact that he was always in trouble. That he was picking a fight with me was par for the course. Perhaps he was just being mean to someone small enough to pick on, or perhaps he was troubled in ways that had everything to do with his own home life and nothing to do with the color of my skin. And though there was indeed at one point an AIM encampment along my route from school to my mother's place of work, I don't know the exact time frame it was there, and I didn't know then exactly what AIM was. My speculation about Bunny's resentment of my whiteness and my mention of the AIM banners have more to do with my own insecurities then and now.

I did understand even then that the AIM protests were somehow directed toward the establishment tribal bureaucracy at the RBC, that my mother was a part of that establishment, and that the establishment did not seem to engage much with the protesters outside their door. My mother, Germaine (Bray) Riegert, is an enrolled member of the White Earth Band of Chippewa (Ojibwe) Indians,[1] but her skin color gives no indication of this. This led, at least in my memory, to resentments among some of her coworkers, an unwillingness to accept her as fully Native and a suspicion of her motives, despite years of successful fundraising for the tribe. Among those suspicious was Vernon Bellecourt, also an enrolled White Earth member and one of the co-founders of AIM in in the late 1960s. Despite AIM's militant campaign against the official government bureaucracy, including its occupation of the Bureau of Indian Affairs headquarters building in Washington, D.C. in November 1972, my memory is that Bellecourt also worked with my mother at the RBC, at least for a time. Bellecourt was tall, wore his long, black hair in braids, and dressed in jeans, cowboy boots, a bolo tie over a button-up shirt and a black leather vest. He had a voice that resounded through the building, and he never had much time for my mother, much less for the skinny "white" kid waiting around for her to finish her workday.

And so, these two disparate figures, one an undersized, impish child, the other an imposing, brash spokesman for a militant brand of Native American identity, represent types that stand in for my experience of going to school in

---

1 The White Earth Band is one of the six member bands that together comprise the Minnesota Chippewa Tribe (MCT). In addition to White Earth, the other bands/reservations include Bois Forte, Fond du Lac, Grand Portage, Leech Lake and Mille Lacs.

White Earth in the early 1970s. This was in part because they stood in such stark contrast to my own shy and deferential demeanor. But more so, I think, because, despite their differences, the two exhibited a combativeness and an unpredictability that characterize many of my memories of that place and time. They have come to represent an ever-present, underlying fear I felt then that the daily routines, which are necessary to any sense of childhood innocence and security, could and would be disrupted at any moment.

In his 2012 book, *Rez Life*, David Treuer writes about the violence that is a part of Indian life, of the positive role a "violent attitude" has played in the Ojibwe defense of their homelands, whether against the Sioux (Lakota and Dakota) from the Great Plains to the west, or against white settlers and the U.S. government that for hundreds of years were intent on stripping the Ojibwe not just of those territories, but of their culture. Treuer also discusses how that same violent attitude is today often directed not outward, in defense, but inward, "against home and place."[2] It is critical to avoid portraying White Earth (or any reservation) solely as destitute and moribund. Native American life flourishes on Indian reservations (and this is, in fact, the main point of Treuer's book), despite ongoing Indian-on-Indian violence, and despite the fact that the drug trade has made this violence much more deadly over the past twenty or so years. But one way to frame the issues here would be to ask whether the violence Bunny directed toward me was because I represented the historical outsider, with each beating I got a comically miniature replay of Little Bighorn, or whether I was a fellow victim of the infighting and domestic abuse that results when people are deprived of resources and turn their anger on one other. To put it reductively, was (and am) I a perpetrator or a victim? Am I white, or am I Indian?

Genealogical documents and ancestry charts put together by my mother's cousin for a family reunion about twenty-five years ago document my mother's family's lineal connection to a woman named O-ge-mau-gee-shi-go-quay, which translates as "Queen of the Skies," born about 1790 and belonging to the crane totem or clan.[3] O-ge-mau-gee-shi-go-quay's great grandchild was my maternal great grandmother, Eugenia (Bisson) Bray (1875–1957). Eugenia Bisson belonged to what was then called the Mississippi Band of Chippewa and moved

---

**2** David Treuer, *Rez Life: An Indian's Journey Through Reservation Life* (New York: Grove Press, 2012), 186–187.

**3** I am ever more grateful to my first cousin, once-removed, Diane Fleischer Preiser, who gathered these genealogical records, along with a tremendous amount of other material regarding the history of my maternal ancestors. Together with my mother, she published/printed them as *The Bray Bunch* in 1993. Most of the information regarding my family's story has been gleaned from this book.

as a young girl with her family from Crow Wing (near Brainerd, MN) to White Earth in 1868, shortly after the founding of the reservation.[4] Her descendants have lived there uninterrupted since. I grew up there until the age of fourteen, when my parents sent me to a Catholic boarding school in central Minnesota.[5] Nonetheless, my mother and four of my seven siblings still live on the reservation, three of my siblings now work for the tribe as had my mother before them, and I still consider it "home." Our foodways, humor, and family structures still very much reflect our Native heritage. At least in terms of genealogical, cultural, and historical ties to the place, it would seem that I belong there as much as Bunny or Bellecourt.

But perhaps this isn't quite so simple. In truth, our family has also always lived somewhat adjacent to reservation life and to Native American culture. First of all, unlike most Minnesota Ojibwe, Eugenia Bisson's family was not forced to move to the reservation, and likely did so because "métis," or "mixed bloods" – as she and her family were known even then – were allowed to receive land allotments at White Earth. The Dawes Act of 1887 had authorized the U.S. government to give allotments, or plots of land, to individual Native Americans living on Indian reservations, dividing lands previously held in common by the tribes. Intended as part of the "civilizing" process for Native Americans, in practice, allotment had little effect in changing the traditional way of life for most Native Americans. More insidiously, however, the allotment program resulted in the loss of huge tracts of tribal land.[6] David Treuer writes that "Native Americans lost more than 90 million acres of tribal land, about two-thirds of the lands held by Native Americans when the Dawes Act was passed; Indians lost, roughly, land that equals the size of the state of California."[7] The result is that most reservations in the United States are a patchwork of lands tribes have managed to maintain, reclaim or repurchase, interspersed between much larger tracts of property owned by non-Natives.

---

**4** The other major historic Minnesota bands include the Pillager, Pembina and Superior.

**5** As far as I am aware, for my parents this had little to do with my Native American background, but rather their hopes that I might become a priest.

**6** On the one hand, this was because Native people had little or no experience in land ownership and consequently lost their property when they were unable to pay mortgages taken out with the land as collateral, or when the land was confiscated because they could not pay illegally charged state property taxes. Most devasting in terms of overall land loss, however, was that land left over after allotment to Natives was put up for sale to non-Natives, including to settlers, railroads and mining and timber interests.

**7** Treuer, *Rez Life*, 134.

This is certainly the case on the White Earth Reservation, the largest in Minnesota, both in terms of land mass and number of members.[8] Located 68 miles east of Fargo and 225 miles northwest of Minneapolis/St. Paul, the original boundaries of the reservation encompass over one-thousand square miles. The White Earth Band today owns only about ten percent of this land; the vast majority of the remainder belongs to non-Indians.[9] State Highway 59 divides the reservation north-south between the fertile farmland of the Red River Valley to the west, an ancient lakebed where the topsoil is six feet deep in places, and the glacially-carved sand and clay hills to the east.[10] The land to the west of the highway is owned almost entirely by wealthy, non-Native grain farmers, their large single-family ramblers nestled in widely dispersed prairie tree groves. While a few wetlands remain scattered among the drained, sprawling fields of "the Valley," innumerable lakes and ponds dot the rolling hills to the East where most of the Indians live, their small homes clustered mainly in and around the reservation villages of White Earth, Nay-Tah-Waush, Rice Lake, and Pine Point.

But Eugenia (Bisson) and her husband, Peter Bray (1857–1948), farmed property that had been allotted to one of their daughters and that would lie today to the west of Highway 59. Their son and my grandfather, Kenneth Bray (1910–2001), and my grandmother, Julia (Simon) Bray (1916–1997), also raised their ten children on a farm, albeit just on the "Indian" side of the highway divide. Neither farm was large-scale like those belonging to today's white grain farmers, but neither was really Native either, because frankly there weren't many Native American farmers at

---

**8** Officially enrolled members of the White Earth Nation currently number roughly 19,000, although according to the 2000 United States census, only about three thousand live on the reservation itself. The majority live off-reservation, with a large population in Minneapolis. The total population of the reservation in the 2000 census was 9,192.

Perhaps the most famous native of White Earth today is Winona LaDuke, who ran for vice president as the nominee of the Green Party of the United States, on tickets headed by Ralph Nader in 1996 and 2000. Other well-known members include Peggy Flanagan, the current Lieutenant Governor of Minnesota, and Gerald Vizenor, a prominent scholar of Native American Studies perhaps best known for his coining and elaboration of the term "survivance."

**9** "Individual enrolled members do hold significant amounts of privately owned fee lands within the reservation and pay property tax to the counties in which the reservation is located." "History of White Earth." No author listed, accessed 12 October 2020, https://whiteearth.com/history. The source does not provide the exact amount of private property owned by tribal members, but I estimate it is far less than an additional ten percent of the original reservation acreage.

**10** In fact, the abundance of white clay in the area is what gives the place its name in both English and Ojibwemowin, the language of the Ojibwe.

White Earth, despite the best efforts of the United States government. My mother and my father, Leo Riegert, Sr. (1932–2018), a white German American with no Native American blood, raised their seven children on a small farm outside the reservation village of Waubun. In other words, my recent Native American ancestors and I have always lived physically somewhat outside the centers of Indian life on the reservation. In some sense, my family represents the "civilized" Native Americans that government policy was intended to produce. In addition, for my siblings and me, our skin color confers advantages that people with a more typically Indian phenotype most often do not have, even as we have the freedom to embrace aspects of our Native American heritage as we wish.

The question am I white or am I Indian? is also complicated by the issue of tribal membership. The MCT changed its membership requirements in 1961, before I was born, to require that future enrollees have, in addition to descendancy from someone on the Indian rolls,[11] a minimum one-quarter Native "blood quantum." Assuming that O-ge-mau-gee-shi-go-quay was a so-called "fullblood," and that no other Indian ancestry entered my gene pool through her husband or subsequently, six generations later my Native American "blood quantum" is 1/64th. I thus do not come close to meeting the blood quantum requirement and, of course, my mother would not have either.

The notion of "blood" as a constitutive of identity will remind many people – and certainly any scholar of German Jewish Studies – of the Nazis and, for those with more detailed knowledge, of their racist definition of Jews in the socalled Nuremberg Laws. But no matter what we may "know" about the construction of race, its power as a concept or ideology is hard to resist. Following – I would argue – colonialist thinking themselves, many enrollees conflate blood quantum and Native phenotype with identity and authenticity, and they resist any changes to membership laws.[12] They do so despite at least one reputable

---

**11** Between roughly 1885 and 1940, the United States Bureau of Indian Affairs undertook censuses to establish official lists of members of Native American tribes. Up to the present moment, descendancy from someone on these lists, known as "rolls," serves as the primary legal basis for formal membership in a Native American tribe that is officially recognized by the United States Government.

**12** The White Earth Band did attempt to adopt a new constitution – written primarily, in fact, by Gerald Vizenor – that would have defined tribal membership solely by descendancy with no regard for blood quantum. In a referendum in November 2013, 79% of tribal members who voted approved the new constitution. Although only about a fifth of total eligible tribal members voted in the referendum, the turnout was more than twice that of most tribal elections. However, for a complicated mix of reasons involving the MCT, the United States Department of the Interior, and newly-elected tribal council members opposed to the revised constitution, it has never taken effect.

study predicting that if the Minnesota Ojibwe do not change their membership requirements, the tribe will mostly disappear within 100 years.[13] This power of race also holds true on a personal level. The majority of my genealogical heritage is European, specifically German. When I told German friends that fraction, they were mildly disappointed, as they had always assumed that I was a "real Indian," and I admit that I myself often make ironic remarks about my Native heritage.

For both the majority of Americans and for myself, race and Native American identity are also complicated by the reified image of "the" Native American. As Gerald Vizenor has pointed out, the tragic, stoic Native, stuck in the past or in a museum, dark-skinned and likely wearing a Plains Indian headdress, is an invention of the non-Indian that bears little or no resemblance to Native Americans living today. The *"indian,"* as Vizenor calls him, presents an immense challenge to the recognition of both tribal specificity *and* diversity, and he remains the image against which living Native Americans are consistently measured, including by many Indians themselves.[14] I would argue that an additional difficulty is the symbolic implications of the indigeneity of Native Americans. Native peoples as the ur-occupants of the Americas calls for an innate authenticity that is perhaps most reflexively embodied in Vizenor's *indian*. On a personal level, the point I am trying to make is that though I may not know what prompted Bunny's aggression or Bellecourt's indifference, I was definitely aware then that I didn't look like they did, and I am aware of that to an even greater degree now. The symbolic power of the "authentic" Native American vis-à-vis my "white" appearance, along with the fact that I lack the stamp of official tribal membership, all contribute to my sense that I'm not "really" Indian in any way that matters.

Either you're in or you're out. Numerous colleagues and friends have encouraged me to speak with my students about my background growing up "on

---

**13** In 2012, the MCT contracted with Wilder Research to conduct a study and produce population projections for its six member bands. Leaving the tribal enrollment criteria as is (one-quarter blood quantum), Wilder Research found that projected MCT membership declines from just over 41,000 in 2013 to just under 9,000 in 2098. During that same 85-year period, White Earth's enrolled citizenship is projected to decrease from around 20,000 to just over 2,000.

**14** Lower-case and italics in the original. Vizenor has proposed the "postindian" as a term which represents the diversity and energy of Indian life today. "The point," says Vizenor, "is that we are long past the colonial invention of the Indian. We come after the invention and we are the postindians . . . . The postindian stands for an active, ironic resistance to dominance, and the good energy of native survivance." In Gerald R. Vizenor, *Postindian Conversations* (Lincoln: University of Nebraska Press, 2003), 84–85.

the rez." I think the idea is that by making students aware of the complexities of who I am, by showing there is more to me than a white male professor of German from someplace up near Canada, it will help them embrace the complexities of their own identities. Or perhaps the idea is that the students, and especially those from disadvantaged backgrounds, will be inspired by the story of someone who grew up in lower-middle-class circumstances on an Indian reservation in the poorest county in Minnesota, but "made good" by getting tenure at a highly selective liberal arts college. For a variety of reasons, not the least of which are the personal insecurities discussed above, I do not discuss these things with my students. (In fact, this text represents the first time I discuss my Native American heritage in print.) And if I were to speak with them, I would prefer to do so in a way that is informed by my research and writing on pre-Holocaust German Jews and that emphasizes and embraces the hybrid position I occupy between two cultures – one Native American and indigenous, one German Catholic and immigrant.

But there seems to be little space for that complexity in 2020. The killing of George Floyd has drawn unprecedented attention to the long history of violence against people of color in the United States, particularly African Americans, to the structures that support ongoing violence and oppression, and to the ways white Americans are implicated in and benefit from those structures. As a scholar of German Jews and of the Holocaust, I resist the term "white supremacy" as a descriptor of those structures, because I believe it covers up at least as much as it reveals in terms of the history of racism and the complexity of racial politics. Nonetheless, the attention to skin color and its literally life-and-death implications places me at this moment squarely on the side of the oppressor. As a white man, at least by appearance, I take seriously the call for me to "sit with my privilege" and to allow more clearly racially marked others to speak about their experiences. I hesitate even to write this much about the current moment, despite the fact that remaining silent about my own experience, at least for the moment, also feels to me like denying a part of who I am, and even as I know that, according to the United States Centers for Disease Control, Native Americans are killed in encounters with police at a rate higher than any other racial or ethnic group, including African Americans. Because can I really speak as a Native American or for them? Isn't it better to leave the speaking to perhaps more authentic voices that have been silenced in the past? The moment demands clear distinctions. Either you're in or you're out. Adjacency is not, it seems, a viable position from which to speak about the exigencies of the current moment.

Jill Doerfler, Associate Professor of American Indian Studies at the University of Minnesota-Duluth, begins to get at a way around blood quantum as a measure of authenticity, arguing persuasively that blood cannot be the "final measure" of Indian identity.[15] She shows how the two issues around which my essay swirls, blood and land, are intimately connected, not because Native Americans have some autochthonous connection to particular pieces of property, but because the two were linked in a deliberate strategy by the United States government to deprive the Ojibwe of thousands of acres of land.[16] Like me, Doerfler is Ojibwe from White Earth and grew up there, but is not enrolled because of insufficient blood quantum. Her writing is motivated by an at times explicit belief in the injustice of this fact and a desire to achieve the official recognition of what she calls "citizenship." And as I write and think about the complexities in the previous pages, I can understand why – it would indeed make things simpler.

But even if I am not as interested in defining who I am once and for all, Doerfler opens up alternative possibilities for thinking about my Native American heritage. She argues that "[t]he Anishinaabeg of White Earth negotiated multifaceted identities both before and after the introduction of blood quantum as a marker of identity and as the sole requirement for tribal citizenship."[17] She describes a complex system of familial connection and continuation that was disconnected from pseudoscientific ideas of blood and race. The Anishinaabeg of White Earth determined who "belonged" based on these familial relationships, in combination with "a variety of lifestyle choices, such as clothing and religious affiliation, as well as economic conditions."[18] Their understandings of identity remained malleable throughout the allotment period, and they resisted federal government attempts to tie their identities to a fixed concept of blood as representative of race.

---

**15** Jill Doerfler, "Postindian Survivance: Gerald Vizenor and Kimberly Blaeser," in *Gerald Vizenor: Texts and Contexts*, eds. Deborah L. Madsen and A. Robert Lee (Albuquerque: University of New Mexico Press, 2010), 189.

**16** The argument is relatively complex. In brief, however, it involved arbitrary, eugenics-based determinations by government officials of who among White Earth Ojibwe were to be considered "mixed-bloods" or "full-bloods," as well as related issues of who was considered "competent" to control their allotments.

**17** Doerfler, *Those Who Belong*, xxiii.

"Anishinaabe(g)" is yet another term used to describe the Ojibwe. I have seen different meanings, but the most common is "the people" or what Doerfler calls the "original people" (xxiii). It is more of an umbrella term referring to a number of tribes or their languages, including the Ojibwe, Odawa, and Potawatomi.

**18** Doerfler, *Those Who Belong*, 29.

I do not know whether other White Earth natives considered my "mixed-blood" ancestors, Eugenia and her family, as belonging to the tribe when they first moved to the reservation, before the establishment of the Indian rolls, or whether even then they were somehow adjacent, even in the eyes of others. But it is possible the question as such never occurred to them or their neighbors. Doerfler shows that while the Ojibwe did adopt and use the terms "mixed-blood" and "full-blood," these designations were based on gradations of the aforementioned markers, not on a racial understanding of blood, and that both groups were considered as belonging to the tribe. As I think this essay makes clear, my own self-understanding as being somehow Native American is based primarily on familial connections. Despite the issues I raise in this reflection, and though we don't often directly talk about it, there is no doubt within my immediate family or among my maternal relatives in general that we are Indian to and for one another.

I would like to close by returning to the issue of land, specifically my connections to White Earth as a place, especially considering that I have not lived there full-time now for forty years. Does that physical distance attenuate my connection to my Native American identity even further, or might there be other ways to consider that connection? In Jewish Studies, for instance, alongside narratives of the "Promised Land," there is a long textual, including scriptural, connection to diaspora as itself also a condition of Jewish collective and individual experience. Jonathan and Daniel Boyarin even argue that diaspora is "paradigmatic of a certain cultural condition in the post-colonial era *tout court*" (x)?[19] But the textual/theological basis for diaspora, its association with divine disruption or justice, is not to my knowledge a part of the Native American tradition. If, as Lilian Friedberg argues, "Native Americans are not, in the strictest sense of the word, a 'diasporic' people,"[20] if diaspora does not provide a model

---

**19** Jonathan and Daniel Boyarin, eds. *Jews and Other Differences: The New Jewish Cultural Studies* (Minneapolis: University of Minnesota Press, 1997), x.

The Boyarins pit their argument for a diasporic Jewish identity against a dangerous ideology that combines race and space and that is used occasionally to legitimize the existence of Israel (or any other nation state, for that matter).

**20** Lilian Friedberg, "Dare to Compare: Americanizing the Holocaust," *American Indian Quarterly* 24, no. 3 (Summer 2000): 371.

Friedberg writes, "The land, 'the Wilderness' or 'the Desert' which has come to signify a 'wasteland' in the symbolic and spiritual orders of other peoples, has never been associated with anything but abundance and eternal sustenance for indigenous peoples because revelation is rooted in the life of reflection on and with the land, not in catastrophic upheaval or divine intervention" (371).

for thinking about my connection to White Earth, are there other alternatives? What is the "cultural condition" of Native Americans in the post-colonial era, if it is not diasporic?

Doerfler hints at another possibility when she writes that "Anishinaabe identity could not be surveyed and divided the way the land had. There were no clear lines and neat allotments; rather, it was a varied and diverse landscape with a mix of lakes, birch, pine and prairie."[21] The implication, I think, is that "land" for Native Americans is in some ways as malleable and shifting as identity. At its core, it does not mean a specific tract of acreage or piece of property, but rather a "landscape" that is itself moving, changeable and in flux. The White Earth Reservation has come to stand in for a much larger landscape throughout the Great Lakes region that is the ancestral home of the Anishinaabe. It is, like most reservations, not a place dispensed to Native Americans by the United States government as some kind of compensation for taking the majority of their homelands, but a place the Indians actively "reserved" for themselves in negotiations as sovereign nations with the United States. It serves as a center of life for the White Earth Ojibwe, even though the vast majority of the members of the White Earth Nation do not actually live on the reservation (see footnote 8) and are dispersed across the United States. When I think of the reservation as my "home," I don't think immediately of the house or property where I grew up, but rather precisely of the "lakes, birch, pine and prairie" that make up the reservation, and that I miss here in Ohio. That decentered center draws me back as often as I can go.

# Works Cited

Boyarin, Jonathan, and Daniel, eds. *Jews and Other Differences: The New Jewish Cultural Studies*. Minneapolis: University of Minnesota Press, 1997.

Doerfler, Jill. "Postindian Survivance: Gerald Vizenor and Kimberly Blaeser." In *Gerald Vizenor: Texts and Contexts*. Edited by Deborah L. Madsen and A. Robert Lee, 186–207. Albuquerque: University of New Mexico Press, 2010.

Doerfler, Jill. *Those Who Belong: Identity, Family, Blood and Citizenship Among the White Earth Anishinaabeg*. East Lansing: Michigan State University Press, 2015.

Friedberg, Lilian. "Dare to Compare: Americanizing the Holocaust." *American Indian Quarterly* 24.3 (Summer 2000): 353–380.

"History of White Earth." No author listed. Accessed 12 October 2020. https://whiteearth.com/history.

---

**21** Doerfler, *Those Who Belong*, 28.

Preiser, Diane Fleischer, and Germaine Bray Riegert. *The Bray Bunch: The Ancestors and Descendants of Peter and Eugenia Bray*. No publisher listed, LOC Catalog Card Number 93-85456, 1993.

Treuer, David. *Rez Life: An Indian's Journey Through Reservation Life*. New York: Grove Press, 2012.

Vizenor, Gerald R. *Postindian Conversations*. Lincoln: University of Nebraska Press, 2003.

Alicia E. Ellis
# Upended

I didn't know that I was born in a third-world country until I was in college and a history professor informed me of that detail. No one in my family had ever told me that Barbados was one of those countries – "island nations" as they are called – that we must feel bad about and send money to along with clothing that was no longer desired. I knew that we were a former colony, the birthplace of British slavery in the Americas in the early seventeenth century.[1] We are called "Little England." It is not a compliment to my mind. It seems aspirational. As if Barbados, itself a small island, wanted to be the adjunct, adjacent to England, that other small island. Barbados now belongs to the commonwealth. We are sovereign, but still bound to the realm. It was only in September 2020 that Barbados decided to become an independent republic. The Prime Minister of Barbados said, "The time has come to fully leave our colonial past behind."[2]

> Wake: the track left on the water's surface by a ship; the disturbance caused by a body swimming or moved, in water; it is the air currents behind a body in flight; a region of disturbed flow.[3]

Rum and tourists keep us afloat. They are the waves we ride round the island (Fig. 1). They are the buoys that sustain us centuries after the end of the slave trade and of slavery. The story of sugarcane and of Black bodies is at the core of the narrative of Barbados. The Black bodies that were bought and traded to cut cane and distill it into sugar and rum for the world's consumption. This extraction is not only about spirits, but it is also about those spirits, those laboring bodies with their worn hands, stiff backs, and knotted feet. Spirits inhabit the floating seascape. That maritime space that is the graveyard of men, women, and children who died so that the world could consume. I also deal in spirits: I

---

**1** Hilary Beckles, *The First Black Slave Society: Britain's "Barbarity Time" in Barbados, 1636–1876* (Kingston, Jamaica: The University of the West Indies Press, 2016) offers a thorough and compelling exploration of that history. Most narrative accounts of Barbados are only to be found on the internet and he was one of the first to chronicle them in Hilary Beckles, *A History of Barbados: From Amerindian Settlement to Caribbean Single Market* (Cambridge: Cambridge University Press, 2006).
**2** Michael Daventry, "Barbados Announces Plans to Remove Elizabeth II and Become a Republic," *Euronews*, 16 September 2020, https://www.euronews.com/2020/09/16/barbados-plans-to-become-a-republic-and-drop-queen-elizabeth-ii-as-head-of-state.
**3** Christina Elizabeth Sharpe, *In the Wake: on Blackness and Being* (Durham: Duke University Press, 2016), 3.

https://doi.org/10.1515/9783110753295-033

**Figure 1:** Silver Sands Beach in Christ Church, 2005. Image in private collection of author.

sometimes put sugar in my tea. I will sometimes have a rum and coke. Sometimes I sucked on sugarcane when I was sitting on the stoop of my grandmother's house. I, too, consume my past, my grandmothers and grandfathers and their children and their children. I am a tourist when I go to Barbados – a consumer of sorts. I am called "Yankee" when there. A visitor. I am not staying. I have only come to see what was once home for just a few years of my life. Is this truly home? Can I ever claim it as such, or will Barbados be the one who will have to claim me as one of its own – as a disturbance or as a region of disturbed flow?

It was March 2005 when I first saw the synagogue in Bridgetown, the capital of Barbados. I was with my niece who was six years old at that time. I was still in residence at Yale not seriously working on the prospectus for my dissertation with the urgency it deserved. Our spring breaks coincided that year. And so, with Schiller's *Wallensteins Tod* by my side and my niece on the other, we flew to Barbados. The Kahal Kadosh Nidhe Israel ("Dispersal of Israel") Synagogue was built in 1654 by a group of three hundred Jews who had fled Brazil (Fig. 2).[4] That first organized Sephardic Jewish community in Dutch Brazil was established

---

4 An interview with a Bajan professor on the website of the *American Jewish Historical Society* reveals that were five hundred to nine hundred Jewish people by the mid-eighteenth century. Laura Leibman, "Rediscovering an Important Link to American Jewish History: Field Notes from the Nidhe Israel Synagogue Complex in Barbados," https://ajhs.org/rediscovering-important-link-american-jewish-history.

**Figure 2:** Nidhe Israel Synagogue (Daniel Berek; used with permission of the photographer).

in the city of Recife in the north-east. The relative peace that they enjoyed changed – even though they had to be inconspicuous about their religious practices under the Dutch – when the Portuguese reconquered Brazil in 1654. Fearing the reenactment of the Inquisition, the Jews of Recife either returned to Holland or fled to Dutch, French, or English colonies in the Caribbean. Barbados was one of those colonies. In *The History of the Marranos*,[5] Cecil Roth writes that "on the island of Barbados, the oldest English colony, the earliest known [Jewish] arrivals were refugees from Brazil . . . In the following year, on August 12, 1656, the Jews were granted full enjoyment of the 'privileges of Laws and Statutes of ye Commonwealth of England.'"[6] What were these privileges? Did they involve sugar and slaves? My search gave me no clear answers. I read about the same events from different sources with competing analyses, and to my growing dismay, I grow impatient enough with what I don't learn that I turn away from my own question.

The Jewish community in Barbados thrived until 1831 when a hurricane devastated the island. By 1848, the Jewish population had declined to only seventy Jews. By 1925, none were left on the island. During the Holocaust, thirty Jewish

---

5 Cecil Roth, *The History of the Marranos* (London: Routledge, 1932).
6 Roth, *The History of the Marranos*, 112–113.

families arrived in Barbados from Eastern Europe, and they were soon followed by several more from Trinidad. By 1968, the Jewish community had begun to rise and had a population of about eighty Jews. In 1987, the synagogue was rededicated in a new location and the Old Jewish cemetery in Bridgetown was restored. The synagogue, one of the oldest such buildings in this hemisphere, features a dark wood ark, European chandeliers, a stained-glass window etched with the Star of David, and black and white marble floors. The former Nidhe Israel building is today used as a library and a museum and is a UNESCO Heritage Site (Fig. 3).

**Figure 3:** Nidhe Israel Synagogue (David G. Molyneaux / TheTravelMavens.com; used with perission of the photographer).

Another significant arrival in Barbados was that of Olaudah Equiano who endured the Middle Passage in the eighteenth century. He was born circa 1745 and died on 31 March 1797 in London. He has no authoritative birthday. Was 1745 the year that he was enslaved as a child? Was it the day that he arrived in the Americas? What does it mean for Equiano to be a "circa"? To only have a beginning that is about or around the time he became unfree? Was it during a sea voyage or even on the auction block where his circa began? His autobiography, *The Interesting Narrative of the Life of Olaudah Equiano, or Gustavas Vassa, the African. Written by Himself* (1789),[7]

---

7 Olaudah Equiano, *The Interesting Narrative of the Life of Olaudah Equiano, or Gustavas Vassa, the African. Written by Himself* (New York: W. W. Norton & Company, 2000).

is a call for abolition and a discursive performance. In the autobiography, Equiano details his survival of the transatlantic slave trade, the traffic in humans as commodities, his arrival in Barbados, his eventual sale to the North American colony of Virginia, and his freedom. He does this with a perfectly executed consideration for his public. Equiano was aware of the heightened stakes of his project – the abolishment of slavery. Equiano's opening gesture in the autobiography is a dedication, an appeal to his readership about the veracity of the document to follow:

> To the Lords Spiritual and
> Temporal, and the Commons
> of the Parliament of Great Britain.
> *My Lords and Gentlemen,*
>
> Permit me, with the greatest deference and respect, to lay at your feet the following genuine Narrative; the chief design of which is to excite in your august assemblies a sense of compassion for the miseries which the Slave-Trade has entailed on my unfortunate countrymen. By the horrors of that trade was I first torn away from all the tender connexions that were naturally dear to my heart; but these, through the mysterious ways of Providence, I ought to regard as infinitely more than compensated by the introduction I have thence obtained to the knowledge of the Christian religion, and of a nation which, by its liberal sentiments, its humanity, the glorious freedom of its government, and its proficiency in arts and sciences, has exalted the dignity of human nature.
>
> I am sensible I ought to entreat your pardon for addressing to you a work so wholly devoid of literary merit; but, as the production of an unlettered African, who is actuated by the hope of becoming an instrument towards the relief of his suffering countrymen, I trust that *such a man*, pleading in *such a cause*, will be acquitted of boldness and presumption.
>
> May the God of heaven inspire your hearts with peculiar benevolence on that important day when the question of Abolition is to be discussed, when thousands, in consequence of your Determination, are to look for Happiness or Misery!
>
> I am,
> MY LORDS AND GENTLEMEN,
> Your most obedient,
> And devoted humble Servant,
>
> OLAUDAH EQUIANO,<br>OR<br>GUSTAVUS VASSA.
>
> Union-Street, Mary-le-bone,
> March 24, 1789.[8]

---

**8** Equiano, *The Interesting Narrative*, 7.

Addressing the many men in the kingdom of this world and of heaven, Equiano's task is to influence and secure freedom for the enslaved. No longer a slave and having settled in London, Equiano's two-volume autobiography is indeed a documentation of great literary merit, for in it, his approach is persuasive and compelling as he works towards the alleviation of the suffering of slaves and the dissolution of slavery. His conciliatory words to his readership – their Christianity, liberalism, humanity, government structure, and aptitude in the arts and sciences – and his confession that he is merely an uneducated African contribute to the flawless articulation of his abolitionist intentions. However, he must indulge his audience with this act of self-abnegation in order to be considered an authentic source. Equiano's conversion to Christianity and display of affection for England signal a man in full possession of the performative. He understands the narrative maneuvers that must be manipulated and shifted in order to have his story move an audience to outrage and even to action. The lack of "literary merit" that Equiano professes is a necessary concession made in order to demonstrate humility and the insignificance of his words. His declaration of the uncultured nature of those trivial and thus harmless words allows his story to be published and read by the public.

The need to authenticate the veracity of any Black story was an accepted feature of every life account of a Black figure. The endorsement of a white, usually male, authenticator is a necessary characteristic of the slave narrative or any document, prose or poetry, composed by Black people who endeavor to express themselves through writing. The works of Phillis Wheatley (*Poems on Various Subjects, Religious and Moral*, 1773)[9] and Harriet Jacobs (*Incidents in the*

---

**9** Wheatley's collection begins in much the same way that Equiano introduces his work. Wheatley writes:

> THE following POEMS were written originally for the Amusement of the Author, as they were the Products of her leisure Moments. She had no Intention ever to have published them; nor would they now have made their Appearance, but at the Importunity of many of her best, and most generous Friends; to whom she considers herself, as under the greatest Obligations. As her Attempts in Poetry are now sent into the World, it is hoped the Critic will not severely censure their Defects; and we presume they have too much Merit to be cast aside with Contempt, as worthless and trifling Effusions. As to the Disadvantages she has laboured under, with Regard to Learning, nothing needs to be offered, as her Master's Letter in the following Page will sufficiently show the Difficulties in this Respect she had to encounter. With all their Imperfections, the Poems are now humbly submitted to the Perusal of the Public.

Phillis Wheatley, *Poems on Various Subjects, Religious and Moral* (New York: Penguin Classics, 2001), 5–6.

*Life of a Slave Girl*, 1861)[10] are just two examples of texts that must show their truth before the telling of their enslavement. Authenticity must be proven before this act of self-representation can be believed and known to be true. Equiano's history is no different. *Written by Himself.*

Equiano captures my attention. I read his autobiography in high school but did not take in its extraordinary nature. Now I know to do so. In chapter two of his autobiography Equiano tells of his first experience of Barbados:

> The clouds appeared to me to be land, which disappeared as they passed along. This heightened my wonder; and I was now more persuaded than ever that I was in another world, and that every thing about me was magic. At last we came in sight of the island of Barbadoes, at which the whites on board gave a great shout, and made many signs of joy to us. We did not know what to think of this; but as the vessel drew nearer we plainly saw the harbour, and other ships of different kinds and sizes; and we soon anchored amongst them off Bridge Town.[11]

Equiano's first glimpse of Barbados from the sea was enchanted. I want to know more about that magic that he observes, but I don't have access to it. I want to know how it is that immediately before this sense of wonder at his arrival in Barbados, Equiano can write about the misery of the slave hold. "Many a time we were near suffocation from the want of fresh air, which we were often without for whole days together."[12] Immediately following this expression of amazement, Equiano's story returns to the truth of slavery, and we are again brought back to his reality that is neither wondrous, magical, nor joyous. He is in another world. One that was built and organized to hold him captive. The anchoring on the harbor is an instance of what Sharpe would call being in the *wake*, a body moved in the water.

When you search for Barbados on the internet, you are most likely to see results related to tourism: its rum, sugarcane, music, cricket, and beautiful beaches – the sea (Fig. 4). If you dig a bit deeper, you'll find slaves. If you want to go subterranean, you'll discover that Barbados was England's first slave colony in the Americas in 1627. If you read the BBC website and their story of Barbados, you will know that:

> Barbados in many respects was England's first experimental tropical agricultural export colony and was successful for a number of related reasons. . . attempts at alternate crops such as indigo and ginger seemed doomed to failure, international affairs conspired to

---

**10** Jacobs, Harriet. *Incidents in the Life of a Slave Girl* (New York: W. W. Norton & Company, 2018).

**11** Equiano, *The Interesting Narrative*, 42.

**12** Equiano, *The Interesting Narrative*, 42.

**Figure 4:** Road with sugarcane, 2005. Image in private collection of author.

> create an economic opening which guaranteed the survival and prosperity of Barbados. The Dutch in north-east Brazil and their allied community of Sephardic Jews were expelled from Recife and Bahia.[13]

The story that the BBC tells is layered. It is encrusted with "Imported manpower" that initially places white indentured servants or prisoners at the center of the economy. With great benevolence, almost without admitting to anything, the BBC tells us of this experiment:

> However, as the cost of white labour in England went up, planters, on the advice of Dutch and Sephardic merchants, turned to West Africa for their source of manpower. Black slaves were imported in large numbers from the Gold Coast region in particular, especially from what is today the country of Ghana. The Asante, Ewe, Fon and Fante peoples provided the *bulk of imports* [emphasis mine] into Barbados. Nigeria also provided slaves for Barbados, the Yoruba, Efik, Igbo and Ibibio being the main ethnic groups targeted.[14]

The specificity of the names and locations of the tribes, the attempt to call it manpower rather than slavery, and the near divorce of the word Black from the word slave as if it was not an established fact that, in this modern era in the Americas, slaves were always Black are appalling. Yet the website tells on itself

---

13 Karl Watson, "Slavery and Economy in Barbados," 2 February 2017 https://www.bbc.co.uk/history/british/empire_seapower/barbados_01.shtml.
14 Watson, "Slavery and Economy in Barbados."

by invoking the well-known place for the purchase of Africans on the continent, the Gold Coast. It knows more than it will freely tell. The Gold Coast is also where I was brought up. It is that part of lower Fairfield County in southwestern Connecticut where enormous wealth is to be found. I often wonder if any of that abundant wealth is derived from the labor of my ancestors in Barbados? Sharpe writes that "The history of capital is inextricable from the history of Atlantic chattel slavery. The disaster and the writing of disaster are never present, are always present."[15] My ancestors came as slaves from the African Gold Coast, and I was raised as a relatively privileged girl on the American Gold Coast. What, if anything, could it change how I think of the word "disaster?" Pierre Nora's concept of "les lieux de mémoire" reminds me that the African Gold Coast is a site of collective memory for Black people – even if it is not recognized as such. I cannot escape its hold on my imagination or my life.

My grandmother raised me for the first five years of my life (Fig. 5). Living with her was boisterous and exciting. Now a vivid silence surrounds that time, even as I recall the joy and safety that shaped my early life with her. It was bright, calming, and beautiful. I imagine sometimes that she still waits for me. She is backlit by the warm sun as I walk toward her, protecting a new insect in my hands – preparing it for her inspection. It was a flash in my life that can never be regained. I didn't stay long enough to appreciate more of her. My

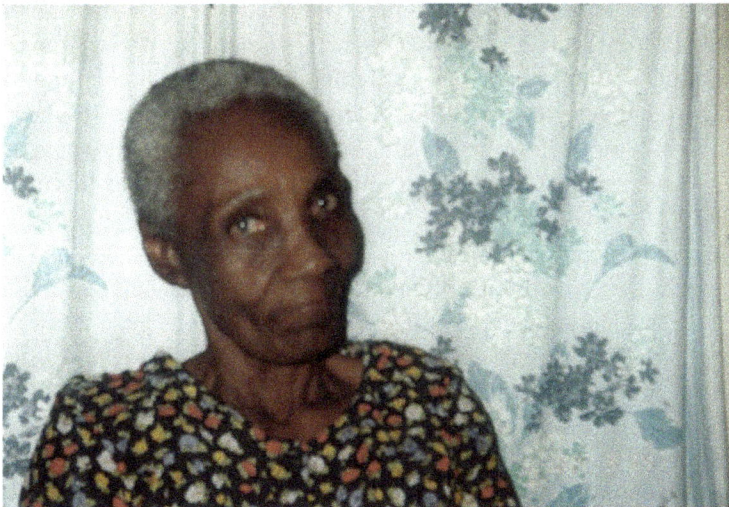

**Figure 5:** Loleta Bowen, ca. 2007. Photographer unknown. Image in private collection of author.

---

**15** Sharpe, *In the Wake*, 5.

grandmother was a tradition. She offered the ethics and values that we all followed with dedication. She was a religious woman who attended church several
times a week and cleaned it after services. She sang in the choir. I would often
walk behind her on the way to church just down the road so that I could watch
other people also on their way to services engage her in conversation, ask for her
advice, tell her their stories. Sharp-tongued with humor and insight, she had a
way of seeing that did the proper work of turning a tragedy into an experience.

I returned to Maine from her funeral in Barbados (how far life takes us).
That was in January 2020 just as COVID-19 hit us all as a global reality. She died
on Christmas Eve in the parish of Christ Church, in an area called Silver Sands. I
extended my time in Barbados after her funeral so that I could discover more
about her, my childhood, and what those things were that I didn't know but
knew had to be there. I wanted to understand more about where I came from
and why my own mother did not take care of me during those formative years.
The death of my grandmother, Loleta Bowen, did not deter me from disturbing
the peace. Asking uncomfortable and indiscreet questions during a time of
mourning. But when would I ever have that chance again? People forget or
claim that they forget. No legible family archive existed. The only things that I
could depend on were what people were willing to tell me. There were few family artifacts. No family bible with the names of relatives and records of their
births and deaths. Very few photos and what was there faded and warped. Few
letters and no diaries or newspaper clippings. Old photos hung on the wall of
my grandmother's front room of people whose names I didn't know. The frames
that contained them hung precariously skewed on the walls, and other photos
had been wedged between the glass and the frame of still other people whose
names I also didn't know. How did I belong to them? What were they to me?
When I ask who the woman with the long white skirt was or who was this man
with the thick glasses, I so often only got "That might have been your great-
auntie, Marva, who died in the hurricane or that was so and so's uncle from St.
Lucia." We were a family without a significant trace – so many circas. There
was no one to tell me stories about my life before I was born. The hush was permanent. There might have never even been a hush, a quick direction to be
quiet, to conceal. Maybe we just never knew. But I know that if I had had the
opportunity to ask, my grandmother would have given me stories.

When I moved to the United States as an almost six-year-old, my conceptual
framework built around the relationship with my grandmother disappeared. I
knew my world, and it was the one that she had built. I knew where to go if I
wanted the best guavas and mangos. I knew not to talk to strange men. I knew
that my grandmother would pick me up from pre-school for lunch, take me home,
and feed me Campbell's vegetarian vegetable soup for lunch, soup that my parents

sent from America to Barbados in a barrel with other non-perishable food, clothes, and the occasional toy. I knew that every morning I would run off from my grand-mother in my underwear to avoid taking my medicine. There would be a fierce chase around the house as she tried to get me to take castor oil that, she insisted, fought off every illness imaginable. It would help me grow, she would say. She took great care. Where was my mother (Fig. 6)? Did she not want me and my two older brothers? I was told that my parents had left Barbados for "opportunity" in the United States. They wanted to make a new life for me and my brothers and thought that the United States would provide that opportunity. Did they appreciate what they were leaving behind? These three children – Carl, Mark, and Alicia? What undercurrents were created in these three small lives when our mother along with our father took off on their flight in search of opportunity? My two older sib-lings have their own stories of this departure. I leave it for them to tell.

Figure 6: My Mother, Photographer, Date Unknown. Image in private collection of author.

In 1982, my brothers and I were brought to the United States rather unceremoni-ously. I recall it with an exactitude that feels like a pin stabbing at the palm of my hand or a faded photograph that I can't find but I know its features down to the look of misery in my oldest brother, a misery that never truly left his eyes. My father came to get us. I didn't know who he was. When he called my name, I ran away from him into my grandmother's house. My brothers had memories of him but to me he was just a strange man. He yelled at me to come to him, and my grandmother told him to leave me alone. I would come around. I suppose I came around. There was no choice. The three of us wore matching polyester outfits on the day we left Barbados on our Pam-Am flight – human cargo on that plane. My brothers looked extremely uncomfortable and disgruntled in their khaki suits, and I wore a khaki skirt with suspenders and a white shirt underneath. Someone had made them for us. Someone took the time to sew these clothes for us. We looked as if we were going to school or to a funeral.

I cannot trace my maternal family's lineage back one hundred years. I am married to a man whose family helped to found the Hartford Colony in 1630. I used to tell a story with an indiscriminate glee about my arrival in America. I had flown into JFK from Barbados. It was Christmas. I knew this because there were so many colorful lights – blues, reds, and greens – and it was cold. Frigid. A warm jacket and a hat were waiting for me when I made it through customs and immigration with my brothers, my father, and my green card. I was bombarded with gifts, and my mother was so happy to hold me because it was Christmas and affection was saved for special times. It wasn't until I was sixteen years old and flipped through my expired passport that I saw the truth, something different than what I had previously known. My entry into America was actually April 1982. It wasn't Christmas at all. The evidence was there, and it had never occurred to me to search for anything else. I was certain of this one thing. My expired passport had ruined my dependable narrative of arrival. Yet I still refuse this story. How do I make sense of my American narrative without giving up my original creation myth? The disruption of my self-perception made me long for stable meanings after discovering that I had invented a past. How can I make myself legible in the future if I couldn't identify something so crucial about myself from the past? The negotiation of this in-betweenness – my artificial story and the real one –put me at odds with my own reflexivity. I needed a usable past and had created an origin story that gave me a certain permanence, marks on the page that could not be erased. An official stamp at the back of the passport that indicated that I had arrived. I had emigrated from Barbados in my polyester suit and it was Christmas and my mother gave me her awkward love. I bring myself back to Sharpe with the air currents behind my body in flight and I was there.

I am a product of two Black diasporas: the transatlantic slave trade from Africa to the Americas in the 17th century and the Caribbean immigration to America in the late twentieth century (Fig. 7). I am also a bicentennial baby. Born

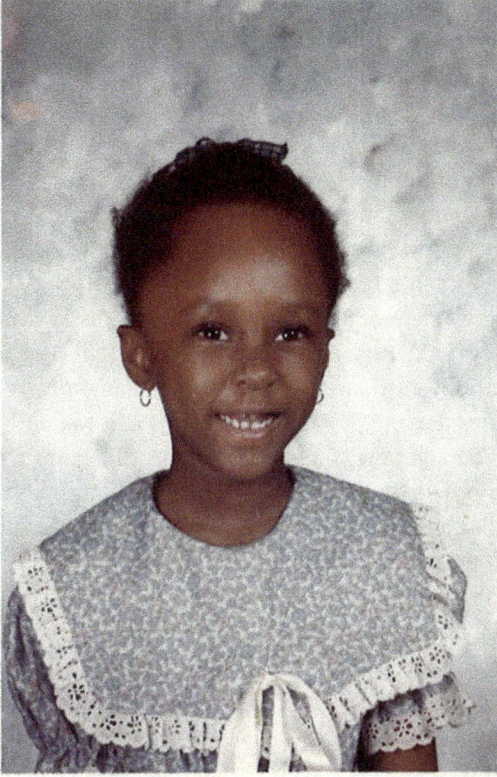

**Figure 7:** Alicia Ellis, 1982. Photographer unknown. Image in private collection of author.

in 1976, I am a two-hundred-year-old replica of free and independent. Stable and now rare like a commemorative coin. I am a celebration of the creation of the United States. A facsimile. I was a slave; I was owned. When I first arrived in America, they asked me at school if they spoke English where I came from. I was confused and wanted to get away from this question. I said, "Sort of." I had this accent that wasn't right, and my words were also incorrect. Instead of saying the number "three" like everyone else in my 1st grade class did, I would say "tree." I was briefly placed in an ESL classroom. My mother fixed that very quickly. I was given psychological tests and my parents were told that I was gifted. I was left in peace. My tongue didn't know how to be an American yet. My small body didn't know how to be warm. My mind didn't know how to say "mum" and "daddy" without the hesitation of newness. I didn't know that I was Black when I came to America. I should say rather that I didn't understand that I was Black in America.

I was young when I left home. So young that when I am inevitably asked that question: "Where are you from? Where is your home?" I have to make a quick decision about how I want to respond. I can offer several answers based on a few simple questions that I quickly ask myself: Do I like you? Can I trust you? Will you understand? I am from New England. I am from Connecticut. I am from Barbados. I am from the continent of Africa. No. I don't know which country in Africa or which tribe I came from. I was an entry in a logbook. I am here because of a ship, because of profit margins, because someone needed a body to labor. I am diaspora. My parents know what "Home" is. For them, it will always be Barbados. When they meet other Bajans, the question is always asked, "When was the last time you were Home?" or "Did you know that Galvin got sick when he left New York and went Home?" There is no confusion there. For them, Home is always Barbados. For me, it is a moving target in lower case. It is context. It is my mood. It is you and how you have asked me. Idle curiosity or the more penetrating need to pin me to a spot on the globe or your imaginary? To uncover the secrets of the cadence of my voice, my private school education, my ivy league "pedigree," my white husband, and my career?

I didn't find any answers in Barbados that January. I went to my grandmother's wake – the viewing, the visitation. What Sharpe describes as the work of mourning is akin to those final moments with my grandmother before her funeral. "*Wake: grief, celebration, memory, and those among the living who, through ritual, mourn their passing and celebrate their life . . .*"[16] I looked at her brown face with its shrunken cheeks and felt a fierce love, a love akin to the love that she gave me with that castor oil almost 40 years earlier. I fixed the small flower attached to the front of her dress. I flew back to my new home and job in Maine. (Actually, I moved to this academic position in 2016 so not entirely new but still full of newness). Another difficult flight. And I sorted through what photographs and letters I had in my possession. I knew that there were other pictures, and I remember them well. My mother insists they never existed. That I had made them up in my head. Perhaps I did make them up so that they could do some recuperative work for me. A way to furnish this nearly empty storehouse of personal memory. In my hand, I hold photographs of the sea and the shoreline, some of me and my brothers, and several of my grandmother. One of my great-grandmother, Marie, pronounced "May-Re" who everyone called Gran-Gran and who spoke to no one save my mother and grandmother. I only knew that she spoke French and was my mother's paternal grandmother who emigrated from St. Lucia to Barbados so many years ago that no one knew for sure anymore when she had arrived. There is no archive there either.

---

16 Sharpe, *In the Wake*, 11.

Black Annotation, Black Redaction. "Black redaction and Black annotation are ways of imagining otherwise."[17] In *In the Wake,* Sharpe defines them for us: to annotate and to redact. Annotation is the making or furnishing (what a perfect word for this) of notes to a literary work or author. And as the OED tells us, metadata is also related to annotation. It is attached to "text, image, or other data."[18] It is data that gives us information about other data: I do not swim. I almost drowned as a small child. I was rescued by my Rastafarian uncle, Dale. I never learned to swim properly after that near death in the sea. I love the sea. I take healing sea baths. I enjoy the saltwater that dries on my skin. I wade in the water whenever I can. I even go as far as to float in the ocean. I am not afraid of the floating away. I have furnished you with some of me. Metadata. The other data that I have about my life is partial. There is no text and few real images. I have no place from which to retrieve them. I can't furnish what doesn't exist. I can't provide what has never been chronicled. I can only imagine otherwise.

I find myself and this life that I live in the full meaning of redaction – even the obsolete definition touches me: the action of driving back; resistance, reaction. I am a makeshift production preoccupied by this desire for a definite form into which I can slot this life. There is something provisional about me. Tentative. I wonder if I can win some of Equiano's literary merit: "Permit me, with the greatest deference and respect, to lay at your feet the following genuine Narrative."

I am "behind" in many ways, but this is how I claim my writing: *Written by Herself.* I remain unsettled. It is this word that materializes when I think of my own story, the personal and the professional. I resist the demand to include what is uninvited. I do not want those sources. I know that my inflections make no sense to some. On occasion, when I am very tired, I will say "tree." My revision of paragraphs and sentences, the inclusion of new footnotes, and the corrections of my interpretations need to be authenticated. I drive back the intentional distribution of harms brought on by that peculiar benevolence which Equiano invokes. I drive back that coercion. I react to the knowledge that I am not the right kind. I am cleaning up the spills of what Sylvia Wynter quoted in Sharpe calls our "narratively condemned status."[19] I take a clean white cloth and wipe away the black ink leaking from the lines of a life that is only faintly annotated and profoundly redacted.

---

**17** Sharpe, *In the Wake,* 124.
**18** Sharpe, *In the Wake,* 112.
**19** Sharpe, *In the Wake,* 13.

In the afterlife of slavery, there are holes that can't be filled. One can only look deeply into the depths and wonder in distress at the true substance of those missing stories with hastily written yet indecipherable names, small collections of undated photographs, and short lists of partial birth dates and deaths. What was very real for Black people – their enslavement – has left a painful trace – the "ongoing disaster of the ruptures of chattel slavery."[20] It is a phantom limb. One can still experience the sensation of what is not there, what no longer exists. Aspiration is the word that Sharpe uses for "keeping and putting breath in the Black body."[21] If aspiration – the breath – is also the word for desire, hope, and wish, then might we not also draw on aspiration to support life in renewed frameworks in which those three needs are also gathered, addressed, and met?

In my upper-level German course on women's writing, we just read Audre Lorde's "A Litany for Survival,"[22] and I find that same meaning of aspiration in Lorde's opening lines but I also find the *wake*:

> For those of us who live at the shoreline
> standing upon the constant edges of decision
> crucial and alone

I cannot help but turn myself toward this "us" that Lorde lyricizes. I feel the urgency of life at the threshold. From that position, I map out how to navigate the persistent burden of being at the limit; it is the fundamental condition of my existence. Roughened for a time by the enduring presence of "the constant edges of decision," I am upended by the sea as I try to imagine how to manage my choices – crucial and alone. I swallow too much saltwater but am still alive and breathing as I right myself.

# Works Cited

Beckles, Hilary. *A History of Barbados: from Amerindian Settlement to Caribbean Single Market*. Cambridge: Cambridge University Press, 2006.
Beckles, Hilary. *The First Black Slave Society: Britain's "Barbarity Time" in Barbados, 1636–1876*. Kingston, Jamaica: The University of the West Indies Press, 2016.

---

**20** Sharpe, *In the Wake*, 5.
**21** Sharpe, *In the Wake*, 130.
**22** Audre Lorde, "A Litany for Survival," in *The Collected Poems of Audre Lorde* (New York: W. W. Norton & Company, 2000), 255.

Daventry, Michael. "Barbados Announces Plans to Remove Elizabeth II and Become a Republic." *Euronews*, 16 September 2020. Accessed 11 October 2020. https://www.euronews.com/2020/09/16/barbados-plans-to-become-a-republic-and-drop-queen-elizabeth-ii-as-head-of-state.

Equiano, Olaudah. *The Interesting Narrative of the Life of Olaudah Equiano, or Gustavas Vassa, the African. Written by Himself*. New York: Norton Critical Editions, 2000.

Jacobs, Harriet. *Incidents in the Life of a Slave Girl*. New York: Norton Critical Edition, 2018.

Leibman, Laura, and Karl Watson. "Rediscovering an Important Link to American Jewish History: Field Notes from the Nidhe Israel Synagogue Complex in Barbados." Accessed 11 October 2020. *American Jewish Historical* Society. https://ajhs.org/rediscovering-important-link-american-jewish-history.

Lorde, Audre. *The Collected Poems of Audre Lorde*. New York: W. W. Norton & Company, 2000.

Sharpe, Christina Elizabeth. *In the Wake: On Blackness and Being*. Durham: Duke University Press, 2016.

Watson, Karl. "Slavery and Economy in Barbados." *BBC History*. 17 February 2011. Accessed 11 October 2020. http://www.bbc.co.uk/history/british/empire_seapower/barbados_01.shtml.

Wheatley, Phillis. *Poems on Various Subjects, Religious and Moral*, 1773. New York: Penguin Classics, 2001.

# A Conversation (Spring 2021)

What follows is based on a conversation that took place with a large percentage of the contributors on March 20, 2021. It has been edited for readability, and additional contributors who were not able to be there on that date have interjected comments. For our readers of this conversation in our book, we remind you that we contributors were talking to one another after the trial of Derek Chauvin in the murder of George Floyd had begun, but well before its conclusion and guilty verdict on 20 April. Present in all our minds were the mass shootings that had occurred on 16 March (just days before this conversation) at three spas or massage parlors in the metropolitan area of Atlanta, Georgia, United States. Eight people were killed, six of whom were Asian women, and one other person was wounded. A 21-year-old male suspect was taken into custody later that day.

We agreed to note full names the first time a person speaks and initials thereafter, since for us the main focus is what we were trying to say to each other and not so much who exactly said what.

**Irene Kacandes** Thank you all for making the time to talk to one another today when we find ourselves meeting again at a moment when several murders have happened in close proximity, murders that yet once again were racially- and gender-motivated.

**Angelika Bammer** It's stunning how this whole volume, from beginning to end, has been accompanied and framed by violences of different kinds, not just in general, but ones that touch our very lives, here and now.

**IK** I agree. We're still dealing with Covid, and it's still the case that certain groups/countries/regions are adversely affected by the pandemic to a much greater extent than others. The immigrant flow at the US southern border and the displacement of large populations all over the world due to famine and war have not slowed down at all. So, in many ways, we're actually not in as different a place as we thought in summer 2020 we might be when we would meet again in 2021. Perhaps, though, others are feeling quite different. There may be things that you're on the lookout for that are quite different. So, let's ask ourselves again, as Ann urged us to ask ourselves when we met last year: are we as some kind of collective or we as individuals any differently attuned to recognizing the connections we aspired to make? Or, to put it otherwise: are we recognizing the aftermaths of historical violence in our present in some other way than when we convened last summer?

https://doi.org/10.1515/9783110753295-034

**AB** One thing that feels different is that institutions, organizations, and businesses across the board have felt compelled to acknowledge the ways that structures of inequality, injustice – and, to use the word that frames our volume, "violence" – are foundational to how they operate. Statements to that effect, and promises to change these structures, abound to an extent that I don't recall seeing yet last year.

**Susan R. Suleiman** Maybe we could put it slightly differently and ask directly about how we feel now in relation to our last meeting. I think that despite the horrors of Atlanta that happened in the last week, I feel a sense of moving toward the end of the tunnel, at least with regard to actual lockdowns. We're beginning to move toward the question: will life resume, begin to resemble what it was before the pandemic? Or, as I've been asking myself, many years from now, will people say: "before Covid" the way we say "before the Great War" or "before Hiroshima"? Or is this experience of the pandemic going to be a more passing thing? Obviously, we can't tell, but I think the fact that we are even beginning to ask that question may show that we're moving out of something. I wouldn't have asked that question in the same way last year.

**Alicia E. Ellis** I can say that I am horrified in broad societal ways about the condition of life in the United States as well as in an existential manner. I feel panic when I hear the news about yet another Black person whose life has been taken. That could be me tomorrow, next month or five years from now. As a tenured professor, I do enjoy the privileges that accompany job security and an academic environment. Tenure offers me a safety net of some sort. But my professional status will not save me from being killed because of my skin color.

**AB** The trauma of racial violence – the "horrors of Atlanta," as you put it, Susan – and the trauma of the pandemic coincide, yet we experience them differently, as Alicia's intervention makes all too clear. The latter we can see an end to. The former, not anytime soon.

**Erin McGlothlin** I've been reading some of the articles about the flu pandemic [of 1918] and how ten years later, nobody was talking about it. It didn't really lodge – at least on a larger global level – in memory in the ways it should have. That's terrifying! That's what I'm thinking about. We're trying to slow this down and look at these traumas and be with them and respond to them and yet, we're being pushed against our will to forget them.

**Marianne Hirsch** This is such an interesting time to have this conversation, especially this week with the events in Atlanta. We're at a threshold moment. It might soon be "after the worst of the pandemic," certainly in the US, but we have no idea what's coming next, and we have no idea how to repair everything that's been broken. Not just what happened this year, but the after-effects of what has been broken, desperately broken, before Covid that we're seeing in so much greater clarity now, in 2020 and 2021. When the killings in Atlanta happened, I just didn't want to turn on the TV or listen or read anything, because after the Biden inauguration and registering that there are some competent people who are trying to move us into vaccines and into a recovery package, possibly a new minimum wage, it seemed like there was progress and a future. I just didn't want to hear any more horror. So now when I hear what's happening at the border and what's happening in Atlanta, it just feels like too much. I think for me it's brought out the exhaustion of the anxious period that we've been living through in the recent past, but also everything that we've been seeing more clearly and learning during this period. For me the question right now is: what will it take, what kind of transitional justice mechanisms can we come up with to memorialize the thousands of people who have been lost? What will it take to repair the communities that have been so fractured by Covid in ways that may or may not affect people on this [Zoom] screen but do affect our neighbors to whom we're adjacent? I don't think we know that yet, but I think it's a very urgent question.

**AB** Your observation, Marianne, that "we're at a threshold moment" gets at something critical. I completely agree. At the same time, I can't help feeling that I've been at this threshold before. Not the same one, of course, but some of what we are experiencing now feels horribly familiar to me: the sense of urgency that we need to do something; the helplessness over what to do; the uncertainty about what exactly it is that we should most immediately and most urgently respond to. Even the uneasiness about how to talk about all of this – what words to use, when to talk and when to stay silent, whom to talk to and whom to talk as – it all feels strangely familiar.

I felt myself at such a threshold moment at the end of the 1960s when I left Germany for the United States. I didn't know how to respond to the history of genocide and antisemitism that I had inherited as a German, from German history. And I didn't know how to respond to the racist history that had led the man I was involved with then – an African American from Birmingham, Alabama – to leave the United States and move to Germany. We lived in the shadow of our respective histories and didn't know how to talk about them with each other.

Years later, I felt myself at such a threshold moment again. It was in Memphis, Tennessee and I had left graduate school to involve myself in community activism around issues of labor rights and racial justice. The issues were not that different from the ones we are facing now: racial and class inequality, poverty, police violence, the disenfranchisement of entire communities, and a system set up to make sure that nothing fundamentally changed. Yet, despite our best-intentioned efforts, I wondered if what we were doing made sense. When a young Black man, whom we were defending against criminal charges, ended up being sentenced to life imprisonment, I was no longer sure if we were helping or making things worse. Perhaps, I thought, I can be more effective in other ways. Perhaps activism is not my way to be useful. So, I returned to graduate school, hoping that this would give me different tools that I could use toward the change we envisioned: conceptual frameworks for understanding how systems work, methods of critical analysis, knowledge of historical connections. But graduate school was also, I realized, an escape from the despairing sense of helplessness we faced trying to change a system we didn't seem to be making a dent in.

I feel as if I'm at such a threshold again now. I feel the same system of compounded injustices. The same sense of urgency to do something, to intervene, to effect change. The same sense of helplessness about how to be effective or even where to start.

**Atina Grossmann** I think that a huge additional factor for me is age and the vulnerabilities of the individual body – which is part of what does make this moment seem more overwhelming than earlier times in our lives when we also despaired of the political situation but still saw ourselves with a long personal horizon for participating in and experiencing change. The combination of political, environmental, and health crises – all linked of course – with the very personal precarities of age and illness (scary despite all our very real and starkly unequal privilege) make the pain of the present as well as the uncertainty of the future feel more urgent and difficult.

**Sylvia Flescher** I don't feel equipped to express myself as eloquently as you academics do. But in terms of Irene's first question, "how are we in a different place now than in the summer of 2020?" I can speak personally that I have been made so much more aware of how I had turned a blind eye to the ongoing racial injustice in this country. That realization feels shameful . . .

I recall watching Bryan Stevenson's TED talk when he objects to people referring to 9/11 as the first terrorist attack on US soil, whereas Black people in this country have been terrorized since they arrived and for sure as citizens ever since Reconstruction. I remember needing time to let that idea really sink in. Last summer I read a piece in the *New York Times*, "It Is Time for Reparations" by Nikole Hannah-Jones, and I realized that I had never thought about what happened after the slaves were freed. I had never previously taken in their abject poverty, their starving for lack of any work. It's easy to justify my reaction and my ignorance by the fact that I attended the French Lycee in New York and was more educated in European than American history. But that's not a very good excuse for ongoing ignorance. I'm glad that the pandemic forced me to think every day about inequality, starting with the unequal access to medical care and the much higher death toll among people of color.

Last summer and into the fall I was feeling extremely lucky but also guilty about my immense privilege. Covid was barely disrupting my life. And the lives of most of my patients in the upper-class suburb where I live and practice psychiatry were equally cushioned. I felt compelled to raise my awareness by reading books I'd never read, watching movies, and through this project talking to all of you. Isabel Wilkerson's *Caste: The Origin of Our Discontents* is probably sticking with me the strongest.

**Tao Leigh Goffe** Before this conversation moves too far in another direction, I want to go back to Atlanta. It seems to me that the murders there already fit into a larger pattern, a kind of cyclic forgetting about violence and Asian communities. I sit with this anti-Asian sentiment and violence in the colonial record of the nineteenth century in both the United States and across the Western Hemisphere. I sit with the forgotten catastrophe and erosion of racial indentureship, an institution designed by European nations to replace racial slavery. This formed a historical adjacency to violence and what Orlando Patterson calls "social death." Hundreds of thousands of East Asian, Southeast Asians, and South Asians were conscripted into this institution in the Americas of brutal backbreaking labor from guano mines in Peru to plantations in Louisiana, but this is not a narrative of violence traditionally found in standard textbooks. I want to know how long before we forget again about Atlanta and "the evidence of things not seen," to give a nod to James Baldwin.

**IK** Tao, thank you so much for that intervention, because for me a huge part of this project is trying to connect up things, to remind each other of pieces of histories precisely so that we can break that cycle of violence and forgetting that you so importantly bring to our attention.

**Leo Spitzer** When I think about the present, what is becoming clearer and clearer but for which I have no resolution, is that we are also living in a moment of ongoing resentment – resentment built upon old resentment. I think this fits in with what Tao is saying about cyclic forgetting. In addition, we are also in the wake of Trump's "Big Lie" which serves to create new resentments. So, in a sense, what's happening is that a large, large group of people in this country feel that they are losing something that was theirs in the past, that they are being shut out of privileges and are being replaced from a dominance they long enjoyed. And I don't know how we can change this brewing resentment. Is it possible to bridge such profound divisions in our national body politic in a relatively non-violent fashion?

**Mark Roseman** Can I pick up on what Leo's saying? I want to share three thoughts on the relevance of our historical work. One thing we're all aware of, whatever similarities or differences there are in terms of the historical periods about which we wrote our contributions to this volume and the present day, is that at least as big as those inherent differences is the difference between looking backwards and looking forward. So, while we're all thinking about the relevance of comparisons, we're all also kind of stuck in the sense of really there's so much open, so much uncertain in our present. The adage that we're 20–20 looking back and blind looking forward really seems to apply to today. In that sense, historical comparisons are always off, except insofar as we are seeking to recapture the uncertainties of past actors that mirror our present state. A second point: I think that as historians, on one way in which we truly can contribute, namely in providing context and complexity, we're often not fulfilling our duty! If I look at colleagues espousing things in social media, it seems to me that much of the care we take as historians when looking backward goes out the window when we talk about contemporary issues. This is often where we fail. My department here at Indiana University has come up with a rather thundering diversity statement. One of the things it talks about is the "white supremacy takeover on January 6." Well, you can just imagine what would happen if you're an historian of Nazism and you argued that the election results in 1932 were an antisemitic takeover. Yes, antisemitism was an extraordinary, powerful theme at the time. But to reduce that moment, that political movement to one motive – even among Nazi members – you'd be laughed out of the court of serious history. So, I feel that kind of care that we use looking back, we should take to understand the complexity of the present. And yet reading and listening to all sorts of things in these last months, I observe that we often don't. Specificity, multiple motives and forces – that's something historians could contribute, even insist on. The third thing I've had on my mind – and this connects to what Marianne was saying – I put in a different direction: I feel very ambivalent about whether what we are doing here in this volume is more or less relevant than it was even last summer when our preoccupation was racist violence. The issue of violence and so much of what we were writing resonated with that experience of Black Lives Matter and continues to do so. Our work in this volume is acutely relevant given the violence that's occurred since we started to talk together and write our essays. Yet in some other ways, what's happened between November and today, as Leo said, the continuation of the Big Lie, makes us aware of the fragility of democracy, and the weakness of institutions. That makes me think that some of what we've been writing, at least what I was writing, is no longer quite expressing my deepest fears which go even beyond the racist violence – though of course that's related to this larger fear but no longer exactly encompassing "it." In other words, the massive and multiple threats to the basic principles of democracy now make "adjacency to violence" only a partial representation of the danger we now face.

**IK** Thanks, Mark. You've taken us to a slightly different question which I might frame as: What are we worried about now? What do we think our role is in such a moment? And how did that shift? In addition to the exhaustion that Marianne mentioned, I do feel differently from last summer when I perceived a kind of solidarity around issues of racial injustice or at least a kind of groundswell to look at it right in the face. Whereas now, just even the discussion of: were these murders in Atlanta caused because the guy was sexually hung up or were they caused because he hates women or because he hates Asians? – just the fact that we as a country could be having such a discussion – seems to indicate that we're further away from a push toward trying to make the world a more just place. That those justifications could even be aired, seems really disgusting to me.

**SRS** But look at Biden and Harris. When they went to Atlanta yesterday, they were really clear about what was the most important thing, which was the inherent and ongoing racism to which everybody has been colluding by not speaking out. And that anti-Asian racism goes all the way back to at least the Chinese Exclusion Act almost one hundred fifty years ago (1882). As Tao and Leo mentioned, it's a long, long story. Still, that doesn't mean that the guy wasn't hung up sexually. We are sophisticated enough to entertain the idea of overdetermination. There are many different possible motives for someone going into a place and shooting it up. He wasn't some kind of ideologue like the guy in New Zealand. He's a 21-year-old shooter with all this going on in his head: the Asians, the women, the sex, his parents, the church were all part of it, to engage in the multiple naming of factors that Mark was just calling for. I don't think naming a single cause will help. On the other hand, you could talk about the most important underlying thing and that would be racism and sexism. Right?

**MH** To stick with Irene's concerns for a moment: I think we need to figure out what that discussion is a symptom of. What are Trump people trying to emphasize or deny when they sort out what a motive might be? Perhaps we are focusing on an individual's motives because we don't want to recognize the systemic and widespread nature of anti-Asian racism right now.

**Claudia Breger** I want to mark the tension in the room. It's very much a tension I'm feeling myself. When Mark and Susan were advocating just now for complexity – and I'm very much in favor of looking at complexity – I find myself worrying about the extent to which my insistence on complexity comes at the expense of solidarity. How do I effectively navigate the relationship between complexity and solidarity in any given moment? I think that that is playing out in this discussion. We are all aware of intersectionality and we can spell it out beautifully in academic jargon. The question is how do we navigate it in specific situations? I find that that really counts for me right now.

**AB** These are such urgent questions. To me, complexity and solidarity would mean that we don't just respond and become "active" when something dramatic happens. Violence doesn't just include the obvious acts, like the murder of George Floyd captured on live video or the racist killing spree of the Atlanta shooter. The condition of poverty – so undramatically common that it comes to be almost invisible – is just as violent; it's just in a different mode. Its violence is hidden in its everydayness, its systemic givenness. It kills you slowly, over time, sucking the life energy out of entire generations. One thing that the events of this year have brought home to me is that violence can be an act, but it can also be experienced as a condition.

**AG** Claudia's identification of that tension between complexity and solidarity strikes me as exactly right. As a historian of National Socialism and the Holocaust who's been thinking a lot about how we find space for teaching those topics – generally identified with a "Eurocentric" approach in the "decolonized curriculum" that my students and my institution, in different ways, have been demanding – I've been trying to grapple with the elision/gap between antisemitism and white supremacy and how we work with it as historians on the one hand and as teachers and aspirational public intellectuals on the other. For example, when my students very attentively read the 1920 Nazi Party Program they had no trouble picking up on the racialized exclusions of the National Socialist vision but where there was explicit reference to antisemitism and exclusion of Jews they immediately read "white supremacy." So, it's one thing to take care with those distinctions (and connections or to put it in terms of this volume, adjacencies) in our scholarship but quite another to invoke them in our everyday lives as part of agitated politicized university communities or other political groups with which we may be involved. We very quickly land in fraught debates about the definitions of antisemitism – for instance look at the controversy and pushback surrounding the IHRA definition – and its relationship to racism and white supremacy [IHRA = International Holocaust Remembrance Alliance founded in 1998 and headquartered in Berlin, Germany]. In those situations, it does not seem to help to say, "oh but a more careful and complex reading of history is necessary." Those kinds of answers didn't go very far with me when I was an impatient radical student and, while I actually think that my students are in many ways more thoughtful than we were in the New Left, they also don't play well in my classroom or college-wide town halls now. Or, to put it otherwise, let's say that, as Claudia notes, I haven't yet figured out how to adequately "navigate" these "specific situations."

**Ann Cvetkovich** In the context of our work on this book, I would link complexity/solidarity to our category of "historical violence." Resonating for me within that term is not so much violence in the past, but historical violence as violence in the present that carries histories with it. That's what we've seen this past week and what we've seen in so many other moments over the last year. It's in the water for me, this work of afterlives. Saidiya Hartman's concept of the "afterlife of slavery" has been so generative for many of us, and it is always with me, along with the afterlives of other histories. A key text, all the more so now, is Lisa Lowe's *The Intimacies of Four Continents*, which explores the interconnections among European colonialisms, Black Atlantic slavery, Indigenous genocides, and the Asian migrations that, as Tao was reminding us, replaced labor that was "lost" when enslaved Africans were set free. I continue to look for tools to do that work, including ways to make gender and sexuality integral to it. One of the major challenges for me of having moved, or rather returned [from Texas] to Canada in this time of pandemic and antiracist struggle is an ongoing and ubiquitous failure to think Blackness and Indigeneity together. Whether in Canada or the USA, that connection often eludes our grasp as settler colonial societies, and it requires both intellectual tools and emotional tools.

**AEE** I think you're right about the connection of historical violence and complexity. We might be able push Hartman's idea of the afterlife of slavery to a more challenging place that embraces Ann's sense of things: the violence in the present and how histories are held within it. The complexity of that condition renders everything that we see and experience as a kind of historical excavation that reveals the current moment as if it were a surprise or unexpected. It's ghostly because you see that the behaviors, words, and beliefs of the past are much too familiar in our contemporary moment. They exist in a different form, but they remain evidence of all the harms that have been done and that still do harm albeit through anti-Asian and anti-Black violence, mass shootings, the white resentment that led to January 6, and systemic racism whose afterlife Hartman reveals is about limited access to health and education, mass incarceration, early death, and poverty. None of what I just said is over. It continues to grow and decimate lives. Slavery is an ongoing brutality whose aftermath still diminishes and suffocates Black lives. The most devastating atrocities from the past are still hovering over us as they play out in the present tense. But then how do we reconcile the idea that life has not changed much with our desire to believe otherwise? Is the desire to believe otherwise a sign of privilege or even blindness when one can read the news and know that Black people are being murdered on an almost daily basis?

**Leo Riegert** I do consider "white supremacy" in its more recent usages a useful term in helping us see the connections between a violent past of colonization and slavery in the United States and Canada and the (often murderous) racism and material deprivation experienced by Blacks and Native Americans in the present. But when everything from the Klan to a Robert Mapplethorpe portrait is described as "white supremacist," what do we lose in terms of understanding the complexities of history and of people's positions? I'm really struck by Atina's anecdote about her students not being capable of bringing themselves to call the Nazi text antisemitic. Is this because Jews are "white" within the current understanding of "white supremacy," despite the fact that Jews were the explicit targets of avowed and proud "white supremacists" in places like Charlottesville and Pittsburgh? So, I worry that terms like "whiteness" and "white supremacy" cover up more than they reveal in terms of privilege and power, as well as in terms of individual lived experiences. Perhaps, in their reductiveness, these terms also limit the possibilities for connection and allyship, as well as our ability to think creatively about the current moment.

**AC** If I could add another topic: this project has also been about acknowledging work at the nexus of Holocaust, Jewish, and trauma studies. The affective networks with people like Irene, Marianne, and Leo (and others such as Nancy Miller and Diana Taylor) link us intellectually and create adjacencies that enable us to come together here. Writing my essay has been a way to honor what that mix has given to me. Can we not challenge the temporal frame that declares things to be over or past? That is: has not the work we've done in queer, feminist, and trauma studies been about challenging the completion of mourning? Intergenerational transmission that Alicia, Atina, Leo and others are referring to works against the idea that anything's over or that we're done with something. Another way to express this is that I'm very worried, unsettled by this invocation of "return to the normal." At the same moment that I was getting my Covid vaccine, my dean was emailing me as a department head about the need to decide in two weeks' time which courses can be offered in person in the fall. Addressing that administrative uncertainty along with trying to think about how to step up the pace on institutional antiracism makes me very worried. We're in a state of ongoing and uneven developments. I don't want to underestimate the significance of the vaccine being available to some people and not others. This work of intimacies and intersecting adjacencies is effectively about thinking contradictions together. And I don't think that's intellectual abstraction. It's a lived reality for me.

**Doris Bergen** I am thinking about what Ann says about feeling unsettled about the idea of "return," and want to talk about returning to normalcy. Because for me, if I think about the question of how do I feel different, what feels so stark compared to a year ago are the multiple ways in which I've become aware of my own privilege, similar to what Sylvia was saying a bit ago. I think, hard as it is, sitting at home doing everything from the computer, I have to compare my difficulties to those of my students. I've been teaching non-stop online during essentially this entire pandemic, and so many of my students are in distress; they're in really, really challenging circumstances, with mental health, with physical health, with social well-being. I've also thought so much about age and aging. My chapter in the book is about the aging Holocaust survivor, Molly Applebaum, and it starts with the plan we had for her to come to my class and then having to do it remotely because of Covid. Well, it's a year later. She is still alive unlike thousands of other people in long-term care homes here in Ontario, but she hasn't even seen her daughter in six months. She's over 90. Her hearing is terrible. She's not getting on zoom with her friends. I could give you so many examples of my privilege compared to others. One of the major sites of Covid transmission here in Canada is meatpacking plants. Who works there? Well, it's not me. It's almost all, what are called here, New Canadians. Temporary workers who are completely exposed. Despite all the thanks to so-called "essential workers," they get no protection. So again, for me, the idea of returning to normalcy is deeply problematic. I get to be the winner. I have my BLM poster from marches last summer right behind the spot I do my zooming from. And there are whatever statements I've signed. But I feel like this event we've gone through – that we're still in the middle of, the idea of returning to normalcy – I don't feel good about that at all. One last thing: I can't remember where, but I read recently: history is the thing that can't be undone. So, if you think about the relationship between history and the present, the minute it can't be undone, it's history. I really appreciated this project, but I feel like in some ways it encapsulates what disturbs me right now in the sense that even to be able to talk and think about these things together–and even be paid for it! – is such a privilege. That's what I find myself thinking about right now: how grateful I am for my job but also how rare it is to have that kind of opportunity and resources.

**AEE** I said this earlier, and I want to try to complicate it for myself a bit more. Normalcy for me means something else entirely than what it might mean for some people in this conversation. Yes, I do enjoy class privilege, but I am still a Black woman living in the predominantly white state of Maine. I am the anomaly here. I am just one member of the 13% of this population that makes up the US. This is said without any anger. It is just my life. My return to normal remains in a space of hypervigilance, while for others it can be a space for forgetting or retreat. It could also just be how the world works, though. There is no ease into which I can relax. No sanctuary where I can stay untouched by the outside world. Because Blackness can't be protected – in coffee shops, while jogging or bird watching, or even while sleeping. Of course, I can hide away for a while, but that doesn't mean that my reality won't be the same as that of every other Black person in this country once I resurface.

**IK** You are so right, Alicia. When I teach about privilege in my Women's and Gender studies courses, I try to get *everybody* to realize that they enjoy privilege in at least one area, by virtue of the fact that we're all connected to an elite university, and often in other respects, too, like being able-bodied, or being tall, or having a loving family. But I think what you're saying is that it's not a math equation. Just because someone enjoys privilege in one or two or even three areas, doesn't mean that their lack of privilege in other areas is overcome or balances out. I feel like I see my privilege in every direction I look. And it starts with getting a paycheck deposited to my bank account every month. What a privilege that is compared to the huge number of unemployed and people who lost their jobs due to lockdowns. I want to touch on the health issue, too. We've been dealing with a major health crisis in my family, as many of you know. Even before that broke in on us, I had been thinking a lot about people in the hospital, especially people who were dying in the hospital alone. But from what I've just gone through, I realize even more profoundly that trying to deal with any health issue with the background of the pandemic is horrifying. I know many of you have had to deal with that, too. And I just want to pause here and acknowledge that. I also want to say on a personal level, my heart goes out to you and your families. Still, my family and perhaps yours have had the financial and mental resources to try to overcome the obstacles and insist on decent care for our family member(s) in crisis. As Marianne said earlier, so many people adjacent to us, near and far, do not have those resources. And again, looking at the big picture, this unevenness in healthcare is very much color-inflected! Privilege is something where I've asked myself: okay, I recognize I have a lot of privilege. What's the next step? I've asked myself that as a teacher, and I've asked myself that as a moral being. To pick up on something Claudia said: we've all thought about intersectionalities. But what do we do with that? To me, it's very similar with privilege. We list our privileges like

Peggy MacIntosh taught us to do. That seems like a really important step. But right now, that is simply, absolutely not enough, right? What do we do with it? How do we let that awareness inflect, grow our sense of solidarity, make us do something?

**AG** Yes. And sometimes this very necessary reminder about our own privilege feels like its own form of evasion, a kind of too reflexive trope that skips over all the ways in which we do feel vulnerable and scared and exhausted. How do we think and feel along these multiple tracks? Even in the classroom? How do we acknowledge our own implication without becoming self-indulgent on the one hand and professorially distant on the other? How do we acknowledge the social and political realities bearing down on our students without simply appropriating and amplifying their own distress and anxiety? In my institution there has actually been a (to me) quite confusing and difficult-to-address push from some of the most activist students to stop discussion of "current events" in the classroom. I sense that they want to control that conversation and not have it be domesticated as part of their academic program and yet at the same time, they are demanding "decolonization" of the curriculum.

**EM** I'd also like to respond to the idea of normalcy. When do we return to normalcy? It's not that we're *not* going to return to normalcy, but rather it's that normalcy will impose itself whether we are ready for it or not. Especially in our public discourse. Especially with regard to the attack on the Capitol. We all have that one event that becomes just too much for us. Marianne mentioned the Atlanta shootings at the beginning of this conversation. For me it was the horrifying events of January 6. Not that we haven't had horrifying events in the last year (particularly the murder of George Floyd), but for me January 6 became a final unbearable addition to the load. I'm particularly devastated by how we've normalized the events of January 6; the public discourse has normalized it. We can read articles in the newspaper about people being arrested, and now we're starting to learn about hearings and trials, but the normalcy is imposing itself through the discourse whether we like it or not. And that's true of all the events we've been talking about. It's Benjamin's angel of history. We're just blown into the future whether we are ready to move on collectively or not, whether social forces are opposing it or not. I don't have an answer for why it's happening. But normalcy is imposing itself on a discursive level and that's what I'm not ready for, for all of these things to be normalized.

**SRS** Could you explain how you think that January 6 has been normalized?

**EM** We have a politician named Ron Johnson who is perpetuating the myth that the perpetrators of January 6 are simply patriots who were a little upset with this election. So, there is this kind of normalization that claims that this was just an affective response by people who were disappointed. That's one way January 6 has been normalized. But January 6 is not really in the media right now either. It's not being talked about in the way it should be. It's not being talked about as a complete break in our system of governance and our democratic principles, to connect up with something Mark said earlier. We're in the middle of this pandemic and so that's what we have to focus on. January 6 is not completely forgotten, but it's not getting the attention that it absolutely deserves. And I would say the fact that the impeachment event came and went is another sign of normalization. I commend the Democratic leadership for try- ing to keep this present. I think Biden himself is trying to sidestep it, so he can govern. But January 6 and the overwhelming evidence of the fragility of democ- racy are not what we're talking about. And the invasion of the Capitol hap- pened only a couple of months ago!

**TLG** This conversation is enlightening as one of the few Black people involved in this project. I'm struck by the connection between what happened at the Capitol and what happened in Atlanta. Whether people see it as connected. In my mind it's deeply connected. So, it's encouraging to me to hear people here talking about the intimacies of continents. We need to be talking about Black and Indigenous people and not just some general Other.

**SRS** Speaking of January 6, there have been a number of reports now that talk about the existence of other white supremacist groups which have gained cour- age from that event. In other words, there have been warnings. *stet* does still report on it. It's not completely forgotten. As for the groups that organized it, well some have been arrested; some have just gone home. But it's the tip of the iceberg, and we know that. I certainly agree that white supremacy has been the elephant in the room for all these years. And now it's all too obvious. If you read Isabel Wilkerson's *Caste,* that Sylvia mentioned earlier, you can't help but realize that the people who came here on January 6 [SRS lives in the DC area] weren't just resentful. They were fearful, and they wanted to hang on to their privilege. They didn't put it in those terms. But they want to hang on to the country, the USA. "The good old days," "MAGA," "make America white again." All those things were not said and are now obvious. We haven't really forgot- ten. At least I hope not. The authorities have not. We're now aware that there are a lot of energized, very dangerous armed groups in this country who feel that they want to take back the country, which means white men run every- thing again.

**Mita Choudhury** I want to address this issue of white supremacy. I think it's really easy, especially for people like us, to focus on January 6 or on the shooter from the murders in Atlanta. The fact is that whiteness exists on a large continuum. I've actually been a little uncomfortable seeing on things like Facebook statements that proclaim: "I stand by my Asian friends and colleagues" and "blah blah blah." Asia is not a country. And Asia itself is so, so varied. That means, Asian is just a stand-in for "oriental." It's the PC person's way of saying "oriental." And I think this is important because in doing this – and this is relevant to the volume and our conversation here – in using the term Asian, you flatten the victim. Asian becomes a kind of imaginary, as opposed to trying to really think about the victim's experience in a textured way. This action itself becomes another form of privilege.

**LR** Mita, your phrase "whiteness exists on a large continuum" jumps out for me. But I hope it's okay that I point out that just now you almost immediately turned from whiteness to thoughts on the flattening of the term "Asian." Earlier, other folks have pointed out the ways that anti-Asian racism and violence is (made) invisible or forgotten in the US, and perhaps that has something to do with a perception of Asians as a more "privileged" minority than African or Native Americans and as therefore somehow subsumed by whiteness. But I'm thinking that you yourself do not see Asians as white, so why do you think you moved away so fast from the observation about whiteness existing on a continuum? I wonder if it has something to do with a larger discomfort that I myself feel about broaching the complexity of whiteness. I wonder if doing so might somehow be a (conscious or unconscious) re-assertion of white privilege, even when the status of my own whiteness is something I have put under scrutiny in my life and in my essay in this volume.

**MC** Hmmm – to be honest, I was trying to say a number of things so quickly that you're assuming a cogency and narrative to my thoughts that probably wasn't there. I do agree with your assertion that whiteness is complex and changing. In fact, I don't think that we interrogate the instability of whiteness enough. Part of the privilege of whiteness is that it is the norm, the default. It'd be great if this conversation – and others – could look at whiteness as something that is always shifting. As for privilege, there are instances when people who are in privileged places can experience marginalization. It depends on positionality and historical circumstances – for instance, this pandemic has left plenty of white people at home or in institutions where they are utterly isolated.

**IK** It felt to me last summer like a really broad swath of people wanted to think about privilege and how uneven and how unjust but again how inequitable life on this planet actually is. In many, many regards. Broader than what we've been talking about so far: North South; in terms of ecological devastation – all those islands that are about to disappear; in terms of ability; in terms of so many issues, access to vaccines to add another. Something Dagmar shared with me last summer just popped back into the forefront of my thoughts: the idea that someone had to prove that they were worthy of a ventilator. I was in Europe at the time, and I was just horrified. The death that was and still is occurring because of Covid is such an obvious sign of these horrible inequalities. Come to think of it, inequality doesn't feel like nearly a big enough word. I do agree with those who are emphasizing the power of words, of language – Erin talking about rhetoric imposing normalcy and Mita about the insidious flattening of "Asian" or "antisemitic" or "white supremacy" and on and on. I feel like that's what's been happening in this last period, partly with all the lies that Leo Spitzer brought up at the beginning of this conversation, lies for the purpose of making it legitimate not to worry about those things, those inequities, for lack of a better word. "We are going to go back to normal again." And "we don't have to think about any of those issues." As Alicia pointed out, "going back to normal" itself means such different things for different people. So, when you asked me, Marianne, and rightfully so, why am I upset about them listing all those reasons for what that young man did in Atlanta, the problem for me right now is a rhetorical one: that so many of those reasons fit into narratives of "that's how men are." One form of going back to normal is that we're going to accept this centuries old idea of patriarchy, that men are just like that and women have to get out of their way. This possible trajectory feels really scary to me, like it's taking us away from that moment when were focused on inequity, when our eyes opened to the fact that African Americans, Native Americans and Latinos were dying of Covid at a rate hallucinatorily higher than white people were. So that's why those arguments, they make me afraid about what's next.

**SRS** I haven't heard anyone say "that's how men are" to talk about this guy. Have you? No, nobody has said that.

**IK** I didn't mean they are saying that per se, I'm saying that when people bring up his sexual hangups in the context of his committing murder, it feels like a slippery slope into those kinds of explanations. It feels like a demonstration of the privilege some (white) people (men) have to NOT talk about it in terms of racism and misogyny.

**SRS** Good point. This is one place where complexity really comes in handy: he had sexual hangups AND his action occurred in a cultural context of racism and misogyny. We could even surmise that his sexual hangups were exacerbated by the racism and misogyny.

**Dagmar Herzog** Well, speaking of the importance of language and what words we use: I want to say more about privilege, even as we've said quite a lot about it already. On the one hand, I understand and share the feeling of having unearned privilege and, on the other, I feel that one thing we've learned in the Covid year is just how powerless people feel. To my mind, powerlessness is a word we need to add to this conversation. Powerlessness is dangerous, especially when masses of people feel it. I think the resentment that we've talked about and the alternate reality in which people were located, as Erin and Leo mentioned, led to the fundamental break that happened on January 6. Many of those human beings still believe they did the right thing. And there are 20–30% of our fellow human beings who still agree with that alternate version of reality. But a lot of that has its roots actually not so much in a sense of privilege – although clearly that's part of it, in the sense that there's an indignant anger that some other group will get what rights these people think are theirs to keep. There's also a real sense of existential insecurity. And that has economic dimensions, but also political-psychological dimensions. The significant proportion of us who study Nazism know all too well what the relationship is between economic destabilization in extreme conditions plus alternate versions of reality proliferating. So, I think we should keep powerlessness and thinking about the ways people manage it in our analytical framework. I mean, when I think about our traumatized students, all of whose futures just got destroyed by the economic consequences of the pandemic and the mental health issues that they're having as a result of that: that's all connected.

**Viktor Witkowski** What we have experienced over the past years of the Trump administration is a steady shift away from agreed-upon language, norms, values, and rules of engagement. For example, as a head of state you do not publicly call Black athletes "sons of b*" if they choose to exercise their right to protest against racism and abuse of power. Those who agreed with the former president on this and other such occasions, find themselves emboldened and encouraged. If they ever felt a sense of powerlessness, it has now been replaced with a sense of righteousness. I'm not saying economics is not a factor. I want to complicate it as a factor. When I was growing up as a Polish kid in a low-income and mostly immigrant neighborhood in Germany, I had a childhood friend who later became one of the local Neo-Nazis. His parents were working-class Germans, and after graduation he started an apprenticeship and became disillusioned about his prospects on the job market and what he perceived as a foreign takeover of his hometown by immigrants, most of whom he knew from his childhood days. In his case, it would be accurate to state that he perceived himself as being in a position of powerlessness. But this was in the early 1990s before we knew people who identify as *Alt-Right* or the *New Right*.

When we look at present-day Germany and at polls from the past few years (including 2020 during Covid), we see a majority of Germans who feel economically secure, yet we also note a spike in antisemitic and Islamophobic crimes. In Poland the situation is even more dire. The far-right ruling party *PiS* has been in direct confrontation with the EU, overhauling Poland's judicial system by appointing party loyalists to key positions, refusing to take in refugees from non-Christian countries, marginalizing what is left of the independent media, severely restricting women's rights, LGBTQ rights, minority rights. At the same time, Poland's economy is booming, spending for social welfare programs – in particular large families – is at a record high, and Poles enjoy a high level of economic stability. Yet some people in my own family are wishing for a Brexit-style exit from the EU, they challenge a woman's right to abortion, they refuse to accept a Polish person who is not white as Polish. I have never seen them this emboldened, so self-assured and without any self-doubt. This worries me greatly.

Germany and Poland cannot serve as foils for what has been happening in the US in particular when we look at the degree of deadly violence, the number of Black and Brown people killed by police, the number of people lost in mass shootings. But in case of these and several other western countries, we are witnessing an alignment of people who share a nativist view, who are – explicitly or implicitly – wishing away democracy, and who are holding on to what they believe is their most precious quality: the superiority of their whiteness no matter their actual economic status.

**DH** It's ridiculous for me to be giving a shout-out to Eric Fromm, because I'm not a huge Fromm fan. But there's this amazing essay from 1937 "Zum Gefühl der Ohnmacht" about the feeling of powerlessness, or the feeling of impotence, which he roots in childhood. He talks about dictatorships but also democracies. In both kinds of systems, people are being told, are fooled into thinking, that they're in charge of their own lives, their own destinies, when really they're not. And that deception is deeply disorienting and has all kinds of emotional and political consequences. I actually think a lot about – I guess it was Leo Spitzer who mentioned the question of how can we reach all of those people? – well, there's Fox News and 4Chan and all these media that maintain alternate views of reality.

**IK** Sorry to interrupt. I really don't follow that kind of media, even though I realize I should be aware of what they're saying. I've never even heard of 4Chan.

**DH** I can put a URL in the chat [rollingstone.com/politics/politics-news/the measure-of-hate-on-4chan-627922/]. All these media maintain these alternate views of reality. At the same time, what I think we need to add to this conversation is that there's economic devastation that is real and that all too many people, also more privileged people as are some of the folks Viktor just brought up, deal with the sense of precarity by being self-protective and having reduced reflexes of solidarity and generosity. It's inevitable.

So, speaking of reduced reflexes of solidarity and generosity: For this volume I wrote on disability, which I was glad Irene encouraged me to do. And during this whole period that we've been trying to consider here together, I've been finding myself coming back to how disability always costs money, someone else's money. And it's therefore in an economic crisis that care for the disabled is always one of the first things to go. The story about the ventilators I shared with Irene last summer? Well, luckily in the United States there are laws that you're not allowed to discriminate against people with disabilities and deny them ventilators. But the minute it comes to what's really going on with disability care and Covid, of course it's a matter of money. Because Medicaid is being cut dramatically, just when people need it most. So, it's suddenly about money again. As it happened in this pandemic year – Irene knows this and some of the rest of you too – my mother was in a nursing home in North Carolina, and things got bad, and I was organizing her daily care from afar. I learned so much about institutional self-protection: the way the institution circles the wagons and wants to save its reputation and its finances. Problematic decisions were made, including the failure to contact-trace appropriately. The experiences forced me – unwillingly – into greater empathy with charity institution directors in the Nazi era. But more generally, the experiences made me think even more about vulnerability and the economics around disability, a crucial dynamic. There's a moment in my essay when I talk about the labor that patients in the asylums had to perform. They were like penal colonies, and there are hallucinatory scenes of psychiatrically ill or cognitively disabled patients having to participate in providing laundry and cooking for the perpetrators, but also in the murder process. I just interviewed last week – so it's not in this essay – a man who was one of the Catholic altar boys who was brought along to swing the incense every time the priest had to go to bury people on the hill above the killing center of Hadamar. This man was born in 1933; he's now 87, and he tells me about what it was like as a child to realize these deaths were unnatural, that the coffins kept being re-used, that sometimes there were two burials a day, and how disconcerting it was to watch the exhausted, starved patients who had to carry the coffins up the slope and so on and so forth. It's all a perverse parody of a funeral. But the participation of patients in the labor, including the labor of killing, is also a perverse parody of work therapy and that has a whole history of vulnerable populations, which we also – to get back to the present – know from our prisons. Here I am in NYC: I come into my building and I use the antibacterial gel, and it was all produced in the NY prisons. The prisoners don't have access to it, or to any kind of proper protection, but they're making it for us! This has been a longwinded way of me putting economic powerlessness on the table.

**AG** Yes, at times, especially at the terrifying beginning of the pandemic, it seemed like discussions got so stuck on the question of ventilator supply that we didn't even get to the question of how and why ventilators became such a critical and presumably scarce item when in fact the very need for ventilators in the first place was partially a result of a starved public health system that didn't even provide the staffing and PPE that would have made often more effective and less invasive interventions possible. Those prized ventilators were in all too many cases (certainly outside the fanciest hospitals) a straight ticket to death, and equity would have required the resources to try alternatives with better or at least equal and less traumatic chances of survival. And now we are confronting the global consequences of basic breathlessness, as in the oxygen shortage in India, Iran and so many other places. Contributors to this volume like Dagmar are experts on these questions, and yet it feels so hard to look deep enough.

**IK** Speaking of breathlessness: there is so much going on in this conversation and I don't want to mess up my role as moderator by missing someone who wants to add a new subject. Amy, did you want to talk about the end of Eric Kligerman's essay in the volume in which he discusses memory being connected to justice?

**AS**: *bound* to justice. The word is bound! To my mind Eric's point is very connected to what Dagmar just said. First of all, in the United States, some people *are* being denied ventilators because of disability. To take another example: people on oxygen weren't considered at all when authorities in Texas and California allowed power to go out. So, the vulnerability of people with disabilities in this country is very significant. And that's related to the larger question that many of you have raised: what makes a deserved life–and in many of your essays: a *grievable* life? That's where Eric's work comes in for me. When we ask, as Dagmar is, if somebody doesn't deserve to have a ventilator, what kind of life is deserved and what kind of future do you have to imagine for yourself, to imagine a future that is deserved? An example from Dagmar's contribution is people peeling potatoes. That used to count as labor and then it doesn't even count any more. So, what is it that makes a life deserved? What kind of productivity is implied and then along with that what is the concomitant implied future? Certain kinds of futures are acceptable. That is, if you have a narrative – that's that overdetermined narrative that we talk about in our essay – that fits a trajectory that fits a deserved life, then you deserve to be saved and rescued. And if not, you don't. The parallel to that then is what makes a life grievable. And that's when what Mita was talking about, the flattening of Asian victims, becomes very important. Because a life can't be grieved in general. You have to grieve an individual, grieve in particular. If you start to associate victims, whether it's George Floyd or it's these women in Atlanta area massage parlors, if you associate them in any way as stigmatized, for any reason, to stigmatize massage parlors – which is what that killer did – then it makes that life not grievable. So that's when, Eric, your point, it's at the end of your essay. And it's just so profound. I'm leaving out the whole argument here by just picking it up at the end, but you do a beautiful job of taking us to this point where you say that the work of memory is bound to justice. And I think that's a lot of what we've been concerned about today. Certain kinds of memory work, many of you said, is about the present, you've said that it's an act of the present, which is true. But I want us to notice that Eric connects memory work to justice: memory work is bound to justice. I feel that captures a lot of what we're talking about here today.

**Eric Kligerman** Thank you, Amy, for your comments. I think you set up a great transition to thinking about the events over the past year in relation to the future, specifically through the lens of justice. When I was first writing on the topic of Holocaust representations over two decades ago, I kept returning to the usual paradigms revolving around psychoanalytic discourses on mourning, repression, and traumatic shocks. I found myself repeating these ideas and couldn't escape them. My movement away from such concepts to the question of justice was precipitated by a brief sentence at the end of Yosef Yerushalmi's book on Jewish memory, *Zakhor,* where he wonders, without any elucidation, if *the opposite of forgetting is not memory but justice.* Since reading this line several years ago, I've tried to reorient my understanding of the intersection between traumatic memory, the Holocaust, and other sites of political violence in relation to justice. Such an uncanny justice would not absolve the past, but rather would haunt the present without becoming pathological. The past transfers to us a debt (or *Schuld*) that requires from us an unremitting engagement and attentiveness (*Aufmerksamkeit)* with the historical as well as personal ruins that we find ourselves adjacent to.

**IK** Thanks for bringing *Zakhor* and that point into the conversation, Eric.

**Karen Remmler** I have been thinking a lot about the relationship of the past to the present, too, and have been wondering when to add my thoughts to this conversation. It seems to me that when we write about the past there are reverberations and affects that travel to the present moment. So, we are in a way always writing about the present moment.

**IK** Karen that is a deeply helpful – and beautiful – way to characterize a lot of the essays in the book. Even the ones that seem to be about the past are saying something about the present. It seems to me we could go in the opposite direction too; that whatever is going on in the present influences what pieces of the past we are capable of thinking or writing about. That's at least a bit related to Michael Rothberg's idea of multidirectional memory. Thank you! Another relevant concept for us to be reminded of here.

**AG** Can we pose the question to ourselves explicitly: what is the work of this conversation which in one way or another we're all having all the time anyway? On the one hand, I find myself thinking about the many ways I could, would rewrite my short personal and historical essay in light of what we are saying here and in light of all the events that continue to swirl around us. On the other, we can't do that at this stage, so I'm not sure I know what we are doing.

**IK** Well, I think I have to take responsibility for why we're here today. For one thing, the way the proposed project was taken up so immediately by so many of you when I first contacted you made me think that there was something about the topic even in its inchoate form that made all of you whom I admire enormously want to contribute and want to talk to each other. I believe that was borne out by the conversations we had last year, the large group conversations, but also smaller ones that many of you had. There was something generated by what was going on, by being together virtually in 2020. As the kind of scholars that we are, we're always working on at least two levels. There's the content of what we're trying to put into words, and there's the performance of doing the work itself. It seemed to me that that was similar with this group: there are the essays that all of you chose to write that were doing work for you yourselves – and that work wasn't the same for all of you; rather, the "assignment" to write on being adjacent to historical violence was doing different work for different people. Some of you articulated that the project was giving you permission to write the kind of piece that you simply wouldn't have written otherwise. So that seems really valuable to me. Some people used it to specifically trespass the boundaries of the discipline they're housed in. Some people used it, like I did myself, to correct a wrong or an omission from the past. There were many different things that people were doing with their essays. But there was also something about being together in this crisis year of 2020. And, so, getting together now is about checking in with each other as people. It's about the basic desire to care for one another that we developed. Also, and perhaps more professionally, I did have a central question for us in relation to the project itself: has the process changed something for us and how we experience what's going on around us? And of course this conversation itself is part of the process. I've admitted to feeling very worried, more anxious about what's next for our country and our society after all these months of stresses and challenges and yet . . . I love discussions; I love meetings; no matter what I think I think when we begin talking, something about the interaction changes what I think, reformulates what I think. Being in dialog with all of you in this group today, in a big group and in subgroups last year, in our one-on-one emails – it's all been an incredible gift to me. Intellectually and emotionally. It's helped me so much with my own sense of isolation and powerlessness. I guess that's a selfish reason to have imposed on your time or at best very personal. So, I do want to say finally, that I called us together because I thought including in the book itself an example of *how* we talk to one another in addition to *what* we want to say to one another right now could model other such discussions that I think we – meaning society writ large – are going to need a lot of to start the process of commemoration and repair Marianne talked about.

**DH** I think I have another answer for Atina about what we're doing here today. When we go back to Amy's beautiful tribute to Eric, that memory is bound to justice, that's demonstrating care and respect for one another, as Irene just suggested. However, I think in addition to that, we're living through a moment of crisis of the relationship of memory to justice–the Covid crisis and the four years of Trump and the January 6 event all are signals that everything that everyone here stands for in terms of thinking intersectionally, thinking about lessons of the past, thinking hard about antiracist pedagogy – all of those reflexes that we've had that have worked pretty damn well for decades – are in crisis. All that making out of Holocaust studies racial justice issues for the United States, all the theories we have about bystanders, everything, everything, everything, is up for grabs. The whole relationship to history. The whole relationship between past and present is in flux right now. And we don't actually know what the value of studying past things is – in whatever discipline we do that in – in relationship to the present and future. I really think that the kind of confidence we had in the 1990s, in the first decade of the twenty-first century and even up to maybe 3–4 years ago, is eroding. I mean all the evidence for *x,y,z* topics and our understanding that they have social justice implications today for students and for the wider public is in doubt now.

We have to rethink our theoretical frameworks for making sense of ambivalence about sexuality. Or how we make sense of the driving motives in some overtly belligerent forms of evangelical Christianity as evident on January 6 with "Jesus is my Savior and Trump is my president" flags being waved. We have a lot of things that need to be theorized with our hearts and our heads and in collaboration with our students and whatever parts of the wider public and activist organizations that we work with. I really think that this issue of the relationship between past and present and hoped-for future is what we're struggling with. That's my long answer to Atina.

**AC** I have so much to say to that! I both strongly agree and strongly disagree with what Dagmar just said. I don't feel like my tools are in crisis. I feel like they're providing an opportunity. People talk about the dialectical relation between crisis and opportunity. This is the moment we've been waiting for. I feel like we get to claim some responsibility for the way in which the conceptual tools that have been generated in the academy can be found in various places. Although there is still work to be done if the logic of the "intimacies of four continents" is not immediately available as a historical analysis. I want to go back to Irene's affective response to Atina's question and how we all wanted to do this work in the midst of so many other things we already had on our plates. I think reminding us of that right now and checking our feeling-temperature right now is so important because insisting that the past and the present are deeply intertwined, as Karen and others have said, bears repeating and being taught and retaught in classrooms. In response to 9/11, I resisted the idea that everything had changed, because I saw it as a spectacular fruition of a set of forces that were already in play. Ultimately, I had to back off from that position and acknowledge the way in which 9/11 became the vehicle for dramatic social change. I was involved in Columbia's September 11, 2001 Narrative Oral History and Memory Project, which under the direction of Mary Marshall Clark, was inspired by testimony and trauma studies. By April 2020, that group already had funding in place to do a Covid-19 oral history project, and I got in touch with Mary Marshall to ask if I could participate. I wanted to record conversations with the HIV/AIDS activists I had interviewed in the 1990s in order to explore the connections between that earlier pandemic and this one. What are the conversations we might have now in order to use the tools and lessons of that earlier history? What do we know about mourning? What do we know about getting drugs into bodies? What do we know about social distancing? And how can we put those tools out there for people to use now? I felt that I needed to connect with my people and make those conversations public.

This group is a bit like that for me, too. What do we already know that we can remind each other about and then put into play? Because again – and maybe I'm thinking too small, but I don't think so – our universities are in crisis at the same time as attention to antiracism is at a level that I haven't seen before. This wouldn't have happened without many decades of work, which is why I feel galvanized even as I feel very worried. To be able to think here as a group is part of the work. I just love the way, Irene, in which you've given us permission to have certain kinds of conversations together that are the food that I need in order to do the work that I know I need to do.

**MH** I think that Dagmar and Ann just beautifully framed the volume and why this volume at this particular moment. To try to put the two statements together: is the present moment a crisis? Is it an opportunity? Well, a crisis is always an opportunity. And, an opportunity can always put us into crisis. Because we want to meet it and if there's pressure, we ask ourselves how can we *best* meet it? That provides a great reason for this conversation. I think there are certain things that wouldn't have happened otherwise. We wouldn't have met on zoom, we wouldn't have had a conversation together if it hadn't been 2020, 2021. We probably wouldn't have written what we've written. Dagmar's idea of the crisis of our disciplines and how she articulated it really spoke to me. Because I think the way so many people in this volume have responded to 2020 and to the invitation to write something is to go back to very small, very personal, but also very urgent stories that are familial, that are in the past, that are engaging people we know and care about in the past but that we never really listened to or wrote for or about or with in the way that a number of you have here. So why did this moment give rise to these essays? I think that's something to think through.

**AB** I love Marianne's emphasis on the transformative power of the "very small, very personal" stories we tell. It reminds me of Doris Lessing, who insisted that the "small, personal voice" of a writer can be a powerful political force.

**IK** I want to pick up on honoring the work that has come before. Some of you have been doing this work for a long time, but there are many many others who are not here whose intellectual and political contributions have been essential to our work. I want to honor and give thanks for the tools others crafted and I got to use.

**Susannah Heschel** I want to say thanks, too. Our project and today's discussion make me feel a tiny bit relieved of my enormous pessimism and I appreciate that. Still, I would like to say something here about my pessimism. I feel that the foundations I've been standing on have collapsed. For me, it is an ethical collapse, the sense that I have lost what I used to feel, when growing up and in college, that one could speak as a Jew, as a civil rights leader, as a minister or in the name of the Bible, or as a Holocaust survivor in the name of justice. Somehow there was an identity that was associated with justice that has slowly crumbled and fallen apart. What disturbs me about January 6 and white supremacy is that they've infiltrated everything: the Congress, the Supreme Court, the military; all of the institutions that we think of as needed for a democracy have already been infiltrated, and I feel the same way about my fields of study for that matter. I feel that way about the Jewish world and about the field of Jewish Studies as well. I just noticed in an article that Israelis ranked #12 in world happiness, and I wonder, what does that mean? How do you associate happiness with people who are living a racist occupation? What does that mean? What is the nature of such happiness? Even certain stances that I used to take, certain alliances that I used to feel, I don't even know who I am. I speak out as what? Representing what? Because all that I have always thought I represent no longer has the moral authority or meaningful voice or resonance or implication that it used to have and that I find very disturbing. I don't know why I wrote an essay for this book about Jews who had converted to Christianity and were treated just horribly when they showed up with a yellow star in church. In part it was because many years ago I had a professor at Hebrew University, David Bankier, who mentioned one day in class that a woman kneeling at the communion rail in Nazi Germany looked over and saw someone else at the communion rail wearing a yellow star and found it disgusting. And it was the disgust that I found important; it wasn't just anger, but disgust. Why I wrote the chapter I did? I couldn't understand why I was writing what I was writing while I was writing it. And it was only after I finished it that I could frame clearer questions about my motivations: why would I care about Jews that converted to Christianity and were badly treated by the Church? What does that have to do with anything? At that moment I realized it was all a metaphor for my own life and how I feel as a Jew. Here I am as a Jew, in the Jewish community or in the synagogue. Or anywhere else and I feel I am wearing a yellow star. And I am a figure of disgust, disgust for my politics, for my views, for my very being. Because today, it's not just about opinions. You express a viewpoint and you yourself become the viewpoint and the object of the disgust. And it is disgust. It's not just political disagreement anymore. You become an object worthy of being murdered, worthy of being silenced. And we know this is nothing new. But for me in my own life, going from the way it was when I was in my teens and college and so on to NOW, that shift, that change, I find it extraordinary. Just unbelievable, what has happened while I do my work, as I sit at my desk, as I teach my classes and watch it all happen and see everything just disintegrate, implode, melt into air? I find it terribly frightening. And I'm terribly pessimistic. I feel terribly sorry for the next generation. I also wonder how this happened. So, believe me, I am grateful for hearing some voices today saying things that alleviate even if only to a small extent my pessimism. I appreciate that.

**IK** Susannah, we could never do without your voice, and you've been saying a lot of vitally important things in social media and elsewhere. Thank you.

**SH** Can I add that yesterday I had a phone call from a woman who was in tears? She's a junior faculty somewhere, and a man, a Jewish sociologist who lost his job a couple of years ago for sexual harassment that he carried out over the course of many decades, is now being rehabilitated. And she as a younger person has the pressure: if she does not accept the rehabilitation will she be shut out of the field? That's just another example of some of the things that are going on. We may think that we've made some progress here and we have. But nonetheless, other forces are coming in, I feel, like a locomotive engine.

**Chunjie Zhang** I would like to echo what Susannah just said. I wrote a piece for this volume about identity politics and identity freedom as something I actually earnestly wish for, but don't realistically believe could happen any time soon. I know this sounds preposterous. Given what's happened in the last four years and given what happened in January, even right now, the last couple of days, what happened in Atlanta to Asian Americans: I could dream about the identity freedom I preached or argued for in the essay, but, really, could that ever become reality or partial reality? A reality that we could really choose to decide who we are on the basis of science, genetics and the fact that we're all connected to each other inside our genes? That the whole of humanity in Africa, Asia, Europe, and America is enmeshed? Even though we know that we have it all within ourselves, that physical appearance doesn't matter that much, would that identity freedom become the actual reality anywhere in this world? This is something I am not optimistic about. Identity politics had been used overtly in a negative way in the past few years, but right now, despite a positive turn, I still feel an impasse. Identity based on physical appearance is still used extensively in daily behavior and mental judgment such as: this person looks like this, then one could expect something and thus behave in a certain way. Such undifferentiated thinking, sometimes unconscious, happens not only on an individual basis, what I experience, what my friends tell me, but also on a policy basis, on a collective basis. So that's something quite disappointing and disquieting as well. Because of this, I'm very interested in the concept of neuroplasticity or mental flexibility. Maybe that would be a way for therapy with pedagogical purposes. I have been conceiving of a course with this concept. I am a literary studies scholar and have been reading and teaching literature, among other things, to discuss textual representations of individual emotions and their correlations to sociocultural transformations. I have been asking myself lately: Is there some way we could use literature to influence or change the brain structure based on the idea of neuroplasticity? Our brain and our neurons have the flexibility to become different; that's a scientific fact. Maybe literature could be used to usher in more positive changes, to make people become more aware of the fact that we're capable of change, of the fact that we share genes from all over the world, that we could move beyond identity barriers based on physical appearance and treat each other with equality, solidarity, and openness? Maybe we could believe in neuroplasticity and use stories to reach out to people and to spread the practice of identity freedom? No matter what we experience on a day-to-day basis, for example, what I have been experiencing as an Asian woman, I still very much appreciate the opportunity to write this piece for this wonderful volume: to imagine, to dream this identity freedom. Maybe one day!

## Coda to the Conversation

[The interventions below were added after the jury found Derek Chauvin guilty of the murder of George Floyd on 20 April 2021]

**Bettina Brandt** In reading through the transcript of this conversation, which I was unable to attend, I am left with two main reactions. For one, the insistence on making connections between the past and the present seems really important to me. Trudie Engel née Frey, the woman about whom I write for this collection, who escaped Vienna with her family as a little girl but who had to leave her grandmother behind in Nazi Europe, had a similar political reflex as those mentioned by Angelika and Atina. As a student in the United States, she joined the civil rights movement and wrote a master's thesis in history about the Harlem Renaissance. When I met Trudie in person in 2013, the parallels she drew included those between her own historical experiences and the predicaments of today's refugees fleeing civil war and other catastrophes around the globe.

My second reaction to your powerful conversation is that especially in the Humanities and Social Sciences, we all too often teach and write in isolation, largely due to systemic academic constraints but also perhaps due to a lack of practice and curiosity? Few of us regularly write with a non-academic audience in mind. Our collaboration in this volume, across generations, across fields, across historical periodsm – and thanks to zoom technology – across geographical distances, was motivated by feelings of empathy with victims of political violence and the desire to translate these feelings into something tangible we could share with each other and through publication with others beyond our group. The whole process leaves me hopeful, but at the same time I worry whether these kinds of empathic collaborations will continue once the global pandemic ends and with it the social isolation that partly brought about the desire for new forms of collaboration and new alliances against old and new racisms. As someone said during your discussion, we academics are not that good about predicting the future, but I do want to register here something about the immediate past: for me and my work it was highly productive to interact with you in this discussion, even if belatedly and virtually, and also in our overall collaboration for this volume. I salute your – our work on the memory of violence and active opposition to its repetition in contemporary shapes and places.

**Christina Matzen** As someone who is only at the beginning of their career, I feel an unusual, although perhaps necessary, amount of hope for the future of academia and for our world. I wrote about prisoner experiences in my essay, and similar to what Chunjie said about identity freedom, I don't think we'll see a criminal justice system that is actually *just* anytime soon (or maybe not ever). But I keep going back to a lecture on antisemitism in which Doris Bergen quoted Abraham Joshua Heschel from *Religion and Race*, "Surrender to despair is surrender to evil. It is important to feel anxiety, it is sinful to wallow in despair." The same day the man charged with killing George Floyd was pronounced guilty, Ma'Khia Bryant – a black teenager from Ohio which is my home – was shot and killed by a police officer. I can't stop thinking about a piece Connie Schultz, a columnist and professional in residence at Kent State University, wrote about meeting with her students the morning after the verdict was pronounced: "It's Wednesday morning. I sign into the eighth one-on-one student video conference but immediately see that, on this call, with this cherished student, there's no oxygen for talking about the final, mundane details of spring semester. The young Black woman looking at me through the computer screen is in almost unspeakable pain." I am tempted to surrender to despair as I contemplate this unspeakable pain. There will be moments when I might. But I will remind myself of Heschel's words, of Ma'Khia's joyful hair tutorial video, of the tools Ann and so many others in this group have developed over the past several decades, and of my future students who will need me as an ally as they navigate a world filled with injustice – and hope.

**Adam Z. Newton** Regrettably, I was unable to attend the virtual meeting during which this colloquy among my fellow contributors unfolded. As my essay in this volume indicates, Atlanta, Georgia – scene of the spa murders that dominated the local news during the previous week – is where I currently reside. Which is only to say, as I recollect that moment more than a month later, that adjacency to historical (racialized) violence in twenty-first century America always happens close to *somebody's* home. Such dislocations will and should *unhome* us, if only momentarily. And to echo Erin McGlothin's remark on the angel of history, truly we are "blown into the future," ahead of our capacity if not our will.

Rather than contrive a voice to insert belatedly into the conversation, I prefer to inscribe those of two others. The first, from a well-known poem composed in the days following September 11, 2001, belongs to the late Adam Zagajewski, whose obituary appeared just a day after the virtual meeting among contributors, March 21, 2021. It feels ever apt.

> "Try to praise the mutilated world.
> Remember June's long days,
> and wild strawberries, drops of rosé wine.
> The nettles that methodically overgrow
> the abandoned homesteads of exiles.
> You must praise the mutilated world.
> You watched the stylish yachts and ships;
> one of them had a long trip ahead of it,
> while salty oblivion awaited others.
> You've seen the refugees going nowhere,
> you've heard the executioners sing joyfully.
> You should praise the mutilated world.
> Remember the moments when we were together
> in a white room and the curtain fluttered.
> Return in thought to the concert where music flared.
> You gathered acorns in the park in autumn
> and leaves eddied over the earth's scars.
> Praise the mutilated world
> and the gray feather a thrush lost,
> and the gentle light that strays and vanishes
> and returns."

The second voice, counterbalancing tempered acceptance with unceasing resistance, belongs to Ta-Nehisi Coates, addressing his son in *Between the World and Me*, a book very much in my thoughts on this day following Derek Chauvin's conviction for the murder of George Floyd.

"But still you must struggle. The Struggle is in your name, Samori – you were named for Samori Touré, who struggled against French colonizers for the right to his own black body. He died in captivity, but the profits of that struggle and others like it are ours, even when the object of our struggle, as is so often true, escapes our grasp . . . . I would not have you descend into your own dream. I would have you be a conscious citizen of this terrible and beautiful world" (68, 108).

**AG** I can't stop thinking about the endless refrain of "I can't breathe" – from Eric Garner to George Floyd and now to the streets and hospital wards of India. I am struck by Chrissy's anecdote about the professor meeting a student after the Chauvin verdict and realizing there wasn't enough oxygen left to talk about what had just happened. There we are again with oxygen – as trope and as basic irreducible need.

**IK** I'm ending this conversation here, with oxygen and with new life. One of our contributors could not converse with us because she was giving birth. When her son reaches our ages, may he find himself living in a very different world than that of April 2020–April 2021 when this book was born and finished.

# Notes on Contributors

**Angelika Bammer** is Professor of Comparative Literature at Emory University. The editor of *Displacements: Cultural Identities in Question* (1994) and co-editor of *The Future of Scholarly Writing: Critical Interventions* (2015), she has published on twentieth-century literature and culture, film and photography, and utopian thought. Her book, *Born After: A German Reckoning* (Bloomsbury Academic, 2019), a study of the transmission of history across four generations in form of a personal narrative, was a PROSE Award Finalist.

**Doris L. Bergen** is the Chancellor Rose and Ray Wolfe Professor of Holocaust Studies at the University of Toronto. She is the author, editor, or co-editor of five books, including *War and Genocide: A Concise History of the Holocaust* (in its 3$^{rd}$ edition) and *Twisted Cross: The German Christian Movement in the Third Reich*. A Fellow of the Royal Society of Canada, Bergen serves on the Academic Committee of the United States Holocaust Memorial Museum in Washington, DC.

**Carol Bohmer**, a lawyer and sociologist, is a Visiting Scholar in the Government Department at Dartmouth College, and a Visiting Senior Research Fellow at King's College, London. She has worked in the area of law and society, examining the way legal and social institutions interact. Her current research interests are in the field of migration and asylum. She is the coauthor with Amy Shuman of *Rejecting Refugees: Political Asylum in the 21$^{st}$ Century* (2007) and *Political Asylum Deceptions: The Culture of Suspicion* (2018), and a number of articles on migration issues.

**Bettina Brandt** is Teaching Professor of German and Jewish Studies at Penn State University. Recent publications include "Taming Foreign Speech: Language Politics in Shadow Plays around 1800" (*German Studies Review* 2018) and "Yoko Tawada's 'Tongue Dance' or the Failed Domestication of a Tongue in Furs" in *Un/Translateables. New Maps for Germanic Literatures* (2016). With V. Glajar, Brandt is editor of *Herta Müller. Politics and Aesthetics* (2013), and, with D. Purdy, of *China in the German Enlightenment* (2016). With D. Schyns, Brandt is Dutch translator of Yoko Tawada's *De Berghollander* (2010). Most recently, Brandt translated Yoko Tawada's German Fukushima poetry cycle into English in *Out of Sight* (2020).

**Claudia Breger** is the Villard Professor of German and Comparative Literature at Columbia University. Having received her PhD and Habilitation from Humboldt University, Berlin, she taught at the University of Paderborn, Germany, and Indiana University, Bloomington, before joining Columbia in 2017. Her research and teaching focus on modern and contemporary culture, with emphases on film, performance, literature, and literary and cultural theory, as well as the intersections of gender, sexuality, and race. Her most recent book, *Making Worlds: Affect and Collectivity in Contemporary European Cinema*, was published by Columbia University Press in Spring 2020.

**Darcy Buerkle** is Professor of History at Smith College. Her research focuses on modern European gender history with an emphasis on German and German Jewish women's intellectual and cultural history. She has worked extensively on the history of emotion and German visual culture of the early 20th century. The author of numerous essays, Buerkle's first

https://doi.org/10.1515/9783110753295-035

book, *Nothing Happened: Charlotte Salomon and an Archive of Suicide,* was published by the University of Michigan Press; a co-edited volume with Skye Doney titled *Mosse's Europe* is forthcoming from the University of Wisconsin Press in 2021.

**Mita Choudhury** works on religion, gender, and politics in early modern France. Her publications include: *Convents and Nuns in French Politics and Culture* (2005); *The Wanton Jesuit and Wayward Saint: A Tale of Sex, Religion, and Politics in Eighteenth-Century France* (2015), and *Belief and Politics in Enlightenment France: Essays in Honor of Dale K. Van Kley* (2019), co-edited with Daniel J. Watkins. Her current project examines sexual assault in the early modern French Church. Choudhury teaches early modern and eighteenth-century European history and Women's Studies at Vassar College.

**Ann Cvetkovich** is Director of the Pauline Jewett Institute of Women's and Gender Studies at Carleton University. She was previously Ellen Clayton Garwood Centennial Professor of English, Professor of Women's and Gender Studies, and founding Director of LGBTQ Studies at the University of Texas at Austin. She is the author of *Mixed Feelings: Feminism, Mass Culture, and Victorian Sensationalism* (Rutgers, 1992); *An Archive of Feelings: Trauma, Sexuality, and Lesbian Public Cultures* (Duke, 2003); and *Depression: A Public Feeling* (Duke, 2012). For additional info, see www.anncvetkovich.com.

**Alicia E. Ellis** is Associate Professor of German at Colby College. Her research interests include German Literature with specializations in nineteenth-century drama and the novella, the European novel in the 20th century, Black Atlantic Studies, Francophone and Anglophone Caribbean Literatures, and Women's and Gender Studies. She has written and lectured on Edwidge Danticat, Franz Grillparzer, Heinrich Heine, ETA Hoffmann, Andrea Levy, Osip Mandelstam, and Sam Selvon. She is the author of *Gender and Identity in Franz Grillparzer's Classical Dramas: Figuring the Female* (2021).

**Sylvia Flescher** is a psychiatrist and psychoanalyst. Her father, a psychiatrist, was a Polish-Jewish survivor of the Holocaust, and her mother was a Swiss Gentile who was honored in 2008 at Yad Vashem as a *Righteous Among the Nations.* Flescher's article, "Googling for Ghosts: A Meditation on Writer's Block, Mourning, and the Holocaust," appeared in *Psychoanalytic Review* in 2012. She is currently at work on a memoir.

**Tao Leigh Goffe** is an assistant professor of literary theory and cultural history at Cornell University. She is a joint hire between Africana Studies and Feminist, Gender, and Sexuality Studies. She specializes in the narratives that emerge from histories of imperialism, technology, and abolition in the Caribbean. Her writing has been published in *Small Axe, Women & Performance: A Journal of Feminist Theory,* and *Boston Review.* Goffe is the founder of the Dark Laboratory, a collective on race, ecology, and creative technology. She is also a founding co-editor for the new peer-reviewed *Journal of Indentureship and Its Legacies.*

**Grazyna Gross** resides in upstate New York. She was born in what was part of prewar Poland. World War II put her family on the run. By war's end she and her mother landed in the American occupation zone of Germany, from where they emigrated to the United States. She is an avid reader of literature, enjoys retrieving historical information from the internet, and writes short fiction.

**Atina Grossmann** is Professor of History at the Cooper Union in New York City. Relevant publications include *Jews, Germans, and Allies: Close Encounters in Occupied Germany*; *Wege in der Fremde: Deutsch-jüdische Begegnungsgeschichte zwischen New York, Berlin, und Teheran* (2012); and co-editor, *Shelter from the Holocaust: Rethinking Jewish Survival in the Soviet Union* (with M. Edele and S. Fitzpatrick 2017), and *The JDC at 100: A Century of Humanitarianism* (with A. Patt, L. Levi, and M. Mandel 2019). Her current research focuses on "Trauma, Privilege, and Adventure: Jewish Refugees in the 'Orient,'" as well as the entanglements of family memoir and historical scholarship.

**Susannah Heschel** is the Eli M. Black Distinguished Professor and chair of the Jewish Studies Program at Dartmouth College. The author of *Abraham Geiger and the Jewish Jesus* (1998); *The Aryan Jesus: Christian Theologians and the Bible in Nazi Germany* (2010); and *Jüdischer Islam: Islam und jüdisch-deutsche Selbstbestimmung* (2018), she and Umar Ryad have recently co-edited, *The Muslim Reception of European Orientalism* (2018). She is currently writing a book with Sarah Imhoff on *Jewish Studies and the Woman Question*. She is a Guggenheim Fellow and has held fellowships at the National Humanities Center and the Wissenschaftskolleg zu Berlin.

**Dagmar Herzog** is Distinguished Professor of History at the Graduate Center, City University of New York. Her books include: *Sex after Fascism: Memory and Morality in Twentieth-Century Germany* (2005); *Sexuality in Europe: A Twentieth-Century History* (2011); *Cold War Freud: Psychoanalysis in an Age of Catastrophes* (2017); and *Unlearning Eugenics: Sexuality, Reproduction, and Disability in Post-Nazi Europe* (2018). She is currently writing on the theology and politics of disability in twentieth-century Germany.

**Marianne Hirsch** writes about the transmission of memories of violence across generations, combining feminist theory with memory studies in global perspective. Recent books include *The Generation of Postmemory: Writing and Visual Culture After the Holocaust* (2012); *Ghosts of Home: The Afterlife of Czernowitz in Jewish Memory* (2010); and *School Photos in Liquid Time: Reframing Difference* (2019), both co-authored with Leo Spitzer; and the co-edited volume *Women Mobilizing Memory* (2020). Hirsch teaches Comparative Literature and Gender Studies at Columbia University and is a former President of the Modern Language Association of America and a member of the American Academy of Arts and Sciences.

**Irene Kacandes,** the Dartmouth College Professor of German Studies and Comparative Literature, teaches also in the fields of Women's, Gender and Sexuality Studies, and Jewish Studies. Her publications include *Teaching the Representation of the Holocaust* (edited with Marianne Hirsch 2005); *Daddy's War. A Paramemoir* (2009); *Let's Talk About Death* (written with Steve Gordon 2015); and *Eastern Europe Unmapped* (edited with Yuliya Komska 2017). She is a past president of the German Studies Association and of the International Society for the Study of Narrative Literature and has edited the series "Interdisciplinary German Cultural Studies" for de Gruyter Verlag since 2005.

**Eric Kligerman** is an Associate Professor of German and Jewish Studies at the University of Florida. His current research focusing on the works of Franz Kafka, examines the intersection between quantum physics and German-Jewish intellectual thought. In addition to his *Sites of the Uncanny: Paul Celan, Specularity and the Visual Arts* (De Gruyter, 2007), he has published on such topics as Nazis in American popular culture and representations of German leftwing terrorism in German film and painting.

**Thomas Kühne** is the Director of the Strassler Center for Holocaust and Genocide Studies and the Strassler Colin Flug Professor in Holocaust History at Clark University. His current research explores the place of the Holocaust and Nazi Germany in the continuities and discontinuities of modern history. He has been awarded the German Bundestag Prize in Parliamentarism Studies and fellowships from the John Simon Guggenheim Memorial Foundation, the Institute for Advanced Study in Princeton, the Deutsche Forschungsgemeinschaft, and others. His monographs include *The Rise and Fall of Comradeship* (2017); *Belonging and Genocide* (2010); and *Dreiklassenwahlrecht und Wahlkultur in Preußen* (1994).

**Priscilla Layne** is Associate Professor of German and Adjunct Associate Professor of African, African American and Diaspora Studies at the University of North Carolina at Chapel Hill. She received her PhD from the University of California at Berkeley. Her research and teaching draw on postcolonial studies, gender studies and critical race theory to address topics like representations of Blackness in literature and film, rebellion, and the concept of the Other in science fiction/fantasy. She is the author of *White Rebels in Black: German Appropriation of African American Culture* (2018), and her current book project is on Afro-German Afrofuturism.

**Erin McGlothlin** is Professor of German and Jewish Studies and Vice Dean of Undergraduate Affairs at Washington University in St. Louis. Her main research interests are in the areas of Holocaust literature and film and German-Jewish literature. She is the author of *Second-Generation Holocaust Literature: Legacies of Survival and Perpetration* (2006) and *The Mind of the Holocaust Perpetrator in Fiction and Nonfiction* (2021). Among the volumes she has co-edited are *Persistent Legacy: The Holocaust and German Studies* (with Jennifer Kapczynski 2016) and *The Construction of Testimony: Claude Lanzmann's* Shoah *and its Outtakes* (with Brad Prager and Markus Zisselsberger 2020).

**Christina Matzen** is a PhD Candidate in History at the University of Toronto. Her research interests include gender, crime, and punishment in modern Germany. She is currently in the process of completing her doctoral dissertation, "Women's Prisons in Twentieth-Century Germany: Gendered Punishment from the Kaiser to the Cold War." Christina has taught courses on European history, the history of the Holocaust, and the history of Jewish humor. She is also committed to prisoner advocacy in the United States and has trained with the Oriana House/North Star Reentry Resource Center in Cleveland, Ohio.

**Leslie Morris** is the Beverly and Richard Fink Professor in Liberal Arts and Professor and Chair of the Department of German, Nordic, Slavic and Dutch at the University of Minnesota, where she served for ten years as Director of the Center for Jewish Studies. She is the author most recently of *The Translated Jew: German Jewish Culture outside the Margins* (Northwestern University Press, 2018) and has co-edited three volumes of essays on German Jewish culture. She is currently writing a hybrid, experimental memoir.

**Adam Zachary Newton** has written and taught at the boundaries of three knowledge practices: literary studies, philosophy, and religion. He has authored six books, beginning with *Narrative Ethics* (1995) and culminating with *Jewish Studies as Counterlife* (2019). Most recently, he edited a special issue of *Humanities* on "Ethics and Literary Practice" (2020); a successor volume, "Refugees and Representation" appears in 2022. Now emeritus, he has held appointments as University Professor and Stanton Chair in Literature and Humanities at Yeshiva University and was the Jane and Rowland Blumberg Centennial Professor at the University of Texas at Austin.

**Eric Niyitunga** is a lecturer at the University of Johannesburg. He obtained his masters in International Peace and Security at Kings College and a doctorate from the University of Johannesburg in Political Studies. Dr Niyitunga is an alumnus of the African Leadership Centre and a member of the American Political Science Association (APSA). He was a United Nations Humanitarian Peace Ambassador in 2018. He is a reader in Africa's international relations, refugee law, practice of diplomacy, foreign policy analysis, and African studies.

**Brad Prager** is Professor of Film Studies and German at the University of Missouri. He specializes in Holocaust film studies and in contemporary German cinema. He is the author of several books including *After the Fact: The Holocaust in Twenty-First Century Documentary Film* (2015) and is co-editor of a volume on Claude Lanzmann's *Shoah* entitled *The Construction of Testimony: Claude Lanzmann's* Shoah *and its Outtakes* (2020). He is also an editor of the journal *New German Critique*.

**Karen Remmler** is the Mary Lyon Professor of Humanities at Mount Holyoke College. Her interdisciplinary research and teaching in English and German focus on the politics and cultures of transnational memory in the aftermath of atrocity and war with an emphasis on memorialization and mourning in the digital age. Her publications include *Waking the Dead: Correspondences between Walter Benjamin's Concept of Remembrance and Ingeborg Bachmann's "Ways of Dying"* (1996); with Sander Gilman, *Reemerging Jewish Culture in Germany: Life and Literature since 1989* (1994); with Leslie Morris, *Contemporary Jewish Writing in Germany* (2002); with Christopher Benfey, *Artists, Intellectuals, and World War II: The Pontigny Encounters at Mount Holyoke College, 1942–1944* (2006). Her current project, "(Im)proper Burial: The Afterlives of the Dead in the Wake of Extreme Violence," focuses on transgenerational residual memories of trauma.

**Leo Riegert, Jr.** is Associate Professor of German at Kenyon College. His work on German-Jewish literature and identity, and particularly on the writings of Karl Emil Franzos, has appeared in the *German Quarterly* and the *Yearbook of the Association of European Jewish Literature Studies*.

**Mark Roseman** is a historian of modern Europe, specializing on the Holocaust. Recent books include *Jewish Responses to Persecution, 1933–1946: Volume I, 1933–1938 (Documenting Life and Destruction: Holocaust Sources in Context)* (with Jürgen Matthäus 2010); *Beyond the racial state* (with Devin Pendas and Richard Wetzell 2017); *Lives, reclaimed. A story of rescue and resistance in Nazi Germany* (2019); *ÜberLeben im Dritten Reich. Handlungsspielräume von Juden und ihren Helfern* (2020). He is general editor of the *Cambridge History of the Holocaust* currently in preparation.

**Amy Shuman** is Professor of folklore and narrative in the Department of English at The Ohio State University. She is the author of *Storytelling Rights: the uses of oral and written texts* *among urban adolescents* (1986) and *Other People's Stories: Entitlement Claims and the Critique of Empathy* (2005). With Carol Bohmer, she has coauthored *Rejecting Refugees: Political Asylum in the 21ˢᵗ Century* (2007) and *Political Asylum Deceptions: The Culture of Suspicion* (2018). She is the recipient of the American Folklore Society Lifetime Achievement Award and a Guggenheim Fellowship.

**Leo Spitzer** is a cultural and comparative historian and writer working in the interdisciplinary field of Memory Studies. His books include *Lives in Between: The Experience of Marginality in a Century of Emancipation* (1998); *Hotel Bolivia: The Culture of Memory in a Refuge from Nazism* (2000); *Ghosts of Home: The Afterlife of Czernowitz in Jewish Memory* (2010) and *School Photos in Liquid Time: Reframing Difference* (2020), both co-authored with Marianne Hirsch. He is currently working on *The Americanization of Poldi*, a memoir about Jewish refugee immigration and acculturation in New York in the 1950s and 1960s.

**Susan Rubin Suleiman** was born in Budapest and emigrated to the U.S. as a child with her parents. A professor emerita of comparative literature at Harvard, she has published many books and articles on contemporary literature and culture, in the U.S. and abroad. Her most recent book is *The Némirovsky Question*: *The Life, Death, and Legacy of a Jewish Writer in 20ᵗʰ-Century France* (2016). Her other books include *Crises of Memory and the Second World War* (2006) and the memoir *Budapest Diary: In Search of the Motherbook* (1996). In 2018 she was awarded France's highest honor, the Légion d'Honneur.

**Viktor Witkowski** is a painter and filmmaker. He was born in Poland and grew up in Germany, where he graduated from the Hochschule für Bildende Künste Braunschweig (HBK Braunschweig, Germany) with a combined master's degree in Studio Art, Art History and Art Education in 2006. The same year, he immigrated to the U.S. where he earned an MFA in Visual Arts from Rutgers University in 2010. He currently splits his time between Vermont (U.S.A.) and Leipzig (Germany). When he is not working on paintings and films, he teaches as a lecturer in Dartmouth College's Studio Art Department.

**Chunjie Zhang** is associate professor of German at the University of California, Davis. She is the author of *Transculturality and German Discourse in the Age of European Colonialism* (Northwestern UP 2017) and the editor of *Composing Modernist Connections in China and Europe* (Routledge 2018). She has co-edited journal issues on Goethe and world literature, aesthetics and politics in the Enlightenment, and Asian German studies and co-edits a book series "Asia, Europe, and Global Connections" at Routledge. She has published articles on Goethe, Herder, Max Weber, travel writings, Chinese gardens in Europe, and the contemporary refugee crisis.

www.ingramcontent.com/pod-product-compliance
Lightning Source LLC
Chambersburg PA
CBHW060017030426

42334CB00019B/2074